Multi-Objective Optimization System Designs and Their Applications

This book introduces multi-objective design methods to solve multi-objective optimization problems (MOPs) of linear/nonlinear dynamic systems under intrinsic random fluctuation and external disturbance. The MOPs of multiple targets for systems are all transformed into equivalent linear matrix inequality (LMI)-constrained MOPs. Corresponding reverse-order LMI-constrained multi-objective evolution algorithms are introduced to solve LMI-constrained MOPs using MATLAB®. All proposed design methods are based on rigorous theoretical results, and their applications are focused on more practical engineering design examples.

Features:

- Discusses multi-objective optimization from an engineer's perspective.
- Contains the theoretical design methods of multi-objective optimization schemes.
- Includes a wide spectrum of recent research topics in control design, especially for stochastic mean field diffusion problems.
- Covers practical applications in each chapter, like missile guidance design, economic and financial systems, power control tracking, minimization design in communication, and so forth.
- Explores practical multi-objective optimization design examples in control, signal processing, communication, and cyber-financial systems.

This book is aimed at researchers and graduate students in electrical engineering, control design, and optimization.

Multi-Objective Optimization System Designs and Their Applications

Bor-Sen Chen

CRC Press
Taylor & Francis Group
Boca Raton London New York

CRC Press is an imprint of the
Taylor & Francis Group, an **informa** business

MATLAB® is a trademark of The MathWorks, Inc. and is used with permission. The MathWorks does not warrant the accuracy of the text or exercises in this book. This book's use or discussion of MATLAB® software or related products does not constitute endorsement or sponsorship by The MathWorks of a particular pedagogical approach or particular use of the MATLAB® software.

First edition published 2024
by CRC Press
2385 NW Executive Center Drive, Suite 320, Boca Raton FL 33431

and by CRC Press
4 Park Square, Milton Park, Abingdon, Oxon, OX14 4RN

CRC Press is an imprint of Taylor & Francis Group, LLC

ISBN: 978-1-032-41564-2 (hbk)
ISBN: 978-1-032-42298-5 (pbk)
ISBN: 978-1-003-36214-2 (ebk)

DOI: 10.1201/9781003362142

Typeset in Times
by Apex CoVantage, LLC

Contents

Part I: General Theory for Multi-Objective Optimization Designs of Stochastic Systems

Part II: Multi-Objective Optimization Designs in Control Systems

Part III: Multi-Objective Optimization Designs in Signal Processing and Systems Communication

Part IV: Multi-Objective Optimization Designs in Cyber-Social Systems

Preface

People always pursue multiple objectives in life. For Chinese-speaking people, the five blessings—longevity, wealth, health, love of virtue, and natural death—are always the five traditional targets to search for in life. These blessings always appear as couplets pasted on doors or doorposts during the lunar new year. Although most existing studies capture characteristics of system engineering designs in the form of single-objective problems (SOPs), multiple objectives arise naturally in real-life scenarios. These objectives conflict with each other normally; in other words, these objectives cannot be optimized simultaneously.

Even mathematicians and engineers could solve the optimal design problems of SOPs for a long time. But it is not easy to solve multi-objective optimization problems (MOPs) when the objectives are somewhat conflicting. MOPs had been inherent in mathematicians and engineers for a long time until Pareto used domination to define the minimization of multi-objective vector in MOPs and consequently developed some theoretical results of MOP. In the last decades, the development of multi-objective evolution algorithms (MOEAs) has significantly improved the efficiency of solving MOPs. Then control engineers combined MOEA methods and linear matrix inequality (LMI) techniques, that is, so-called LMI-constrained MOEAs, to solve LMI-constrained MOPs of multi-objective controls and filter design problems of linear stochastic systems via the help of MOEAs and the LMI toolbox in MATLAB.

In general, it is still very difficult to solve MOPs of nonlinear systems because it is necessary to solve a highly nonlinear Hamilton-Jacobi inequality (HJI)–MOP. Recently several interpolation methods like the Takagi-Sugeno fuzzy method, global linearization scheme, and gain schedule method have been proposed to interpolate several local linearized systems to efficiently approximate a nonlinear system. With these interpolation methods, HJI-constrained MOPs for the multi-objective design of nonlinear stochastic systems can be transformed into equivalent LMI-constrained MOPs and therefore can be efficiently solved with the help of MOEAs and the LMI toolbox in MATLAB with more practical applications.

This book is divided into four parts. In the first part of this book, the concepts and general theories of MOPs will be introduced. The applications of MOPs to the multi-objective designs of control systems are given in the second part. In the third part of this book, I will introduce multi-objective optimal design problems in signal processing and communication systems. Finally, in the fourth part of this book, multi-objective optimization design problems in cyber-social systems will be introduced to meet the recent rapid development of MOPs in cyber-social or financial systems via a smart wireless network in the future 5G or 6G era.

Last but not least, I would like to thank Dr. Gagandeep Singh, an editor at CRC, who has given me a lot of help during the course of this project. The author is also grateful to Ms. Ci-Jun Wang for her careful typing of this book.

About the Author

Dr. Bor-Sen Chen received a B.S. in electrical engineering from the Tatung Institute of Technology, Taipei, Taiwan, in 1970; an M.S. in geophysics from the National Central University, Chungli, Taiwan, in 1973; and a Ph.D. from the University of Southern California, Los Angeles, California, USA, in 1973. From 1973 to 1987, he was a lecturer, associate professor, and professor at the Tatung Institute of Technology. From 1987, he has been a professor, chair professor, and Tsing Hua distinguished chair professor with the Department of Electrical Engineering at National Tsing Hua University, Hsinchu, Taiwan. His research interests include robust control theory and engineering design, robust signal processing and communication system design, and systems biology. He has published more than 340 journal papers, including 150 papers in control, 80 papers in signal processing and communication, and 110 papers in systems biology. He has also published 10 monographs. He has been the recipient of numerous awards for his academic accomplishments in robust control, fuzzy control, H∞ control, stochastic control, signal processing, and systems biology, including four Outstanding Research Awards of the National Science Council, the Academic Award in Engineering from the Ministry of Education, National Chair Professor of the Ministry of Education, and Best Impact Award of IEEE Taiwan Section for having the most SCI citations of IEEE members in Taiwan. His current research interest focuses on the H∞ team formation network tracking control of large-scale unmanned aerial vehicles, large-scale biped robots and their team cooperation, deep neural network (DNN)-based control design of nonlinear dynamic systems, systems medicine design based on design specifications, and deep learning schemes. He is a life fellow of IEEE.

Part I

General Theory for Multi-
Objective Optimization
Designs of Stochastic Systems

1 Introduction to Multi-Objective Optimization Problems

1.1 INTRODUCTION

In this chapter, a more general multi-objective optimization problem (MOP) under non-linear inequality constraints is introduced for algebraic systems. In general, it is very difficult to solve a nonlinear inequality-constrained MOP. Based on the linearization scheme at some operation points, the nonlinear inequality-constrained MOP can be transformed into a solvable linear matrix inequality (LMI)–constrained MOP. Since the number of design parameters of MOPs in control and signal processing design of nonlinear stochastic systems is always much larger than the number of multi-objectives, the conventional multi-objective evolution algorithms (MOEAs), which always update the design parameters by evolution algorithm to search for the optimal multi-objectives, are not suitable for practical designs of MOPs in the field of control and signal processing even if they have excellent results in some simple design examples of MOP. In this chapter, a reverse-order LMI-constrained MOEA is proposed to directly update the multi-objective vector of an MOP for an algebraic system first and then to search for the multi-objective optimal design parameters of an MOP indirectly with the help of powerful convex search algorithm from the LMI toolbox in MATLAB. If we become familiar with the design procedure of nonlinear inequality-constrained MOPs of algebraic systems in this chapter, it will be easier to solve the MOPs of nonlinear stochastic systems in the following chapters.

Recently, the Takagi-Sugeno (T-S) fuzzy model, global linearization scheme, and gain scheduling method have been utilized to interpolate several local linearized systems to approximate nonlinear stochastic systems. With these interpolation methods, the MOPs of control and filter design in nonlinear stochastic systems can be transformed into LMI-constrained MOPs. Therefore, based on the reverse-order MOEA in this chapter, the MOPs of control and signal processing in nonlinear stochastic systems can be easily solved by the proposed reverse-order LMI-constrained MOEA method in the following chapters.

1.2 MULTI-OBJECTIVE OPTIMIZATION PROBLEMS IN ALGEBRAIC SYSTEMS

Let us consider the following general MOP of an algebraic system with the objective functions $\min_{x \in \Omega}(f_1(x), \cdots, f_i(x), \cdots, f_N(x))$ within a feasible convex region Ω [1–3]

$$\min_{x \in \Omega}(f_1(x), \cdots, f_i(x), \cdots, f_N(x)) \tag{1.1}$$

DOI: 10.1201/9781003362142-2

subject to the following nonlinear inequality constraints:

$$g_i(x) \leq 0, i = 1, ..., J \tag{1.2}$$

$$h_k(x) = 0, k = 1, ..., K \tag{1.3}$$

where $x = (x_1, ..., x_i, ..., x_n)^T \in \mathbb{R}^n$ are design parameters, all the objective functions $f_i(x), i = 1, ..., N$ are convex, all inequality constrained functions $g_i(x), i = 1, ..., J$ are convex, and all equality functions $h_k(x), k = 1, ..., K$ are linear in Ω.

Remark 1.1: The design parameter x could be a vector or matrix, that is, $x \in \mathbb{R}^{n \times m}$.

In general it is not easy to solve the MOP in (1.1)–(1.3) directly [4, 5]. It is appealing to solve the MOP in (1.1)–(1.3) indirectly by minimizing the upper bounds via the following MOP

$$\min_{x \in \Omega} (\alpha_1, \cdots, \alpha_i, \cdots, \alpha_N) \tag{1.4}$$

$$\text{subject to} \quad f_i(x) \leq \alpha_i, i = 1, ..., N \tag{1.5}$$

$$g_i(x) \leq 0, i = 1, ..., J \tag{1.6}$$

$$h_k(x) = 0, k = 1, ..., K \tag{1.7}$$

where α_i denotes the upper bound of $f_i(x)$ for $i = 1, ..., N$ and Ω denotes the feasible set of x.

Before further the discussion of MOP in (1.4)–(1.7), we have to introduce the following definitions of Pareto optimal solutions of the previous MOP [1–4].

Definition 1.1 (Pareto Dominance [1, 2]): Let $x^1 \in \Omega$ and $x^2 \in \Omega$ be the feasible solution of the MOP in (1.4)–(1.7) corresponding to the objective values $(\alpha_1^1, ..., \alpha_i^1, ..., \alpha_N^1)$ and $(\alpha_1^2, ..., \alpha_i^2, ..., \alpha_N^2)$, respectively. The solution $(\alpha_1^1, ..., \alpha_i^1, ..., \alpha_N^1)$ is said to dominate $(\alpha_1^2, ..., \alpha_i^2, ..., \alpha_N^2)$ if $\alpha_1^1 \leq \alpha_1^2, \ldots, \alpha_i^1 \leq \alpha_i^2, \ldots, \alpha_N^1 \leq \alpha_N^2$ and at least one inequality is a strict inequality.

Definition 1.2 (Pareto Optimality [1, 2]): A solution $x^* \in \Omega$ with the set of objective values $(\alpha_1^*, ..., \alpha_i^*, ..., \alpha_N^*)$ is said to be a Pareto optimal solution if and only if there does not exist another feasible solution that dominates it; that is, we use domination to define the "simultaneous minimization" in the MOP in (1.4)–(1.7).

Definition 1.3 (Pareto Optimal Solution Set [1, 2]): For the given MOP in (1.4)–(1.7), in general, the Pareto optimal solution is not unique. There exists a set of Pareto optimal solutions. The Pareto optimal solution set ρ^* is defined as follows:

$$\rho^* = \left\{ x^* \in \Omega \middle| x^* \text{ is Pareto optimal} \right\} \tag{1.8}$$

Definition 1.4 (Pareto Front [1, 2]): For the given MOP in (1.4)–(1.7), the Pareto front F_p is defined as follows:

$$F_p = \{(\alpha_1^*, ..., \alpha_i^*, ..., \alpha_N^*) \big| x^* \in \rho^*\} \tag{1.9}$$

After the introduction of the previous definitions, we will prove that the solutions of the MOP in (1.4)–(1.7) are also the solutions of the MOP in (1.1)–(1.3). In this situation, the MOPs in (1.1)–(1.3) can be solved by the equivalent MOP in (1.4)–(1.7) indirectly.

Theorem 1.1: The MOPs in (1.4)–(1.7) are equivalent to the MOP in (1.1)–(1.3).

Proof: The proof is simple. If we can prove that all the inequalities in (1.5) and (1.6) become equalities when the optimization objective vector $(\alpha_1^*, ..., \alpha_i^*, ..., \alpha_N^*)$ is achieved. It will be proved by contradiction. Suppose the inequality constraints in (1.5) are strictly held for a Pareto optimal solution; that is, there exists a Pareto optimal solution $(\alpha_1^*, ..., \alpha_i^*, ..., \alpha_N^*)$ with $f_i(x^*) < \alpha_i^*$, for some i. As a result, we have $f_i(x^*) = \alpha_i^1 < \alpha_i^*$. In this situation $(\alpha_1^*, ..., \alpha_i^*, ..., \alpha_N^*)$ is dominated by $(\alpha_1^*, ..., \alpha_i^1, ..., \alpha_N^*)$. This will contradict the fact that the Pareto optimal solution $(\alpha_1^*, ..., \alpha_i^*, ..., \alpha_N^*)$ cannot be dominated by another solution in Definition 1.2.

1.3   REVERSE-ORDER LMI-CONSTRAINED MOEAS FOR MOPs

Based on Theorem 1.1, the MOP in (1.1)–(1.3) is equivalent to the equivalent MOP in (1.4)–(1.7). However, it is very difficult to solve the MOP in (1.4) under nonlinear constraints (1.5)–(1.7). In this situation, interpolation methods like the T-S fuzzy interpolation method [6–8], global linearization method [9], and gain scheduling method [10] can be employed to simplify the MOP in (1.4) under the nonlinear inequality constraints in (1.5)–(1.7). In this book, the LMI-constrained MOEA will be employed to solve MOPs of nonlinear stochastic systems. Therefore, before using the proposed algorithm, the nonlinear constraints should be linearized in some local regions. In the following, some linearized constraints are used to approximate the nonlinear constraints in (1.5)–(1.7) in some local regions to simplify the MOP in (1.4)–(1.7). This leads to the following local linearizations of nonlinear constraints in different local regions [9]

$$\begin{bmatrix} \dfrac{\partial f_i'(x)}{\partial x} \\[2mm] \dfrac{\partial g_j'(x)}{\partial x} \\[2mm] \dfrac{\partial h_k'(x)}{\partial x} \end{bmatrix}_{x=x_l} = \begin{pmatrix} A_{il} \\[2mm] B_{jl} \\[2mm] C_{kl} \end{pmatrix}, l = 1, ..., L \tag{1.10}$$

that is, the nonlinear constraints are linearized in L local regions, where $f_i'(x)$, $g_j'(x)$, and $h_k'(x)$ are the nonlinear parts of $f_i(x)$, $g_j(x)$, and $h_k(x)$, respectively, and $A_{il} \in \mathbb{R}^n$, $B_{jl} \in \mathbb{R}^n$, and $C_{kl} \in \mathbb{R}^n$ are local linearizations of $f_i'(x)$, $g_j'(x)$, and $h_k'(x)$ at local operation point x_l.

Based on the linearized constraints in L local regions in (1.10), the MOP in (1.4)–(1.7) can be simplified to the following MOP with linearized constraints in L local regions around the operation points $x_l, l = 1, ..., L$.

$$\min_x (\alpha_1, \cdots, \alpha_i, \cdots, \alpha_N) \tag{1.11}$$

$$\text{subject to } \left.\begin{array}{c} a_{i0} + (A_{i0}^T + A_{i1}^T)x \le \alpha_i \\ \vdots \\ a_{i0} + (A_{i0}^T + A_{iL}^T)x \le \alpha_i \end{array}\right\} i = 1, ..., N \tag{1.12}$$

$$\left.\begin{array}{c} b_{j0} + (B_{j0}^T + B_{j1}^T)x \le 0 \\ \vdots \\ b_{j0} + (B_{j0}^T + B_{jL}^T)x \le 0 \end{array}\right\} j = 1, ..., J \tag{1.13}$$

$$\left.\begin{array}{c} c_{k0} + (C_{k0}^T + C_{k1}^T)x = 0 \\ \vdots \\ c_{k0} + (C_{k0}^T + C_{kL}^T)x = 0 \end{array}\right\} k = 1, ..., K \tag{1.14}$$

where $a_{i0} + A_{i0}^T x$, $b_{j0} + B_{j0}^T x$, and $c_{k0} + C_{k0}^T x$ are the linear parts of $f_i(x)$, $g_j(x)$, and $h_k(x)$ in (1.5)–(1.7), respectively, which are globally linear and don't need the linearization in (1.10) at local regions.

Remark 1.2: The nonlinear constraints (1.5)–(1.7) of MOP in (1.4) are replaced by L linearized constraints at L local regions around the operation points $\{x_l\}_{l=1}^L$. The linear-constrained MOP in (1.11)–(1.14) can be represented by an LMI-constrained MOP as follows:

$$\min_x (\alpha_1, \cdots, \alpha_i, \cdots, \alpha_N) \tag{1.15}$$

$$\text{subject to } a_{i0} - \alpha_i + (A_{i0}^T + A_{il}^T)x \le 0, i = 1, ..., N \tag{1.16}$$

$$b_{j0} + (B_{j0}^T + B_{jl}^T)x \le 0, j = 1, ..., J \tag{1.17}$$

$$c_{k0} + (C_{k0}^T + C_{kl}^T)x = 0, k = 1, ..., K \tag{1.18}$$

$$l = 1, ..., L$$

After the MOP in (1.4) under nonlinear constraints (1.5)–(1.7) is simplified to the MOP in (1.11) with linear constraints in (1.12)–(1.14) or LMI-constrained MOP in (1.15)–(1.18), the MOEA will be employed to treat the MOP in (1.11) or (1.15). In general, the conventional MOEAs update the design parameters $\{x_i\}_{i=1}^n$ by an evolution algorithm (EA) to search the multi-objective vector $(\alpha_1, ..., \alpha_i, ..., \alpha_N)$ in (1.11) or (1.15) for the Pareto optimal solutions $(\alpha_1^*, ..., \alpha_i^*, ..., \alpha_N^*)$ via the nondominated sorting and crowd tournament selection schemes in the parametric space [1–3]. In general, the number n of design parameters $x = [x_1, ..., x_n]^T$ is always much larger than the number N of the multi-objective vectors $(\alpha_1,...,\alpha_i,...,\alpha_N)$ in (1.11) in practical applications. Especially in the following chapters, the design parameters include fuzzy controller parameters and observer parameters of complex fuzzy systems with a large number of local controllers and observers. It is almost impossible to employ EA for updating these fuzzy controller and observer parameters to achieve an MOP of complex T-S fuzzy systems, such as the multi-objective H_2/H_∞ fuzzy observer-based nonlinear stochastic systems in Chapter 3. Further, for conventional MOEAs, the corresponding objective vector of a child population randomly generated by crossover and mutation operations may exceed the real Pareto front F_p if the current population is close to the real Pareto optimality. Clearly, if the generated child population is infeasible, it will be directly discarded. In this situation, it is very difficult to generate a feasible child population while the current population is close to the real Pareto optimality [3, 4]. This is why the conventional MOEAs have not been addressed on the MOP of control or filter design in nonlinear dynamic systems based on the T-S fuzzy model method and global linearization method, even they are very powerful in MOPs of other more simple designed systems.

In this book, a reverse-order LMI-constrained MOEA is proposed to solve the MOP in (1.11)–(1.14) or LMI-constrained MOP in (1.15)–(1.18) and the LMI-constrained MOPs in the following chapters. First, instead of updating the design parameters in $x = (x_1,...,x_i,...,x_n)^T \in \mathbb{R}^n$ by EA to search the multi-objective vector $(\alpha_1,...,\alpha_i,...,\alpha_N)$ for the Pareto optimal solutions $(\alpha_1^*,...,\alpha_i^*,...,\alpha_N^*)$, we update the multi-objective vector $(\alpha_1,...,\alpha_i,...,\alpha_N)$ by EA to search for the Pareto optimal solutions $(\alpha_1^*,...,\alpha_i^*,...,\alpha_N^*)$ directly and then find the corresponding feasible solutions $\{x_i^*\}_{i=1}^n$ of design parameters indirectly by solving the LMIs in (1.12)–(1.14) with the help of the LMI toolbox in MATLAB. The proposed reverse-order LMI-based MOEA can also avoid the previously mentioned shortage of exceeding the real Pareto front F_p if the current population is close to the real Pareto optimality due to child populations randomly generated by crossover and mutation operations. In the proposed reverse-order LMI-constrained MOEA, if there exists a child population $(\tilde{\alpha}_1,...,\tilde{\alpha}_i,...,\tilde{\alpha}_N)$ that is not feasible for the LMI constraints in (1.12)–(1.14), then a population $(\alpha_1,...,\alpha_i,...,\alpha_N)$ from the parent set that is closest to $(\tilde{\alpha}_1,...,\tilde{\alpha}_i,...,\tilde{\alpha}_N)$ is selected, and the child population is replaced by $\frac{1}{2}(\alpha_1 + \tilde{\alpha}_1,...,\alpha_i + \tilde{\alpha}_i,...,\alpha_N + \tilde{\alpha}_N)$. The mechanism will be repeated until the fixed child population is feasible for the linear constraints in (1.12)–(1.14).

Based on these analyses, the reverse-order MOEA is proposed for the MOP in (1.11)–(1.14):

Step 1: Choose the search region $R = [\underline{\alpha}_1, \bar{\alpha}_1] \times \cdots \times [\underline{\alpha}_i, \bar{\alpha}_i] \times \cdots \times [\underline{\alpha}_N, \bar{\alpha}_N]$ of the multi-objective vector $(\alpha_1,...,\alpha_i,...,\alpha_N)$ and the feasible region $\Omega = [\underline{x}_1, \bar{x}_1] \times \cdots \times [\underline{x}_i, \bar{x}_i] \times \cdots \times [\underline{x}_n, \bar{x}_n]$ of design parameters $\{x_i^*\}_{i=1}^n$. Give the population number N_p, iteration number N_i, crossover rate C_r, and mutation rate c_m of the EA in the proposed reverse-order LMI-constrained MOEA. Set iteration index $j = 1$.

Step 2: Randomly generate feasible objective vectors $\{(\alpha_1^i,...,\alpha_N^i)\}_{i=1}^{N_p}$ from the search region R by examining whether $\{(\alpha_1^i,...,\alpha_N^i)\}_{i=1}^{N_p}$ are feasible in R and the solutions $\{x_i^*\}_{i=1}^n$ of LMIs in (1.12)–(1.14) are feasible in Ω. If not, they are infeasible and should be deleted. Define the parent set as $P_{parent}^j = \{(\alpha_1^i, \cdots, \alpha_N^i)\}_{i=1}^{N_p}$.

Step 3: Employ crossover and mutation operators for the parent set P_{parent}^j and produce N_p child population, that is, $\{\tilde{\alpha}_1^i, \cdots, \tilde{\alpha}_N^i\}_{i=1}^{N_p}$. Define the child set as $P_{parent}^j = \{(\tilde{\alpha}_1^i, \cdots, \tilde{\alpha}_N^i)\}_{i=1}^{N_p}$. If there exists a child population $(\tilde{\alpha}_1, \cdots, \tilde{\alpha}_N)$ that is not feasible for the linear constraints in (1.12)–(1.14), then a population $(\alpha_1, ..., \alpha_i, ..., \alpha_N)$ from the parent set that is closest to $(\tilde{\alpha}_1, \cdots, \tilde{\alpha}_N)$ is selected, and the child population P_{child}^j is feasible for the linear constraints in (1.12)–(1.14); that is, the solutions $\{x_i^*\}_{i=1}^n$ of linear constraints via the LMI toolbox in MATLAB should be in feasible region Ω.

Step 4: Apply the nondominated sorting operator to the set $P_{parent}^j \cup P_{child}^j$ to obtain the corresponding nondominated front $F_p^j = \{F_1^j, F_2^j, ...\}$.

Step 5: Apply the crowded comparison assignment operator to the sets F_i^j and generate the corresponding crowding distance of each element in F_i^j, for $i = 1, ...$ Based on their crowding distance, the set $\{F_i^j\}_{i \in N}$ in F_p^j can be sorted in descending order.

Step 6: Let $t_j \in N$ be the minimum positive integer such that $\sum_{i=1}^{t_j} |F_i^j| > N_p$, where $|S|$ denotes the cardinality of the set S, that is, $t_j = \arg\min_{t_j \in N} \sum_{i=1}^{t_j} |F_i^j| > N_p$.

Update the iteration index $j = j+1$ and let the parent set $P_{parent}^j = \{F_i^j\}_{i=1}^{t_j-1} \cup \tilde{F}_i^{t_j}$ for the next iteration where $\tilde{F}_i^{t_j}$ is the set containing the first G^j elements in $F_{t_j}^j$ and $G^j = N_p - \sum_{i=1}^{t_j} |F_i^j|$.

Step 7: Repeat Steps 3 to 6 until the iteration index $j = N_i$ and set the final population $P_{parent}^{N_i} = F_p$ as the Pareto front.

Step 8: Choose a desired Pareto optimal objective vector $(\alpha_1^*, ..., \alpha_i^*, ..., \alpha_N^*) \in F_p$ according to the one's own reference with the corresponding Pareto optimal solution $\{x_i^*\}_{i=1}^n$ of the MOP in (1.11)–(1.14).

Remark 1.3: The detailed crowded comparison assignment operator and nondominated sorting operator are described as follows [1–4]:

(a) Crowded comparison assignment operator:
  (i) Given a finite set I with the cardinality $l = |I|$ set $\{I_i = 0\}_{i=1}^l$.
  (ii) Set the objective index $j = 1$.
  (iii) Sort I in descending order according to the jth objective value.
  (iv) Assign the values of the first element I_1 and the last element I_l in the sorted set I as follows

$$I_1 = \infty, I_l = \infty$$

  (v) Assign the values to the element $\{I_i\}_{i=2}^{l-1}$ in the sorted set I:

$$I_i = I_i + (f_j^{i+1} - f_j^{i-1})/(f_j^{max} - f_j^{min}), i = 2, ..., l-1$$

where f_j^{i+1} and f_j^{i-1} are the jth objective value of the ith element and $(i-1)$th element in the sorted set I, respectively, and $f_j^{\max}$ and $f_j^{\min}$ are the maximum jth objective value and minimum jth objective value in the sorted set I, respectively.

(vi) Update the objective index $j = j+1$ and repeat steps iii–v.

(b) Nondominating sorting operator:

(i) Given a set P, for each element $p \in P$, generate the corresponding counter index n_p and domination set S_p as follows:

$$S_p = \{q \| q \in P, p < q\}, n_p = \{q \| q \in P, q \le p\}$$

(ii) Define the first domination front F_1 as

$$F_1 = \{p \| p \in P, n_p = 0\}$$

and the jth domination front F_j as

$$F_j = \left\{ q \Big\| p \in F_{j-1}, q \in S_p, n_p = \sum_{i=1}^{j-1} |F_i| \right\}, \text{ for } j \in N$$

where $|S|$ denotes the cardinality S.

1.4 SIMULATION EXAMPLE

After the reverse-order LMI-constrained MOEA is proposed for solving the MOP under a class of linear inequality constraints, with some modification, the proposed algorithm can be used to solve nonlinear constrained MOPs of algebraic systems. In the following, an MOP of the Zitzler-Debi test function in [5] is given to illustrate the design procedure and then validate the performance of the proposed reverse-order LMI-constrained MOEA. Consider the following test problem [5]

$$\min_{\{x_i\}_{i=0}^{30}} (f_1(x), f_2(x)) \tag{1.19}$$

$$\text{subject to } 0 \le x_i \le 1, \text{ for } i = 1,...,30 \tag{1.20}$$

where $x = (x_1,..., x_{30})^T$ is the augmented vector of design variables and $f_1(x)$ and $f_2(x)$ are objective functions with

$$f_1(x) = x_1$$

$$f_2(x) = g(x)\left(1 - \left(\frac{x_1}{g(x)}\right)^2\right)$$

where

$$g(x) = 1 + 9\left(\sum_{i=2}^{30} x_i\right)\Big/29$$

Then, by using the indirect method in (1.4)–(1.7), the MOP in (1.19) and (1.20) can be transformed into the following equivalent MOP

$$\min_{\{x_i\}_{i=0}^{30}}(\alpha_1,\alpha_2) \tag{1.21}$$

$$\text{subject to } 0 \le x_i \le 1, \quad i = 1,...,30 \tag{1.22}$$

$$x_1 \le \alpha_1 \tag{1.23}$$

$$g(x) - x_1^2 / g(x) \le \alpha_2 \tag{1.24}$$

By applying the local linearization in (1.10) of nonlinear constraint in (1.24) at the operation point $x^0 = (x_1,...,x_i,...,x_{30})^T$ with $x_i = 0.5$ for $i = 1,...,30$. Then MOP in (1.21)–(1.24) can be transformed into the following MOP

$$\min(\alpha_1,\alpha_2) \tag{1.25}$$

$$\text{subject to } 0 \le x_i \le 1, \quad i = 1,...,30 \tag{1.26}$$

$$0 \le x_1 \le \alpha_1 \tag{1.27}$$

$$29 - \alpha_2 - 29x_1 + (99 - 9\alpha_2)\sum_{i=2}^{30} x_i \le 0 \tag{1.28}$$

Remark 1.4: (i) The linear inequality–constrained MOP in (1.11)–(1.14) or the LMI-constrained MOP in (1.15)–(1.18) is obtained based on L local linearized constraints in L local regions around L operation points from the nonlinear inequality–constrained MOP in (1.4)–(1.7). For simplicity, the linearized constraint in (1.28) is obtained with only one operation point at $x^0 = (0.5,...,0.5,...,0.5)^T$, and $0 \le x_i \le 1, i = 1,...,30$ is considered a feasible region. (ii) The nonlinear constraint $g^2(x)$ in (1.24) can be obtained as $g^2(x) = 1 + \frac{18}{29}\sum_{i=2}^{30} x_i + (\frac{9}{29})^2\sum_{i=2}^{30} x_i x_j$. Therefore, the linearized $g^2(x)$ at the operation point $x^0 = (0.5,...,0.5,...,0.5)^T$ can be obtained as $1 + \frac{99}{29}\sum_{i=2}^{30} x_i$.

Based on the reverse-order MOEA, setting the population number $N_r = 80$, the iteration number $N_i = 120$, crossover rate $C_r = 0.8$, and mutation rate $C_m = 0.2$ in the proposed LMI-constrained MOEA, the Pareto front of the MOP in (1.25)–(1.28) is shown in Figure 1.1. Compared with the Pareto front in [5], due to only one local linearization, the Pareto front obtained by the proposed reverse-order MOEA is similar but with a more linear curvature. However, it can be seen that the Pareto optimal solutions are uniformly distributed on the Pareto form under the nondominating sorting and crowding comparison. In general, the results can be further improved by increasing the number of operation points for the local linearization of nonlinear constraints as shown in (1.10), that is, with more linear inequality constraints as shown in (1.12)–(1.14) based on the more local linearizations at more operation points. More sophisticated design examples of MOPs of nonlinear systems will be given in the following chapters.

Pareto Front

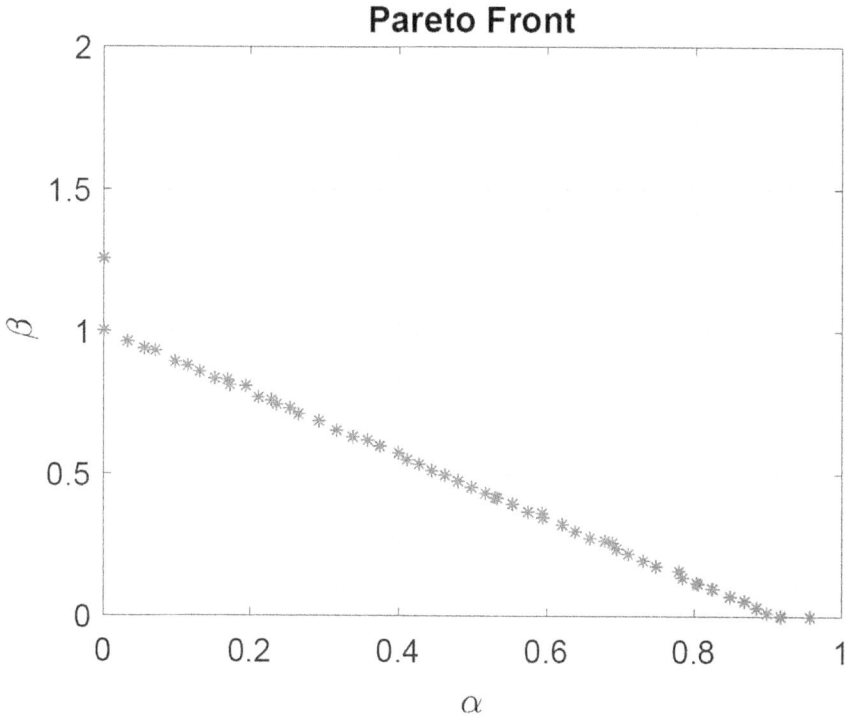

FIGURE 1.1 The Pareto front of the MOP in (1.25)–(1.28) by the proposed LMI-constrained MOEA with the linearized constraint (1.28) at one operation point.

1.5 CONCLUSION

In this chapter, a more general MOP of algebraic systems under nonlinear inequality constraints is introduced first. With the help of local linearized techniques in different local regions, the nonlinear inequality–constrained MOP can be transformed into an MOP under a set of LMI constraints. A reverse-order LMI-constrained MOEA is also proposed to efficiently solve the LMI-constrained MOP. Recently, MOP designs [6–10] of nonlinear stochastic systems in the field of control and signal processing have been more appealing. Several interpolation methods have been utilized to interpolate several local linearized systems to approximate a nonlinear system, such as the fuzzy interpolation method [6, 8], globate linearization method [9], gain scheduling method [10], and so on. Based on these local linearization methods, the proposed reverse-order LMI-constrained MOEA can be further extended to solve dynamic system-constrained MOPs of control and filter design of nonlinear stochastic systems in the following chapters.

2 Multi-Objective Optimization Design for Linear and Nonlinear Stochastic Systems

2.1 INTRODUCTION

After the introduction of MOPs under algebraic inequality constraints [1, 2] in Chapter 1, for more practical applications [6, 7, 9, 11–18], MOPs of the stabilization control of linear and nonlinear stochastic systems will be introduced in this chapter. Some fundamental properties of MOPs of multi-objective stabilization control strategies of stochastic systems are introduced first. For the convenient illustration of MOPs of multi-objective stabilization control design in linear stochastic systems, the MOPs of multi-objective H_2/H_∞ stabilization control design of linear stochastic systems are introduced as a design example. MOPs of multi-objective H_2/H_∞ stabilization control design problems can be transformed into linear matrix inequality–constrained MOPs. A reverse-order MOEA in [13] is also introduced to efficiently solve LMI-constrained MOPs for the multi-objective H_2/H_∞ stabilization control design of linear stochastic systems.

MOPs of multi-objective H_2/H_∞ stabilization on control of nonlinear stochastic systems need to solve a Hamilton-Jacobi inequality (HJI)–constrained MOP. At present, there exists no analytical method to efficiently solve HJIs. Based on the global linearization technique [9, 19], a nonlinear stochastic system can be approximated by an interpolated system consisting of several local linearized stochastic systems through some interpolation functions. Then the HJI-constrained MOP of the multi-objective nonlinear stochastic stabilization control system can be transformed into an LMI-constrained MOP, which can be also solved by a reverse-order LMI-constrained MOEA [2, 13], with some modifications. These results are useful for the development of MOPs of control and filter designs for linear and nonlinear stochastic systems in the following chapters.

2.2 MULTI-OBJECTIVE OPTIMIZATION CONTROL DESIGN PROBLEMS OF LINEAR STOCHASTIC SYSTEMS

Consider the following linear stochastic jump-diffusion system.

$$dx(t) = (Ax(t) + Bu(t) + v(t))dt + Cx(t)dw(t) + Dx(t)dp(t)$$
$$x(0) = x_0 \tag{2.1}$$

DOI: 10.1201/9781003362142-3

where $x(t) \in \mathbb{R}^n$ denotes the state vector; $u(t) \in \mathbb{R}^m$ denotes control input; $v(t) \in \mathbb{R}^n$ denotes the external disturbance; $w(t) \in \mathbb{R}$ denotes the standard Wiener process with zero mean and unit variance; and $p(t) \in \mathbb{R}$ denotes the Poisson counting process with $Ep(t) = \lambda t$, where λ denotes the Poisson jump intensity. $A \in \mathbb{R}^{n \times n}$, $B \in \mathbb{R}^{n \times m}$, $C \in \mathbb{R}^{n \times n}$, and $D \in \mathbb{R}^{n \times n}$ are system matrices.

Remark 2.1: Consider the following linear dynamic system with intrinsic random fluctuation and external disturbance

$$\frac{d}{dt}x(t) = (A + Cn(t) + Dq(t))x(t) + Bu(t) + v(t)$$

$$x(0) = x_0 \tag{2.2}$$

where $n(t) \in \mathbb{R}$ and $q(t) \in \mathbb{R}$ are white noise and Poisson process, respectively. $Cn(t)$ and $Dq(t)$ denote the continuous intrinsic random fluctuation and the discontinuous intrinsic random fluctuation of system matrix A, respectively. By the fact that $dw(t) = n(t)dt$ and $dp(t) = q(t)dt$, then the linear dynamic system with intrinsic fluctuation and external disturbance in (2.2) can be represented by the linear stochastic jump-diffusion system in (2.1).

Some important properties of the Wiener process $w(t)$ and Poisson counting process $p(t)$ are given as follows [11]:

$$
\begin{array}{ll}
(i) E\{w(t)\} = 0 & (ii) E\{dw(t)\} = 0 \\
(iii) E\{d^2 w(t)\} = dt & (iv) E\{dp(t)\} = \lambda dt \\
(v) E\{dp(t)dw(t)\} = 0 & (vi) E\{dp(t)dt\} = 0
\end{array}
\tag{2.3}
$$

For the linear stochastic jump-diffusion system in (2.1), the multi-objective optimization problem for the control design is formulated as follows:

$$\min_{u(t) \in U} (J_1(u(t)), ..., J_i(u(t)), ..., J_N(u(t)))$$

$$\text{subject to (2.1)} \tag{2.4}$$

where U denotes the feasible solution set of $u(t)$.

It is not easy to solve the MOP in (2.4) directly. The following suboptimal method is employed to solve it indirectly.

$$(\alpha_1^*, ..., \alpha_i^*, ..., \alpha_N^*) = \min_{u(t) \in U} (\alpha_1, ..., \alpha_i, ..., \alpha_N) \tag{2.5}$$

$$\text{subject to (2.1) and } J_i(u(t)) \le \alpha_i, i = 1, ..., N \tag{2.6}$$

Before the solution of the MOP in (2.5) and (2.6), some properties of the MOP in (2.5) are given in the following remark.

Remark 2.2: Some properties of the MOP in (2.5) are given as follows [1]:

(1) Pareto dominance: For the MOP in (2.5), suppose there exist two multi-objective feasible solutions $u_1(t)$ and $u_2(t)$ with the corresponding objective

values $(\alpha_1^1,...,\alpha_i^1,...,\alpha_N^1)$ and $(\alpha_1^2,...,\alpha_i^2,...,\alpha_N^2)$, respectively. The solution $u_1(t)$ is said to dominate $u_2(t)$ if $\alpha_1^1 \le \alpha_1^2,...,\alpha_i^1 \le \alpha_i^2,...,\alpha_N^1 \le \alpha_N^2$ and at least one of the inequalities is a strict inequality.

(2) Pareto optimality: The feasible solution $u^*(t)$ with the corresponding objective values $(\alpha_1^*,...,\alpha_i^*,...,\alpha_N^*)$ of the MOP in (2.5) is said to be of Pareto optimality with respect to the feasible solution set U if and only if there does not exist another feasible solution that dominates it.

Based on the properties of the MOP in (2.5) in Remark 2.2, we get the following result.

Theorem 2.1: The suboptimal solution of the MOP in (2.5) for linear stochastic jump-diffusion system in (2.1) is equivalent to the optimal solution of the MOP in (2.4).

Proof: If we can prove that all inequalities in (2.6) will approach equalities when the solution of the MOP in (2.5) and (2.6) is achieved, then the theorem is true. It could be proved by contradiction. If the equalities in (2.6) are violated, then the solution of the MOP is achieved, that is, $J_i(u^*(t)) = \alpha_i' < \alpha_i^*$ for any i. In this solution, $(\alpha_1^*,...,\alpha_i',...,\alpha_N^*)$ will dominate $(\alpha_1^*,...,\alpha_i^*,...,\alpha_N^*)$. According to the Pareto optimality of the MOP, it will violate that $(\alpha_1^*,...,\alpha_i^*,...,\alpha_N^*)$ is the Pareto optimal solution. Therefore, when the solution of the MOP in (2.5)–(2.6) is achieved, $J_i(u^*(t)) = \alpha_i^*$.

Based on Theorem 2.1, the MOP in (2.4) is equivalent to the MOP in (2.5)–(2.6). For the simplicity of illustration of the MOP in (2.4), the multi-objective H_2/H_∞ control strategy for linear stochastic jump-diffusion systems in (2.1) is considered a design example of the MOP in (2.4). Let us denote

$$J_2(u(t)) = E \int_0^\infty (x^T(t)Q_1 x(t) + u^T(t)Ru(t))dt \qquad (2.7)$$

$$J_\infty(u(t)) = \frac{E \int_0^\infty (x^T(t)Q_2 x(t) + u^T(t)Ru(t))dt - x^T(0)Px(0)}{E \int_0^\infty v^T(t)v(t)dt} \qquad (2.8)$$

Remark 2.2: The H_2 control performance in (2.7) is the mean linear quadratic control performance without considering the external disturbance $v(t)$; that is, in the H_2 optimal control design, we always assume $v(t) = 0$ because the H_2 optimal control cannot treat the corruption problem of external disturbance. The H_∞ performance is the effect of external disturbance on the quadratic control performance from the energy perspective. The term $x^T(0)Px(0)$ in the numerator of (2.8) is to extract the effect of initial $x(0)$ on the H_∞ control performance.

Then the multi-objective H_2/H_∞ control design problem is equivalent to the following MOP

$$\min_{u(t) \in U} (J_2(u(t)), J_\infty(u(t))) \qquad (2.9)$$

subject to (2.1)

Based on Theorem 2.1, the MOP of the H_2/H_∞ control strategy in (2.9) is equivalent to the following MOP

$$(\alpha_2^*, \alpha_\infty^*) = \min_{u(t) \in U} (\alpha_2, \alpha_\infty) \qquad (2.10)$$

subject to

$$J_2(u(t)) = E \int_0^\infty (x^T(t)Q_1 x(t) + u^T(t)Ru(t))dt \leq \alpha_2 \qquad (2.11)$$

$$J_\infty(u(t)) = \frac{E \int_0^\infty (x^T(t)Q_2 x(t) + u^T(t)Ru(t))dt - x^T(0)Px(0)}{E \int_0^\infty v^T(t)v(t)dt} \leq \alpha_\infty \qquad (2.12)$$

Before we solve the MOP of the H_2/H_∞ control strategy for the linear stochastic jump diffusion system in (2.1), the Itô-Lévy formula of a linear stochastic jump diffusion system is given for the Lyapunov function $V(x(t)) \in C^2(0, \infty)$ with $V(0) = 0$ and $V(x(t)) > 0$ as follows [7, 11]:

$$\begin{aligned} dV(x(t) &= V_x^T(x)[Ax(t) + Bu(t) + v(t)]dt \\ &+ \tfrac{1}{2}x^T(t)C^T V_{xx}(x)Cx(t)dt \\ &+ V_x^T(x)Cx(t)dw(t) + [V(x(t) + Dx(t)) - V(x(t))]dp(t) \end{aligned} \qquad (2.13)$$

where $V_x(x) = \frac{\partial V(x(t))}{\partial x(t)}$ and $V_{xx}(x) = \frac{\partial^2 V(x(t))}{\partial x^2(t)}$.

Remark 2.3: Since the Wiener process $w(t)$ is continuous but not differential and the Poisson process $p(t)$ is jumping and discontinuous in (2.1), the Itô term $\tfrac{1}{2}x^T(t)C^T V_{xx}(x)Cx(t)$ is due to the differential result by the Itô derivative, and the Lévy term $[V(x(t) + Dx(t)) - V(x(t))]dp$ is due to the derivative by the change of energy via these jumps in the Poisson process.

Based on the Itô-Lévy formula and with the Lyapunov function $V(x(t)) = x^T(t)Px(t)$ for a positive symmetric matrix $P > 0$, then we get the following result.

Theorem 2.2: The MOP of the H_2/H_∞ control strategy in (2.10)–(2.11) can be solved by the following optimal H_2/H_∞ control and the worst-case external disturbance

$$u^*(t) = -R^{-1}B^T P^* x(t) \qquad (2.14)$$

$$v^*(t) = \frac{1}{\alpha_\infty} P^* x(t) \qquad (2.15)$$

where $P^* > 0$ is the solution of the following MOP.

$$\min_{P>0}(\alpha_2, \alpha_\infty) \qquad (2.16)$$

subject to

$$\begin{aligned} &PA + A^T P + Q_1 + C^T PC + \lambda(D^T P + PD + D^T PD) \\ &- PBR^{-1}B^T P \leq 0 \end{aligned} \qquad (2.17)$$

$$TrPR_0 \le \alpha_2 \tag{2.18}$$

$$PA + A^T P + Q_2 + C^T PC + \lambda(D^T P + PD + D^T PD)$$
$$-PBR^{-1}B^T P + \tfrac{1}{\alpha_\infty} PP \le 0 \tag{2.19}$$

where $R_0 = E[x(0)x^T(0)]$.

Proof: Please refer to Appendix 2.5.1.

Since (2.17) and (2.19) are Riccati-like inequalities, it is still not easy to solve the Riccati-like constrained MOP (2.16)–(2.19). Let us denote $W = P^{-1}$ and multiply W to both sides of (2.17) and (2.19). Then we get

$$AW + WA^T + WQ_1 W + WC^T W^{-1} CW$$
$$+ \lambda(WD^T + DW + WD^T W^{-1} DW) - BR^{-1}B^T \le 0 \tag{2.20}$$

and

$$AW + WA^T + WQ_2 W + WC^T W^{-1} CW$$
$$+ \lambda(WD^T + DW + WD^T W^{-1} DW) - BR^{-1}B^T + \tfrac{1}{\alpha_\infty} I \le 0 \tag{2.21}$$

respectively. By several times of operation of Schur complement [9], (2.20) and (2.21) can be represented by the following LMIs, respectively.

$$\begin{bmatrix} AW + WA^T + \lambda(WD^T + DW) & W & WC^T & WD^T \\ -BR^{-1}B^T & & & \\ W & -Q_1^{-1} & 0 & 0 \\ CW & 0 & -W & 0 \\ DW & 0 & 0 & -\tfrac{1}{\lambda}W \end{bmatrix} \le 0 \tag{2.22}$$

$$\begin{bmatrix} AW + WA^T + \lambda(WD^T + DW) & W & WC^T & WD^T \\ -BR^{-1}B^T + \tfrac{1}{\alpha_\infty} I & & & \\ W & -Q_2^{-1} & 0 & 0 \\ CW & 0 & -W & 0 \\ DW & 0 & 0 & -\tfrac{1}{\lambda}W \end{bmatrix} \le 0 \tag{2.23}$$

and (2.18) is equivalent to

$$R_0 - \tfrac{\alpha_2}{n} W \le 0 \tag{2.24}$$

Therefore, the MOP in (2.16)–(2.19) of the multi-objective H_2/H_∞ control design in (2.10)–(2.12) can be transformed into the following LMI-constrained MOP,

$$\min_{W>0}(\alpha_2, \alpha_\infty) \tag{2.25}$$

$$\text{subject to LMIs in (2.22)} - (2.24) \tag{2.26}$$

After solving the LMI-constrained MOP in (2.25)–(2.26), we get the optimal $W^* > 0$, let $P^* = (W^*)^{-1}$, and obtain the multi-objective H_2/H_∞ control

$$u^*(t) = -R^{-1} B^T P^* x(t) \tag{2.27}$$

for the linear stochastic jump-diffusion system in (2.1).

According to the analyses in Chapter 1, the reverse-order LMI-constrained MOEA to solve the MOP in (2.25)–(2.26) of the H_2/H_∞ control design of the linear stochastic jump diffusion system in (2.1) is proposed as follows:

The reverse-order LMI-constrained MOEA for the MOP in (2.25)–(2.26) for the multi-objective H_2/H_∞ control design of the linear stochastic jump diffusion system in (2.1).

Step 1: Select the search region $[\underline{\alpha}_2, \overline{\alpha}_\infty] \times [\underline{\beta}_2, \overline{\beta}_\infty]$ for the feasible objective vector $(\alpha_2, \alpha_\infty)$ and set the iteration number N_i, the population number N_p, the crossover rate C_r, and mutation ratio m_r in the LMI-constrained MOEA. Set iteration number $i = 1$.

Step 2: Select N_p feasible individuals (chromosomes) from the feasible chromosome set randomly to be the initial population P_1.

Step 3: Operate the EA with crossover rate C_r and mutation ratio m_r, and generate $2N_p$ feasible chromosomes by examining whether their corresponding objective vectors $(\alpha_2, \alpha_\infty)$ are feasible objective vectors for the LMIs (2.22)–(2.24).

Step 4: Set the iteration index $i = i + 1$ and select N_p chromosomes from the $2N_p$ feasible chromosomes in Step 3 through the nondominated sorting method to be the population P_{i+1}.

Step 5: Repeat Steps 3 and 4 until the iteration number N_i is reached. If the iteration number N_i is satisfied, then set $P_{N_i} = P_F$ as the Pareto front.

Step 6: Select a "preferable" objective individual $(\alpha_2^*, \alpha_\infty^*) \in P_F$ according to the designer's preference with the optimal $W^* = (P^*)^{-1}$. Then the multi-objective H_2/H_∞ control $u^*(t) = -R^{-1} B^T P^* x(t)$.

Remark 2.4: The previous reverse-order LMI-constrained MOEA for multi-objective H_2/H_∞ control design in (2.10)–(2.12) could be easily extended to the more general MOP in (2.4) or (2.5)–(2.6) if objective functions $J_i(u), i = 1,..., N$ are given.

2.3 MULTI-OBJECTIVE OPTIMIZATION CONTROL DESIGN PROBLEMS OF NONLINEAR STOCHASTIC SYSTEMS

Since most physical systems are nonlinear, it is more appealing to discuss the MOPs of control design of nonlinear stochastic systems for more practical applications.

Consider the following nonlinear stochastic jump diffusion systems [6, 7, 11–13]

$$\begin{aligned} dx(t) &= f(x(t)) + g(x(t))u(t) + h(x(t))v(t) + l(x(t))dw(t) \\ &\quad + d(x(t))dp(t) \end{aligned} \tag{2.28}$$

where the vector $x(t) \in \mathbb{R}^n$, control input $u(t) \in \mathbb{R}^m$, and external disturbance $v(t) \in \mathbb{R}^l$. $w(t) \in \mathbb{R}$ is the standard Wiener process, and $l(x(t))dw(t)$ denotes continuous intrinsic random fluctuation. $p(t)$ denotes the Poisson counting process with mean λ in a unit of time, and $d(x(t))dp(t)$ is regarded as intrinsic discontinuous random fluctuation (jumping process). The nonlinear functional vector $f(x(t)) \in \mathbb{R}^n$, input functional matrix $g(x(t)) \in \mathbb{R}^{n \times m}$, and disturbance coupling functional matrix $h(x(t)) \in \mathbb{R}^{n \times l}$ $l(x(t)) \in \mathbb{R}^n$ and $d(x(t)) \in \mathbb{R}^n$ must satisfy the Lipschitz continuity to guarantee the unique solution of (2.28).

For the nonlinear stochastic jump-diffusion system in (2.28), the multi-objective optimization problem for the control design can be formulated as follows

$$\min_{u(t) \in U} (J_1(u), ..., J_i(u), ..., J_N(u)) \tag{2.29}$$

$$\text{subject to (2.28)}$$

Since it is not easy to solve the MOP in (2.29) directly, the following suboptimal method is employed to solve it indirectly.

$$(\alpha_1^*, ..., \alpha_i^*, ..., \alpha_N^*) = \min_{u(t) \in U} (\alpha_1, ..., \alpha_i, ..., \alpha_N) \tag{2.30}$$

$$\text{subject to (2.28) and } J_i(u(t)) \le \alpha_i, i = 1, ..., N \tag{2.31}$$

Lemma 2.1: The suboptimal solution of the MOP in (2.30) and (2.31) for the non-linear stochastic jump diffusion system in (2.28) is equivalent to the optimal solution of the MOP in (2.29).

Proof: Similar to the proof in Theorem 2.1.

From Lemma 2.1, the MOP in (2.29) for the nonlinear stochastic jump diffusion system in (2.28) can be solved via the MOP in (2.30) and (2.31). For a simple illustration of the MOP of the nonlinear stochastic jump diffusion system in (2.28), the following multi-objective H_2/H_∞ control design is given as a design example of the MOP in (2.30) and (2.31).

$$(\alpha_2^*, \alpha_\infty^*) = \min_{u(t) \in U} (\alpha_2, \alpha_\infty) \tag{2.32}$$

$$\text{subject to } J_2(u) = E \int_0^\infty (x^T(t)Q_1 x(t) + u^T(t)Ru(t))dt \le \alpha_2 \tag{2.33}$$

$$J_\infty(u) = \frac{E \int_0^\infty (x^T(t)Q_2 x(t) + u^T(t)Ru(t))dt - V(x(0))}{E \int_0^\infty v^T(t)v(t))dt} \le \alpha_2 \tag{2.34}$$

Before solving the MOP in (2.32)–(2.34), the Itô-Lévy formula of a nonlinear stochastic jump diffusion system is given as follows [12]:

$$
\begin{aligned}
dV(x(t)) = \; & V_x^T(x(t))[f(x(t)) + g(x(t))u(t) + h(x(t))v(t)]dt \\
& + \tfrac{1}{2}l^T(x(t))V_{xx}(x(t))l(x(t))dt \\
& + V_x^T(x(t))l(x(t))dw(t) \\
& + [V(x(t) + d(x(t))) - V(x(t))]dp(t)
\end{aligned}
\tag{2.35}
$$

Based on the Itô-Lévy formula in (2.35), we get the following result:

Theorem 2.3: The MOP of the H_2/H_∞ control strategy in (2.32)–(2.34) can be solved as follows

$$
u^*(t) = -\tfrac{1}{2}R^{-1}g^T(x(t))V_x^*(x(t))
\tag{2.36}
$$

$$
v^*(t) = \tfrac{1}{2\alpha_\infty}R^{-1}g^T(x(t))V_x^*(x(t))
\tag{2.37}
$$

where $V^*(x(t))$ is the solution of the following MOP.

$$
\min_{V(x(t))} (\alpha_2, \alpha_\infty)
\tag{2.38}
$$

subject to the following Hamilton Jacobi inequalities,

$$
\begin{aligned}
& V_x^T(x(t))f(x(t)) + x^T(t)Q_1x(t) + \tfrac{1}{2}l^T(x(t))V_{xx}(x(t))l(x(t)) \\
& + \lambda[V(x(t) + d(x(t))) - V(x(t))] \\
& + \tfrac{1}{4}V_x^T(x(t))g(x(t))R^{-1}g^T(x(t))V_x(x(t)) \leq 0
\end{aligned}
\tag{2.39}
$$

$$
EV(x(0)) \leq \alpha_2
\tag{2.40}
$$

$$
\begin{aligned}
& V_x^T(x(t))f(x(t)) + x^T(t)Q_2x(t) + \tfrac{1}{2}l^T(x(t))V_{xx}(x(t))l(x(t)) \\
& + \lambda[V(x(t) + d(x(t))) - V(x(t))] \\
& + \tfrac{1}{4}V_x^T(x(t))g(x(t))R^{-1}g^T(x(t))V_x(x(t)) \\
& + \tfrac{1}{4\alpha_\infty}V_x^T(x(t))h(x(t))h^T(x(t))V_x(x(t)) \leq 0
\end{aligned}
\tag{2.41}
$$

Proof: Please refer to Appendix 2.5.2.

Since the HJIs in (2.39) and (2.41) are nonlinear partial differential inequality equations, it is very difficult to solve the HJI-constrained MOP in (2.38)–(2.41) for the multi-objective H_2/H_∞ control strategy for the nonlinear stochastic jump diffusion system in (2.28). In this chapter, the global linearization method [9, 19] is employed to interpolate with a set of local linearized systems at the vertices of the polytope to approximate the nonlinear stochastic jump diffusion system in (2.28). This can overcome the difficult in solving the HJI-constrained MOP in (2.38)–(2.41). This leads to bounding all local linearized systems by a polytope C_0 with J vertices as follows [9, 19]:

$$\begin{bmatrix} \dfrac{\partial f(x(t))}{\partial x(t)} \\[2mm] \dfrac{\partial g(x(t))}{\partial x(t)} \\[2mm] \dfrac{\partial h(x(t))}{\partial x(t)} \\[2mm] \dfrac{\partial l(x(t))}{\partial x(t)} \\[2mm] \dfrac{\partial d(x(t))}{\partial x(t)} \end{bmatrix} \in C_0 \left\{ \begin{bmatrix} A_1 \\ B_1 \\ H_1 \\ C_1 \\ D_1 \end{bmatrix} \cdots \begin{bmatrix} A_i \\ B_i \\ H_i \\ C_i \\ D_i \end{bmatrix} \cdots \begin{bmatrix} A_J \\ B_J \\ H_J \\ C_J \\ D_J \end{bmatrix} \right\}, \forall x(t) \tag{2.42}$$

Normally, we can say that the trajectory $x(t)$ of the nonlinear stochastic jump diffusion system in (2.28) can be represented by the convex combination of the trajectories of the following J local linearized stochastic systems at J vertices of the polytope if the convex hull C_0 consists of all local linearized systems at all [9, 19]

$$dx(t) = (A_i x(t) + B_i u(t) + H_i v(t))dt + C_i x(t)dw(t)$$
$$+ D_i x(t)dp(t), i = 1,...,J \tag{2.43}$$

Based on the global linearization theory [19], the trajectory $x(t)$ of the nonlinear stochastic jump diffusion system in (2.28) can be represented by a convex combination of the trajectories of the J local linearized stochastic jump diffusion systems in (2.43) as follows:

$$dx(t) = \sum_{i=1}^{J} h_i(x(t))(A_i x(t) + B_i u(t) + H_i v(t))dt + C_i x(t)dw(t)$$
$$+ D_i x(t)dp(t) \tag{2.44}$$

where $h_i(x(t)), i = 1,...,J$ denotes the interpolation functions with $0 \le h_i(x(t)) \le 1$ and $\sum_{i=1}^{J} h_i(x(t)) = 1$; that is, we can replace the trajectory of the nonlinear stochastic jump diffusion system in (2.28) with the trajectory of the interpolated stochastic system in (2.44).

Remark 2.5: Except the global linearization method, there are other interpolation methods to interpolate local linearized stochastic systems by different interpolation functions to efficiently approximate a nonlinear stochastic system, for example, the T-S fuzzy method through the interpolatory fuzzy bases [6, 7, 11, 12].

Lemma 2.2 ([7, 11]): For any matrix S_i with appropriate dimensions and interpolation function $h_i(x(t))$ with $0 \le h_i(x(t)) \le 1$ and $\sum_{i=1}^{J} h_i(x(t)) = 1$, then for any matrix $P > 0$, we have

$$\left(\sum_{i=1}^{J}h_i(x(t))S_i^T\right)P\left(\sum_{i=1}^{J}h_i(x(t))S_i\right)\leq\sum_{i=1}^{J}h_i(x(t))S_i^T PS_i \qquad (2.45)$$

Theorem 2.4: If the nonlinear stochastic jump diffusion system in (2.28) is represented by the global linearization in (2.44), then Theorem 2.3 for the MOP of the H_2/H_∞ control strategy in (2.36)–(2.41) can be solved with $V(x(t))=x^T(t)Px(t)$ by the following

$$u^*(t)=\sum_{i=1}^{J}h_i(x(t))R^{-1}B_iP^*x(t) \qquad (2.46)$$

$$v^*(t)=\frac{1}{\alpha_\infty}\sum_{i=1}^{J}h_i(x(t))B_iP^*x(t) \qquad (2.47)$$

where $P^*>0$ is the solution of the following Riccati-like inequality-constrained MOP

$$\min_{P>0}(\alpha_2,\alpha_\infty) \qquad (2.48)$$

subject to $A_i^TP+PA_i+Q_1-PB_iR^{-1}B_i^TP+C_i^TPC_i$
$$+\lambda(D_i^TP+PD_i+D_i^TPD_i)\leq 0 \qquad (2.49)$$

$$TrPR_0\leq\alpha_2 \qquad (2.50)$$

$$A_i^TP+PA_i+Q_1-PB_iR^{-1}B_i^TP+C_i^TPC_i$$
$$+\lambda(D_i^TP+PD_i+D_i^TPD_i)-PB_iR^{-1}B_i^TP+\frac{1}{\alpha_\infty}PH_iH_i^TP\leq 0 \qquad (2.51)$$

for $i=1,...,J$

Proof: Please refer to Appendix C.

Let $W=P^{-1}$. Following a similar procedure as in (2.20) to (2.24), the MOP in (2.48)–(2.51) is equivalent to the following MOP

$$\min_{W>0}(\alpha_2,\alpha_\infty) \qquad (2.52)$$

subject to

$$\begin{bmatrix} A_iW+WA_i^T+\lambda(WD_i^T+D_iW) & & & \\ -B_iR^{-1}B_i^T & W & WC_i^T & WD_i^T \\ & & & \\ W & -Q_1^{-1} & 0 & 0 \\ C_iW & 0 & -W & 0 \\ D_iW & 0 & 0 & -\frac{1}{\lambda}W \end{bmatrix}\leq 0 \qquad (2.53)$$

$$
\begin{bmatrix}
\begin{matrix} A_i W + W A_i^T + \lambda (W D_i^T + D_i W) \\ -B_i R^{-1} B_i^T + \frac{1}{\alpha_\infty} H_i H_i^T \end{matrix} & W & W C_i^T & W D_i^T \\
W & -Q_2^{-1} & 0 & 0 \\
C_i W & 0 & -W & 0 \\
D_i W & 0 & 0 & -\frac{1}{\lambda} W
\end{bmatrix} \le 0 \qquad (2.54)
$$

$$
R_0 - \frac{\alpha_2}{n} W \le 0 \qquad (2.55)
$$

Based on the previous analysis, if we could solve W^* from the MOP in (2.52)–(2.55), then we can let $P^* = (W^*)^{-1}$ and obtain the multi-objective H_2/H_∞ control $u^*(t)$ in (2.46). Further, the proposed reverse-order LMI-constrained MOEA in the previous section could be employed to solve the MOP in (2.52)–(2.55) with some modifications.

2.4   CONCLUSION

In this chapter, we investigate the MOPs of multi-objective stabilization control strategies of linear and nonlinear stochastic jump diffusion systems. Further, the MOPs of multi-objective H_2/H_∞ stabilization control strategies of linear and nonlinear stochastic jump diffusion systems with external disturbance are given as design examples to illustrate the design procedure of MOPs in the multi-objective optimization control design of linear and nonlinear jump diffusion systems with external disturbance. A reverse-order LMI-constrained MOEA is also proposed to treat the MOPs of multi-objective stabilization control of linear and nonlinear stochastic jump diffusion systems with external disturbance. These results can be easily extended to the MOPs of multi-objective filter design in signal processing [9, 14–16], power control of wireless communication networks [17, 18], or multi-objective observer-based tracking control design of linear and nonlinear stochastic jump diffusion systems in the following chapters.

2.5   APPENDIX

2.5.1   Proof of Theorem 2.2

From (2.11) and without consideration of $v(t)$ in (2.1), we get

$$
\begin{aligned}
& J_2(u(t)) \\
& = E \int_0^\infty (x^T(t) Q_1 x(t) + u^T(t) R u(t)) dt \\
& = E x^T(0) P x(0) - E x^T(\infty) P x(\infty) + E \int_0^\infty (x^T(t) Q_1 x(t) \\
& \quad + u^T(t) R u(t) + \frac{dx^T(t) P x(t)}{dt}) dt
\end{aligned} \qquad (2.56)
$$

By the Itô-Lévy formula in (2.13) with $V(x(t)) = x^T(t)Px(t)$, we get

$$
\begin{aligned}
&J_2(u(t))\\
&= Ex^T(0)Px(0) - Ex^T(\infty)Px(\infty) + E\int_0^\infty (x^T(t)Q_1 x(t)\\
&\quad + u^T(t)Ru(t) + x^T(t)P[Ax(t) + Bu(t)]Px(t)\\
&\quad + x^T(t)C^T PCx(t) + \lambda x^T(t)(D^T P + PD + D^T PD)x(t))dt\\
&= Ex^T(0)Px(0) - Ex^T(\infty)Px(\infty)\\
&\quad + E\int_0^\infty [(Ru(t) + B^T Px(t))^T R^{-1}(Ru(t) + B^T Px(t))\\
&\quad + x^T(t)(Q_1 + PA + A^T P + C^T PC\\
&\quad + \lambda(D^T P + PD + D^T PD) - PBR^{-1}B^T P)x(t)]dt
\end{aligned}
\tag{2.57}
$$

By (2.14) and (2.17), we get

$$
\begin{aligned}
J_2(u(t)) &\le Ex^T(0)Px(0) - Ex^T(\infty)Px(\infty)\\
&\le Ex^T(0)Px(0) = TrPR_0 \le \alpha_2
\end{aligned}
\tag{2.58}
$$

By the Itô-Lévy formula in (2.13) with $V(x(t)) = x^T(t)Px(t)$, the numerator of (2.11) is equivalent to the following

$$
\begin{aligned}
&-Ex^T(0)Px(0) + E\int_0^\infty (x^T(t)Q_2 x(t) + u^T(t)Ru(t))dt\\
&= -Ex^T(\infty)Px(\infty) + E\int_0^\infty (x^T(t)Q_2 x(t) + u^T(t)Ru(t))dt\\
&\quad + dx^T(t)Px(t)\\
&= -Ex^T(\infty)Px(\infty) + E\int_0^\infty [x^T(t)Q_2 x(t)\\
&\quad + u^T(t)Ru(t) + x^T(t)P[Ax(t) + Bu(t) + v(t)]\\
&\quad + [x^T(t)A^T + u^T(t)B^T + v^T(t)]Px(t)\\
&\quad + x^T(t)C^T PCx(t) + \lambda x^T(t)(D^T P + PD + D^T PD)x(t)]dt\\
&= -Ex^T(\infty)Px(\infty)\\
&\quad + E\int_0^\infty \{(Ru(t) + B^T Px(t))^T R^{-1}(Ru(t) + B^T Px(t))\\
&\quad - (\alpha_\infty^{\frac{1}{2}}v(t) - \alpha_\infty^{-\frac{1}{2}}Px(t))^T (\alpha_\infty^{\frac{1}{2}}v(t) - \alpha_\infty^{-\frac{1}{2}}Px(t))\\
&\quad + x^T(t)(Q_2 + PA + A^T P + C^T PC\\
&\quad + \lambda(D^T P + PD + D^T PD) - PBR^{-1}B' P + \tfrac{1}{\alpha_\infty}PP)x(t)\\
&\quad + \alpha_\infty v^T(t)v(t)]dt\\
&\le E\int_0^\infty v^T(t)v(t)dt \quad \left(by\ (2.14),(2.15)\ and\ (2.19)\right)
\end{aligned}
\tag{2.59}
$$

Therefore, we get

$$
\frac{E\int_0^\infty (x^T(t)Q_1 x(t) + u^T(t)Ru(t))dt - Ex^T(0)Px(0)}{E\int_0^\infty v^T(t)v(t)dt} \le \alpha_\infty
\tag{2.60}
$$

which is (2.11). Therefore, the MOP in (2.10)–(2.12) can be transformed into the MOP in (2.16)–(2.19).

2.5.2 Proof of Theorem 2.3

From (2.33), we get

$$
\begin{aligned}
J_2(u(t)) &= EV(x(0)) - EV(x(\infty)) \\
&\quad + E\int_0^\infty (x^T(t)Q_1x(t) + u^T(t)Ru(t))dt + dV(x(t))
\end{aligned}
\tag{2.61}
$$

By the Itô-Lévy formula in (2.35), we get

$$
\begin{aligned}
&J_2(u(t)) \\
&= EV(x(0)) - EV(x(\infty)) + E\int_0^\infty \{x^T(t)Q_1x(t) + u^T(t)Ru(t) \\
&\quad + V_x^T(x(t))[f(x(t)) + g(x(t))u(t)] + \tfrac{1}{2}l^T(x(t))V_{xx}(x(t))l(x(t) \\
&\quad + \lambda[V(x(t) + d(x(t)) + V(x(t))]\}dt \\
&= EV(x(0)) - EV(x(\infty)) + E\int_0^\infty \{x^T(t)Q_1x(t) + V_x^T(x(t))^T f(x) \\
&\quad + \tfrac{1}{2}l^T(x(t))V_{xx}(x(t))l(x(t) + \lambda[V(x(t) + d(x(t)) - V(x(t))] \\
&\quad - \tfrac{1}{4}V_x^T(x(t))g(x(t))R^{-1}g^T(x(t))V_x(x(t)) \\
&\quad + (Ru(t) + \tfrac{1}{2}g^T(x(t))V_x(x(t)))R^{-1}(Ru(t) + \tfrac{1}{2}g^T(x(t))V_x(x(t)))\}dt
\end{aligned}
\tag{2.62}
$$

By (2.36) and HJI on (2.39), we get $J_2(u(t)) = EV(x(0)) - EV(x(\infty)) \le EV(x(0)) \le \alpha_2$, which is (2.40).

From the numerator of (2.34), by the Itô-Lévy formula in (2.35)

$$
\begin{aligned}
&E\int_0^\infty (x^T(t)Q_2x(t) + u^T(t)Ru(t))dt - V(x(0)) \\
&= -V(x(\infty)) + E\int_0^\infty (x^T(t)Q_2x(t) + u^T(t)Ru(t)) + dV(x(t)) \\
&= -EV(x(\infty)) + E\int_0^\infty \{x^T(t)Q_2x(t) + V_x^T(x(t))^T f(x) \\
&\quad + \tfrac{1}{2}l^T(x(t))V_{xx}(x(t))l(x(t) + \lambda[V(x(t) + d(x(t)) - V(x(t))] \\
&\quad - \tfrac{1}{4}V_x^T(x(t))g(x(t))R^{-1}g^T(x(t))V_x(x(t)) \\
&\quad + \tfrac{1}{4\alpha_\infty}V_x(x(t))h(x(t))h^T(x(t))V_x(x(t)) + \alpha_\infty v^T(t)v(t) \\
&\quad + (Ru(t) + \tfrac{1}{2}g^T(x(t))V_x(x(t)))R^{-1}(Ru(t) + \tfrac{1}{2}g^T(x(t))V_x(x(t))) \\
&\quad - (\alpha_\infty^{\frac{1}{2}}v(t) - \tfrac{1}{2}\alpha_\infty^{-\frac{1}{2}}h^T(x(t))V_x(x(t)))^T (\alpha_\infty^{\frac{1}{2}}v(t) - \tfrac{1}{2}\alpha_\infty^{-\frac{1}{2}}h^T(x(t))V_x(x(t)))\}dt
\end{aligned}
\tag{2.63}
$$

By (2.30), (2.37), and HJI in (2.41), we get

$$
E\int_0^\infty (x^T(t)Q_1x(t) + u^T(t)Ru(t))dt - V(x(0)) \le \alpha_\infty E\int_0^\infty v^T(t)v(t)dt
\tag{2.64}
$$

which is (2.34). Therefore the multi-objective H_2/H_∞ control design in (2.32)—(2.34) can be transformed into the MOP in (2.36)—(2.41).

2.5.3 PROOF OF THEOREM 2.4

Based on Theorem 2.3, we select the Lyapunov function $V(x(t)) = x^T(t)Px(t)$. By the global linearization method in (2.42)–(2.44), we get

$$f(x(t)) = \sum_{i=1}^{J} h_i(x(t))A_i x(t), g(x(t)) = \sum_{i=1}^{J} h_i(x(t))B_i x(t)$$

$$l(x(t)) = \sum_{i=1}^{J} h_i(x(t))C_i x(t), h(x(t)) = \sum_{i=1}^{J} h_i(x(t))H_i x(t) \qquad (2.65)$$

$$d(x(t)) = \sum_{i=1}^{J} h_i(x(t))D_i x(t)$$

Then

$$u^*(t) = -\frac{1}{2}R^{-1}g^T(x(t))V_x(x(t)) = -\sum_{i=1}^{J} h_i(x(t))R^{-1}B_i Px(t) \qquad (2.66)$$

$$v^*(t) = \frac{1}{2\alpha_\infty}g^T(x(t))V_x(x(t)) = \frac{1}{\alpha_\infty}\sum_{i=1}^{J} h_i(x(t))B_i Px(t) \qquad (2.67)$$

which are (2.46) and (2.47), respectively.

Substituting (2.65) into HJIs in (2.39) and (2.41), we get

$$\sum_{i=1}^{J} h_i(x(t))x_i^T(t)[A_i^T P + PA_i + Q_1 - PB_i R^{-1}B_i^T P + C_i^T PC_i$$

$$+\lambda(D_i^T P + PD_i + D_i^T PD_i)x_i(t)] \le 0 \qquad (2.68)$$

$$\sum_{i=1}^{J} h_i(x(t))x_i^T(t)[A_i^T P + PA_i + Q_2 - PB_i R^{-1}B_i^T P + C_i^T PC_i$$

$$+\lambda(D_i^T P + PD_i + D_i^T PD_i) + \frac{1}{\alpha_m}PH_i H_i^T P] \le 0 \qquad (2.69)$$

respectively. Then from the inequalities in (2.68) and (2.69), we get the Riccati-like inequalities in (2.49) and (2.51), respectively. And with $V(x(t)) = x^T(t)Px(t)$, then (2.40) becomes (2.50).

Part II

Multi-Objective Optimization Designs in Control Systems

3 Multi-Objective H_2/H_∞ Stabilization Control Strategies of Nonlinear Stochastic Systems

3.1 INTRODUCTION

Since many stochastic systems in engineering, economics, and biology contain both continuous and Poisson jump processes—an Itô-Lévy type system [10, 20–24]—H_2 and H_∞ control design for stochastic Poisson jump-diffusion systems is an active field. The H_2 control strategy is an important optimal control design for minimizing a desired control performance index, such as a cost function [25–27]. The H_∞ control strategy is an important robust control design for eliminating the effects of external disturbances on control performance, and it has been employed to handle robust control problems with uncertain disturbances [28–31]. H_2 and H_∞ control design problems for linear stochastic Poisson jump-diffusion systems have been widely discussed in the last decade [32–35], but few studies discuss the control design problems of nonlinear stochastic Poisson jump-diffusion systems. In practice, designers always expect the system controller to be not only robust but also optimal. Combining the advantages of both the H_2 and H_∞ controls, a stochastic mixed H_2/H_∞ control design has been proposed [36, 37] and realized in practical linear stochastic Poisson jump-diffusion system control problems [35]. However, the stochastic mixed H_2/H_∞ control design is employed to design a controller that minimizes the H_2 performance index under the constraint of a prescribed H_∞ attenuation level (i.e., a constrained single-objective optimization problem (SOP)), thus providing a suboptimal solution for robust controller design. In this chapter, we provide a multi-objective control design for nonlinear stochastic Poisson jump-diffusion systems with external disturbances. Unlike the conventional mixed H_2/H_∞ control design, the multi-objective control problem is defined as how to design a controlled input $u(t)$ to simultaneously minimize the H_2 and H_∞ control performance indices in the Pareto optimal sense for nonlinear stochastic Poisson jump-diffusion systems.

The multi-objective control problem appears in [38, 39]. Since the H_2 and H_∞ control performance indices generally compete for the admissible control input $u(t)$, they can be regarded as conflicting with one another in a multi-objective optimization problem. In fact, for a multi-objective control problem, it is impossible to improve the H_2 performance index without deteriorating the H_∞ performance index, and vice versa. Moreover, there exists no unique solution $u(t)$ corresponding to both the minimum values of the H_2 and H_∞ performance indices. Rather, there exists a set

DOI: 10.1201/9781003362142-5

of feasible optimal solutions for the multi-objective control design problem in the Pareto optimal sense. The "optimal" design of the MOP for the nonlinear stochastic Poisson jump-diffusion system can be seen as how to select a preferred solution from the Pareto front to obtain the Pareto optimal solution.

There are two main difficulties in solving the multi-objective control problem for nonlinear stochastic Poisson jump-diffusion systems. The first issue is that the Pareto front of the multi-objective control problem is difficult to obtain through direct calculation. The second is that the Hamilton-Jacobi inequalities for the nonlinear stochastic Poisson jump-diffusion system are also difficult to solve. For the first issue, an indirect method is proposed for solving the MOP of multi-objective control by simultaneously minimizing the upper bounds (α, β) of the H_2 and H_∞ performance indices. By applying the proposed indirect method, the multi-objective control problem can be transformed into an HJI-constrained MOP (i.e., the solutions of the multi-objective control for a nonlinear stochastic Poisson jump-diffusion system are constrained by two HJIs). However, the HJIs are still difficult to solve except in some special cases. For the second issue, the T-S fuzzy interpolation method [2, 40, 41] is introduced to approximate the nonlinear stochastic jump-diffusion system by interpolating a set of linear stochastic jump-diffusion systems. Based on the T-S fuzzy interpolation scheme, a set of linear matrix inequalities can replace the HJIs. Thus, the HJI-constrained MOP can be replaced by an LMI-constrained MOP. Therefore, the multi-objective control for nonlinear stochastic Poisson jump-diffusion systems becomes an issue of how to specify a set of T-S fuzzy control gains $\{K_i\}_{i=1}^m$ to minimize (α, β), simultaneously subject to a set of LMI constraints, where m is the number of local linear models used to approximate the nonlinear stochastic Poisson jump-diffusion system via the T-S fuzzy interpolation method and α and β are the upper bounds of H_2 and H_∞ control performance, respectively.

The LMI-constrained MOP is quite different from the conventional MOP with algebraic constraints in Chapter 1. Recently, many MOEAs for Pareto optimality have been proposed for solving MOPs [38, 42], but most have been limited to static systems, which lead to algebraic functional constraints. Hence, much work is necessary to apply the MOEA to the multi-objective control problem for nonlinear stochastic Poisson jump-diffusion systems. In this chapter, based on the indirect method and the T-S fuzzy interpolation method, we propose a reverse-order LMI-constrained MOEA for the multi-objective control design problem of nonlinear stochastic jump-diffusion systems. The reverse-order LMI-constrained MOEA searches for the Pareto front of the multi-objective control strategy for the nonlinear stochastic jump-diffusion system via evolution operators and a nondominated sorting scheme. By gradually decreasing (α, β) simultaneously and maintaining its domination via a nondominating sorting scheme, the Pareto optimal solution set of the multi-objective control for the nonlinear stochastic Poisson jump-diffusion system can be approached indirectly. As long as the Pareto front is obtained, the multi-objective control design problem for the nonlinear stochastic Poisson jump-diffusion system can be solved, since each vector (α^*, β^*) of the Pareto front P_{front}^* corresponds to a group of T-S fuzzy control gains $\{K_i^*\}_{i=1}^m$, guaranteeing that the H_2 and H_∞ control performance indices (α, β) are minimized simultaneously. Finally, we give a multi-objective missile pursuit system with cheating jets and a wind-gust disturbance as an example to illustrate the design procedure and confirm the performance of the proposed design method.

This chapter provides an efficient method to transform the multi-objective control problem of nonlinear Poisson jump-diffusion systems into an LMI-constrained MOP, which can be efficiently solved with the help of the proposed LMI-constrained MOEA search method. Once the Pareto front is obtained, the designer can select an T-S fuzzy controller from the Pareto optimal solution set P_{set}^* according to their preference.

This chapter is divided into five sections. In Section 3.2, we describe a nonlinear stochastic Poisson jump system with jump noise. We define the multi-objective control problem in Section 3.3, where we apply an indirect method for solving the multi-objective control problem of a nonlinear stochastic Poisson jump-diffusion system by minimizing the upper bounds (α, β) simultaneously. In Section 3.4, we propose multi-objective control via the T-S fuzzy method to transform the HJI-constrained MOP into an LMI-constrained MOP. The multi-objective controller design via the LMI-constrained MOEA is introduced in Section 3.5. A simulation example of guidance control of a missile of a missile pursuit system is given in Section 3.6, and conclusions are given in Section 3.7.

Notation: A^T: the transpose of matrix A; $A \geq 0$ $(A > 0)$: symmetric positive semi-definite (symmetric positive definite) matrix A; I: identity matrix; $\|x\|_2$: Euclidean norm for the given vector $x \in \mathbb{R}^n$; c^2: class of functions $V(x)$ twice continuously differential with respect to x; f_x: gradient column vector of n_x-dimensional twice continuously differentiable function $f(x)$ (i.e., $\partial f(x)/\partial x$); f_{xx}: Hessian matrix with elements of second partial derivatives of n_x-dimensional twice continuously differentiable function $f(x)$, (i.e., $\partial^2 f(x)/\partial x^2$); $L_F^2\left(\mathbb{R}^+, \mathbb{R}^{n_y}\right)$: space of nonanticipative stochastic processes $y(t) \in \mathbb{R}^l$ with respect to an increasing σ-algebra $F_t (t \geq 0)$ satisfying $\|y(t)\|_{L^2(\mathbb{R}^+, \mathbb{R}^{n_y})} \triangleq E\{\int_0^\infty y^T(t)y(t)dt\}^{\frac{1}{2}} < \infty$; $B(\Theta)$: Borel algebra generated by Θ; E: expectation operator; $P\{\cdot\}$: probability measure function; and $\bar{\lambda}(M)$: maximum eigenvalue of real-value matrix M.

3.2 PRELIMINARIES

For increased versatility of theories in this chapter, we assume the Poisson jump process to be a marked Poisson point process, and its definition is given as follows:

Definition 3.1 ([41]): A stochastic process $\Pi(t)$ is called a marked Poisson jump process (also called a marked Poisson point process) if

$$\Pi(t) = \int_0^t \int_\Theta T(x(s^-), \theta) N(ds, d\theta) \tag{3.1}$$

where Θ is the Poisson mark space, the jump amplitude $T(x(s^-), \cdot)$ is assumed to be bounded continuous function and satisfied with $T(x(0), \cdot) = 0$, and $N(\cdot, \cdot)$ is the Poisson random measure on $\mathbb{R}^+ \times \Theta$.

Let (Ω, F_t, F, P) be a filtration probability space. Here, F_t is the σ-algebra generated by an one-dimensional standard Wiener process $W(t)$ and an one-dimensional Poisson jump process $\Pi(t)$ as:

$$F_t(t) = \sigma\left\{\iint_{(0,S]\times A} N(dt, d\theta)\big| 0 < s \leq t, A \in B(\Theta)\right\} \vee \left\{W(s)\big| 0 < s \leq t\right\} \vee \sigma_{null}$$

where σ_{null} contains all of the P-null set of F.

Its corresponding nonlinear stochastic Poisson jump-diffusion system is defined as:

$$\begin{cases} dx(t) = [f(x(t)) + g(x(t))u(t) + v(t)]dt + \sigma(x(t))dW(t) \\ \qquad + \int_\Theta T(x(t^-),\theta)N(dt,d\theta) \\ y(t) = m(x(t)), x(0) = x_0, \text{ and } T(x_0,\cdot) = 0 \end{cases} \qquad (3.2)$$

where $f : \mathbb{R}^{n_x} \to \mathbb{R}^{n_x}$, $g : \mathbb{R}^{n_x} \to \mathbb{R}^{n_x \times n_u}$, $\sigma : \mathbb{R}^{n_x} \to \mathbb{R}^{n_x}$, $T : \mathbb{R}^{n_x} \times \Theta \to \mathbb{R}^{n_x}$, and $m : \mathbb{R}^{n_x \times n_u} \to \mathbb{R}^{n_y}$ are nonlinear Borel measurable continuous functions, which are satisfied with Lipschitz continuity.

The càdlàg process $x(t) \in \mathbb{R}^{n_x}$ is the state vector; the initial state vector $x(0) = x_0$; $y(t) \in \mathbb{R}^{n_y}$ is the controlled output vector; the input vector $u(t) \in L_F^2(\mathbb{R}^+; \mathbb{R}^{n_u})$ is the admissible control law with respect to $\{F_t\}_{t \ge 0}$; $v(t) \in L_F^2(\mathbb{R}^+; \mathbb{R}^{n_v})$ is regarded as an unknown finite-energy stochastic external disturbance; and the term $\Theta \in \mathbb{R}^{n_\Theta}$ is the mark space of the Poisson random measure $N(dt,d\theta)$. Since a one-dimensional standard Wiener process $W(t)$ is a continuous but nondifferentiable stochastic process, the term $\sigma(x(t))dW(t)$ can be regarded as a continuous state dependent internal disturbance. The integral term $\int_\Theta T(x(t^-),\theta)N(dt,d\theta)$, caused by a nonlinear discontinuous change of the system in (3.2) at the time instant t, is used to denote the Poisson jump, where the jump amplitude depends on the jump amplitude coefficient function $T(x(t^-),\theta)$. Moreover, the Wiener process $W(t)$ and Poisson jump process $\Pi(t)$ are assumed to be mutually independent for all $t \ge 0$ (i.e., $E\{[\int_\Theta N(dt,d\theta)]dW(t)\} = 0$).

Some important properties of the Poisson jump process $\Pi(t)$ are given as follows ([10]):

(i) $\int_0^t \int_\Theta N(dt,d\theta) = \int_0^t dN(s;\theta) = N(t;\theta)$, where $N(t;\theta)$ is the mark-generated simple Poisson process, and the F_t-measurable independent and identically distributed random vector $\theta \subset \Theta$ is called the Poisson mark vector.

(ii) $E\{\int_0^t \int_\Theta N(dt,d\theta)\} = \lambda t$, where the finite scalar number $\lambda > 0$ is the Poisson jump intensity.

(iii) $E\{N(dt,d\theta)\} = \pi(d\theta)$, where $\pi(d\theta) \triangleq \lambda \int_\Theta \phi_\Theta(\theta)$ and $\phi_\Theta(\theta)$ is the mark probability density function.

(iv) $E\{[\int_\Theta N(dt,d\theta)]dt\} = 0$.

In this chapter, without loss of generality, we assume that (a) the system states of the nonlinear stochastic Poisson jump-diffusion system in (3.2) are available, and (b) (3.2) has an equilibrium point at the origin of $\mathbb{R}^{n_x}$ (i.e., $x_{eq} = 0$). If the equilibrium point of the nonlinear stochastic system of interest with a Poisson jump process is not at the origin point, then the designer can shift it to the origin by changing variables.

Definition 3.2: For the nonlinear stochastic Poisson jump-diffusion system in (3.2) with $v(t) = 0$, its H_2 performance index $J_2(u(t))$ is defined as follows:

$$J_2(u(t)) = E\left\{\int_0^\infty (x^T(t)Q_1 x(t) + u^T(t)R_1 u(t))dt\right\} \qquad (3.3)$$

Definition 3.3: For the given stochastic Poisson jump-diffusion system in (3.2) with $x_0 = 0$, the H_∞ robustness performance index $J_\infty(u(t))$ of the given stochastic system is defined as follows:

$$J_\infty(u(t)) = \sup_{\substack{v(t) \in L^2(\mathbb{R}_+, \mathbb{R}^{n_v}) \\ v \neq 0, x_0 = 0}} \frac{E\left\{\int_0^\infty (y^T(t)Q_2 y(t) + u^T(t)R_2 u(t))dt\right\}}{E\left\{\int_0^\infty v^T(t)v(t)dt\right\}} \tag{3.4}$$

where $Q_2, R_2 > 0$, that is, the worst-case effect from the exogenous disturbance signal $v(t) \in L^2(\mathbb{R}_+, \mathbb{R}^{n_v})$ to the controlled output $y(t)$ and $u(t)$ from the average energy point of view.

Remark 3.1: If the initial condition $x_0 \neq 0$, then the H_∞ performance index $J_\infty(u(t))$ should be rewritten as

$$J_\infty(u(t))$$
$$= \sup_{\substack{v(t) \in L^2(\mathbb{R}_+, \mathbb{R}^{n_v}), \\ v \neq 0, x_0 \neq 0}} \frac{E\left\{\int_0^\infty (y^T(t)Q_2 y(t) + u^T(t)R_2 u(t))dt - V(x_0)\right\}}{E\left\{\int_0^\infty v^T(t)v(t)dt\right\}} \tag{3.5}$$

where $V(\cdot) \in C^2$ and $V(\cdot) \geq 0$; that is, the effect of the initial condition x_0 should be deleted in order to obtain the real effect of $v(t)$ on the controlled output.

The Itô-Lévy formula of the nonlinear stochastic jump-diffusion system in (3.2) is given as follows:

Lemma 3.1 ([20]): Let $V(\cdot) \in C^2(\mathbb{R}^{n_x})$ and $V(\cdot) \geq 0$. For the nonlinear stochastic system with Wiener and Poisson jump process in (3.2), the Itô-Lévy formula of $V(x(t))$ is:

$$dV(x(t)) = V_x^T [f(x(t)) + g(x(t))u(t) + v(t)]$$
$$+ \tfrac{1}{2}\sigma^T(x(t))V_{xx}\sigma(x(t))dt + V_x^T\sigma(x(t))dW(t) \tag{3.6}$$
$$+ \int_\Theta \{V(x(t) + T(x(t^-), \theta)) - V(x(t))\}N(dt, d\theta)$$

3.3 MULTI-OBJECTIVE STATE FEEDBACK CONTROL FOR THE NONLINEAR STOCHASTIC POISSON JUMP-DIFFUSION SYSTEM

Roughly speaking, an optimization problem is defined as an MOP if there does not exist a single solution that is simultaneously optimal with respect to every objective function. The H_2 control design focuses on keeping the energy of system states and admissible control laws as low as possible. However, with respect to the H_∞ control design, a better H_∞ performance index is always coupled with a higher input energy cost. As a result, the aims of H_2 and H_∞ control design conflict with each other. Thus, the concurrent optimization problem for the H_2 and H_∞ performance indices is an MOP with a dynamic constraint (dynamic-constrained MOP).

Definition 3.4 ([22]): The multi-objective control of a given nonlinear stochastic Poisson jump-diffusion system is to design an admissible control law $u(t)$ that can minimize the H_2 and H_∞ performance indices in the Pareto optimal sense simultaneously, that is:

$$\min_{u(t) \in U} (J_2(u(t)), J_\infty(u(t)))$$

subject to (3.2)   (3.7)

where U is the set of all the admissible control laws for the given nonlinear stochastic system in (3.2); the objective functions $J_2(u(t))$ and $J_\infty(u(t))$ are defined in (3.3) and (3.5), respectively; and the vector of the objective functions $(J_2(u(t)), J_\infty(u(t)))$ is called the objective vector of $u(t)$.

For an MOP, the domination concept is employed to decide which admissible control law is better; that is, we use domination to define the "simultaneous minimization" in the MOP in (3.7). Thus, the solution of the MOP in (3.7) is a set of admissible control laws with a nondominated objective vector. The definition of domination is given as follows:

Definition 3.5 ([2, 22]): For a multi-objective control problem, an objective vector (α, β) of $u(t)$ is said to dominate another objective vector $(\dot{\alpha}, \dot{\beta})$ of $\dot{u}(t)$ if and only if both the following conditions are true: (I) $\alpha \le \dot{\alpha}$ and $\beta \le \dot{\beta}$, and (II) at least one of the two inequalities in (I) with strict inequality. Moreover, the admissible control law $u(t)$ is regarded as a better solution than $\dot{u}(t)$.

Unfortunately, there is no efficient method to directly solve the control $u(t)$ of the MOP in (3.7). Thus, we introduce an indirect method to help us solve the MOP in (3.7).

Lemma 3.2 ([15, 43]): Suppose α and β are the upper bounds of the H_2 and H_∞ performance indices, respectively; that is, $J_2(u(t)) \le \alpha$ and $J_\infty(u(t)) \le \beta$. The MOP in (3.7) is equivalent to the following MOP:

$$\min_{u(t) \in U} (\alpha, \beta)$$

subject to $J_2(u(t)) \le \alpha$ and $J_\infty(u(t)) \le \beta$ (3.8)

Lemma 3.3 ([10]): For any two real matrices A and B with appropriate dimensions, we have:

$$A^T B + B^T A \le \gamma^2 A^T A + \gamma^{-2} B^T B$$

where γ is any nonzero real number.

By applying Lemmas 3.2 and 3.3 and the Itô-Lévy formula in (3.6), we can obtain the sufficient condition to design an admissible control law $u(t)$ to minimize (α, β) under the constraints $J_2(u(t)) \le \alpha$ and $J_\infty(u(t)) \le \beta$ for the MOP in (3.8) of the nonlinear stochastic Poisson jump-diffusion system in (3.2). For convenience, we use J_2 and J_∞ to denote $J_2(u(t))$ and $J_\infty(u(t))$ in (3.3) and (3.5), respectively.

Theorem 3.1: The multi-objective control problem in (3.8) for the nonlinear stochastic Poisson jump-diffusion system in (3.2) can be solved if the following MOP:

$$\min_{u(t)\in U}(\alpha,\beta)$$

$$\text{subject to (3.13), (3.14) and (3.15)}$$

(3.9)

can be solved for some Lyapunov function $V(x(t))$ and $V(x_0)\neq 0$.

Proof: We first derive the sufficient condition for $J_\infty \leq \beta$ of the MOP in (3.8). Add and subtract the term $dV(x(t))$ to the integrand of $E\left\{\int_0^\infty (y^T(t)Q_2 y(t)+u^T(t)R_2 u(t))dt\right\}$. Using the Itô-Lévy formula of $V(x(t))$ in (3.6), the term $E\{\int_0^\infty dV(x)\}$ can be expressed as

$$
\begin{aligned}
E\{\int_0^\infty dV(x(t))\} = E\int_0^\infty \{ & V_x^T f(x(t)) + V_x^T g(x(t))u(t) \\
& + V_x^T v(t) + \tfrac{1}{2}\sigma^T(x(t))V_{xx}\sigma(x(t)) \\
& + \int_\Theta \{V(x(t)+T(x(t^-),\theta)) \\
& - V(x(t))\}\pi(d\theta)\}dt
\end{aligned}
$$

(3.10)

Since $\lim_{t\to\infty} V(x(t)) \geq 0$, we have:

$$
\begin{aligned}
& E\{\int_0^\infty (y^T(t)Q_2 y(t)+u^T(t)R_2 u(t))dt\} \\
& \leq E\{V(x_0)\} + E\{\int_0^\infty [(y^T(t)Q_2 y(t)+u^T(t)R_2 u(t))dt + dV(x(t))]\}
\end{aligned}
$$

According to Lemma 3.3, we obtain:

$$V_x^T v(t) \leq \beta v^T(t)v(t) + (4\beta)^{-1}V_x^T V_x$$

(3.11)

Using the inequality in (3.11), we have

$$
\begin{aligned}
& E\{\int_0^\infty (y^T(t)Q_2 y(t)+u^T(t)R_2 u(t))dt\} \\
& \leq E\{V(x_0)\} + E\{\int_0^\infty (y^T(t)Q_2 y(t)+u^T(t)R_2 u(t) \\
& + V_x^T f(x(t)) + V_x^T g(x(t))u(t)) + \tfrac{1}{2}\sigma^T(x(t))V_{xx}(x(t))\sigma(x(t)) \\
& + \int_\Theta [V(x(t)+T(x(t^-),\theta)) - V(x(t))\pi(d\theta)] \\
& + \tfrac{1}{4\beta}V_x^T V_x + \beta v^T(t)v(t)\}dt
\end{aligned}
$$

(3.12)

If the inequality:

$$
\begin{aligned}
\Psi_\infty(V(x(t))) \triangleq \; & y^T(t)Q_2 y(t) + u^T(t)R_2 u(t) + V_x^T f(x(t)) \\
& + V_x^T g(x(t))u(t)) + \tfrac{1}{2}\sigma^T(x(t))V_{xx}(x(t))\sigma(x(t)) \\
& + \int_\Theta \{V(x(t)+T(x(t^-),\theta)) - V(x(t))\}\pi(d\theta) \\
& + \tfrac{1}{4\beta}V_x^T V_x \leq 0
\end{aligned}
$$

(3.13)

holds, then we have $J_\infty \leq \beta$. Second, we discuss the sufficient condition for the nonlinear stochastic system with the Poisson jump process in (3.2) with $J_2 \leq \alpha$. By adding and subtracting the term $dV(x(t))$ into the integrand of J_2, we obtain the following inequality by the Itô-Lévy formula in (3.6):

$$E\{\int_0^\infty (x^T(t)Q_1 x(t) + u^T(t)R_1 u(t))dt\}$$
$$\leq E\{V(x_0)\} + E\{\int_0^\infty (x^T(t)Q_1 x(t) + u^T(t)R_1 u(t)$$
$$+ V_x^T f(x(t)) + V_x^T g(x(t))u(t) + \tfrac{1}{2}\sigma^T(x(t))V_{xx}(x(t))\sigma(x(t))$$
$$+ \int_\Theta \{V(x(t) + T(x(t^-),\theta)) - V(x(t))\}\pi(d\theta))dt\}$$

If the following two inequalities:

$$\Psi_2(V(x(t))) \triangleq x^T(t)Q_1 x(t) + V_x^T f(x(t)) + u^T(t)R_1 u(t)$$
$$+ \tfrac{1}{2}\sigma^T(x(t))V_{xx}(x(t))\sigma(x(t)) + \int_\Theta \{V(x(t) + T(x(t^-),\theta)) \qquad (3.14)$$
$$- V(x(t))\}\pi(d\theta) + V_x^T g(x(t))u(t)) \leq 0$$

and

$$E\{V(x_0)\} \leq \alpha \qquad (3.15)$$

hold, then we have $J_2 \leq \alpha$. Based on this analysis, if the control law $u(t)$ is a feasible solution of the MOP in (3.9) under the constraints of (3.13), (3.14), and (3.15), then $u(t)$ will ensure that $J_2 \leq \alpha$ and $J_\infty \leq \beta$ in (3.8) simultaneously. Therefore, the solutions of the MOP in (3.9) are also the solutions of the MOP in (3.8).

Note that the inequalities in (3.13) and (3.14) are the Hamilton-Jacobi-Isaacs inequality (HJII) and Hamilton-Jacobi-Bellman inequality (HJBI) of the nonlinear stochastic Poisson jump-diffusion system in (3.2), respectively.

Remark 3.2: In many practical applications, the nonlinear stochastic Poisson jump-diffusion system may only have finite marks (i.e., $\Theta = \{\theta_1, \theta_2, ..., \theta_m\}$). In this situation, the integral of the Poisson random measure with respect to Θ can be rewritten as:

$$\int_{\Theta = \{\theta_1,\theta_2,...,\theta_m\}} N(dt, d\theta) = \sum_{i=1}^m dN(t;\theta_i) \qquad (3.16)$$

with $E\{dN(t;\theta_i)\} = \lambda_i dt$, for $i = 1, 2, ..., m$.

Now (3.2) can be rewritten as:

$$dx(t) = \{f(x(t)) + g(x(t))u(t) + v(t)\}dt + \sigma(x(t))dW(t)$$
$$+ \sum_{i=1}^m T(x(t^-),\theta_i)dN(t;\theta_i) \qquad (3.17)$$

And the term $\int_{\Theta}\{V(x(t)+T(x(t^-),\theta))-V(x(t))\}\pi(d\theta)$ in (3.13) and (3.14) can be replaced by:

$$\sum_{i=1}^{m}\lambda_i\{V(x(t)+T(x(t^-)))-V(x(t))\}$$

To treat the multi-objective control problem in (3.9), we also need to define the stability of the nonlinear stochastic Poisson jump-diffusion system in (3.2).

Definition 3.6 ([44]): Suppose $x_{eq}=0$ is an equilibrium point of the following nonlinear stochastic Poisson jump-diffusion system:

$$\begin{aligned}dx(t) &= \{f(x(t))+g(x(t))u(t)\}dt+\sigma(x(t))dW(t)\\ &\quad+\int_{\Theta}T(x(t^-),\theta)N(dt,d\theta)\end{aligned} \tag{3.18}$$

The nonlinear stochastic Poisson jump-diffusion system in (3.18) is said to be exponentially mean square stable (i.e., stable in the mean square sense) if, for some positive constants C_1 and C_2, the following inequality holds:

$$E\left\{\|x(t)\|_2^2\right\}\le C_1\exp(-C_2 t),\forall t>0 \tag{3.19}$$

Theorem 3.2: For the nonlinear stochastic Poisson jump-diffusion system in (3.2), suppose the given Lyapunov function $V(x(t))$, which satisfies the following two inequalities:

$$m_1\|x(t)\|_2^2\le V(x(t))\le m_2\|x(t)\|_2^2 \tag{3.20}$$

where $m_1>0$ and $m_2>0$. If the external noise $v(t)=0$, and $u(t)$ is a solution of MOP in (3.9), then $u(t)$ stabilizes (3.2) exponentially in the mean square sense.
Proof: By the Itô-Lévy formula of $V(x(t))$ in (3.6) and the fact that $u(t)$ is a feasible solution of (3.14), we obtain:

$$\begin{aligned}dE\{V(x(t))\} &= E\{dV(x(t))\}\\ &= E\{V_x^T f(x(t))+V_x^T g(x(t))u(t)+\tfrac{1}{2}\sigma^T(x(t))V_{xx}(x(t))\sigma(x(t))\\ &\quad+\int_{\Theta}[V(x(t)+T(x(t^-),\theta))-V(x(t))]\pi(d\theta)\}dt\\ &\le E\{-x^T(t)Qx(t)\}dt\le \tfrac{-m_3}{m_2}E\{V(x(t))\}dt<0\end{aligned} \tag{3.21}$$

where m_3 is the smallest eigenvalue of the positive definite matrix Q. We have:

$$\frac{d}{dt}E\{V(x(t))\}\le\frac{-m_3}{m_2}E\{V(x(t))\}dt<0 \tag{3.22}$$

By using inequalities in (3.20), we get:

$$E\left\{\|x(t)\|^2\right\}\le E\{m_1^{-1}V(x_0)\}\exp(-m_3 t\times m_2^{-1}) \tag{3.23}$$

We obtain $\lim_{t\to\infty} x(t) = 0$ (e.g., $x(t)$ exponentially in the mean square sense).

Based on Theorems 3.1 and 3.2, the multi-objective control design problem is equivalent to how to solve the MOP in (3.9) for (3.2) with HJII and HJBI constraints in (3.13) and (3.14). Although the multi-objective control for the nonlinear stochastic Poisson jump-diffusion system in (3.2) can be reformulated as an MOP in (3.9) via the proposed indirect method, the HJII and HJBI are still difficult to solve except in special cases.

3.4 MULTI-OBJECTIVE STATE-FEEDBACK CONTROL FOR THE NONLINEAR STOCHASTIC T-S FUZZY JUMP-DIFFUSION SYSTEM

To accomplish the multi-objective state-feedback control design for the nonlinear stochastic jump-diffusion system in (3.2), one needs to solve the HJII in (3.13) and the HJBI in (3.14), which are difficult to solve analytically. To overcome this problem, the T-S fuzzy interpolation method is introduced [6, 40, 45]. A fuzzy dynamic model will be applied to interpolate several local linearized stochastic systems around some operation points to approximate the nonlinear stochastic jump-diffusion system in (3.2). This fuzzified model is described by a group of if-then rules and is used to simplify the multi-objective H_2 / H_∞ state-feedback control design for the nonlinear stochastic jump-diffusion system in (3.2).

The ith rule of the T-S fuzzy model for approximating the nonlinear stochastic jump-diffusion system in (3.2) is given as follows:

System Rule i : for $i = 1,...,m$

If $z_1(t)$ is $G_{i,1}$ and $z_2(t)$ is $G_{i,2}$ &, and $z_g(t)$ is $G_{i,g}$,

then

$$\begin{cases} dx(t) = [A_i x(t) + B_i u(t) + v(t)]dt + D_i x(t)dW(t) \\ \qquad + \int_\Theta E_i(\theta)x(t^-)N(dt,d\theta) \\ y(t) = M_i x(t) \end{cases}$$

where A_i, B_i, C_i, D_i, and M_i are constant matrices with appropriate dimensions, and $E_i(\theta)$ is a matrix function of mark vector θ with appropriate dimensions, that is, $E_i : \mathbb{R}^{n_\theta} \to \mathbb{R}^{n_x}$; $z_1(t),...,z_g(t)$ are fuzzy premise variables, and $G_{i,j}$ is the membership function with respect to $z_j(t)$ for $j = 1,2,...,g$.

Thus, the overall fuzzy nonlinear stochastic Poisson jump-diffusion system can be inferred as follows:

$$\begin{cases} dx(t) = \sum_{i=1}^m h_i(z(t))\{[A_i x(t) + B_i u(t) + v(t)]dt + D_i x(t)dW(t) \\ \qquad + \int_\Theta E_i(\theta)x(t^-)N(dt,d\theta) \\ y(t) = \sum_{i=1}^m h_i M_i x(t) \end{cases} \qquad (3.24)$$

where

$$z(t) = [z_1(ti),...,z_g(t)]^T$$

$$\mu_i(z(t)) = \prod_{j=1}^{g} G_{i,j}(z_j(t)) \geq 0, \text{ for } i = 1,2,...,m$$

$$h_i(z(t)) \triangleq \frac{\mu_i(z(t))}{\sum_{i=1}^{m} \mu_i(z(t))} \text{ and } \sum_{i=1}^{m} h_i(z) = 1$$

Similarly, the control input $u(t)$ for the nonlinear stochastic jump-diffusion system in (3.2) can be approximated by the following fuzzy control input:

The ith T-S fuzzy control law for the corresponding local system is given as

Control Rule i : for $i = 1,...,m$

If $z_1(t)$ is $G_{i,1}$ and $z_2(t)$ is $G_{i,2}$ & , and $z_g(t)$ is $G_{i,g}$,

Then $u_i(t) = K_i x(t)$

where K_i, $i = 1,...,m$ are constant matrices with appropriate dimensions.

The overall T-S fuzzy controller for the nonlinear stochastic T-S fuzzy Poisson jump-diffusion system in (3.24) can be interpolated as:

$$u(t) = \sum_{i=1}^{m} h_i(z(t))(K_i x(t)) \tag{3.25}$$

Substituting (3.25) into the T-S fuzzy system in (3.24) yields the closed-loop system of the following form:

$$dx(t) = \sum_{i=1}^{m} \sum_{j=1}^{m} h_i(z(t))h_j(z(t))\{[(A_i + B_i K_j)x(t) + v(t)]dt$$
$$+ D_i x(t)dW(t) + \int_\Theta E_i(\theta)x(t^-)N(dt,d\theta)$$

Even though the T-S fuzzy interpolation method can efficiently reflect the system behavior, we still need to consider the effect of approximation errors of the T-S fuzzy interpolation method because they can affect the stability and control performance of the system.

After counting the approximation errors of T-S fuzzy interpolation method, (3.2) can be represented as:

$$\begin{cases} dx(t) = \sum_{i=1}^{m} \sum_{j=1}^{m} h_i(z(t))h_j(z(t))\{[A_i x(t) + \Delta f(x(t)) + \\ (B_i + \Delta g(x(t))K_j)x(t) + v(t)]dt + (D_i x(t) + \Delta \sigma(x(t))) \cdot \\ dW(t) + \int_\Theta [E_i(\theta)x(t^-) + \Delta E_{i,\theta}(x(t^-),\theta)]N(dt,d\theta)\} \\ y(t) = \sum_{i=1}^{m} h_i M_i x(t) + \Delta m(x(t)) \end{cases} \tag{3.26}$$

where $f(x(t)) = \sum_{i=1}^{m} h_i(z)A_i x(t) + \Delta f(x(t))$, $g(x(t)) = \sum_{i=1}^{m} h_i(z)B_i + \Delta g(x(t))$, $\sigma(x(t)) = \sum_{i=1}^{m} h_i(zs)D_i x(t) + \Delta\sigma(x(t))$, and $T(x(t^-,\theta)) = \sum_{i=1}^{m} h_i(z)E_i(\theta)x(t) + \Delta E_{i,\theta}(x(t),\theta)$ and the terms Δf, Δg, $\Delta\sigma$, $\Delta E_{i,\theta}$, and Δm are used to denote the T-S fuzzy approximation errors in the terms $f(x(t))$, $g(x(t))$, $\sigma(x(t))$, $\int_\Theta T(x(t^-),\theta)$, and $m(x(t))$, respectively.

Assumption: We assume the upper bounds of the T-S fuzzy approximation errors in the Euclidean norm for the nonlinear stochastic Poisson jump-diffusion system in (3.26) are given as:

$$\left\|\Delta f(x(t))\right\|_2 \le \varepsilon_1 \left\|x(t)\right\|_2, \left\|\Delta\sigma(x(t))\right\|_2 \le \varepsilon_3 \left\|x(t)\right\|_2,$$

$$\Delta g(x(t))\Delta g^T(x(t)) \le \varepsilon_2 I, \left\|\Delta m(x(t))\right\|_2 \le \varepsilon_5 \left\|x(t)\right\|_2, \tag{3.27}$$

$$\text{and } \left\|\Delta E_{i,\theta}(x(t),\theta)\right\|_2 \le \varepsilon_4 \left\|x(t)\right\|_2$$

Lemma 3.4: For any matrix M_i with appropriate dimension and the weighted functions $h_i(z)$ with $0 \le h_i(z) \le 1$, for $i \in N^+$, $1 \le i \le m$, and $\sum_{i=1}^{m} h_i(z) = 1$, we have

$$\left(\sum_{j=1}^{m} h_i(z)M_j\right)^T P\left(\sum_{j=1}^{m} h_i(z)M_j\right) \le \sum_{j=1}^{m} h_i(z)M_i^T P M_i \tag{3.28}$$

Based on the T-S fuzzy interpolation method in (3.26), the HJI-constrained MOP in (3.9) for the nonlinear stochastic Poisson jump-diffusion system in (3.2) can be transformed into an LMI-constrained MOP as follows:

Theorem 3.3: If the following LMI-constrained MOP can be solved, then the multi-objective stabilization control problem in (3.7) for the nonlinear stochastic Poisson jump-diffusion system in (3.2) can be solved.

$$\min_{\xi=\{W,Y_1,Y_2,\ldots,Y_m\}} (\alpha,\beta) \tag{3.29}$$

$$\text{subject to LMIs in } (3.30)-(3.32)$$

$$W \ge \alpha^{-1} Tr(R_{x_0})I \tag{3.30}$$

$$\begin{bmatrix} \Xi_{i,j} & Y_j^T & WD_i^T & W & W & W \\ * & -(R_1+sI)^{-1} & 0 & 0 & 0 & 0 \\ * & * & -\dfrac{W}{2} & 0 & 0 & 0 \\ * & * & * & -Q_1^{-1} & 0 & 0 \\ * & * & * & * & -(\hat{\Delta})^{-1} & 0 \\ * & * & * & * & * & -\dfrac{Tr(R_{x_0})}{\alpha}\Delta^{-1} \end{bmatrix} \le 0 \tag{3.31}$$

$$\begin{bmatrix} \Xi_{i,j} + \frac{1}{\beta}I & Y_j^T & WD_i^T & WM_i^T & W & W \\ * & -(R_2 + sI)^{-1} & 0 & 0 & 0 & 0 \\ * & * & -\frac{W}{2} & 0 & 0 & 0 \\ * & * & * & -(2Q_2)^{-1} & 0 & 0 \\ * & * & * & * & -(\tilde{\Delta})^{-1} & 0 \\ * & * & * & * & * & -\frac{Tr(R_{x_0})\Delta^{-1}}{\alpha} \end{bmatrix} \leq 0 \qquad (3.32)$$

$$i, j = 1,...,m$$

where

$$\Delta \triangleq \int_\Theta \{3\varepsilon_4^2 I + 2E_i^T(\theta)E_i(\theta)\}, \widehat{\Delta} \triangleq \alpha Tr\{R_{x_0}\}^{-1}(\varepsilon_1^2 + 2\varepsilon_3^2)I,$$

$$\tilde{\Delta} \triangleq (2\varepsilon_5^2\overline{\lambda}(Q_2)I + \widehat{\Delta}), R_{x_0} \triangleq E\{x_0 x_0^T\}, E\{Tr(x_0 x_0^T)\} \triangleq Tr(R_{x_0}),$$

$$Y_j \triangleq K_j W, s > 0$$

and

$$\Xi_{i,j} \triangleq (A_i W + WA_i^T + W + (B_i Y_j)^T + B_i Y_j + \frac{\varepsilon_2}{s}I$$
$$+ \int_\Theta \{WE_i^T(\theta) + E_i(\theta)W + W\}\pi(d\theta))$$

Proof: We choose $V(x(t)) = x^T(t)Px(t)$ as a Lyapunov function for the nonlinear stochastic Poisson jump system in (3.2), where $P = P^T > 0$ is a positive definite matrix. We know that

$$E\{V(x_0)\} \leq \overline{\lambda}(P)E\{Tr(x_0 x_0^T)\} = \overline{\lambda}(P)TrR_{x_0}$$

If we set $\overline{\lambda}(P)Tr(R_{x_0}) < \alpha$, then we get:

$$P \leq \alpha Tr\{R_{x_0}\}^{-1}I \text{ and } E\{x_0^T Px_0\} \leq \alpha \qquad (3.33)$$

Based on (3.26) and (3.27), applying Lemmas 3.2 and 3.3 and (3.33), we have the following inequalities:

$$y^T(t)Q_2 y(t) \leq \sum_{i=1}^{m} h_i(z)x^T(t)(2\varepsilon_5^2 I + 2M_i^T Q_2 M_i)x(t) \qquad (3.34)$$

$$2x^T(t)Pf(x(t)) \leq$$
$$\sum_{i=1}^{m} h_i(z)x^T(t)(A_i^T P + PA_i + P + \alpha Tr\{R_{x_0}\}^{-1}\varepsilon_1^2 I)x(t) \tag{3.35}$$

$$2x^T(t)Pg(x(t))u(t) \leq$$
$$\sum_{i=1}^{m}\sum_{j=1}^{m} h_i(z)h_j(z)x^T(t)[2PB_iK_j + \tfrac{\varepsilon_2}{s}PP + sK_j^T K_j]x(t) \tag{3.36}$$

$$\tfrac{1}{2}\sigma^T(x(t))V_{xx}(x(t))\sigma(x(t)) \leq$$
$$\sum_{i=1}^{m} h_i(z)x^T(t)[2D_i^T PD_i + \alpha Tr\{R_{x_0}\}^{-1}(2\varepsilon_3^2 I)]x(t) \tag{3.37}$$

and

$$\int_\Theta \{[x(t) + T(x(t^-),\theta)]^T P[x(t) + T(x(t^-),\theta)]$$
$$-x^T(t)Px(t)\}\pi(d\theta) \leq$$
$$x^T(t)\{\sum_{i=1}^{m} h_i(z)\int_\Theta(\alpha Tr\{R_{x_0}\}^{-1}[3\varepsilon_4^2 I + 2E_i^T(\theta)E_i(\theta)]$$
$$+P + E_i^T(\theta)P + PE_i(\theta))\pi(d\theta)\}x(t) \tag{3.38}$$

where $s > 0$ in (3.36). Now, apply the inequalities to (3.13), and we have:

$$\Psi_\infty(x^T(t)Px(t)) \leq \sum_{i=1}^{m}\sum_{j=1}^{m} h_i(z)h_j(z)x^T(t)(2M_i^T Q_2 M_i$$
$$+K_j^T R_2 K_j + 2\varepsilon_5^2 I + A_i^T P + PA_i + P + \tfrac{1}{\beta}PP + PB_i K_j$$
$$+(PB_i K_j)^T + 2D_i^T PD_i + \alpha Tr\{R_{x_0}\}^{-1}(2\varepsilon_3^2 + \varepsilon_1^2)I + \tfrac{\varepsilon_2}{s}PP \tag{3.39}$$
$$+sK_j^T K_j + \int_\Theta\{\alpha Tr\{R_{x_0}\}^{-1}[3\varepsilon_4^2 I + 2E_i^T(\theta)E_i(\theta)] + P$$
$$+E_i^T(\theta)P + PE_i(\theta)\}\pi(d\theta))x(t)$$

If the following inequalities hold:

$$2M_i^T Q_2 M_i + K_j^T R_2 K_j + 2\varepsilon_5^2 I + A_i^T P + PA_i + P + \tfrac{1}{\beta}PP$$
$$+PB_i K_j + (PB_i K_j)^T + 2D_i^T PD_i + \alpha Tr\{R_{x_0}\}^{-1}(2\varepsilon_3^2 + \varepsilon_1^2)I$$
$$+\tfrac{\varepsilon_2}{s}PP + sK_j^T K_j + \int_\Theta\{\alpha Tr\{R_{x_0}\}^{-1}[3\varepsilon_4^2 I + 2E_i^T(\theta)E_i(\theta)] \tag{3.40}$$
$$+P + E_i^T(\theta)P + PE_i(\theta)\}\pi(d\theta) \leq 0$$

then we have $J_\infty \leq \beta$. Consider the HJBI in (3.14):

$$\Psi_2(x^T(t)Px(t)) \triangleq x^T(t)Q_1x(t) + u^T(t)R_1u(t) + 2x^T(t)P$$
$$\cdot f(x(t)) + 2x^T(t)Pg(x(t))u(t) + [D_ix(t) + \Delta\sigma(x(t))]^T P$$
$$\{[D_ix(t) + \Delta\sigma(x(t))] + \int_\Theta \{[x(t) + T(x(t^-),\theta)]^T P[x(t)$$
$$+T(x(t^-),\theta)] - x^T(t)Px(t)\}\pi(d\theta)$$

(3.41)

By using the inequalities in (3.34)–(3.38), we obtain:

$$\Psi_2(x^T(t)Px(t)) \leq \sum_{i=1}^m h_i(z)h_j(z)x^T(t)(Q_1 + K_j^T R_2 K_j + A_i^T P_i$$
$$+PA + P + (PB_iK_j)^T + PB_iK_j + \tfrac{\varepsilon_2}{s}PP + sK_j^T K_j + 2D_i^T PD_i$$
$$+\alpha Tr\{R_{x_0}\}^{-1}(2\varepsilon_3^2 + \varepsilon_1^2)I + \int_\Theta \{\alpha Tr\{R_{x_0}\}^{-1}[3\varepsilon_4^2 I + 2E_i^T(\theta)$$
$$\cdot E_i(\theta)] + P + E_i^T(\theta)P + PE_i(\theta)\}\pi(d\theta))x(t)$$

(3.42)

If the following inequalities hold:

$$Q_1 + K_j^T R_2 K_j + A_i^T P_i + PA + P + (PB_iK_j)^T + PB_iK_j$$
$$+\tfrac{\varepsilon_2}{s}PP + sK_j^T K_j + 2D_i^T PD_i + \alpha Tr\{R_{x_0}\}^{-1}(2\varepsilon_3^2 + \varepsilon_1^2)I$$
$$+\int_\Theta \{\alpha Tr\{R_{x_0}\}^{-1}[3\varepsilon_4^2 I + 2E_i^T(\theta)\cdot E_i(\theta)] + P + E_i^T(\theta)P$$
$$+PE_i(\theta)\}\pi(d\theta) \leq 0$$

(3.43)

then we obtain $J_2 \leq \alpha$. In general, the algebraic Riccati-like inequalities in (3.40) and (3.43) are not easy to solve, as they are bilinear matrix inequalities (BMIs) of P and K_j. Fortunately, they can be reformulated as a group of LMIs by introducing new variables. Let $W = P^{-1}$ and $Y_j = K_j W$, then (3.40) and (3.43) are equivalent to the following two inequalities, respectively:

$$A_iW + WA_i^T + W + (B_iY_j)^T + B_iY_j + \int_\Theta \{W + WE_i^T(\theta) + E_i(\theta)$$
$$\cdot W\}\pi(d\theta) + \tfrac{1}{\beta}I + \tfrac{\varepsilon_2}{s}I + W(2\varepsilon_5^2 I + \alpha Tr\{R_{x_0}\}^{-1}(2\varepsilon_3^2 + \varepsilon_1^2)I)$$
$$\cdot W + 2WD_i^T W^{-1}D_iW + \alpha Tr\{R_{x_0}\}^{-1}W(\int_\Theta [3\varepsilon_4^2 I + 2E_i^T(\theta)$$
$$\cdot E_i(\theta)]\pi(d\theta))W + Y_j^T(sI + R_2)Y_j + 2WM_i^T Q_2 M_i W \leq 0$$

(3.44)

$$A_iW + WA_i^T + W + (B_iY_j)^T + B_iY_j + \int_\Theta \{W + WE_i^T(\theta) + E_i(\theta)$$
$$\cdot W\}\pi(d\theta) + \tfrac{\varepsilon_2}{s}I + 2WD_i^T W^{-1}D_iW + W[\alpha Tr\{R_{x_0}\}^{-1}(2\varepsilon_3^2$$
$$+\varepsilon_1^2)I]W + WQ_1W + \alpha Tr\{R_{x_0}\}^{-1}W(\int_\Theta [3\varepsilon_4^2 I + 2E_i^T(\theta)E_i(\theta)]$$
$$\cdot \pi(d\theta))W + Y_j^T(sI + R_2)Y_j \leq 0$$

(3.45)

Applying the Schur complement ([46]) to these inequalities, we obtain the LMIs in (3.31) and (3.32). Since $P \leq \alpha Tr\{R_{x_0}\}^{-1} I$ implies:

$$W \geq \alpha^{-1} Tr\{R_{x_0}\}I \tag{3.46}$$

we obtain the LMIs in (3.30). The T-S fuzzy control gain $\{K_i\}_{i=1}^m$ is also obtained where $K_i = Y_i W^{-1}$ for $i = 1, ..., m$ when the MOP in (3.29) can be solved.

Remark 3.3: If the Poisson mark is finite (i.e., $\Theta = \{\theta_1, ..., \theta_m\}$, then the terms $\Xi_{i,j}$ and Δ of the LMI in (3.31) and (3.32) can be rewritten as:

$$\Xi_{i,j} = A_i W + WA_i^T + W + (B_i Y_j)^T + B_i Y_j + \frac{\varepsilon_2}{s} I + \sum_{i=1}^m \lambda_i \{\int_\Theta \{WE_i^T + E_i W + W\}$$

and $\Delta = \sum_{i=1}^m \lambda_i \{3\varepsilon_4^2 I + 2E_i^T E_i\}$.

Theorem 3.4: For the nonlinear stochastic Poisson jump-diffusion system in (3.2), if the external noise $v(t) = 0$, and $(\dot{\alpha}, \dot{\beta})$ is a feasible solution of (3.29), then the T-S fuzzy control signal $u(t) = \sum_{i=1}^m h_i(x)(K_i x(t))$ stabilizes the nonlinear stochastic Poisson jump-diffusion system in (3.2) exponentially in the mean square sense.

Proof: Because the proof is similar to that of Theorem 3.2, we omit it.

Now the multi-objective control design problem for the nonlinear stochastic Poisson jump-diffusion system in (3.2) can be regarded as how to search for a feasible solution $\xi = \{W, Y_1, Y_2, ..., Y_m\}$ that satisfies the LMIs in (3.30)–(3.32) and minimizes (α, β) in (3.29) simultaneously in the Pareto optimality sense. Although many MOEAs have been proposed for solving MOPs with algebraic function constraints, it is very difficult to search $\xi = \{W, Y_1, Y_2, ..., Y_m\}$ by MOEA to solve the LMI-constrained MOP in (3.29)–(3.32). Few researchers have discussed the LMI-constrained MOP in (3.29). To overcome the LMI-constrained MOP in (3.29), in the next section, we will introduce the reverse-order LMI-constrained MOEA.

3.5 MULTI-OBJECTIVE STATE FEEDBACK CONTROLLER DESIGN BY USING THE PROPOSED REVERSE-ORDER LMI-CONSTRAINED MOEA

The conventional MOEA is used to solve multi-objective problems via a stochastic search method based on a "survival of the fittest" law. The MOEA derives a set of compromise solutions called Pareto optimal solutions through the evolution algorithm, including crossover, mutation, and a nondominated sort operator [2, 38, 42]. Since the MOEA can search the Pareto optimal solutions in parallel globally and escape from local optima, it is particularly suitable for solving MOPs of nonlinear stochastic Poisson jump-diffusion systems. However, the constraints in (3.30)–(3.32) on the MOP in (3.29) are a set of LMIs. To solve the LMI-constrained MOP, some modifications to the MOEA are necessary. Prior to further discussion, we note some important definitions about Pareto optimality for the LMI-constrained MOEA.

Definition 3.7: For the given LMI-constrained MOP in (3.29), (α, β) is called a feasible objective vector if, for the given (α, β) with the LMIs in (3.30), (3.31), and (3.32), there exists a feasible solution $\xi_{(\alpha,\beta)} = \{W, Y_1, Y_2, ..., Y_m\} \in U$ such that the corresponding T-S fuzzy-controlled input $u_{(\alpha,\beta)}$ in (3.25) can guarantee $(J_2(u_{(\alpha,\beta)}), J_\infty(u_{(\alpha,\beta)})) \leq (\alpha, \beta)$.

Definition 3.8: A feasible solution $\xi_{(\alpha,\beta)} \in U$ is said to be Pareto optimal with respect to the given LMI-constrained MOP in (3.29) if and only if there is no other feasible objective vector solution $(\dot\alpha, \dot\beta)$ dominating (α, β).

Definition 3.9: For the given LMI-constrained MOP in (3.29), the Pareto optimal set P^*_{set} is defined as:

$$P^*_{set} \triangleq \{\xi_{(\alpha,\beta)} \in U \text{ there does not exist another feasible solution}$$

$$\xi_{(\alpha,\beta)} \in U \text{ such that } (\dot\alpha, \dot\beta) \text{ dominates } (\alpha, \beta)\}$$

Definition 3.10: For a given LMI-constrained MOP, the Pareto front P^*_{front} is defined as:

$$P^*_{front} \triangleq \{(\alpha, \beta) | \xi_{(\alpha,\beta)} \in P^*_{set}\}$$

It is obvious that if a feasible objective vector (α, β) is said to dominate another feasible objective vector $(\dot\alpha, \dot\beta)$, the feasible solution $\xi_{(\alpha,\beta)} = \{W, Y_1, Y_2, ..., Y_m\}$ is better than $\xi_{(\dot\alpha,\dot\beta)} = \{\dot W, \dot Y_1, \dot Y_2, ..., \dot Y_m\}$ in the Pareto optimal sense. By applying the concept of domination to the comparison of the feasible objective vectors, the proposed LMI-constrained MOEA can efficiently approach the Pareto front P^*_{front}.

Definition 3.11: The ideal point (α, β) for the lower bound of the LMI-constrained MOP in (3.29) for the nonlinear stochastic Poisson jump-diffusion system in (3.2) is defined as:

$$\underline{\alpha} = \min_{\xi \in U}$$

subject to LMIs (3.30) and (3.31)

and

$$\underline{\beta} = \min_{\xi \in U} \beta$$

subject to LMIs in (3.32)

Note that the ideal point $(\underline{\alpha}, \underline{\beta})$ is always an infeasible solution of the given MOP; otherwise, the given problem is not an MOP.

The detailed design procedure of the proposed LMI-constrained MOEA approach for the multi-objective T-S fuzzy controlled design problem of the nonlinear stochastic Poisson jump-diffusion system is given as follows.

3.5.1 THE LMI-CONSTRAINED MOEA PROCEDURE FOR MULTI-OBJECTIVE T-S FUZZY-CONTROL DESIGN

Step 1 Initialization

> **Step 1.1** Set the search region $\Gamma = (\alpha, \bar{\alpha}) \times (\beta, \bar{\beta})$, where $\bar{\alpha}$ and $\bar{\beta}$ are sufficiently large real numbers with $\bar{\alpha} > \alpha$ and $\bar{\beta} > \beta$, respectively. Set the maximum number of individuals N_g in the evolution algorithm, the iteration number N, the crossover rate M_c, and the mutation rate M_r.
> **Step 1.2** Randomly select N_g feasible objective individuals from the search region Γ to be the population P_1 and set the iteration index $t = 1$.

Step 2 Update: If the iteration index $t \leq N$

> **Step 2.1** Make the P_t perform an EA (including crossover and the mutation operator) and examine N feasible processes to produce $2N_g$ feasible individuals for the MOP in (3.29).
> **Step 2.2** For each iteration number index t, perform the elitist strategies to select N_g elitist individuals from the $2N_g$ feasible individuals obtained in **Step 2.1**; that is, select N_g individuals from the $2N_g$ feasible individuals via nondominated sort scheme.
> **Step 2.3** Set iteration index $t = t + 1$, and the N_g elitist individuals in **Step 2.2** are set to be the population P_{t+1}.

Step 3 Stop Criteria: Is the iteration index $t > N$?

> **Step 3.1** If $t > N$, then stop and set $P_{front}^* = P_{N+1}$, where P_{front}^* is called the Pareto front. Otherwise, go to **Step 2.1**.

Step 4 Select T-S fuzzy control gains K_i

> **Step 4.1** Select a preferred feasible objective individual $(\alpha^\dagger, \beta^\dagger) \in P_{front}^*$ according to the designer's preference. Once the preferred feasible objective individual is selected, the corresponding $\xi^\dagger = \{W, Y_1^\dagger, Y_2^\dagger, ..., Y_m^\dagger\} \in P_{set}^*$ is obtained. By using $\xi^\dagger$, the proposed T-S fuzzy controller $u(t) = \sum_{i=1}^m h_i(x)(K_i^\dagger x(t))$ in (3.25) can be constructed and the multi-objective control problem for nonlinear stochastic Poisson jump-diffusion system in (3.2) can be solved with $J_2 = \alpha^\dagger$ and $J_\infty = \beta^\dagger$ simultaneously.

3.6 SIMULATION EXAMPLE

In this section, we give a simulation example to verify the proposed multi-objective control scheme for a nonlinear stochastic Poisson jump-diffusion system. A LMI-constrained MOEA is employed to search a set of Pareto optimal solutions for the multi-objective control problem of the nonlinear stochastic Poisson jump-diffusion system in (3.2).

For the guidance control of a missile pursuit system [47] in polar coordinates, the relative dynamic motion between a homing missile and a target can be described by a nonlinear stochastic Poisson jump-diffusion system as follows:

$$
\begin{cases}
dx(t) = [f(x(t)) + g(x(t))u(t) + v(t)]dt + \sigma(x(t))dW(t) \\
\quad\quad + \int_\Theta T(x(t^-), \theta)N(dt, d\theta) \\
y(t) = m(x(t)),\ x(0) = x_0,\ \text{and}\ T(x_0, \cdot) = 0
\end{cases}
\tag{3.47}
$$

where
$x(t) = [x_1(t), x_2(t), x_3(t)]^T$, $u(t) = [u_1(t), u_2(t)]^T$, $v(t) = [0, v_1(t), v_2(t)]^T$, $f(x(t)) = [x_2(t), x_3^2(t)x_1^{-1}(t), -x_2(t)x_3(t)x_1^{-1}(t)]^T$, $m(x(t)) = x_3(t)$, $\sigma(x(t)) = diag(c_1, c_2, c_3)f(x(t))$, $f(x(t))$, $g(x(t)) = [0,0;-1,0;0,-1]^T$, and $T(x(t^-), \theta) = [0\ \ 0\ \ I_\theta(\theta_i)\ \ x_3(t_i^-)$ $x_1^{-1}(t_i^-)]^T$.

Here, $x_1(t)$ is the relative distance between the missile and target; $x_2(t)$ is the radial relative velocity; $x_3(t)$ is the tangential relative velocity; and $u_1(t)$ and $u_2(t)$ are the missile's guidance control along the $x_1(t)$ and $x_2(t)$ aces, respectively. The terms $v_1(t) = 3\sin(40t)$ and $v_2(t) = 8\sin(40t)$ are the external disturbances corresponding to the $x_2(t)$ and $x_3(t)$ axes due to windgusts and the target acceleration, respectively; the state-dependent internal fluctuation generated by the Wiener process is used to denote the effect of aerodynamic fluctuation and jet engine vibration where $c_1 = c_3 = 0.3$ and $c_2 = 0.1$. The discontinuous changes in the nonlinear stochastic system generated by Poisson process $N(t, \Theta)$ can be regarded as the deceptive behavior of the missile to avoid being shot down by shooting out gas from two side jets randomly. Suppose $\Theta = \{\theta_1, \theta_2\}$ with $E\{N(dt, \theta_1)\} = 0.1dt$ and $E\{N(dt, \theta_2)\} = 0.1dt$, where θ_1 and θ_2 represent gas shooting out from the two side jets, respectively, and the marked function I_θ is given by $I_\theta(\theta_1) = 1$ and $I_\theta(\theta_1) = -1$, respectively.

The multi-objective H_2/H_∞ guidance control design problem for the nonlinear stochastic missile pursuit system with the Poisson jump maneuver in (3.47) can be defined as:

$$
\min_{u(t) \in U} (J_2(u(t)), J_\infty(u(t)))
$$

$$
subject\ to\ (3.47)
$$

In this simulation, we set the weighting matrices to be $R_1 = diag(0.003, 0.003)$, $R_2 = diag(0.005, 0.005)$, $Q_1 = diag(0.1, 0.003, 0.003)$, $Q_2 = diag(0.008, 0.007)$, and $R_{x_0} = [3500, -550, -100]^T \times [3500, -550, -100]$.

The fuzzy premise vector $z(t)$, fuzzy operation points $x_i(t)$, and fuzzy interpolation matrices A_i, B_i, D_i, and $E_i(\theta)$ are given as follows:

$z(t) = [z_1(t), z_2(t)]^T = [x_1(t), x_3(t)]^T$; $x_1^{(1)}(t) = 335$; $x_1^{(2)}(t) = 1100$; $x_1^{(3)}(t) = 2500$; $x_1^{(4)}(t) = 6000$; $x_1^{(5)}(t) = 9000$; $x_3^{(1)}(t) = -350$; $x_3^{(2)}(t) = 350$.

$$A_i = \begin{bmatrix} 0 & 1 & 0 \\ 0 & 0 & \frac{x_3^{(i_3)}}{x_1^{(i_1)}} \\ 0 & \frac{-x_3^{(i_3)}}{x_1^{(i_1)}} & 0 \end{bmatrix}, \ B_i = \begin{bmatrix} 0 & 0 \\ -1 & 0 \\ 0 & -1 \end{bmatrix}, \ D_i = diag(0.3,0.3,0.1)A_i, \ E_1(\theta) = diag(0,0,\tfrac{1}{x_1^{(i_1)}}),$$

$$\text{and } E_2(\theta) = diag(0,0,\tfrac{-1}{x_1^{(i_1)}}), \text{ with } \begin{cases} i = 5(i_3 - 1) + i_1 \\ i_1 \in \{1,2,...,5\} \\ i_3 \in \{1,2\} \end{cases}.$$

The fuzzy membership functions are given in Figure 3.1. Moreover, the upper bounds of the T-S fuzzy approximation errors in the Euclidean norm are also given in the following:

$$\begin{cases} \left\| \Delta f(x(t)) \right\|_2 \le 10^{-2} \left\| x(t) \right\|_2 ; \left\| \Delta\sigma(x(t)) \right\|_2 \le 10^{-4} \left\| x(t) \right\|_2 \\ \left\| \Delta E_{i,\theta}(x(t),\theta) \right\|_2 \le 10^{-3} \left\| x(t) \right\|_2, \text{ and } \left\| \Delta m(x(t)) \right\|_2 \le 10^{-2} \left\| x(t) \right\|_2 \end{cases}$$

For the LMI-constrained MOEA to solve the MOP of the H_2/H_∞ guidance control design of the missile, the search region Γ is set as $\Gamma = [1 \times 10^6, 5 \times 10^6] \times [0,1]$, the maximum number of individuals $N_g = 50$, the iteration number $N_s = 70$, the crossover rate $M_c = 0.8$, and the mutation rate $M_r = 0.2$.

Once the iteration number N_s is achieved, the Pareto front P_{front}^* of the MOP for the nonlinear stochastic Poisson jump-diffusion system of a missile pursuit system in (3.47) can be obtained as shown in Figure 3.2.

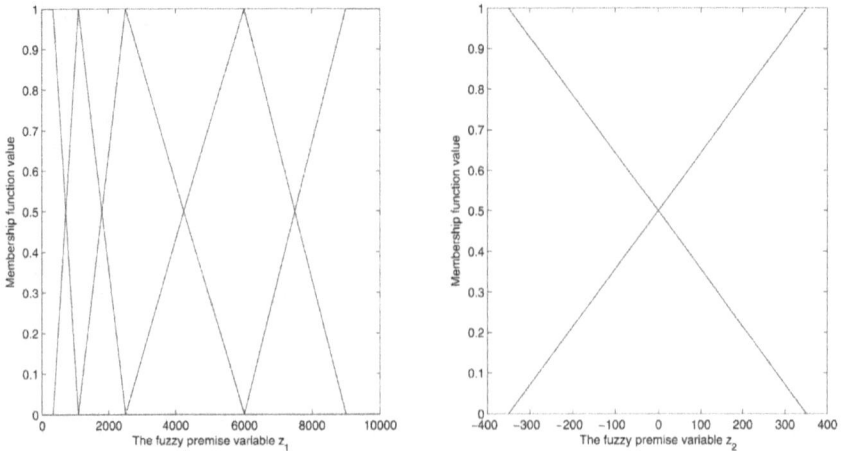

FIGURE 3.1   The fuzzy membership functions of the fuzzy premise variables $z_1(t)$ and $z_2(t)$.

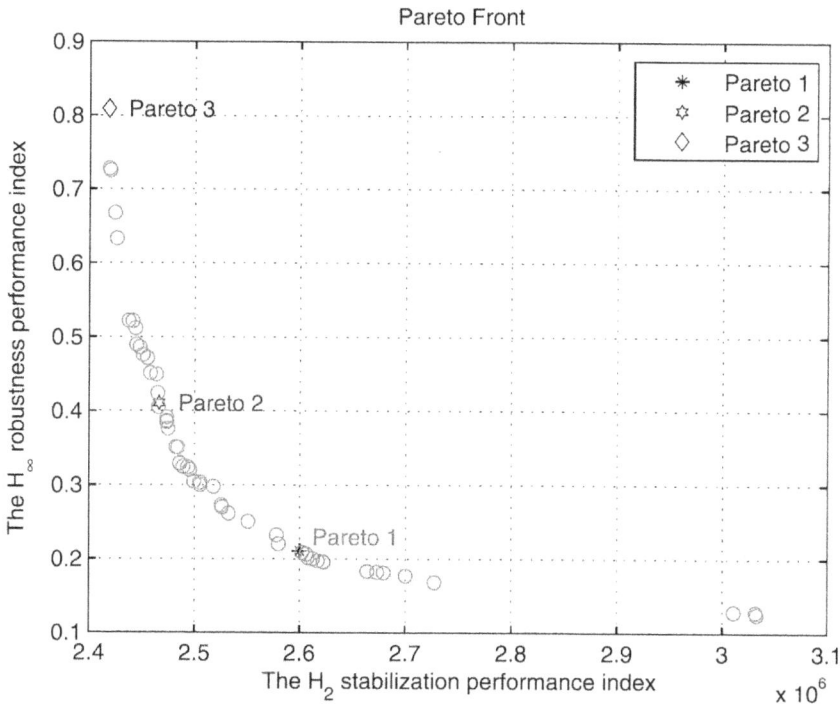

FIGURE 3.2 The Pareto front P^*_{front} of the MOP in (3.47) obtained by the proposed LMI-constrained MOEA.

To illustrate how to select the preferred solution, we choose three Pareto optimal solutions from the Pareto front for comparison; their Pareto objective vectors are given as follows:

Pareto Solution 1	Pareto Solution 2	Pareto Solution 3
$(2.62\times10^6, 0.21)$	$(2.46\times10^6, 0.45)$	$(2.44\times10^6, 0.63)$

Moreover, the T-S fuzzy control gains of the three chosen Pareto optimal solutions are given in Tables 3.1–3.3 in the Appendix, respectively. The simulation results in Figures 3.3 to 3.7 are given to illustrate the performance of the multi-objective T-S fuzzy control of the three chosen Pareto optimal solutions. In Figure 3.3, Pareto optimal solution 1 has the largest J_2 performance index of the three chosen Pareto optimal solutions because the integral value of its trajectory with respect to t is the largest of the three. However, with respect to Pareto optimal solution 3, its trajectory has the minimum integral value with respect to time. Thus, it has the smallest J_2 performance index of the three. In Figure 3.5, since Pareto optimal solution 1 has the best J_∞ performance index of the three chosen Pareto optimal solutions, its trajectory is robust with less vibration of the tangential relative velocity when the Poisson jumps

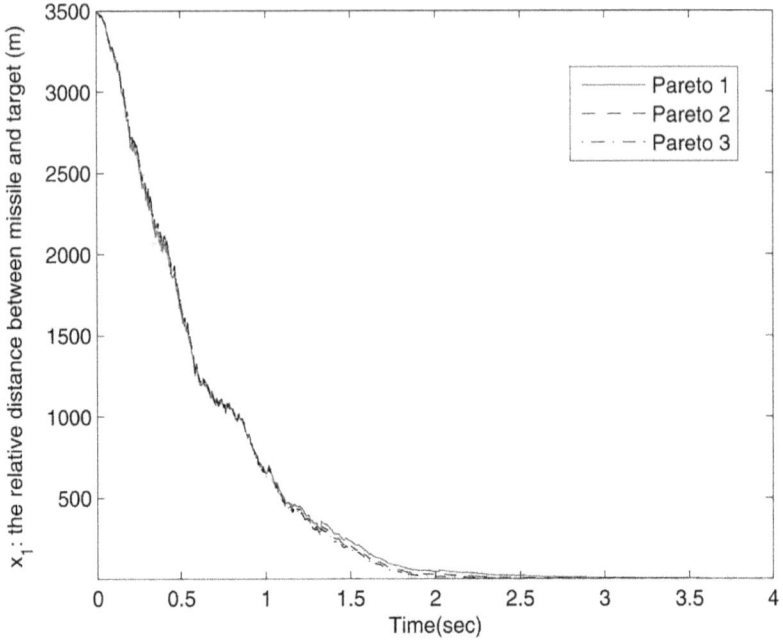

FIGURE 3.3 Radial relative velocities $x_1(t)$ between the middle and target of the three chosen Pareto solutions.

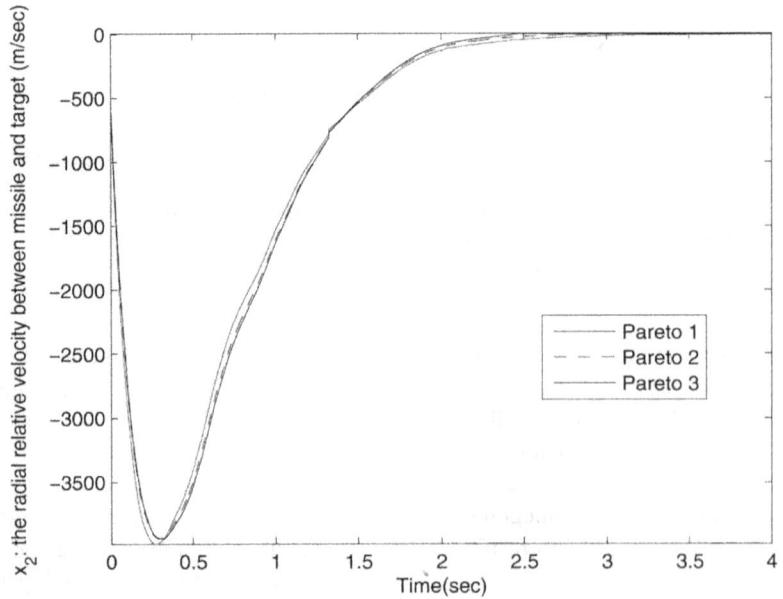

FIGURE 3.4 Tangential relative velocities $x_2(t)$ between the missile and target of the three chosen Pareto solutions.

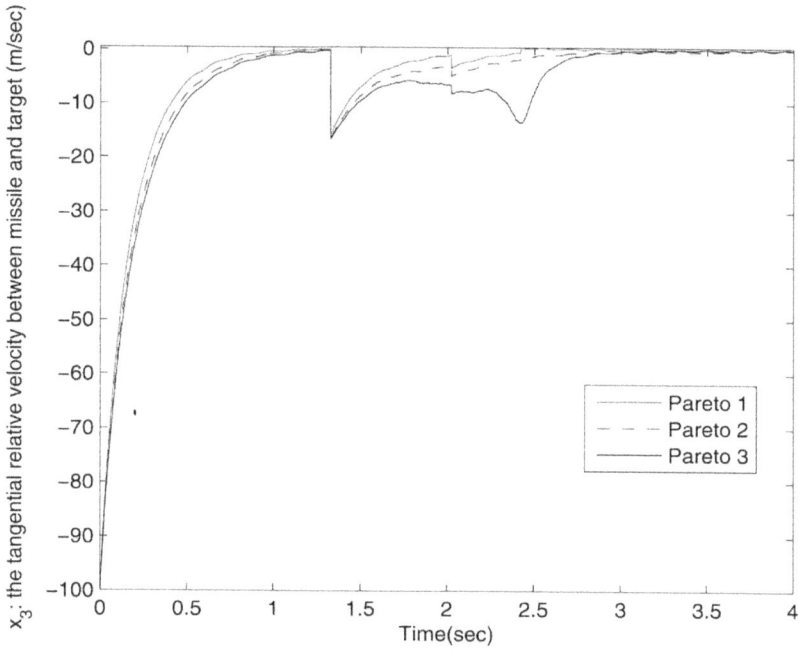

FIGURE 3.5 Tangential relative velocities $x_3(t)$ between the missile and target of the three chosen Pareto solutions.

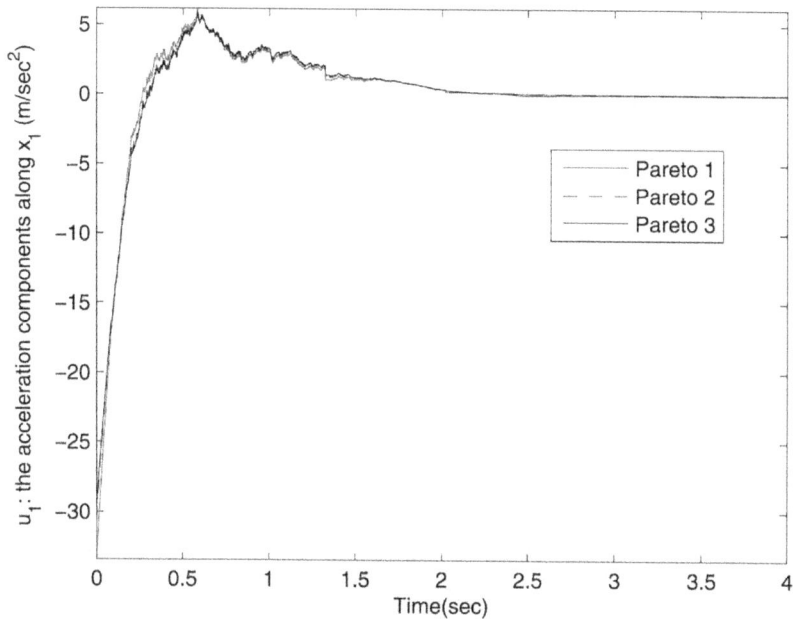

FIGURE 3.6 Trajectory of the control input signal $u_1(t)$ of the three chosen Pareto solutions.

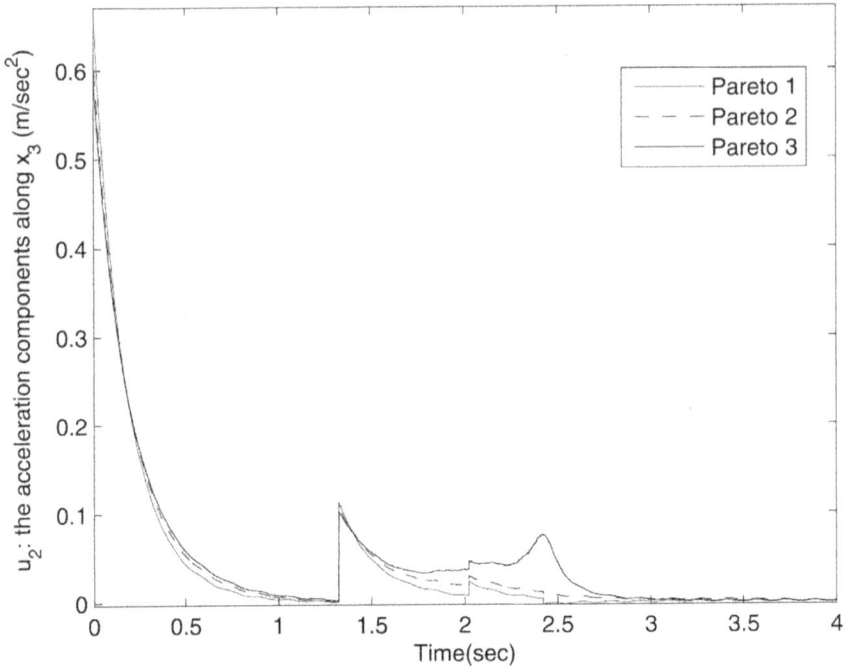

FIGURE 3.7   Trajectory of the control input signal $u_2(t)$ of the three chosen Pareto solutions.

occur. In contrast, because Pareto optimal solution 3 has the worst J_∞ performance index, it is less robust and has the most serious vibration of the three solutions. It is easy to observe that Pareto solution 2 is the preferred solution of the three chosen Pareto solutions because Pareto optimal vector 2 provides a compromise between J_2 and J_∞. Finally, the trajectories of the marked Poisson jump processes in (3.47) are given in Figure 3.8.

3.7   CONCLUSION

This chapter investigated multi-objective control for nonlinear stochastic Poisson jump-diffusion systems via the T-S fuzzy interpolation method. Unlike most MOPs that only focus on algebraic systems, the proposed multi-objective control design method can handle dynamically constrained MOPs and simultaneously achieve the H_2 optimal cost and H_∞ robustness performance indices of a nonlinear stochastic Poisson jump-diffusion system. To avoid solving the HJIs, the T-S fuzzy interpolation method is employed, and these HJI constraints can be replaced by two sets of LMI constraints (i.e., the HJI-constrained MOP can be transformed into an LMI-constrained MOP). If there exist some feasible solutions for the LMI constraints in (3.30)–(3.32), then the LMI-constrained MOP can be solved by the proposed reverse-order LMI-constrained MOEA. Thus, the multi-objective control design problem of the nonlinear stochastic Poisson jump-diffusion systems can be efficiently solved with the help of the LMI toolbox of the commercial MATLAB

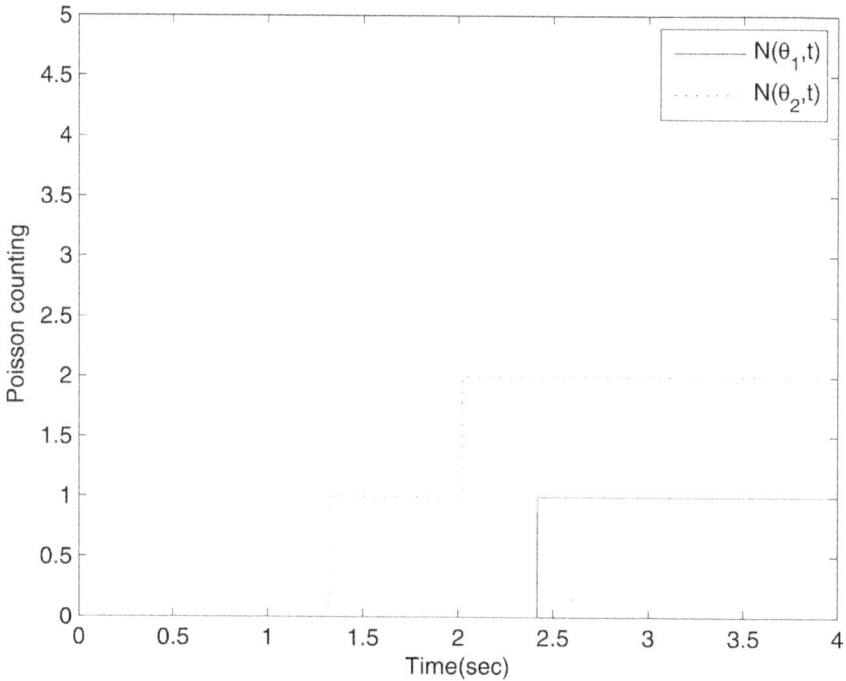

FIGURE 3.8   The trajectory of marked Poisson jump process.

software. When the Pareto front is obtained, the designer can select a preferred T-S fuzzy controller from the set of Pareto optimal controllers according to their preference and complete the multi-objective T-S fuzzy controller design. Finally, we gave an example of stochastic homing missile control design to confirm the performance of the proposed multi-objective control design for nonlinear stochastic Poisson jump-diffusion systems through a computer simulation.

3.8   APPENDIX

TABLE 3.1
The T-S fuzzy control gains for Pareto solution 1

$$K_1^1 = \begin{bmatrix} 10.49 & 6.42 & 0.06 \\ -0.17 & 0.24 & 5.42 \end{bmatrix} \qquad K_2^1 = \begin{bmatrix} 10.52 & 6.44 & 0.01 \\ -0.07 & 0.04 & 5.48 \end{bmatrix}$$

$$K_3^1 = \begin{bmatrix} 10.52 & 6.44 & 0.00 \\ -0.03 & 0.02 & 5.49 \end{bmatrix} \qquad K_4^1 = \begin{bmatrix} 10.52 & 6.44 & 0.00 \\ -0.01 & 0.00 & 5.49 \end{bmatrix}$$

$$K_5^1 = \begin{bmatrix} 10.52 & 6.44 & 0.00 \\ 0.00 & 0.00 & 5.49 \end{bmatrix} \qquad K_6^1 = \begin{bmatrix} 10.49 & 6.42 & -0.06 \\ 0.17 & -0.24 & 5.42 \end{bmatrix}$$

(Continued)

TABLE 3.1   (Continued)

$$K_7^1 = \begin{bmatrix} 10.52 & 6.44 & -0.01 \\ 0.07 & -0.04 & 5.48 \end{bmatrix} \qquad K_8^1 = \begin{bmatrix} 10.52 & 6.44 & 0.00 \\ 0.03 & -0.02 & 5.49 \end{bmatrix}$$

$$K_9^1 = \begin{bmatrix} 10.52 & 6.44 & 0.00 \\ 0.01 & 0.00 & 5.49 \end{bmatrix} \qquad K_{10}^1 = \begin{bmatrix} 10.52 & 6.44 & 0.00 \\ 0.00 & 0.00 & 5.49 \end{bmatrix}$$

TABLE 3.2
The T-S fuzzy control gains for Pareto solution 2

$$K_1^2 = \begin{bmatrix} 9.51 & 5.74 & 0.12 \\ -0.16 & 0.19 & 5.96 \end{bmatrix} \qquad K_2^2 = \begin{bmatrix} 9.52 & 5.80 & 0.03 \\ -0.05 & 0.04 & 6.02 \end{bmatrix}$$

$$K_3^2 = \begin{bmatrix} 9.52 & 5.80 & 0.01 \\ -0.02 & 0.02 & 6.02 \end{bmatrix} \qquad K_4^2 = \begin{bmatrix} 9.52 & 5.80 & 0.00 \\ 0.00 & 0.00 & 6.02 \end{bmatrix}$$

$$K_5^2 = \begin{bmatrix} 9.52 & 5.80 & 0.00 \\ 0.00 & 0.00 & 6.02 \end{bmatrix} \qquad K_6^2 = \begin{bmatrix} 9.51 & 5.74 & -0.12 \\ 0.16 & -0.19 & 5.96 \end{bmatrix}$$

$$K_7^2 = \begin{bmatrix} 9.52 & 5.80 & -0.03 \\ 0.05 & -0.04 & 6.02 \end{bmatrix} \qquad K_8^2 = \begin{bmatrix} 9.52 & 5.80 & -0.01 \\ 0.02 & -0.01 & 6.02 \end{bmatrix}$$

$$K_9^2 = \begin{bmatrix} 9.51 & 5.74 & 0.00 \\ 0.00 & 0.00 & 6.02 \end{bmatrix} \qquad K_{10}^2 = \begin{bmatrix} 9.52 & 5.80 & 0.00 \\ 0.00 & 0.00 & 6.02 \end{bmatrix}$$

TABLE 3.3
The T-S fuzzy control gains for Pareto solution 3

$$K_1^3 = \begin{bmatrix} 9.47 & 5.68 & 0.11 \\ -0.26 & 0.166 & 5.07 \end{bmatrix} \qquad K_2^3 = \begin{bmatrix} 9.49 & 5.75 & 0.02 \\ -0.08 & 0.03 & 5.11 \end{bmatrix}$$

$$K_3^3 = \begin{bmatrix} 9.49 & 5.76 & -0.008 \\ 0.048 & 0.030 & 0.944 \end{bmatrix} \qquad K_4^3 = \begin{bmatrix} 9.49 & 5.76 & 0.00 \\ -0.01 & 0.00 & 5.12 \end{bmatrix}$$

$$K_5^3 = \begin{bmatrix} 9.49 & 5.76 & 0.00 \\ 0.00 & 0.00 & 5.12 \end{bmatrix} \qquad K_6^3 = \begin{bmatrix} 9.47 & 5.68 & -0.11 \\ 0.26 & -0.17 & 5.07 \end{bmatrix}$$

$$K_7^3 = \begin{bmatrix} 9.49 & 5.75 & -0.02 \\ 0.08 & -0.03 & 5.11 \end{bmatrix} \qquad K_8^3 = \begin{bmatrix} 9.49 & 5.75 & -0.01 \\ 0.04 & -0.01 & 5.12 \end{bmatrix}$$

$$K_9^3 = \begin{bmatrix} 9.49 & 5.76 & 0.00 \\ 0.01 & 0.00 & 5.12 \end{bmatrix} \qquad K_{10}^3 = \begin{bmatrix} 9.49 & 5.76 & 0.00 \\ 0.01 & 0.00 & 5.12 \end{bmatrix}$$

4 Multi-Objective Tracking Control Design of T-S Fuzzy Systems
Fuzzy Pareto Optimal Approach

4.1 INTRODUCTION

After the MOP of H_2/H_∞ stabilization control of nonlinear stochastic systems has been introduced in Chapter 3, the MOP of H_2/H_∞ tracking control design of nonlinear T-S fuzzy systems will be introduced in this chapter. In system control design, the tasks of stabilization and tracking are two typical control design problems. Many control design techniques have been developed for nonlinear system control designs, for example, exact feedback linearization [48, 49], sliding mode control [48], and adaptive control [48]. Fuzzy control has been employed successfully to solve many nonlinear control problems [50, 51]. Recently, many robust fuzzy tracking control schemes have been developed for nonlinear tracking control design problems [52–61]. In general, tracking control design problems are more difficult than fuzzy stabilization control design problems because they need to track a prescribed reference signal.

In [52], the feedback linearization technique was proposed to design a fuzzy tracking controller for discrete time systems. As pointed out in [58, 61], fuzzy controllers based on feedback linearization may be not guaranteed to be stable for nonminimum phase systems because of their inverse system design method. Recently, the optimal H_2 tacking control design was proposed for nonlinear stochastic systems to achieve the least mean square tracking error [58, 61–64]. In general, the optimal H_2 tracking control design can be achieved only when the statistics of stochastic disturbances are available. Otherwise, its performance will be degraded if the statistics of stochastic disturbance are incorrect. To remedy the shortcomings of H_2 tracking, a robust control design based on H_∞ tracking has received great attention for its robustness properties against uncertain disturbance without the knowledge of the statistics of disturbances. In [53], based on the T-S fuzzy model, a robust H_∞ tracking control design was proposed for nonlinear dynamic systems with external disturbance. An effective matrix decoupling technique was developed to derive a single-step LMI condition for the observer-based H_∞ control design [65]. In [29, 66–69], the H_∞ fuzzy control was studied for many different systems to achieve H_∞ control performance. Recently, mixed H_2/H_∞ tracking designs [34, 70–75] were proposed to minimize the H_2 tracking error with the consideration of a prescribed H_∞ attenuation level to

DOI: 10.1201/9781003362142-6

eliminate the influence of uncertain disturbances. However, the conventional mixed H_2/H_∞ tracking control design is an optimal H_2 tracking control under a prescribed H_∞ attenuation constraint and basically is only a constrained single-objective problem for tracking control design. In this chapter, the proposed multi-objective tracking control design can achieve the optimal H_2 and robust H_∞ tracking performance for fuzzy systems simultaneously.

Although most existing studies capture characteristics of tracking control designs in the form of SOPs, multiple objectives arise naturally in real-life scenarios. These objectives normally conflict with each other; in other words, these objectives cannot be optimized simultaneously. In general, the optimal solution of multiple objectives is not unique, and there may exist many optimal solutions for multiple objectives. In contrast to finding the global optimal solution in SOPs, Pareto optimal solutions have been proposed for MOPs [76–78]. Classical optimization methods suggest converting an MOP to SOP by emphasizing one particular Pareto optimal solution by a weighted sum method at a time [78]. When such a method is used to find multiple solutions of an MOP, it has to be applied with different weightings many times, hopefully finding a different solution in each simulation run. Recently, a number of multi-objective evolutionary algorithms have been discussed for their ability to find multiple Pareto optimal solutions for algebraic systems in one single simulation run [2, 76–79]. From a recent review paper about multi-objective fuzzy control [80], it is seen that the work of multi-objective control of fuzzy logical systems can be divided into two categories: (i) identification of control parameters and/or rules (e.g. tuning of membership function parameters and rule selection as a post-processing method) and (ii) learning controller structure (e.g. learning rule bases); that is, at present, a MOEA is employed to learn the fuzzy controller based on a fuzzy rule-based system and still cannot be applied to the multi-objective control design problems of T-S fuzzy dynamic systems [80–82]. Even though MOEAs have been employed widely to solve MOPs of algebraic or rule-based systems, it is still not easy to apply them to solve MOPs of nonlinear dynamic systems at present. More effort is still needed to apply them to multi-objective H_2/H_∞ tracking control design problems for nonlinear dynamic systems. In general, MOPs for nonlinear dynamic systems are more difficult than how to specify variables to minimize the multi-objective functions in the conventional MOPs of algebraic or rule-based systems.

Recently, T-S fuzzy systems have been used to efficiently approximate nonlinear dynamic systems [70, 83–88]. In this chapter, the H_2 and H_∞ tracking control problems are formulated as an MOP for fuzzy systems with uncertain measurement noises and disturbances. The H_2 tracking performance needs to be minimized for the minimum tracking error. Similarly, the robust H_∞ tracking performance also needs to be minimized for the minimum worst-case effects of measurement noise and disturbance on the tracking error. Therefore, the multi-objective H_2/H_∞ tracking control design needs to minimize H_2 tracking and robust H_∞ tracking performance simultaneously. In general, it is not easy to solve this multi-objective tracking control design problem for fuzzy systems directly. In this chapter, the MOP for H_2/H_∞ tracking control design of fuzzy systems is solved from the indirect perspective. Based on three LMI constraints and the proposed indirect technique, we find the H_2 tracking and robust H_∞ tracking performance have the upper bounds α and β, respectively. Therefore, the indirect method of multi-objective H_2/H_∞ tracking control design becomes how to specify the fuzzy control gains to minimize the upper bounds $(\alpha\ \beta)$ simultaneously subject to sets of LMIs. We also show that the proposed indirect MOP is equivalent to the original MOP for multi-objective H_2/H_∞ tracking control design of fuzzy systems when Pareto optimal solutions are achieved. In summary, as compared with the conventional mixed H_2/H_∞ tracking control designs, which solve the optimal H_2 tracking control problem under the constraint of a prescribed robust H_∞ tracking performance, that is, to minimize α under a prescribed H_∞ attenuation level β, the proposed multi-objective H_2/H_∞ tracking control design needs to minimize α and β simultaneously.

In this chapter, in order to solve the simultaneous optimization of the MOP for the multi-objective H_2/H_∞ tracking control design for fuzzy systems, the LMI technique and an MOEA based on nondominating sorting scheme [2, 76, 79] are combined together. Hence, the LMI-constrained MOP can be solved by an LMI-based MOEA search method via a nondominating sorting scheme to achieve optimal H_2 and robust H_∞ tracking performance simultaneously. The feasible solutions and the computational complexities of LMI-constrained MOP are also discussed in this chapter. Since an LMI is convex, the LMI-constrained MOP for fuzzy systems is always feasible and can be solved efficiently by the convex optimization scheme [89] via the proposed LMI-based MOEA search method [2, 76, 79] with the help of the LMI toolbox in MATLAB. Further, for comparison, the multi-objective H_2/H_∞ tracking control design based on the weighted sum method and the conventional mixed H_2/H_∞ tracking control method are also solved for fuzzy systems as an alternative choice.

Because there exist sets of feasible optimal solutions, the LMI and MOEA techniques are employed to solve the LMI-constrained MOP. An LMI-based MOEA search via a number of genetic operators, such as selection, mutation, and crossover, is proposed with the help of the LMI toolbox in MATLAB to search for the sets of Pareto optimal solutions for fuzzy control gains from the LMI constraints. Hence, one of Pareto optimal solutions is selected to achieve a multi-objective H_2/H_∞ tracking performance by the designer according to their preference. Finally, a simulation example of an MOP of H_2/H_∞ tracking control of a two-link robot system is provided to illustrate the design procedure and confirm the performance of the proposed results for nonlinear robotic systems with external disturbance.

Notations and Definitions: I and 0 denote the identity and zero matrices with appropriate dimensions, respectively. Asymmetric matrix $P >, <, \geq$, and ≤ 0 means that it is positive definite, negative definite, positive semidefinite, and negative semidefinite, respectively. A block diagonal matrix is shown by $diag(\cdot)$.

In reality, there exist many solutions for an MOP, called Pareto optimal solutions. Comparison among Pareto optimal solutions requires an order relation, called domination, between different solutions. Before further discussion, take the following example to easily understand [76–78]. Consider an MOP with two objectives $(\alpha\ \beta)$ and the decision variable $W \in \Omega$, where Ω is the feasible set. Some properties of the Pareto optimal solutions of the MOP are given as follows [76–78].

Definition 4.1 (Dominance): There are two solutions, $W_1 \in \Omega$ and $W_2 \in \Omega$, for the two objective values $(\alpha_1\ \beta_1)$ and $(\alpha_2\ \beta_2)$ of the MOP, respectively. $(\alpha_1\ \beta_1)$ is said to dominate $(\alpha_2\ \beta_2)$ if $\alpha_1 \leq \alpha_2$ and $\beta_1 \leq \beta_2$.

Definition 4.2 (Pareto optimal solution): A solution W^* is the Pareto optimal solution of the MOP with respect to Ω if there does not exist another feasible solution $W^\circ \in \Omega$ such that the objective values $(\alpha^\circ \, \beta^\circ)$ dominate $(\alpha^* \, \beta^*)$.

Definition 4.3 (Pareto front): For the MOP, the Pareto front defined as $\Gamma \triangleq \{(\alpha^* \, \beta^*) | W^* \text{ is the Pareto optimal solution of the MOP and } (\alpha^* \, \beta^*) \text{ is generated by } W^*\}$.

4.2   SYSTEM DESCRIPTION AND PROBLEM FORMULATION

A nonlinear system can be represented by the T-S fuzzy model [83], which is a fuzzy piecewise interpolation of several local linear systems to approximate a nonlinear system via fuzzy bases. The T-S fuzzy model is described by fuzzy if-then rules as follows [70, 84–86]

$$
\begin{aligned}
&\text{Plant Rule } i \\
&\text{If } z_1(t) \text{ is } F_{i1} \text{ and } \cdots \text{ and } z_g(t) \text{ is } F_{ig} \\
&\text{then } \dot{x}(t) = A_i x(t) + B_i u(t) + E\omega(t), \text{for } i = 1,...,l
\end{aligned} \tag{4.1}
$$

where $x(t) \triangleq \begin{bmatrix} x_1(t) & x_2(t) & \cdots & x_n(t) \end{bmatrix}^T$ denotes the state vector; $u(t) \triangleq \begin{bmatrix} u_1(t) & u_2(t) & \cdots & u_m(t) \end{bmatrix}^T \in R^{m \times 1}$ denotes the control input; $\omega(t) \triangleq \begin{bmatrix} \omega_1(t) & \cdots & \omega_p(t) \end{bmatrix}^T \in R^{p \times 1}$ is the external disturbance with the influence matrix $E \in R^{n \times p}$; F_{ij} is the fuzzy set; $A_i \in R^{n \times n}$, $B_i \in R^{n \times m}$ are the fuzzy system parameters; l is the number of if-then rules; and $z_1(t), z_2(t),...,z_g(t)$ are the premise variables. The T-S fuzzy system in (4.1) is employed to represent a nonlinear dynamic system, and the fuzzy system parameters A_i and B_i can be identified by the optimal parameter estimation method [62, 64, 68, 69, 83, 86, 88]. Hence, the fuzzy system in (4.1) is inferred as follows [70, 84–86]

$$
\dot{x}(t) = \sum_{i=1}^{l} h_i(z(t)) \big[A_i x(t) + B_i u(t) \big] + E\omega(t) \tag{4.2}
$$

where $\mu_i(z(t)) = \Pi_{j=1}^{g} F_{ij}(z_j(t))$, $h_i(z(t)) = \mu_i(z(t)) / \sum_{i=1}^{l} \mu_i(z(t))$ with the property $\sum_{i=1}^{l} h_i(z(t)) = 1$, and $z(t) \triangleq [z_1(t) \cdots z_g(t)]$. The fuzzy model in (4.2) is an interpolation of local linear systems for general nonlinear dynamic systems and can be used to model the behaviors of complex dynamic systems [54]. Consider the following reference model to be tracked [54]

$$
\dot{x}_r(t) = \sum_{i=1}^{l} h_i(z(t)) A_{ir} x_r(t) + r(t) \tag{4.3}
$$

where $x_r(t)$ is the desired reference state to be tracked, A_{ir} is the matrix to be specified for the transient behavior of $x_r(t)$, and $r(t)$ is the bounded reference input to be specified by the user for any desired reference signal at the steady state. It is assumed that A_{ir} and $r(t)$ are selected so that $x_r(t)$ represents a desired trajectory for $x(t)$ to track. Let us denote the tracking error $e(t) \triangleq x(t) - x_r(t)$. It is more appealing to use state feedback control to achieve the multi-objective H_2/H_∞ tracking performance; therefore, we have the following fuzzy controller

$$u(t) = \sum_{j=1}^{l} h_j(z(t))\left[K_j(x(t) - x_r(t))\right] = \sum_{j=1}^{l} h_j(z(t))K_j e(t) \qquad (4.4)$$

where $K_j \in R^{m \times n}$ is the fuzzy control gain to be designed.

Remark 4.1: (i) In the previous fuzzy systems, the premise variable $z(t)$ can be measurable state variables, outputs, or a combination of measurable state variables. For the T-S fuzzy model, using state variables as premise variables is common but not always done [84, 85]. The limitation of this approach is that some state variables must be measurable to construct the fuzzy observer and controller [85]. This is common limitation for T-S fuzzy control design [7, 35]. (ii) The problem of how to construct T-S fuzzy models for nonlinear systems can be found in [83, 88].

After constructing the fuzzy system in (4.2) and the reference system in (4.3), the fuzzy augmented system can be expressed as follows

$$\dot{\bar{x}} = \sum_{i=1}^{l}\sum_{j=1}^{l} h_i(z(t))h_j(z(t))\left[\bar{A}_{ij}\bar{x}(t) + \bar{E}\bar{\omega}(t)\right] \qquad (4.5)$$

where

$$\bar{x} \triangleq \begin{bmatrix} e(t) \\ x_r(t) \end{bmatrix}, \; \bar{\omega}(t) \triangleq \begin{bmatrix} \omega(t) \\ r(t) \end{bmatrix}, \; \bar{A}_{ij} \triangleq \begin{bmatrix} A_i + B_i K_j & A_i - A_{ir} \\ 0 & A_{ir} \end{bmatrix}, \; \bar{E} \triangleq \begin{bmatrix} E & -I \\ 0 & I \end{bmatrix}$$

Since the reference signal $r(t)$ is to be specified by the user in the future and can be treated as an uncertain external input by designer, it can be included in the general disturbance $\bar{\omega}(t)$. Additionally, the tracking error $e(t)$ can be represented by $\bar{x}(t)$ as follows

$$e(t) = \begin{bmatrix} I & 0 \end{bmatrix}\bar{x}(t) \qquad (4.6)$$

Based on the fuzzy augmented system in (4.5), the H_2 tracking performance without considering the effect of general disturbance $\bar{\omega}(t)$ can be represented by

$$J_2(u) \triangleq \int_0^{t_f} e^T(t)Q_1 e(t) + u^T(t)Ru(t)dt$$

$$= \int_0^{t_f} \bar{x}^T(t)\bar{Q}_1\bar{x}(t) + u^T(t)Ru(t)dt \qquad (4.7)$$

where $Q_1 \in R^{n \times n}$ and $R \in R^{n \times n}$ are the weighting matrices for $e(t)$ and $u(t)$, respectively, and $\bar{Q}_1 = \begin{bmatrix} I & 0 \end{bmatrix}^T Q_1 \begin{bmatrix} I & 0 \end{bmatrix}$. For the optimal tracking control design, the H_2 tracking performance in (4.7) is to consider the penalty on both the quadratic tracking error and control effort. Further, in order to efficiently attenuate the effect of general disturbance $\bar{\omega}(t)$, the robust H_∞ tracking performance is given by

$$
J_\infty(u) \triangleq \frac{\int_0^{t_f} e^T(t)Q_2 e(t)dt}{\int_0^{t_f} \bar{\omega}^T(t)\bar{\omega}(t)dt} = \frac{\int_0^{t_f} \bar{x}^T(t)\bar{Q}_2\bar{x}(t)dt}{\int_0^{t_f} \bar{\omega}^T(t)\bar{\omega}(t)dt}, \text{ with } \bar{x}(0) = 0 \tag{4.8}
$$

where $Q_2 \in R^{n \times n}$ is the weighting matrix for $e(t)$. In (4.7) and (4.8), t_f denotes the terminal time of control, and $\bar{Q}_2 = \begin{bmatrix} I & 0 \end{bmatrix}^T Q_2 \begin{bmatrix} I & 0 \end{bmatrix}$.

Then the multi-objective H_2/H_∞ tracking control problem is given as follows:

$$
\min_{u(t) \in U}(J_2(u) \quad J_\infty(u)) \text{ subject to (4.5)} \tag{4.9}
$$

Where U denotes the feasible set of control laws, and $J_2(u)$ and $J_\infty(u)$ are defined in (4.7) and (4.8), respectively. At present, most MOPs focus on static systems, and it is not easy to solve the MOP in (4.9) for the fuzzy augmented system in (4.5) directly.

Remark 4.2: The conventional mixed H_2/H_∞ control designs are always considered multi-objective H_2/H_∞ control designs by some authors [34, 70–75]. For the conventional mixed H_2/H_∞ tracking control design, one always minimizes the H_2 tracking performance $J_2(u)$ in (4.7) under a specified robust H_∞ tracking performance constraint $J_\infty(u) \leq \beta$ with a given β in (4.8); that is, the conventional mixed H_2/H_∞ tracking control design is only a single-objective H_2 tracking control design under a prescribed robust H_∞ tracking performance constraint. More detailed discussion will be presented in the following sections.

4.3 MULTI-OBJECTIVE H_2/H_∞ TRACKING CONTROL DESIGN

Since there is some conflict between J_2 in (4.7) and J_∞ in (4.8), there may not exist a unique solution for the MOP in (4.9). Further, because of the complexity of J_2 in (4.7), J_∞ in (4.8), and the fuzzy augmented system in (4.5), it is very difficult to solve the MOP in (4.9) directly. Therefore, an indirect method is proposed to solve the MOP in (4.9) subject to J_2 and J_∞ indirectly. Suppose $J_2 \leq \alpha$ and $J_\infty(u) \leq \beta$; that is, α and β are the upper bounds of the H_2 tracking and robust H_∞ tracking performance, respectively. Then, the MOP in (4.9) is transformed into the following suboptimal MOP for the multi-objective H_2/H_∞ tracking control design

$$
(\alpha^* \quad \beta^*) \triangleq \min_{u(t) \in U}(\alpha \quad \beta) \tag{4.10}
$$

subject to

$$
J_2(u) = \int_0^{t_f} \bar{x}^T(t)\bar{Q}_1\bar{x}(t) + u^T(t)Ru(t)dt \leq \alpha \tag{4.11}
$$

$$
J_\infty(u) = \frac{\int_0^{t_f} \bar{x}^T(t)\bar{Q}_2\bar{x}(t)dt}{\int_0^{t_f} \bar{\omega}^T(t)\bar{\omega}(t)dt} \leq \beta \tag{4.12}
$$

The following result will prove that the suboptimal MOP in (4.10) is equivalent to the MOP in (4.9) for the multi-objective H_2/H_∞ tracking control design when the Pareto optimal solutions of MOP are achieved.

Theorem 4.1: The MOP in (4.10) is equivalent to the MOP in (4.9).

Proof: The proof of Theorem 4.1 is straightforward. One only needs to prove that both inequality constraints in (4.11) and (4.12) become equality for the Pareto optimal solutions for the MOP in (4.10). Given a 3-tuple Pareto optimal solution $(u^*(t), \alpha^*, \beta^*)$, we assume that either one of the inequalities in (4.11) or (4.12) remains a strict inequality at the optimal solution. Without loss of generality, suppose $J_2(u^*) < \alpha$. As a result, there exists α_1 such that $\alpha_1 < \alpha^*$ and that $J_2(u^*) = \alpha_1$. Now, for the same $u^*(t)$, the solution (α_1, β^*) dominates the Pareto optimal solution (α^*, β^*), leading to a contradiction [79, 89, 90]. This implies that both inequality constraints in (4.11) and (4.12) indeed become equalities for Pareto optimal solutions. The optimization problem in (4.10) is hence equivalent to the MOP in (4.9).

Remark 4.3: (i) Based on Theorem 4.1, the MOP of the multi-objective H_2/H_∞ tracking control design problem in (4.9) can be replaced by the MOP in (4.10) with the inequality constraints in (4.11)–(4.12), which are much easier to solve. Therefore, we focus on the MOP in (4.10) in sequence. (ii) For the robust H_∞ tracking performance in (4.12), it is equivalent to

$$\int_0^{t_f} \overline{x}^T(t)\overline{Q}_2\overline{x}(t)dt \le \beta \int_0^{t_f} \overline{\omega}^T(t)\overline{\omega}(t)dt \tag{4.13}$$

If the effect of initial condition $\overline{x}(0) \ne 0$ is considered in the input/output ratio in (4.12) or (4.13), then the robust H_∞ tacking performance in (4.13) should be modified as follows [70, 86]

$$\int_0^{t_f} \overline{x}^T(t)\overline{Q}_2\overline{x}(t)dt \le \overline{x}^T(0)P\overline{x}(0) + \beta \int_0^{t_f} \overline{\omega}^T(t)\overline{\omega}(t)dt \tag{4.14}$$

for some matrix $P > 0$.

Based on this analysis, we get the following result.

Theorem 4.2: If the following MOP is solved,

$$(\alpha^* \quad \beta^*) = \min_{(W_{11}, P_{22}, Y_j) \in \Omega} (\alpha \quad \beta) \tag{4.15}$$

subject to the following LMIs

$$\begin{bmatrix} W_{11}A_i^T + A_iW_{11} + B_iY_j + Y_j^TB_i^T & (A_i - A_{ir}) & W_{11} & Y_j^T \\ (A_i - A_{ir})^T & A_{ir}^TP_{22} + P_{22}A_{ir} & 0 & 0 \\ W_{11} & 0 & -Q_1^{-1} & 0 \\ Y_j & 0 & 0 & -R^{-1} \end{bmatrix} \le 0 \tag{4.16}$$

$$
\begin{bmatrix} x_r^T(0)P_{22}x_r(0)-\alpha I & e^T(0) \\ e(0) & -W_{11} \end{bmatrix} \le 0
\tag{4.17}
$$

$$
\begin{bmatrix} W_{11}A_i^T + A_iW_{11} + B_iY_j + Y_j^T B_i^T & (A_i - A_{ir}) & E & -I & W_{11} \\ (A_i - A_{ir})^T & A_{ir}^T P_{22} + P_{22}A_{ir} & 0 & P_{22} & 0 \\ E^T & 0 & -\beta I & 0 & 0 \\ W_{11} & 0 & 0 & 0 & -Q_2^{-1} \end{bmatrix} \le 0
\tag{4.18}
$$

where $W_{11} \triangleq P_{11}^{-1} > 0$ and $Y_j \triangleq K_jW_{11}$; that is, $K_j = Y_jW_{11}^{-1}$, then the MOP in (4.10) for the multi-objective H_2/H_∞ tracking control design can be solved. In (4.15), Ω denotes the feasible set in which all elements of Ω must be satisfied with $W_{11} > 0$, $P_{22} > 0$, and (4.16)–(4.18).

Proof: The proof is divided into two parts. The first is for H_2 tracking performance, and the second is for robust H_∞ tracking performance. Initially, we give the following Lyapunov function for the fuzzy augmented systems in (4.5) as

$$
V(\bar{x}) \triangleq e^T(t)P_{11}e(t) + x_r^T(t)P_{22}x_r(t) = \bar{x}^T(t)P\bar{x}(t)
\tag{4.19}
$$

where $P_{11} > 0$, $P_{22} > 0$, and $P \triangleq diag(P_{11}, P_{22}) > 0$. For the H_2 tracking performance part with $\bar{\omega}(t) = 0$, we have

$$
\begin{aligned}
J_2 &= \int_0^{t_f} \bar{x}^T(t)\bar{Q}_1\bar{x}(t) + u^T(t)Ru(t)dt \\
&\le \bar{x}^T(0)P\bar{x}(0) + \int_0^{t_f} (\bar{x}^T(t)\bar{Q}_1\bar{x}(t) + u^T(t)Ru(t) \\
&\quad + \dot{\bar{x}}^T(t)P\bar{x}(t) + \bar{x}^T(t)P\dot{\bar{x}}(t))dt
\end{aligned}
\tag{4.20}
$$

due to $\bar{x}^T(t_f)P\bar{x}(t_f) \ge 0$. Next, by substituting (4.5) into (4.20), we get

$$
\begin{aligned}
J_2 &\le \bar{x}^T(0)P\bar{x}(0) + \int_0^{t_f} \Big\{ \bar{x}^T(t)\bar{Q}_1\bar{x}(t) \\
&\quad + \sum_{i=1}^{L}\sum_{j=1}^{L} h_i(z(t))h_j(z(t))\bar{x}^T(t)\Big[\bar{A}_{ij}^T P \\
&\quad + P\bar{A}_{ij} + [I \quad 0]^T K_j^T RK_j[I \quad 0]\Big]\bar{x}(t) \Big\} dt
\end{aligned}
\tag{4.21}
$$

Obviously, if

$$
\bar{A}_{ij}^T P + P\bar{A}_{ij} + [I \quad 0]^T K_j^T RK_j[I \quad 0] + \bar{Q}_1 \le 0
\tag{4.22}
$$

and

$$
\bar{x}^T(0)P\bar{x}(0) \le \alpha
\tag{4.23}
$$

hold, then we can conclude

$$J_2 \le \alpha \tag{4.24}$$

That is, if the inequalities in (4.22)–(4.23) hold, then J_2 is bounded by α. However, the inequality in (4.22) is the bilinear matrix inequality of P, K_j. In general, it is not easy to solve a BMI, which could not be solved efficiently by the convex optimization technique with the help of the LMI toolbox in MATLAB. To solve the BMI in (4.22), we first pre- and post-multiply the BMI in (4.22) by the matrix $diag(W_{11}, I)$ and further apply the Schur complement [90] to them; thus, we obtain the LMI in (4.16). Similarly, the inequality in (4.23) can be transformed into the LMI in (4.17).

Consider the robust H_∞ tracking performance part with $\bar{x}(0) = 0$. Similar to the analysis in (4.20)–(4.21) with $\bar{x}^T(t_f)P\bar{x}(t_f) \ge 0$ and $\bar{x}^T(0)P\bar{x}(0) = 0$, we obtain

$$\int_0^{t_f} \bar{x}^T(t)\bar{Q}_2\bar{x}(t)dt \le \int_0^{t_f} \sum_{i=1}^L \sum_{j=1}^L h_i(z(t))h_j(z(t)) \Big[\bar{x}^T(t)\bar{Q}_2\bar{x}(t)$$
$$+ \bar{x}^T(t)\bar{A}_{ij}^T P\bar{x}(t) + \bar{x}^T(t)P\bar{A}_{ij}\bar{x}(t) \tag{4.25}$$
$$+ \bar{\omega}^T(t)\bar{E}^T P\bar{x}(t) + \bar{x}^T(t)P\bar{E}\bar{\omega}(t) \Big]dt$$

By the fact that $a^T b + b^T a \le (1/\beta)a^T a + \beta b^T b$ for a positive scale value β and any vector or matrix a and b [90], we get

$$\int_0^{t_f} \bar{x}^T(t)\bar{Q}_2\bar{x}(t)dt \le \int_0^{t_f} \sum_{i=1}^L \sum_{j=1}^L h_i(z(t))h_j(z(t)) \Big[\bar{x}^T(t)\bar{Q}_2\bar{x}(t)$$
$$+ \bar{x}^T(t)\bar{A}_{ij}^T P\bar{x}(t) + \bar{x}^T(t)P\bar{A}_{ij}\bar{x}(t) \tag{4.26}$$
$$+ \frac{1}{\beta}\bar{x}^T(t)P\bar{E}\bar{E}^T P\bar{x}(t) + \beta\bar{\omega}^T(t)\bar{\omega}(t) \Big]dt$$

Apparently, if the following inequality holds

$$\bar{Q}_2 + \bar{A}_{ij}^T P + P\bar{A}_{ij} + \frac{1}{\beta}P\bar{E}\bar{E}^T P \le 0 \tag{4.27}$$

which is equivalent to

$$\begin{bmatrix} \bar{A}_{ij}^T P + P\bar{A}_{ij} + \bar{Q}_2 & P\bar{E} \\ \bar{E}^T P & -\beta I \end{bmatrix} \le 0 \tag{4.28}$$

by using the Schur complement, then the upper bound β for robust H_∞ tracking performance can be obtained as follows

$$J_\infty(u) = \frac{\int_0^{t_f} \bar{x}^T(t)\bar{Q}_2\bar{x}(t)dt}{\int_0^{t_f} \bar{\omega}^T(t)\bar{\omega}(t)dt} \le \beta \tag{4.29}$$

In order to transform the BMI in (4.28) into the LMI in (4.18), we first pre- and post-multiply the BMI in (4.28) by the matrix $diag(W_{11}, I, I, I)$ and then use the Schur complement. In summary, if the LMIs in (4.16)–(4.18) hold, then J_2 and J_∞ have upper bounds α and β, respectively. Therefore, the MOP in (4.15) subject to the LMIs (4.16)–(4.18) can solve the MOP in (4.10) indirectly.

Remark 4.4: Unlike the conventional MOP, which is only for the algebraic systems in Chapter 1 [2, 76–79], we develop the MOP for the optimal tracking control design in (4.15) for the fuzzy system in (4.2) with the fuzzy controller in (4.4) via the LMI approach. That is, the multi-objective H_2/H_∞ tracking control design problem becomes how to solve the following MOP

$$(\alpha^* \quad \beta^*) = \min_{(W_{11}, P_{22}, Y_j) \in \Omega} (\alpha \quad \beta) \tag{4.30}$$

subject to $(4.16) - (4.18)$

Remark 4.5: (i) If we only consider the optimal H_2 tracking control design problem, then the MOP in (4.30) is reduced to the following SOP

$$\min_{(W_{11}, P_{22}, Y_j) \in \Omega} \alpha \tag{4.31}$$

subject to $(4.16) - (4.17)$

(ii) If we only consider the optimal H_∞ tracking control design problem [6, 7, 12, 13], then the MOP in (4.30) is reduced to the following SOP

$$\min_{(W_{11}, P_{22}, Y_j) \in \Omega} \beta \tag{4.32}$$

subject to (4.18)

(iii) In other studies [34, 52, 70–75], for example, the conventional mixed H_2/H_∞ tracking control design needs to solve the following SOP under the constraint of a prescribed attenuation level β of the robust H_∞ tracking performance

$$\alpha_0 \triangleq \min_{(W_{11}, P_{22}, Y_j) \in \Omega} \alpha \tag{4.33}$$

subject to $(4.16) - (4.18)$

As compared with the SOPs in (4.31)–(4.33), in which only one performance is minimized, the MOP in (4.30) needs to minimize α and β simultaneously. Obviously, the MOP in (4.30) for the multi-objective H_2/H_∞ tracking control design is different from the SOPs in (4.31)–(4.33) in their design target and optimal solution method. In contrast to the SOPs in (4.31)–(4.33), the global optimal solution for the MOP in (4.30) may not exist. In other words, there exists no unique solution K_j^* such that α and β are minimized simultaneously. Hence, more effort is required for the MOP in (4.30) to seek a set of Pareto optimal solutions to achieve simultaneous minimization by nondominating sorting schemes [76–79].

Remark 4.6: (i) The weighted-sum optimization method is to convert an MOP to an SOP by emphasizing one optimal solution of weighted sum, that is, to minimize $(\omega_1 \alpha + \omega_2 \beta)$ for some weightings $\omega_1 \geq 0$ and $\omega_2 \geq 0$ with $\omega_1 + \omega_2 = 1$ instead of directly minimizing $(\alpha \quad \beta)$. For example, by following a similar procedure in the proof of Theorem 4.2 with the weighted sum $(\omega_1 \alpha + \omega_2 \beta)$ to replace $(\alpha \quad \beta)$ for all $\omega_1 \geq 0$, $\omega_2 \geq 0$, and $\omega_1 + \omega_2 = 1$, the MOP for the multi-objective H_2/H_∞ tracking control design can be solved by

$$\min_{(W_{11}, P_{22}, Y_j) \in \Omega} \omega_1 \alpha + \omega_2 \beta$$

subject to $(4.16)-(4.18)$ \hfill (4.34)

(ii) The weighted sum method in (4.34) provides an alternative solution for the MOP of the multi-objective H_2/H_∞ tracking control design for fuzzy systems. The shortage of the weighed sum method for the MOP in (4.34) is that you need to solve all solutions for the MOP with different weightings ω_1 and ω_2 with $\omega_1 \geq 0$, $\omega_2 \geq 0$, and $\omega_1 + \omega_2 = 1$. The MOP by the weighted sum method in (4.34) can be solved directly with the help of LMI toolbox in MATLAB if the weightings ω_1 and ω_2 are given. When such a method is used to find multiple solutions of an MOP, it has to be applied with different weightings many times, hopefully finding a different solution at each computational run. However, the Pareto solutions for the MOP in (4.30), which is proposed in this study, can be obtained by the LMI-based MOEA in a single run.

Remark 4.7: (i) Considering the optimal H_2 tracking performance, the MOP (4.30) is reduced to the following SOP

$$\underline{\alpha} = \min_{(W_{11}, P_{22}, Y_j, \beta) \in \Omega_\alpha} \alpha$$

subject to $(4.16)-(4.18)$ \hfill (4.35)

where Ω_α denotes the feasible set in which all elements of Ω_α must be satisfied with $W_{11} > 0$, $P_{22} > 0$, $\beta > 0$, and (4.16)–(4.18). When $\alpha = \underline{\alpha}$, the corresponding $\beta = \underline{\beta}$ is denoted as the upper bound of β in the feasible region (see Figure 4.2).

(ii) Considering the optimal robust H_∞ tracking performance, the MOP in (4.30) is reduced to the following SOP

$$\underline{\beta} = \min_{(W_{11}, P_{22}, Y_j, \alpha) \in \Omega_\beta} \beta$$

subject to $(4.16)-(4.18)$ \hfill (4.36)

where Ω_β denotes the feasible set in which all elements of Ω_β must be satisfied with $W_{11} > 0$, $P_{22} > 0$, $\alpha > 0$, and (4.16)–(4.18). When $\beta = \underline{\beta}$, the corresponding $\alpha = \underline{\alpha}$ is denoted as the upper bound of α in the feasible region.

(iii) In general, the solutions $(\alpha^*\ \ \beta^*)$ of the MOP in (4.30) should be under the following constraint

$$\underline{\alpha} \le \alpha^* < \bar{\alpha} \text{ and } \underline{\beta} \le \beta^* < \bar{\beta} \tag{4.37}$$

Remark 4.8: In contrast to the conventional EA algorithm to search one parameter for the SOPs in (4.31)–(4.33), we need to search the optimal parameters α^* and β^* simultaneously for the optimal MOP. In the search process, one parameter cannot dominate another parameter or it will violate simultaneous minimization. In this situation, the Pareto optimal solutions for the MOPs in (4.30) should be searched via an MOEA scheme through a nondominating search scheme in the following section.

4.4   REVERSE-ORDER LMI-BASED MOEA APPROACH FOR MULTI-OBJECTIVE H_2/H_∞ TRACKING CONTROL DESIGN

Based on the Pareto optimal solution properties that we mentioned in Section 4.1, the reverse-order LMI-based MOEA scheme is proposed to search for the optimal parameters α^* and β^* simultaneously to solve the MOP in (4.30) for fuzzy tracking control design [76–78]. For the property of the dominance for the MOP in (4.30), consider two solutions $(W_{11}^1, P_{22}^1, Y_j^1)$ and $(W_{11}^2, P_{22}^2, Y_j^2)$ in Ω for two objective values $(\alpha_1\ \ \beta_1)$ and $(\alpha_2\ \ \beta_2)$ subject to the LMIs in (4.16)–(4.18), respectively. $(\alpha_1\ \ \beta_1)$ is said to dominate $(\alpha_2\ \ \beta_2)$ if $\alpha_1 \le \alpha_2$ and $\beta_1 \le \beta_2$, with at least one inequality being a strict inequality. In relation to the property of the Pareto optimal solution for the MOP in (4.30), a solution $(W_{11}^*, P_{22}^*, Y_j^*)$ is the Pareto optimal solution of the MOP in (4.30) with respect to Ω if there does not exist another feasible solution $(W_{11}^\circ, P_{22}^\circ, Y_j^\circ)$ such that objective values $(\alpha^\circ\ \ \beta^\circ)$ dominate $(\alpha^*\ \ \beta^*)$. The Pareto front for the MOP in (4.30) defined as $\Gamma \triangleq \{(\alpha^*\ \ \beta^*) | (W_{11}^*, P_{22}^*, Y_j^*)$ is the Pareto optimal solution of the MOP in (4.30), and $(\alpha^*\ \ \beta^*)$ is generated by $(W_{11}^*, P_{22}^*, Y_j^*)$ subject to the LMIs in (4.16)–(4.18)}.

An evolutionary algorithm is a stochastic search method based on natural selection and is also particularly suitable for solving the MOP in (4.30) due to its population-based nature, which allows a set of Pareto optimal solutions to be obtained in a single run. A number of MOEAs have been proposed to solve MOPs with simple algebraic constraints. In the multi-objective H_2/H_∞ tracking control design problem, a reverse-order LMI-based MOEA scheme is proposed to search the optimal parameters α^* and β^* to solve the LMI-constrained MOP in (4.30) through a nondominating sorting scheme. If appropriate matrices W_{11}^*, P_{22}^*, and Y_j^* are specified to minimize both α and β in the MOP in (4.30) by the proposed LMI-based MOEA scheme, then the corresponding fuzzy control gains are obtained as $K_j^* = Y_j^* P_{11}^*$ for the multi-objective H_2/H_∞ tracking control design. We will discuss some important steps for the reverse-order LMI-based MOEA and some characteristics of the Pareto optimal solutions of the MOP in (4.30) as follows:

(1) Objective value encoding:

The objective values α and β are encoded into chromosomes $C_k = (\alpha_k \quad \beta_k)$. The proposed MOEA approach uses decimal-valued objective values for chromosome construction to avoid a long binary string and large search space in the conventional genetic algorithm (GA).

Definition 4.4: Each chromosome C_k has an individual $(\alpha_k \quad \beta_k)$ in terms of two objective values, and a set of chromosomes is called a population.

(2) Initial solution by the LMI method:

From the MOPs in (4.30) and (4.37), it is seen that the selection range of the Pareto front is within $\underline{\alpha} \le \alpha^* < \bar{\alpha}$ and $\underline{\beta} \le \beta^* < \bar{\beta}$. The traditional MOEA methods select candidate individuals $(\alpha \quad \beta)$ from the selection region as initial population. In the study, we need to impose the LMI constraints in (4.16)–(4.18) into the traditional MOEA method. First, a number of initial population C_k, $k = 1,...,N$ (or chromosomes) will be generated from the selection region. Then, in order to ensure some of these solutions will be feasible for the MOP in (4.30), each initial solution will be checked by the existence of the matrices W_{11}, P_{22}, and Y_j in the LMIs in (4.16)–(4.18). If not, these initial individuals have to be deleted and replaced by other randomly selected feasible initial individuals so that the initial populations are all in the feasible set Ω for the MOP in (4.30). Note that the LMI constraints in (4.16)–(4.18) create more restrictions on the searching of feasible solutions than the traditional MOEA method. The LMI toolbox in MATLAB can help these individuals (chromosomes) very efficiently check whether they satisfy the LMIs in (4.16)–(4.18). Therefore, the previous initialization scheme can help accelerate convergence speed.

(3) Nondominated sorting:

Since we want to minimize two parameters $(\alpha \quad \beta)$ simultaneously, one parameter cannot dominate another one in the search process to guarantee simultaneous minimization in the MOP in (4.30). Therefore, we could only select the parameters through a nondominating sorting method in the LMI-based MOEA search process. Before the selection step in the process, all individuals of the population are assigned ranks based on the property of dominance. All individuals are classified into one category, which is ranked in level (1 is the best level, 2 is the next-best level, and so on). At the beginning, the first nondominated set in the current population is constituted based on the property of dominance, and each individual of the first nondominated set is classified. Next, those classified individuals of the first nondominated set are ignored temporarily in order to classify the residual individuals of the population in the same way for the second nondominated set. The sorting process continues until all individuals in the population are ranked. In the following multi-objective optimization search process with a nondominating sorting scheme, the individuals with better ranks will have more copies than other individuals, which creates the elitist characteristic in the evolution algorithm and allows better convergence to Pareto optimal solutions [2].

(4) Fitness assignment:

Each individual is assigned a crowding distance to preserve uniformity of population. For a particular individual $(\alpha_i \quad \beta_i)$ with nondominated level j, we first find all other individuals $(\alpha_k \quad \beta_k)$, which are assigned a rank equal to nondominated level j, and individual distance is defined as the distance between two individuals $(\alpha_k \quad \beta_k)$ on either side of this individual $(\alpha_i \quad \beta_i)$ along each objective. Then, the crowding distance of individual $(\alpha_i \quad \beta_i)$ is calculated as the sum of individual distances corresponding to each objective. If the individual $(\alpha_i \quad \beta_i)$ is a boundary individual, the crowding distance is assigned an infinite value. After performing selection, those individuals with higher crowding distances have more survivability because those individuals with higher value make up the uniform distribution of the population.

(5) Crossover and mutation:

In this study, the basic mechanisms of the crossover and mutation operators are described in [76, 77]. After generating offspring population, they need to be checked by the three LMIs in (4.16)–(4.18) to find whether $W_{11} > 0$, $P_{22} > 0$, and Y_j so that these offspring populations are feasible.

Based on this analysis, a detailed design procedure of the proposed LMI-based MOEA approach for the multi-objective H_2/H_∞ tracking control design is as follows.

Design procedure:

Step 1: Select membership functions, construct fuzzy plant rules in (4.1), and specify the desired reference model in (4.3).

Step 2: Set the selection range $\underline{\alpha} \le \alpha^* < \bar{\alpha}$ and $\underline{\beta} \le \beta^* < \bar{\beta}$, the maximum number M of generations, total population size L, crossover ratio, and mutation ratio for the MOEA.

Step 3: Initialize a number of random individual $(\alpha_i \quad \beta_i)$ within the selection range, and take those individuals $(\alpha_i \quad \beta_i)$ that satisfy the LMI constraints in (4.16)–(4.18) as the initial population.

Step 4: Sort the current population into different fronts by a nondominating sorting scheme, and assign a crowding distance to each individual.

Step 5: Perform the selection step to select half the size of the current population as the parent population based on rank and crowding distance.

Step 6: Perform the crossover and mutation operators by the MOEA, and check the three LMIs in (4.16)–(4.18) to construct feasible offspring for the MOP in (4.30).

Step 7: Combine the parent population with the offspring, do steps 4 and 5, and select individuals according to their ranks and crowding distances until the population size reaches the total population size. A new generation is created for optimization search after doing step 7.

Step 8: Repeat steps 6 and 7 until the maximum generation M is reached.

When the previous LMI-based MOEA processing is finished, the Pareto front $(\alpha^* \quad \beta^*)$ is obtained, and the corresponding set of Pareto optimal solutions $(W_{11}^*, P_{22}^*, Y_j^*)$ will be also obtained. Then the corresponding fuzzy control gains $K_j^* = Y_j^* P_{11}^* = Y_j^* W_{11}^{-1}$ are obtained for the multi-objective H_2/H_∞ tracking control design.

The total computational complexity of the proposed LMI-based MOEA scheme is analyzed as follows: the computational complexity of solving the LMIs in (4.16)–(4.18) is about the other $O(l^2 n(n+1)/2)$, where $2n$ is the order of $P \in R^{2n \times 2n}$, and l is the number of fuzzy rules. The computational complexity of the MOEA scheme is about the order $O(2ML^2)$, where L is the size of population, and M is the maximum number of generation of the MOEA scheme. Thus, the total computational complexity of the proposed LMI-based MOEA scheme is about the order $O(n^2 l^2 ML^2)$.

4.5 SIMULATION EXAMPLE

Consider a two-link robot system as shown in Figure 4.1. The dynamic equation of the two-link robot system is given as follows [53]:

$$M(q)\ddot{q} + C(q,\dot{q})\dot{q} + G(q) = \tau \tag{4.38}$$

where

$$M(q) \triangleq \begin{bmatrix} (m_1 + m_2)l_1^2 & m_2 l_1 l_2 (s_1 s_2 + c_1 c_2) \\ m_2 l_1 l_2 (s_1 s_2 + c_1 c_2) & m_2 l_2^2 \end{bmatrix}, G(q) \triangleq \begin{bmatrix} -(m_1 + m_2)l_1 g s_1 \\ m_2 l_2 g s_2 \end{bmatrix}, C(q,\dot{q}) \triangleq$$

$$m_2 l_1 l_2 (c_1 s_2 - s_1 c_2) \begin{bmatrix} 0 & -\dot{q}_2 \\ -\dot{q}_1 & 0 \end{bmatrix}, q \triangleq \begin{bmatrix} q_1 \\ q_2 \end{bmatrix}.$$

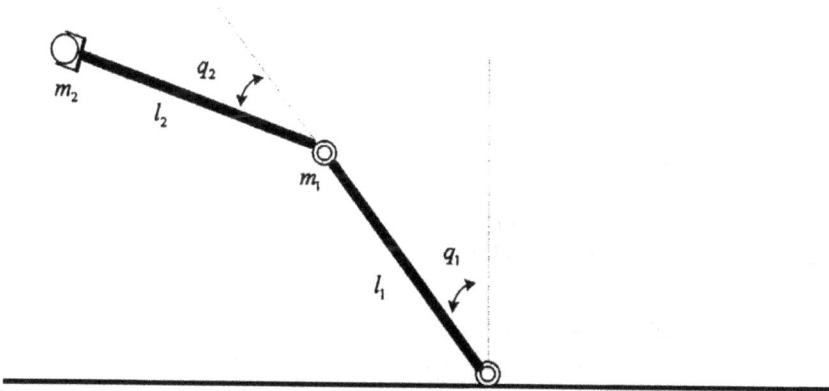

FIGURE 4.1   The configuration of two-link robot systems.

q_1, q_2 are generalized coordinates; $M(q)$ is the moment of inertia; $C(q,\dot{q})$ includes Coriolis and centripetal forces; and $G(q)$ is the gravitational force. Other quantities are: link mass $m_1 = 1$, $m_2 = 1$ (kg), link length $l_1 = 1$, $l_2 = 1$ (m), angular position $s_1 \triangleq \sin(q_1)$, $s_2 \triangleq \sin(q_2)$, $c_1 \triangleq \cos(q_1)$, and $c_2 \triangleq \cos(q_2)$.

Let $x_1 \triangleq q_1$, $x_2 \triangleq \dot{q}_1$, $x_3 \triangleq q_2$, and $x_4 \triangleq \dot{q}_2$; then (4.38) can be written as the following state-space form including external disturbances:

$$
\begin{aligned}
\dot{x}_1 &= x_2 + \omega_1 \\
\dot{x}_2 &= f_1(x) + g_{11}(x)\tau_1 + g_{12}(x)\tau_2 + \omega_2 \\
\dot{x}_3 &= x_4 + \omega_3 \\
\dot{x}_4 &= f_2(x) + g_{21}(x)\tau_1 + g_{22}(x)\tau_2 + \omega_4
\end{aligned}
\tag{4.39}
$$

where the disturbances are given as $\omega_1 \triangleq 0.1\sin(2t)$, $\omega_2 \triangleq 0.1\cos(2t)$, $\omega_3 \triangleq 0.1\cos(2t)$, and $\omega_4 \triangleq 0.1\sin(2t)$.

$$
f_1(x) = \frac{(s_1 c_2 + c_1 s_2)[m_2 l_1 l_2 (s_1 s_2 + c_1 c_2)x_2^2 - m_2 l_2^2 x_4^2]}{l_1 l_2[(m_1 + m_2) - m_2(s_1 s_2 + c_1 c_2)^2]} + \frac{[(m_1 + m_2)l_2 gs_1 - m_2 l_2 gs_2(s_1 s_2 + c_1 c_2)]}{l_1 l_2[(m_1 + m_2) - m_2(s_1 s_2 + c_1 c_2)^2]},
$$

$$
f_2(x) = \frac{(s_1 c_2 + c_1 s_2)[-(m_1 + m_2)l_1^2 x_2^2 + m_2 l_1 l_2(s_1 s_2 + c_1 c_2)x_4^2] + [-(m_1 + m_2)l_1 gs_1(s_1 s_2 + c_1 c_2) + (m_1 + m_2)l_1 gs_2]}{l_1 l_2[(m_1 + m_2) - m_2(s_1 s_2 + c_1 c_2)^2]},
$$

$$
g_{11}(x) = \frac{m_2 l_2^2}{m_2 l_1^2 l_2^2[(m_1 + m_2) - m_2(s_1 s_2 + c_1 c_2)^2]}, \quad g_{12}(x) = \frac{-m_2 l_1 l_2(s_1 s_2 + c_1 c_2)}{m_2 l_1^2 l_2^2[(m_1 + m_2) - m_2(s_1 s_2 + c_1 c_2)^2]},
$$

$$
g_{21}(x) = \frac{-m_2 l_1 l_2(s_1 s_2 + c_1 c_2)}{m_2 l_1^2 l_2^2[(m_1 + m_2) - m_2(s_1 s_2 + c_1 c_2)^2]}, \quad g_{22}(x) = \frac{(m_1 + m_2)l_1^2}{m_2 l_1^2 l_2^2[(m_1 + m_2) - m_2(s_1 s_2 + c_1 c_2)^2]}.
$$

To use the fuzzy control approach, we suppose that x_1, x_2, x_3, and x_4 are all available through the optical encoder attached on the robot. In this example, angular positions q_1, q_2 are constrained within $[-\pi/2, \pi/2]$. To minimize the design complexity, we use the following nine fuzzy if-then rules to construct the T-S fuzzy model for the systems in (4.39):

Rule 1: If x_1 is about $-\pi/2$ and x_3 is about $-\pi/2$, then $\dot{x} = A_1 x + B_1 u + E\omega$.
Rule 2: If x_1 is about $-\pi/2$ and x_3 is about 0, then $\dot{x} = A_2 x + B_2 u + E\omega$.
Rule 3: If x_1 is about $-\pi/2$ and x_3 is about $\pi/2$, then $\dot{x} = A_3 x + B_3 u + E\omega$.
Rule 4: If x_1 is about 0 and x_3 is about $-\pi/2$, then $\dot{x} = A_4 x + B_4 u + E\omega$.
Rule 5: If x_1 is about 0 and x_3 is about 0, then $\dot{x} = A_5 x + B_5 u + E\omega$.
Rule 6: If x_1 is about 0 and x_3 is about $\pi/2$, then $\dot{x} = A_6 x + B_6 u + E\omega$.
Rule 7: If x_1 is about $\pi/2$ and x_3 is about $-\pi/2$, then $\dot{x} = A_7 x + B_7 u + E\omega$.
Rule 8: If x_1 is about $\pi/2$ and x_3 is about 0, then $\dot{x} = A_8 x + B_8 u + E\omega$.
Rule 9: If x_1 is about $\pi/2$ and x_3 is about $\pi/2$, then $\dot{x} = A_9 x + B_9 u + E\omega$.

where $x \triangleq [x_1 \quad x_2 \quad x_3 \quad x_4]^T$, $u \triangleq [\tau_1 \quad \tau_2]^T$, and $\omega \triangleq [\omega_1 \quad \omega_2 \quad \omega_3 \quad \omega_4]^T$ with $E \triangleq I$. Moreover, we have

$$A_1 = \begin{bmatrix} 0 & 1 & 0 & 0 \\ 5.927 & -0.001 & -0.315 & -8.4 \times 10^{-6} \\ 0 & 0 & 0 & 1 \\ -6.859 & 0.002 & 3.155 & 6.2 \times 10^{-6} \end{bmatrix}, A_2 = \begin{bmatrix} 0 & 1 & 0 & 0 \\ 3.043 & -0.001 & 0.179 & -2 \times 10^{-4} \\ 0 & 0 & 0 & 1 \\ 3.544 & 0.031 & 2.561 & 1.14 \times 10^{-5} \end{bmatrix},$$

$$A_3 = \begin{bmatrix} 0 & 1 & 0 & 0 \\ 6.273 & 0.003 & 0.434 & -10^{-4} \\ 0 & 0 & 0 & 1 \\ 9.104 & 0.016 & -1.057 & -3.2 \times 10^{-5} \end{bmatrix}, A_4 = \begin{bmatrix} 0 & 1 & 0 & 0 \\ 6.454 & 0.002 & 1.243 & 2 \times 10^{-4} \\ 0 & 0 & 0 & 1 \\ -3.187 & -0.031 & 5.191 & -1.8 \times 10^{-5} \end{bmatrix},$$

$$A_5 = \begin{bmatrix} 0 & 1 & 0 & 0 \\ 11.134 & 0 & -1.815 & 0 \\ 0 & 0 & 0 & 1 \\ -9.092 & 0 & 9.164 & 0 \end{bmatrix}, A_6 = \begin{bmatrix} 0 & 1 & 0 & 0 \\ 6.170 & -0.001 & 1.687 & -2 \times 10^{-4} \\ 0 & 0 & 0 & 1 \\ -2.356 & 0.031 & 4.530 & 1.1 \times 10^{-5} \end{bmatrix},$$

$$A_7 = \begin{bmatrix} 0 & 1 & 0 & 0 \\ 6.121 & -0.004 & 0.621 & 10^{-4} \\ 0 & 0 & 0 & 1 \\ 8.879 & -0.019 & -1.012 & 4.4 \times 10^{-5} \end{bmatrix}, A_8 = \begin{bmatrix} 0 & 1 & 0 & 0 \\ 3.642 & 0.002 & 0.072 & 2 \times 10^{-4} \\ 0 & 0 & 0 & 1 \\ 2.429 & -0.031 & 2.983 & -1.9 \times 10^{-5} \end{bmatrix},$$

$$A_9 = \begin{bmatrix} 0 & 1 & 0 & 0 \\ 6.293 & -0.001 & -0.219 & -1.2 \times 10^{-5} \\ 0 & 0 & 0 & 1 \\ -7.465 & 0.002 & 3.269 & 9.2 \times 10^{-6} \end{bmatrix}, B_1 = \begin{bmatrix} 0 & 0 \\ 1 & -1 \\ 0 & 0 \\ -1 & 2 \end{bmatrix}, B_2 = \begin{bmatrix} 0 & 0 \\ 0.5 & 0 \\ 0 & 0 \\ 0 & 1 \end{bmatrix}, B_3 = \begin{bmatrix} 0 & 0 \\ 1 & 1 \\ 0 & 0 \\ 1 & 2 \end{bmatrix}.$$

where $B_5 = B_9 = B_1$, $B_4 = B_6 = B_8 = B_2$, and $B_7 = B_3$. For convenience of design, triangle-type membership functions are adopted for the previous nine if-then rules. More fuzzy details can be found in [53]. Additionally, the reference model is given as

$$A_r \triangleq \begin{bmatrix} 0 & 1 & 0 & 0 \\ -6 & -5 & 0 & 0 \\ 0 & 0 & 0 & 1 \\ 0 & 0 & -6 & -5 \end{bmatrix} \text{ and } r(t) \triangleq \begin{bmatrix} 0 \\ 8\sin(t) \\ 0 \\ 8\cos(t) \end{bmatrix}.$$

As for the rest of settings, we give the weighting matrices $Q_1 \triangleq 0.01 \times I$, $Q_2 \triangleq 0.1 \times I$, and $R \triangleq 0.001 \times I$.

For the proposed LMI-based MOEA to solve the MOP in (4.30), we give the selection range $\underline{\alpha} = 0.01$, $\bar{\alpha} = 3$, $\underline{\beta} = 0.019$, $\bar{\beta} = 10$, the number of generations $M = 100$, total population size $L = 100$, crossover ratio 0.9 and mutation ratio 0.15. Following the design procedure, Figure 4.2 shows the final population for the Pareto optimal solutions of the MOP in (4.30) for the two-link robot system. In Figure 4.2, we choose three Pareto optimal solutions for comparison with each other. Moreover, the optimal objectives $(\alpha^* \quad \beta^*)$ and the tracking error performance

$$\|e\|^2 = \int_0^{20} e^T(t)e(t)dt$$

of the three Pareto optimal solutions and the conventional solution of the SOP in (4.33) for the conventional mixed H_2/H_∞ tracking control design are given in Table 4.1. Based on the simulation results, Pareto solutions 1 and 2 are close to the optimal H_2 and H_∞ solutions, respectively. In addition, Pareto solution 3 is a compromise of the optimal H_2 solution with the optimal H_∞ solution. To solve the SOP in (4.33) and compare its results with the three Pareto optimal solutions, we give a prescribed H_∞ disturbance attenuation level $\beta = 0.2436$, which is equal to one of the optimal H_∞ solutions. After solving the SOP in (4.33), we obtain $\alpha_0 = 0.9417$ in Table 4.1. We can easily see that the tracking error performance in Pareto solution 1 is the smallest in Table 4.1. Since Pareto solution 2 is close to the optimal H_∞ solution, it has the worst tracking error performance from the H_2 tracking performance perspective. Since Pareto solution 3 has a compromise between H_2 optimal tracking and optimal H_∞ disturbance rejection, it has smaller tracking error performance

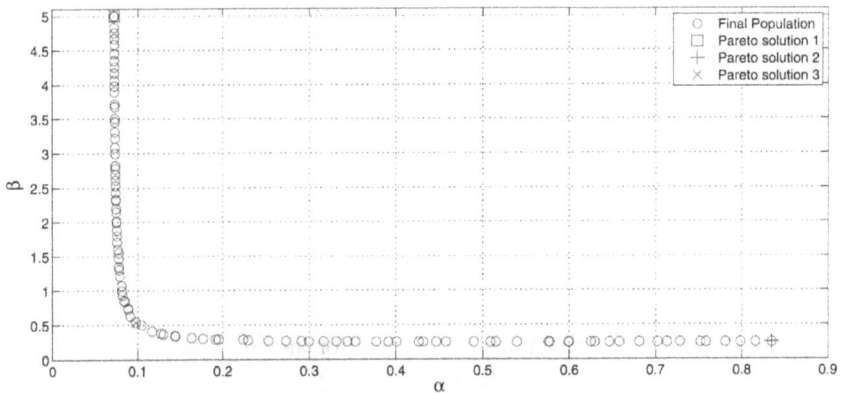

FIGURE 4.2 The Pareto Γ obtained by solving the MOP in (4.30) for the two-link robot system, and three Pareto solutions are chosen for comparison in Table 4.1.

TABLE 4.1

Optimal Objectives $(\alpha* \quad \beta*)$ and Tracking Error Performance $\|e\|^2$ of Three Pareto Optimal Solutions and Conventional Solution of the SOP in (4.33)

	$\alpha*$	$\beta*$	$\|e\|^2$
Pareto solution 1	0.0722	5	0.7397
Pareto solution 2	0.8348	0.2436	0.9571
Pareto solution 3	0.0978	0.5391	0.8336
Conventional solution of SOP (4.33)	0.9417	0.2436	1.4682

than Pareto solution 2, and it has a lower H_∞ disturbance attenuation level β than Pareto solution 1. Moreover, the tracking error performance of all Pareto optimal solutions is less than the conventional solution of the SOP in (4.33). Further, Figures 4.3 and 4.4 illustrate the simulation results of fuzzy tracking control designs with the initial condition $(x_1(0), x_2(0), x_3(0), x_4(0), x_{r1}(0), x_{r2}(0), x_{r3}(0), x_{r4}(0))^T \triangleq (0.5, 0, -0.5, 0, 0, 0, 0, 0)^T$ for

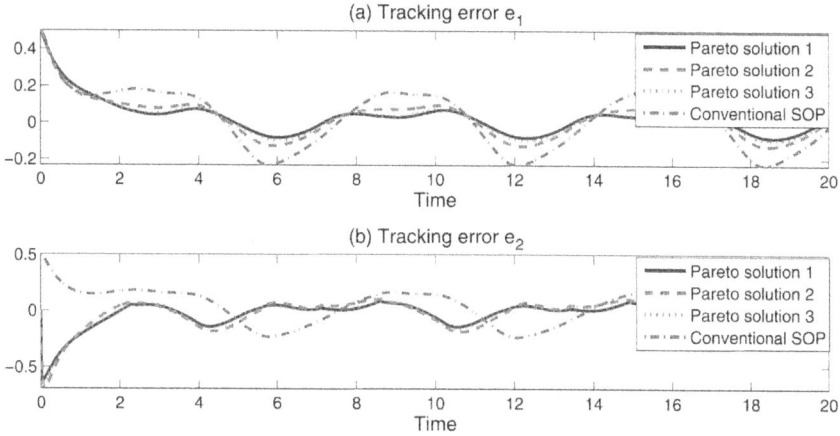

FIGURE 4.3 The trajectories of tracking errors e_1 and e_2 corresponding to (a) and (b), respectively. The solid line is of Pareto solution 1, the dashed line is of Pareto solution 2, the dotted line is of Pareto solution 3, and the dash-dot line is of conventional solution of the SOP in (4.33).

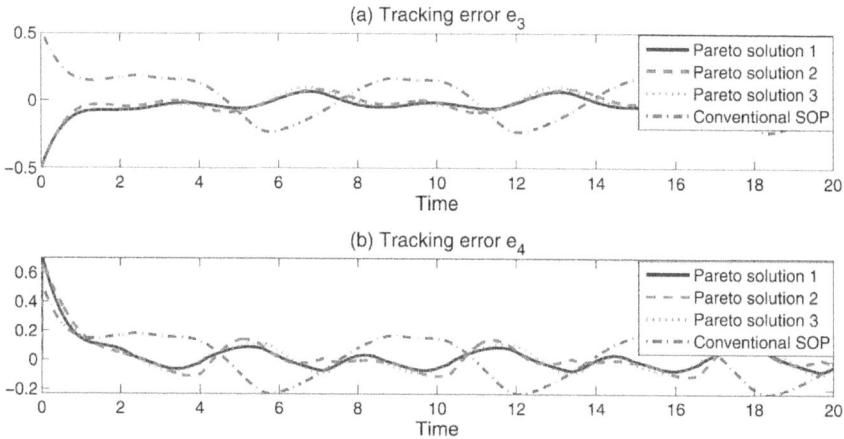

FIGURE 4.4 The trajectories of tracking errors e_3 and e_4 corresponding to (a) and (b), respectively. The solid line is of Pareto solution 1, the dashed line is of Pareto solution 2, the dotted line is of Pareto solution 3, and the dash-dot line is of conventional solution of the SOP in (4.33).

each Pareto solution and the conventional one. Figures 4.3 and 4.4 show the trajectories of the tracking errors e_1, e_2, e_3, and e_4.

4.6 CONCLUSION

This chapter studied the multi-objective robust fuzzy tracking control design of Takagi and Sugeno fuzzy systems. The purposes of this chapter are the following: (1) Based on the T-S fuzzy model, we formulated the multi-objective robust fuzzy tracking control design problem of nonlinear dynamic systems as an MOP with two conflicting objectives, H_2 tracking performance and robust H_∞ tracking performance, which can be optimized simultaneously. At present, researchers have focused more attention on the MOP of algebraic systems, and most control engineers have paid more attention to conventional mixed H_2/H_∞ control design problems. This chapter introduced an efficient method to solve the MOP for the optimal H_2/H_∞ tracking control design of nonlinear dynamic systems. (2) By the proposed indirect technique and linear matrix inequality scheme, the multi-objective H_2/H_∞ tracking control design problem can be transformed into an equivalent LMI-constrained MOP. Therefore, the multi-objective H_2/H_∞ tracking control design problem can be efficiently solved in a single run by way of the proposed LMI-based MOEA search method with the help of the LMI toolbox in MATLAB and the advantage of a nondominating sorting scheme for the Pareto optimal solutions. A simulation example of a two-link robot is given to illustrate the design procedure and to confirm the robust tracking performance of the proposed multi-objective H_2/H_∞ tracking control design. For more practical applications, multi-objective H_2/H_∞ fuzzy observer-based tracking control design or a more general multi-objective control design problem (i.e., a MOP with three or four design objectives) are given in the following chapters.

5 Multi-Objective Missile Guidance Control with Stochastic Continuous Wiener and Discontinuous Poisson Noises

5.1 INTRODUCTION

As far as the national military is concerned, missile guidance to hit a target is quite important. Missile guidance has always been a very popular topic in the aerospace field and military affairs. The control principles of missile guidance are well known to control engineers. Many guidance controls have been discussed in [91–93], and many guidance technologies have been developed to improve guidance control performance and to accommodate environmental disturbances. These guidance techniques are mainly based on classical control theory. Various guidance laws have been exploited with different control concepts over the years [92]. Currently, most popular terminal guidance laws are based on linear quadratic regulatory (LQR) theory [94], linear quadratic Gaussian (LQG) theory [95], linear exponential Gaussian theory, sliding mode theory [96–98], H_∞ robust theory, and fuzzy logic control theory [84, 99–102]. Because the dynamic system can represent relative motion between a missile and target, which is highly nonlinear and uncertain due to unmodeled dynamics and parameter perturbations resulting from missile modeling, it is more appealing to design a guidance law to achieve multi-objective (MO) H_2/H_∞ guidance performance to make the missile optimally track the target and attenuate the effect of continuous Wiener noise, discontinuous Poisson jump noise, and external disturbance simultaneously. Since it is still very difficult to solve the MO H_2/H_∞ guidance problem directly, in this chapter, an equivalent indirect MO H_2/H_∞ guidance problem is proposed by minimizing the upper bounds of H_2 and H_∞ performance simultaneously. However, it is not easy to efficiently solve the complex MO H_2/H_∞ guidance problem because we need to solve a nonlinear Hamilton-Jacobin inequality–constrained MO H_2/H_∞ guidance problem. Unfortunately, at present, there still exists no good method to solve such HJI-constrained MO H_2/H_∞ guidance problems for nonlinear stochastic missile guidance systems. Therefore, in this chapter, using this fuzzy interpolation approach, an HJI-constrained MO H_2/H_∞ guidance strategy for missile guidance design can be

DOI: 10.1201/9781003362142-7

transformed into a linear matrix inequality–constrained MO H_2/H_∞ guidance problem. Then a reverse-order LMI-constrained MOEA is also developed to solve the MO H_2/H_∞ guidance control design problem through a nondominated sorting scheme with the help of the MATLAB LMI toolbox. On the whole, the H_2/H_∞ guidance control design problem does not have a unique solution, but a set of Pareto optimal solutions could be obtained by the proposed reverse-order LMI-constrained MOEA, from which the designer could select one according to their own preference. Finally, a simulation example is given to validate the effectiveness of the proposed MO H_2/H_∞ guidance control strategy of the stochastic missile guidance system with intrinsic continuous and discontinuous random fluctuation as well as external distance.

5.2 THE 3-D SPHERICAL COORDINATE STOCHASTIC MISSILE GUIDANCE SYSTEM

The kinematic equation between the missile and target, as shown in Figure 5.1, can be recast into the following nonlinear state space missile guidance system [92, 93]:

$$\dot{x}(t) = F(x(t)) + Bu(t) + Dv(t) \tag{5.1}$$

where $x(t) = \begin{bmatrix} r \\ \varphi \\ \theta \\ V_r \\ V_\varphi \\ V_\theta \end{bmatrix}$, $F(x(t)) = \begin{bmatrix} V_r \\ \dfrac{V_\varphi}{r\cos\theta} \\ \dfrac{V_\theta}{r} \\ \dfrac{V_\varphi^2 + V_\theta^2}{r} \\ \dfrac{-V_r V_\varphi}{r} + \dfrac{V_\varphi V_\theta \tan\theta}{r} \\ \dfrac{-V_r V_\theta}{r} - \dfrac{V_\varphi^2 \tan\theta}{r} \end{bmatrix}$, $u(t) = \begin{bmatrix} u_r \\ u_\varphi \\ u_\theta \end{bmatrix}$, $v(t) = \begin{bmatrix} v_r \\ v_\varphi \\ v_\theta \end{bmatrix}$,

$B = \begin{bmatrix} 0 & 0 & 0 \\ 0 & 0 & 0 \\ 0 & 0 & 0 \\ -1 & 0 & 0 \\ 0 & -1 & 0 \\ 0 & 0 & -1 \end{bmatrix}$, $D = \begin{bmatrix} 0 & 0 & 0 \\ 0 & 0 & 0 \\ 0 & 0 & 0 \\ 1 & 0 & 0 \\ 0 & 1 & 0 \\ 0 & 0 & 1 \end{bmatrix}$, where $V_r = \dot{r}$, $V_\varphi = r\dot{\varphi}\cos\theta$, and $V_\theta = r\dot{\theta}$.

Let us denote the controlled state variable $\eta(t) = [r \quad V_\varphi \quad V_\theta]^T$ for missile guidance control as

$$\eta(t) = Lx(t) \tag{5.2}$$

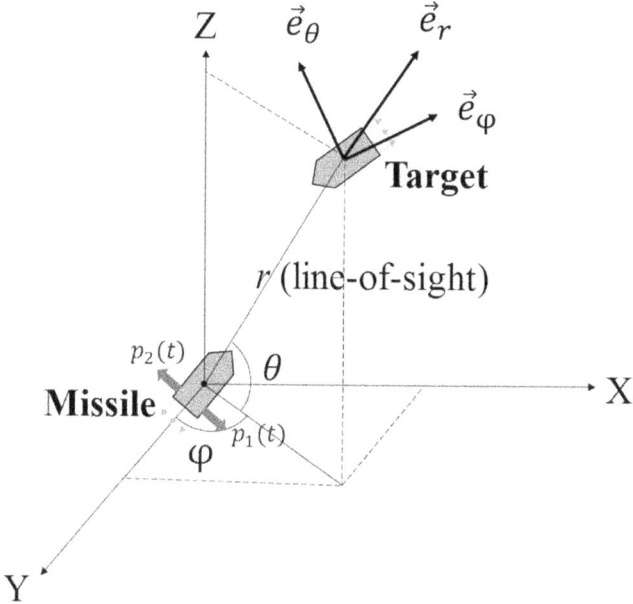

FIGURE 5.1   3-D pursuit evasion geometry in missile guidance system.

where $L = \begin{bmatrix} 1 & 0 & 0 & 0 & 0 & 0 \\ 0 & 0 & 0 & 0 & 1 & 0 \\ 0 & 0 & 0 & 0 & 0 & 1 \end{bmatrix}$.

Remark 5.1: When $r \to 0$, $V_\varphi \to 0$, and $V_\theta \to 0$, it means a missile and target in the head-on condition. Consequently, as $\eta(t) \to 0$, it represents that the missile will hit the target in the head-on condition. Among three relative velocities, V_r, V_φ, and V_θ, only the relative velocity along to the line of sight (LOS) (i.e., V_r) could decrease the relative distance between missile and target. Therefore, it can be decreased to zero.

In real situations, the missile dynamic system in (5.1) should be modified as follows:

$$\dot{x}(t) = F(x(t)) + Bu(t) + Dv(t) + H(x(t))n(t) + \sum_{k=1}^{2} G_k(x(t))p_k(t) \qquad (5.3)$$

where $H(x(t))n(t)$ denotes the effect of intrinsic random continuous fluctuations, and $n(t)$ is zero-mean white noise; the term $\sum_{k=1}^{2} G_k(x(t))p_k(t)$ denotes the effect of intrinsic stochastic discontinuous Poisson processes, $p_k(t)$ represents the suddenly cheating side-stop maneuvering through the two-side jets of the target, and $v(t)$ denotes the external disturbance due to the target's acceleration.

The nonlinear stochastic missile guidance system in (5.3) could be represented by [10, 15]

$$dx(t) = (F(x(t)) + Bu(t) + Dv(t))dt + H(x(t))dW(t)$$
$$+ \sum_{k=1}^{2} G_k(x(t))dN_k(t)$$

(5.4)

where F, H, and $G_k : \mathbb{R}^{n_x} \rightarrow \mathbb{R}^{n_x}$ are nonlinear Borel measurable continuous functions, which are satisfied with local Lipschitz continuity. Wiener noise $W(t)$ is a continuous but nondifferentiable stochastic process with $dW(t) := n(t)dt$, and $H(x(t))dW(t)$ denotes the effect of continuous stochastic intrinsic noise. $N(t)$ is a Poisson counting process with $dN_k(t) := p_k(t)dt$, and $\sum_{k=1}^{2} G_k(x(t))dN_k(t)$ denotes the effect of discontinuous stochastic noises. Assume $dW(t)$ and $dN(t)$ are independent, then $E\{[dN(t)][dW(t)]\} = 0$.

Some important properties of the Wiener process $W(t)$ and Poisson jump process $N_k(t)$ are given as follows: (1) $E\{W(t)\} = E\{dW(t)\} = 0$, (2) $E\{dW(t)dW(t)\} = dt$, (3) $E\{dN_k(t)\} = \lambda dt$, (4) $E\{dN(t)dt\} = 0$.

Lemma 5.1 ([10, 15]): Let the Lyapunov function $V(x(t)) \geq 0$. For the nonlinear stochastic missile guidance system in (5.4), the Itô-Lévy formula of $V(x(t))$ is given as:

$$dV(x(t)) = [V_x^T (F(x(t)) + Bu(t) + Dv(t))$$
$$+ \tfrac{1}{2} H^T (x(t))V_{xx}H(x(t))]dt + V_x^T H(x(t))dW(t)$$
$$+ \sum_{k=1}^{2} \{V(x(t) + G_k(x(t))) - V(x(t))\}dN_k(t)$$

(5.5)

where the $\hat{Ito}$ term $\tfrac{1}{2} H^T (x(t))V_{xx}H(x(t))$ is used to compensate for the derivative of the Wiener process $dW(t)$.

5.3   MULTI-OBJECTIVE H_2/H_∞ GUIDANCE CONTROL DESIGN FOR NONLINEAR STOCHASTIC MISSILE SYSTEMS

In the nonlinear stochastic missile guidance system in (5.4), a good guidance law $u(t)$ must guarantee a decreasing relative distance $r(t)$ and at the same time keep the pitch and yaw LOS rates $V_\varphi(t)$ and $V_\theta(t)$ as small as possible, that is, in the head-on condition. Therefore, our first design objective is to design the guidance command $u(t)$ so that the controlled variable $\eta(t) = [r(t), V_\varphi(t), V_\theta(t)]^T$ is reduced to as close to zero as possible with minimal control effort. In this situation, the first design objective is to specify $u(t)$ to minimize the following H_2 guidance performance

$$J_1(u) = E\left\{\int_0^{t_f} [\eta^T (t)Q\eta(t) + u^T (t)Ru(t)]dt\right\}$$

(5.6)

However, the H_2 guidance control in (5.6) could not treat the uncertain external disturbance $v(t)$; that is, it assumes $v(t) = 0$. Hence, the following H_∞ guidance performance is given as the second design objective to efficiently attenuate the effect of external disturbance $v(t)$ on guidance performance with the initial condition $x(0) \triangleq x_0 \neq 0$:

$$J_2(u) = \frac{E\{\int_0^{t_f} [\eta^T(t)Q\eta(t)]dt - V(x_0)\}}{E\{\int_0^{t_f} [v^T(t)v(t)]dt\}} \tag{5.7}$$

That is, the second design objective is to specify $u(t)$ to minimize the effect of external disturbance on the missile guidance simultaneously.

Since the simultaneous minimization of both H_2 guidance performance in (5.6) and H_∞ guidance performance in (5.7) are more appealing for guidance control design, the following MO H_2/H_∞ guidance control design is formulated as follows:

$$\min_{u(t)}(J_1(u), J_2(u)) \tag{5.8}$$
$$s.t. (5.4)$$

where $\min_{u(t)}(J_1(u), J_2(u))$ means to specify $u(t)$ to minimize $J_1(u)$ and $J_2(u)$ at the same time. The vector $(J_1(u), J_2(u))$ is called the objective vector of $u(t)$. As a whole, the solution of the MO H_2/H_∞ guidance problem is not unique. There exist a set of Pareto optimal solutions in the Pareto optimal sense.

In general, it is very difficult to solve the MO H_2/H_∞ guidance problem in (5.8) directly. Consequently, we propose an indirect method to minimize their upper bounds simultaneously as follows:

$$\min_{u(t)}(\alpha, \beta) \tag{5.9}$$
$$\text{subject to (5.4), } J_1(u) \leq \alpha \text{ and } J_2(u) \leq \beta$$

that is, we specify guidance law $u(t)$ to minimize the upper bounds α and β simultaneously. Based on the following lemma, we will prove the MO H_2/H_∞ guidance controls in (5.8) and (5.9) are equivalent when Pareto optimal solutions are achieved.

Lemma 5.2: The MO H_2/H_∞ guidance problems in (5.8) and (5.9) are equivalent when Pareto optimal solutions of these two MO H_2/H_∞ optimal guidance problems are achieved, that is, $(J_1^*, J_2^*) = (\alpha^*, \beta^*)$.

Proof: Please refer to Appendix 5.8.1.

Lemma 5.2 provides an indirect method to solve the MO H_2/H_∞ guidance problems of the nonlinear stochastic missile guidance system in (5.4). Then, we get the following results.

Theorem 5.1: The MO H_2/H_∞ guidance problem in (5.9) can be solved by the following HJI-constrained MO problem:

$$\min_{u(t)}(\alpha, \beta) \tag{5.10}$$
$$\text{subject to } \Gamma^{(1)} \leq 0, \Gamma^{(2)} \leq 0 \text{ and } E\{V(x_0)\} \leq \alpha$$

where

$$\Gamma^{(1)} = V_x^T F(x(t)) + V_x^T Bu(t) + \tfrac{1}{2} H^T(x(t))V_{xx}H(x(t)) + \lambda \sum_{k=1}^{2}[V(x(t))$$
$$+ G_k(x(t)) - V(x(t))] + x^T(t)\bar{Q}x(t) + u^T(t)Ru(t) \le 0,$$

$$\Gamma^{(2)} = \Gamma^{(1)} + \tfrac{1}{4\beta}V_x^T DD^T V_x - u^T(t)Ru(t) \le 0,$$

$$\bar{Q} = \bar{Q}^T = L^T QL \ge 0, \text{ and } V(x) > 0.$$

Proof: Please refer to Appendix 5.8.2.

On the whole, it is almost impossible to solve a closed-form guidance law $u(t)$ from the MO H_2/H_∞ guidance problem in (5.10) because HJIs $\Gamma^{(1)} \le 0$ and $\Gamma^{(2)} \le 0$ are nonlinear partial differential Hamilton Jacobi inequalities. At present, there is still no good method to solve HJIs of $\Gamma^{(1)}$ and $\Gamma^{(2)}$ for the MO H2/H∞ guidance problem in (5.10). Thus, the Takagi-Sugeno fuzzy interpolation method is employed to simplify these HJIs for the control design procedure of the MO H_2/H_∞ guidance problem in (5.10).

5.4   REVERSE-ORDER LMI-BASED MOEA APPROACH FOR MULTI-OBJECTIVE H_2/H_∞ TRACKING CONTROL DESIGN

The MO H_2/H_∞ guidance law of the nonlinear stochastic missile system in (5.4) needs to solve the HJI-constrained MO H_2/H_∞ guidance problem in (5.10), where the solutions of HJIs are very difficult to obtain even with numerical methods. To overcome this HJI-constrained MO H_2/H_∞ guidance problem in (5.10), a T-S fuzzy interpolation method is employed to approximate the nonlinear stochastic missile guidance system in (5.4) by interpolating several locally linearized systems via the fuzzy bases. This fuzzified linear system is described by a group of if-then rules and is employed to deal with the MO H_2/H_∞ guidance control problem.

The ith rule of the T-S fuzzy model for the nonlinear stochastic missile guidance system in (5.4) is described by:

Plant Rule i :

If $z_1(t)$ is F_{i1}, and, ..., and $z_g(t)$ is F_{ig} (5.11)

$$dx(t) = (A_i x(t) + Bu(t) + Dv(t))dt$$

Then
$$+ H_i x(t)dW(t) + \sum_{k=1}^{2} G_{ik}x(t)dN_k(t)$$

where F_{ij} is the fuzzy set. A_i, H_i, and G_i are the locally linearized system matrices of $F(x(t))$, $H(x(t))$, and $G(x(t))$ at the corresponding fuzzy set, respectively. $z_1(t),....,z_g(t)$ are the premise variables, $F_{ij}(z_j)$ is the membership grade of $z_j(t)$ in F_{ij}, and $h_i(z(t))$ is the ith interpolation function with $i = 1, 2, ..., l$. The overall fuzzy system in (5.11) can be inferred as follows [10]:

$$dx(t) = \sum_{i=1}^{l} h_i(z(t))[(A_i x(t) + Bu(t) + Dv(t))dt$$
$$+ H_i x(t)dW(t) + \sum_{k=1}^{2} G_{ik} x(t)dN_k(t)]$$

(5.12)

where $z(t) = [z_1(t),...,z_g(t)]^T$, $u_i(z(t)) = \prod_{j=1}^{g} F_{ij}(z_j(t))$, $h_i(z) = \dfrac{u_i(z(t))}{\sum\limits_{i=1}^{l} u_i(z(t))}$, and $\sum_{i=1}^{l} h_i$

$(z(t)) = 1$, for $i = 1, 2, ..., l$.

It is natural to assume $u_i(z(t)) \geq 0$, $h_i(z(t)) \geq 0$, and $\sum_{i=1}^{l} h_i(z(t)) = 1$, for $i = 1, 2, ..., l$.

The physical meaning of the stochastic fuzzy model in (5.12) is that the locally linearized stochastic systems in (5.11) at a different fuzzy set F_{ij} are interpolated piecewise by interpolation function $h_i(z(t))$ to approximate the original nonlinear stochastic missile guidance system in (5.4). Note that the identifications of A_i, H_i, and G_i, can be obtained by the fuzzy system identification method in the MATLAB toolbox. Consequently, the nonlinear stochastic missile guidance system in (5.4) could be represented by

$$dx(t) = \sum_{i=1}^{l} h_i(z(t))[(A_i x(t) + Bu(t) + Dv(t))dt$$
$$+ H_i x(t)dW(t) + \sum_{k=1}^{2} G_{ik} x(t))dN_k(t)] + \Delta F(x(t))$$
$$+ \Delta H(x(t))dW(t) + \sum_{k=1}^{2} \Delta G(x(t))dN_k(t)$$

(5.13)

where $\Delta F(x(t))$, $\Delta H(x(t))$, and $\Delta G(x(t))$, denote the fuzzy approximation error between the stochastic missile guidance system in (5.4) and the T-S fuzzy interpolation system in (5.12) with regard to the system in (5.13).

The following fuzzy guidance law is employed to deal with the MO H_2/H_∞ missile guidance problem:

Guidance Law Rule i:

If $z_1(t)$ is F_{i1}, and, ...,and $z_g(t)$ is F_{ig}

(5.14)

Then $u(t) = K_i x(t)$, for $i = 1, ..., l$

Then, the overall fuzzy guidance law can be expressed as

$$u(t) = \sum_{i=1}^{l} h_i(z(t))K_i x(t)$$

(5.15)

where $h_i(z(t))$ is defined in (5.12), and K_i, $i = 1,...,l$ are the fuzzy control parameters. Substituting (5.15) into (5.13), we get the closed-loop system of the missile guidance system as:

$$dx(t) = \sum_{i=1}^{l}\sum_{j=1}^{l} h_i(z(t))h_j(z(t))[((A_i + BK_j)x(t)$$

$$+ Dv(t))dt + H_i x(t)dW(t) + \sum_{k=1}^{2} G_{ik} x(t)dN_k(t)] \qquad (5.16)$$

$$+ \Delta F(x(t)) + \Delta H(x(t))dW(t) + \sum_{k=1}^{2} \Delta G_k(x(t))dN_k(t)$$

Suppose there exist some positive scalars r_1, r_2, and r_3 such that the fuzzy approximation errors are bounded as follows:

$$\begin{aligned}
\left\| \Delta F(x(t)) \right\|_2 &\le r_1 \left\| x(t) \right\|_2 \\
\left\| \Delta H(x(t)) \right\|_2 &\le r_2 \left\| x(t) \right\|_2 \\
\left\| \Delta G_1(x(t)) \right\|_2 &\le r_{31} \left\| x(t) \right\|_2 \\
\left\| \Delta G_2(x(t)) \right\|_2 &\le r_{32} \left\| x(t) \right\|_2
\end{aligned} \qquad (5.17)$$

Based on this analysis, the MO H_2/H_∞ guidance problem becomes how to specify fuzzy guidance law in (5.15) or K_i, $i = 1,...,l$ for the missile guidance system in (5.13) so that the MO H_2/H_∞ guidance problem in (5.10) could be solved for K_i, $i = 1,...,l$ in (5.15); that is, with the fuzzy missile system in (5.13) and fuzzy guidance law (5.15), the MO H_2/H_∞ guidance problem in (5.10) in Theorem 5.1 could be simplified significantly and solved easily.

Lemma 5.3 ([15]): For any matrix M_i with appropriate dimensions and the fuzzy interpolation functions $h_i(z)$ with $0 \le h_i(z) \le 1$, for $i \in \mathbb{N}^+$, $1 \le i \le l$, $P > 0$, and $\sum_{i=1}^{l} h_i(z) = 1$, we have

$$\left(\sum_{j=1}^{l} h_j(z)M_j \right)^T P\left(\sum_{i=1}^{l} h_i(z)M_i \right) \le \sum_{i=1}^{l} h_i(z)M_i^T P M_i$$

Let us first consider a Lyapunov function candidate as:

$$V(x(t)) = x^T(t)Px(t) \qquad (5.18)$$

where the weighting matrix P is a positive-definitive symmetric matrix; that is, $P = P^T > 0$.

Theorem 5.2: The fuzzy guidance law in (5.15) for the multi-objective H_2/H_∞ guidance control of nonlinear stochastic missile guidance system in (5.13) could be solved by the following MO problem:

$$\min_{K_i, P>0} (\alpha, \beta) \text{ subject to}$$

$$P \le \alpha[Tr(R_{x_0})]^{-1} I, \Phi_{ij}^{(1)} \le 0, \Phi_{ij}^{(2)} \le 0 \qquad (5.19)$$

where

$$
\begin{aligned}
\Phi_{ij}^{(1)} = {}& [0.5(A_i^T P + PA_i + A_j^T P + PA_j) + 0.5(PBK_i + (PBK_i)^T \\
& + PBK_j + (PBK_j)^T) + (H_i^T PH_i + H_j^T PH_j)] \\
& + \lambda \sum_{k=1}^{2} \{(G_{i,k}^T PG_{i,k} + G_{j,k}^T PG_{j,k}) + 0.5(G_{i,k}^T P + PG_{i,k} \\
& + G_{j,k}^T P + PG_{j,k})\} + (r_1^2 I + PP + 2Pr_2^2 + \lambda P + \\
& + 3\lambda P(r_{3,1}^2 + r_{3,2}^2 + \bar{Q}) + 0.5(K_i^T RK_i + K_j^T RK_j))
\end{aligned}
\tag{5.20}
$$

$$
\Phi_{ij}^{(2)} = \Phi_{ij}^{(1)} - 0.5(K_i^T RK_i + K_j^T RK_j) + \tfrac{1}{\beta} PDD^T P
\tag{5.21}
$$

By Schur complement, the MO H_2/H_∞ guidance problem could be transformed into the following LMI-constrained MO H_2/H_∞ guidance problem

$$
\min_{K_i}(\alpha, \beta)
\tag{5.22}
$$

$$
\Lambda \geq \alpha^{-1} Tr(R_{x_0}) I
\tag{5.23}
$$

$$
\begin{bmatrix}
\Xi_{ij}^{(1)} & \Lambda G_i^T & \Lambda G_j^T & \Lambda H_i^T & \Lambda H_j^T & \Lambda & Z_i^T & Z_j^T \\
* & -\frac{\Lambda}{\lambda} & 0 & 0 & 0 & 0 & 0 & 0 \\
* & * & -\frac{\Lambda}{\lambda} & 0 & 0 & 0 & 0 & 0 \\
* & * & * & -2\Lambda & 0 & 0 & 0 & 0 \\
* & * & * & * & -2\Lambda & 0 & 0 & 0 \\
* & * & * & * & * & \bar{J} & 0 & 0 \\
* & * & * & * & * & * & -R^{-1} & 0 \\
* & * & * & * & * & * & * & -R^{-1}
\end{bmatrix} \leq 0
\tag{5.24}
$$

$$
\begin{bmatrix}
\Xi_{ij}^{(2)} & \Lambda G_i^T & \Lambda G_j^T & \Lambda H_i^T & \Lambda H_j^T & \Lambda \\
* & -\frac{\Lambda}{\lambda} & 0 & 0 & 0 & 0 \\
* & * & -\frac{\Lambda}{\lambda} & 0 & 0 & 0 \\
* & * & * & -2\Lambda & 0 & 0 \\
* & * & * & * & -2\Lambda & 0 \\
* & * & * & * & * & \bar{J}
\end{bmatrix} \leq 0
\tag{5.25}
$$

$$
\forall i \leq j(i, j = 1, ..., l)
$$

where J is any nonzero real number,

$$\bar{J} = -(Jr_1^2 I + \bar{Q})^{-1}, \ \Lambda = P^{-1}, \ Z_j = K_j \Lambda,$$

$$\Xi_{ij}^{(1)} = 0.5(A_i\Lambda + \Lambda A_i^T + A_j\Lambda + \Lambda A_j^T) + 0.5(BZ_i + (BZ_i)^T + BZ_j + (BZ_j)^T)$$

$$+ 0.5\lambda(\Lambda G_{i,k}^T + G_{i,k}\Lambda + \Lambda G_{j,k}^T + G_{j,k}\Lambda) + \lambda\Lambda + 3\lambda r_3^2 \Lambda + J^{-1}I + r_2^2\Lambda \ ,$$

and $\Xi_{ij}^{(2)} = \Xi_{ij}^{(1)} + \dfrac{1}{\beta} DD^T$. Thus, the HJI-constrained MO problem in (5.10) could be
replaced by the LMI-constrained MO problem in (5.22).

Proof: Please refer to Appendix 5.8.3.

5.5   MO H_2/H_∞ GUIDANCE CONTROL OF NONLINEAR STOCHASTIC MISSILE SYSTEM DESIGN VIA REVERSE-ORDER LMI-CONSTRAINED MOEA

It is not easy to solve the MOP in Theorem 2 directly. In this section, a reverse-order
LMI-constrained MOEA search is developed to help us solve the MOP in Theorem
5.2 iteratively.

The reverse-order LMI-constrained MOEA procedure for MO fuzzy guidance
controlled design:

Step 1: Set the feasible range $[\alpha^*, \bar{\alpha}] \times [\beta^*, \bar{\beta}]$ for the feasible objective vectors
   (α, β), the maximum number of generations N_g, the total population size N_p,
   the crossover ratio N_c, and the mutation ratio N_m in the LMI-constrained
   MOEA.

Step 2: Generate a population such that all chromosomes (solutions) (α, β) in
   the population are within the feasible range, that is, those chromosomes
   (α, β) that satisfy the LMI constraints in (5.23)–(5.25), as the initial population and set the iteration index $k = 1$.

Step 3: Perform crossover and mutation operations in evolution algorithm to
   $2N_p$ number feasible solutions for the population, and examine three LMIs
   (5.23)–(5.25) to construct feasible solutions for the MOP in (5.22).

Step 4: Select N_p solutions based on the survival of the fittest from the $2N_p$ feasible solutions in Step 3 through the nondominated sorting method. Sort the current population into different fronts and combine the parent population with
   solution, and perform Step 3 to select chromosomes according to their rank
   and fitness value until the population size reaches the total population size.

Step 5: Set the iteration index $k = k + 1$ and repeat Steps 3, 4, and 5 until the
   maximum number of generations is reached. It is worth mentioning that the
   Pareto form in the last generation is the optimal Pareto front.

Step 6: If the LMIs in (5.23)–(5.25) have a $P > 0$, then the nonlinear stochastic missile guidance system in (5.16) with $K_i = Z_i P$ $(i = 1, 2, ..., l)$ would be
   stable, and the MO H_2/H_∞ missile guidance problem in (5.4) can be solved
   with $J_1 = \alpha^*$ and $J_2 = \beta^*$ simultaneously.

5.6 SIMULATION EXAMPLE AND RESULT

The following example is given to verify the performance of the proposed fuzzy nonlinear MO H_2/H_∞ missile guidance law. The maneuvering strategies of the target in the 3-D model [92] are utilized as external disturbance $v(t)$ to confirm the tracking performance and the robustness of the fuzzy MO H_2/H_∞ missile guidance laws; that is, for the external disturbance $v(t)$, we used step target:

$$v(t) = [v_{r,st}, v_{\varphi,st}, v_{\theta,st}]^T = \left[\Omega_T \vec{e}_r, \Omega_T \frac{-\dot{\theta}}{\sqrt{\dot{\theta}^2 + \dot{\varphi}\cos^2\theta}} \vec{e}_\psi, \Omega_T \frac{-\dot{\varphi}\cos\theta}{\sqrt{\dot{\theta}^2 + \dot{\varphi}\cos^2\theta}} \vec{e}_\theta \right]^T$$

where Ω_T is the target's navigation gain, which we set as a random value within 0–4G in our simulation. To demonstrate the tracking performance robustness of the proposed method, the following scenarios are considered in the missile guidance system in (5.1):

Target is toward to missile ($v_r < 0$): $r = 9\,\text{km}$, $\varphi = \pi/4$, $\theta = \pi/3$, $V_r = -500\,\text{m/s}$, $V_\varphi = 200\,\text{m/s}$, $V_\theta = 150\,\text{m/s}$.

Applying the T-S fuzzy approach considering the approximation error, the nonlinear stochastic missile guidance system in (5.4) can be rewritten as (5.16). The upper bounds of the fuzzy approximation errors in (5.17) are also known to be $r_1 = 5.3101 \times 10^{-6}$, $r_2 = 5.6970 \times 10^{-4}$, and $r_3 = 7.5638 \times 10^{-5}$. We used 12 rules based on the premise variables $z(t) = (z_1(t), z_2(t), z_3(t)) = [r, V_\varphi, V_\theta] = [x_1(t), x_5(t), x_6(t)]$ to approximate the nonlinear nonlinear stochastic missile guidance system in (5.4).

The operation points of $x_1(t)$, $x_5(t)$, and $x_6(t)$ are respectively given at: $x_1^{(1)} = 4723$, $x_1^{(2)} = 7000$, $x_1^{(3)} = 9000$, $x_5^{(1)} = 187.4$, $x_5^{(2)} = 427.8$, $x_6^{(1)} = -231.6$, and $x_6^{(2)} = 297.4$, and we selected the weighting matrices $Q = R = diag(10^{-5}, 10^{-5}, 10^{-5})$. By solving the MO H_2/H_∞ missile guidance problem in (5.19)–(5.22), we can get Λ and Z_i, with $i = 1, 2, ..., 12$. Therefore, we can obtain $K_i = Z_i \Lambda^{-1}$, where $N_p = 30$, iteration number $k = 50$, crossover ratio $N_c = 0.9$, and mutation ratio $N_m = 0.1$. Once the iteration number $k = 50$ is achieved, the Pareto front for the Pareto optimal solutions of the MOP for the nonlinear stochastic missile guidance system in (5.4) can be obtained as given in Figure 5.2.

We choose three Pareto optimal solutions, A, B, and C, in Table 5.1 from Figure 5.2 to illustrate how to select the preferable solution. In Figure 5.2, Pareto optimal solution A has the best J_2 performance among the three Pareto optimal solutions so that the trajectory of Pareto optimal solution A has the minimum perturbation but the maximum J_1 performance. However, for Pareto optimal solution C, its trajectory has the maximum perturbation among the three Pareto optimal solutions because it has the minimum J_1 performance. The J_1 performance of Pareto solution B is between those of Pareto solution A and Pareto solution C. The state trajectories of $x(t) = [x_1(t), x_2(t) \cdots x_6(t)]^T$ of Pareto optimal solutions A, B, and C of the MO H_2/H_∞ guidance control law are shown in Figures 5.3–5.6 with intrinsic continuous random fluctuation caused by the Wiener process in Figure 5.6 and intrinsic discontinuous fluctuations caused by the Poisson

TABLE 5.1

The Pareto Objective Vectors of the Three Chosen Pareto Optimal Solutions in Figure 5.2

	Pareto Objective Vector (α, β)
Pareto optimal solution A	$(\alpha_A, \beta_A) = (1.5688 \times 10^4, 5.8045 \times 10^{-5})$
Pareto optimal solution B	$(\alpha_B, \beta_B) = (1.1601 \times 10^4, 3.3740 \times 10^{-4})$
Pareto optimal solution C	$(\alpha_C, \beta_C) = (1.0971 \times 10^4, 4.7511 \times 10^{-3})$

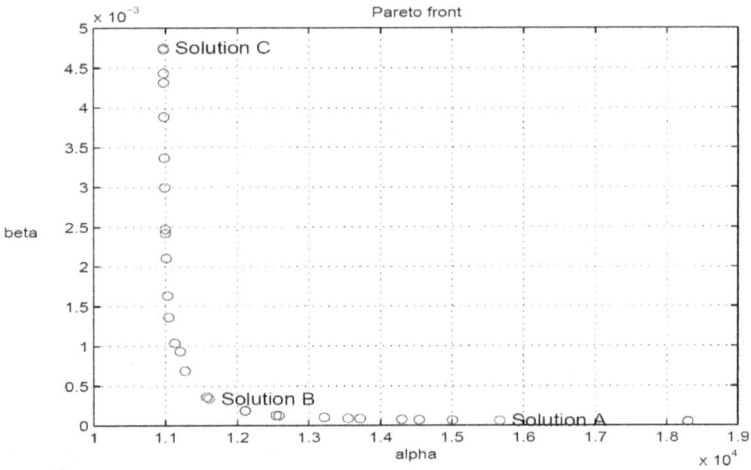

FIGURE 5.2 The Pareto front for Pareto optimal solutions of the MO H_2/H_∞ missile guidance problem in (5.22) can be obtained by the proposed LMI-constrained MOEA. The three marked Pareto optimal solutions in Table 5.1 illustrate how to select the preferable solution with the tracking results in Figures 5.3–5.5, respectively.

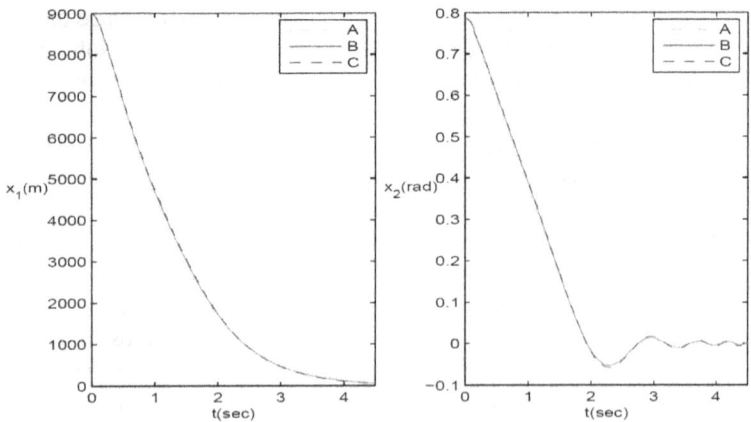

FIGURE 5.3 The trajectories $x_1(t)$ and $x_2(t)$ of Pareto optimal solutions A, B, and C.

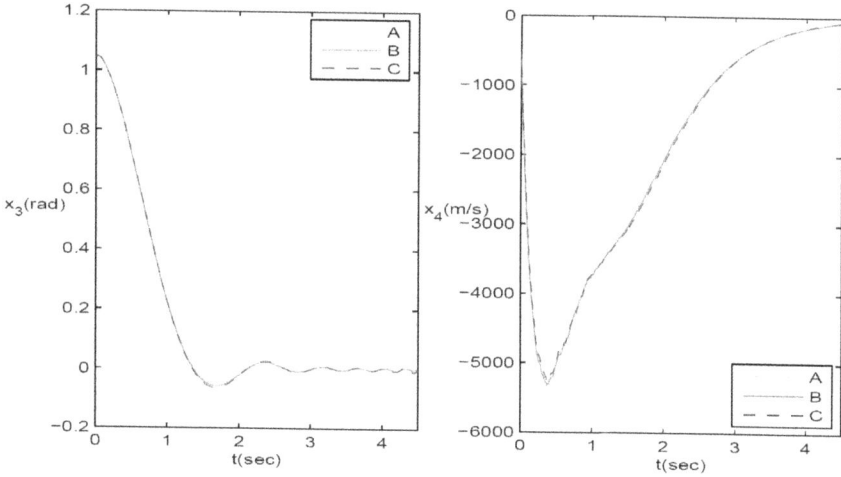

FIGURE 5.4 The trajectories $x_3(t)$ and $x_4(t)$ of Pareto optimal solutions A, B, and C.

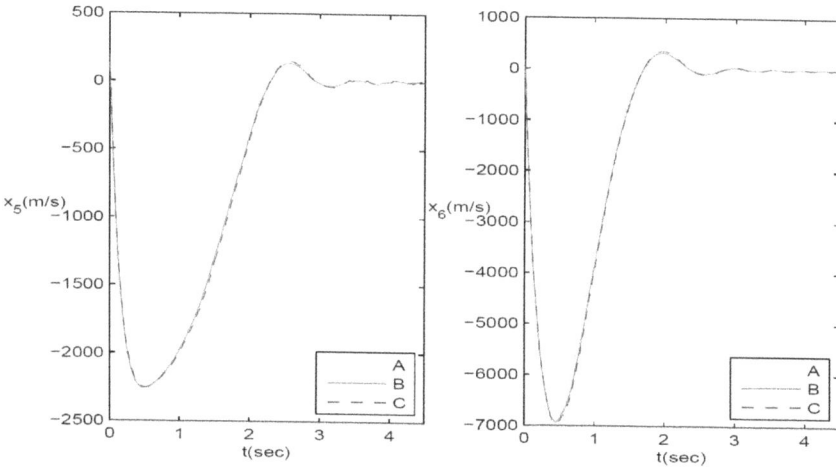

FIGURE 5.5 The trajectories $x_5(t)$ and $x_6(t)$ of Pareto optimal solutions A, B, and C.

process due to the two side jets of target as side-stop maneuvering in Figure 5.8 and 5.9, respectively. From Figure 5.9, we can see the 3-D trajectory of Pareto solution A is the smoothest one among the three solutions. Hence, J_1 performance had a more significant effect than J_2 performance. It is obvious that Pareto optimal solution A is the preferable guidance control solution of the three Pareto optimal solutions because Pareto solution A is a compromise solution in the MO H_2/H_∞ missile guidance problem of the nonlinear stochastic missile guidance system. However, these three chosen Pareto optimal solutions have satisfactory regulation results to achieve the desired steady states of the missile guidance system in spite of intrinsic continuous and discontinuous random fluctuation and external disturbance.

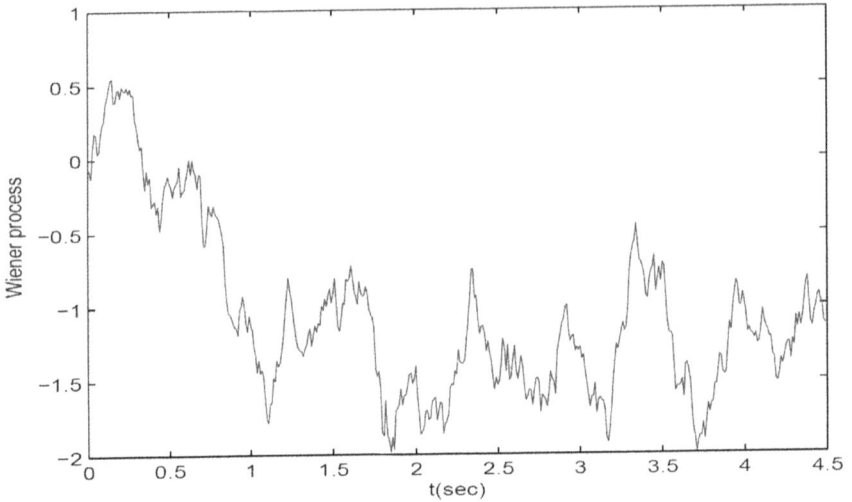

FIGURE 5.6 The trajectory of continuous Wiener process $W(t)$ caused by the modeling uncertainty of the missile and the accumulated angle error of the gyroscope.

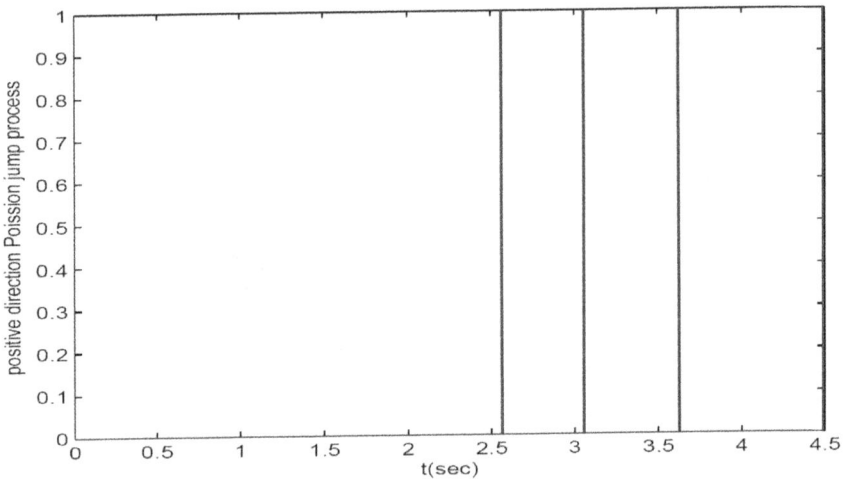

FIGURE 5.7 The trajectory of discontinuous positive Poisson processes $P_1(t)$ caused by the inaccurate radar measurement of the missile because of the target's sudden side-step maneuver.

5.7 CONCLUSION

This study investigated the MO H_2/H_∞ missile guidance law for the nonlinear stochastic missile guidance system via the T-S fuzzy model interpolation method. The proposed MO H_2/H_∞ missile guidance law could simultaneously solve the optimal

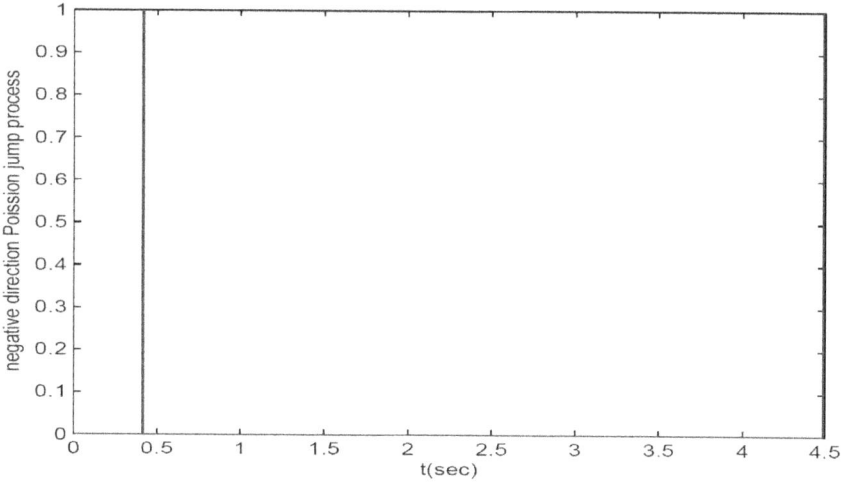

FIGURE 5.8 The trajectory of discontinuous negative Poisson processes $P_2(t)$ caused by the inaccurate radar measurement of the missile because of the target's sudden side-step maneuver.

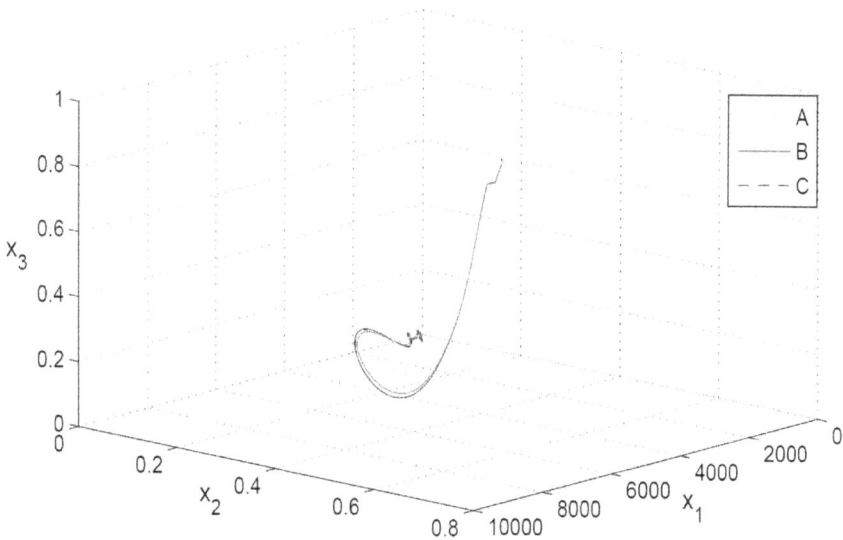

FIGURE 5.9 The 3-D view for the relative position of solutions A, B, and C.

robustness and optimal target tracking problems of a stochastic missile guidance system with continuous Wiener noise as well as discontinuous Poisson jumping noise. To avoid dealing with a nonlinear HJI-constrained MOP, the T-S fuzzy model interpolation method is applied such that the HJI-constrained MOP of the MO H_2/H_∞ missile guidance problem can be replaced by an LMI-constrained MOP of the MO

H_2/H_∞ missile guidance problem. If there exist some feasible solutions for the LMI-constrained MOP in (5.22)–(5.25), the LMI-constrained MOP could be solved simply by the proposed reverse-order LMI-constrained MOEA. Hence, the MO H_2/H_∞ missile guidance law for the nonlinear stochastic missile guidance system can be solved efficiently with the help of the MATLAB LMI toolbox. When the Pareto front in Figure 5.2 is obtained for Pareto optimal solutions of the MOP in (5.22)–(5.25), the designer can select a preferable fuzzy missile guidance law from the set of Pareto optimal fuzzy missile guidance laws according to their own preference to finish the MO H_2/H_∞ missile guidance control design of the nonlinear stochastic missile guidance system.

5.8   APPENDIX

5.8.1   PROOF OF LEMMA 5.2

The proof of Lemma 5.2 is straightforward. One only needs to prove that both inequality constraints in the MO problem in (5.9) become equality constraints for Pareto optimal solutions. We will show this by contradiction. Given a 3-tuple Pareto optimal solution $(u^*(t), \alpha^*, \beta^*)$ of the MO problem in (5.9), we assume that either of the inequalities in (5.9) remains a strict inequality at the Pareto optimal solution. Without loss of generality, suppose that $J_1^*(u^*(t)) \le \alpha^*$. As a result, there exists α_1 such that $\alpha_1 < \alpha^*$ and $J_1^*(u^*(t)) \le \alpha_1$. Now, for the same $u^*(t)$, the solution (α_1, β^*) dominates the Pareto optimal solution (α^*, β^*), leading to a contradiction. This implies that both inequality constraints in problem (5.9) indeed become equality for constraint-optimal solutions. Thus, the MOP in (5.9) is equivalent to the MOP described in (5.8).

5.8.2   PROOF OF THEOREM 5.1

Without consideration of $v(t)$, by using the fact of $V(x_{t_f}) \ge 0$ and the Itô-Lévy formula in Lemma 5.1, we have

$$
\begin{aligned}
J_1(u(t)) &= E\left\{\int_0^{t_f} [\eta^T(t)Q\eta(t) + u^T(t)Ru(t)]dt\right\} \\
&= E\{V(x_0) - V(x_{t_f})\} + E\left\{\int_0^{t_f} [dV(x) + \eta^T(t)Q\eta(t) \right. \\
&\quad + u^T(t)Ru(t)]dt\} \\
&\le E\{V(x_0)\} + E\left\{\int_0^{t_f} [dV(x) + x^T(t)\bar{Q}x(t) + u^T(t)Ru(t)]dt\right\} \qquad (5.26) \\
&= E\{V(x_0)\} + E\left\{\int_0^{t_f} [V_x^T F(x(t)) + V_x^T Bu(t) \right. \\
&\quad + \tfrac{1}{2}H^T(x(t))V_{xx}H(x(t)) + \lambda \sum_{k=1}^{2}[V(x(t) + G_k(x(t)) \\
&\quad - V(x(t))] + x^T(t)\bar{Q}x(t) + u^T(t)Ru(t)]dt\}
\end{aligned}
$$

If both the following Hamilton-Jacobi Bellman inequality

$$V_x^T F(x(t)) + V_x^T Bu(t) + \tfrac{1}{2} H^T(x(t)) V_{xx} H(x(t)) + \lambda \sum_{k=1}^{2} [V(x(t))$$

$$+ G_k(x(t)) - V(x(t))] + x^T(t)\bar{Q}x(t) + u^T(t)Ru(t) \leq 0 \qquad (5.27)$$

and the initial inequality $E\{V(x_0)\} \leq \alpha$ hold, we have

$$J_1(u(t)) = E\{\int_0^{t_f} [\eta^T(t)Q\eta(t) + u^T(t)Ru(t)]dt\} \leq E\{V(x_0)\} \leq \alpha$$

Lemma 5.4 ([10]): For any matrices (or vectors) X and Y with appropriate dimensions, we have

$$X^T Y + Y^T X \leq J X^T X + J^{-1} Y^T Y \qquad (5.28)$$

where J is any nonzero real number.

Considering the disturbance $v(t)$, the fact of $V(x_{t_f}) \geq 0$, Lemma 5.1, and Lemma 5.4, we have

$$E\{\int_0^{t_f} [\eta^T(t)Q\eta(t) + u^T(t)Ru(t)]dt\}$$

$$= E\{V(x_0) - V(x_{t_f})\} + E\{\int_0^{t_f} [dV(x) + \eta^T(t)Q\eta(t)$$

$$+ u^T(t)Ru(t)]dt\}$$

$$\leq E\{V(x_0)\} + E\{\int_0^{t_f} [dV(x) + x^T(t)\bar{Q}x(t) + u^T(t)Ru(t)]dt\} \qquad (5.29)$$

$$\leq E\{V(x_0)\} + E\{\int_0^{t_f} [V_x^T F(x(t)) + V_x^T Bu(t) + \tfrac{1}{4\beta} V_x^T DD^T V_x$$

$$+ \beta v^T(t)v(t) + \tfrac{1}{2} H^T(x(t)) V_{xx} H(x(t)) + \lambda \sum_{k=1}^{2} [V(x(t)$$

$$+ G_k(x(t))) - V(x(t))] + x^T(t)\bar{Q}x(t) + u^T(t)Ru(t)]dt\}$$

If the following Hamilton-Jacobi-Bellman inequality holds,

$$V_x^T F(x(t)) + V_x^T Bu(t) + \tfrac{1}{4\beta} V_x^T DD^T V_x$$

$$+ \tfrac{1}{2} H^T(x(t)) V_{xx} H(x(t)) + \lambda \sum_{k=1}^{2} [V(x(t)) + G_k(x(t)) - V(x(t))] \qquad (5.30)$$

$$+ x^T(t)\bar{Q}x(t) + u^T(t)Ru(t) \leq 0$$

we can get the H guidance performance

$$J_2(u(t)) = \frac{E\{\int_0^{t_f} [\eta^T(t)Q\eta(t) + u^T(t)Ru(t)]dt - V(x_0)\}}{E\{\int_0^{t_f} [v^T(t)v(t)]dt\}} \le \beta$$

5.8.3 PROOF OF THEOREM 5.2

We choose $V(x(t)) = x^T(t)Px(t)$ as a Lyapunov function for the nonlinear stochastic missile guidance system in (5.4), where $P = P^T > 0$ is a positive definite matrix. It is clear that

$$E\{V(x_0)\} \le \bar{\xi}(P)E\{Tr(x_0 x_0^T)\} = \bar{\xi}(P)Tr(R_{x_0}) \tag{5.31}$$

If we set $\bar{\xi}(P)Tr(R_{x_0}) \le \alpha$, then we get $P \le \alpha Tr(R_{x_0})^{-1}I$, which is in (5.23), and it implies $E\{x_0^T Px_0\} \le \alpha$. By using Lemma 5.3, Lemma 5.4, and (5.17), we can get the following inequalities

$$\begin{aligned}
&V_x^T F(x(t)) \\
&= \sum_{i=1}^l h_i(z)x^T(t)[A_i^T P + PA_i]x(t) + x^T(t)P\Delta F(x(t)) \\
&\quad + x(t)\Delta F^T(x(t))P \\
&\le \sum_{i=1}^l h_i(z)x^T(t)[A_i^T P + PA_i]x(t) + \Delta F^T(x(t))\Delta F(x(t)) \\
&\quad + x^T(t)PPx(t) \\
&\le \sum_{i=1}^l h_i(z)x^T(t)[A_i^T P + PA_i + r_1^2 I + PP]x(t)
\end{aligned} \tag{5.32}$$

$$\begin{aligned}
&V_x^T Bu(t) \\
&= \sum_{j=1}^l h_j(z)x^T(t)2PBK_j x(t) \\
&= \sum_{j=1}^l h_j(z)x^T(t)[PBK_j + (PBK_j)^T]x(t)
\end{aligned} \tag{5.33}$$

$$\begin{aligned}
&\tfrac{1}{2}H^T(x(t))V_{xx}(x(t))H(x(t)) \\
&= \left(\sum_{i=1}^l h_i(z)H_i x(t) + \Delta H(x(t))\right)^T P \left(\sum_{i=1}^l h_i(z)H_i x(t) + \Delta H(x(t))\right) \\
&\le \sum_{i=1}^l \sum_{j=1}^l h_i(z)h_j(z)x^T(t)[2H_i^T PH_j]x(t) \\
&\quad + 2\Delta H^T(x(t))P\Delta H(x(t)) \\
&\le \sum_{i=1}^l h_i(z)x^T(t)[2H_i^T PH_i + 2Pr_2^2]x(t)
\end{aligned} \tag{5.34}$$

$$\sum_{k=1}^{2}[V(x(t))+G_k(x(t))-V(x(t))]$$

$$=\sum_{k=1}^{2}[x(t)+G_k(x(t))]^T P[x(t)+G_k(x(t))]-x^T(t)Px(t)$$

$$=\sum_{k=1}^{2}[x(t)+\sum_{i=1}^{l}h_i(z)G_{i,k}x(t)+\Delta G_k(x(t))]^T P[x(t)$$

$$+\sum_{i=1}^{l}h_i(z)G_{i,k}x(t)+\Delta G_k(x(t))]-x^T(t)Px(t)$$

$$\leq\sum_{i=1}^{l}h_i(z)x^T(t)\{\sum_{k=1}^{2}[(2G_{i,k}^T PG_{i,k})+(G_{i,k}^T P)+(PG_{i,k})]$$

$$+P+3Pr_{3,k}^2\}x(t)$$

(5.35)

$$u^T(t)Ru(t)\leq\sum_{j=1}^{l}h_j(z)x^T(t)(K_j^T RK_j)x(t)$$

(5.36)

Substituting (5.32)–(5.36) into (5.26), we get

$J_1(u(t))$

$$\leq E\{V(x_0)\}+E\left\{\sum_{i=1}^{l}\sum_{j=1}^{l}h_i(z)h_j(z)x^T(t)\{[(A_i^T P+PA_i)+r_1^2 I+PP+(PBK_j\right.$$

$$+(PBK_j)^T)+2H_i^T PH_i+2Pr_2^2]+\lambda\sum_{k=1}^{2}\{(2G_{i,k}^T PG_{i,k})+(G_{i,k}^T P)+(PG_{i,k})+P$$

$$+3Pr_{3,k}^2\}+\bar{Q}+K_j^T RK_j\}x(t)\bigg\}$$

$$=E\{V(x_0)\}+E\left\{\sum_{i=1}^{l}h_i^2(z)x^T(t)\{[(A_i^T P+PA_i)+r_1^2 I+PP+(PBK_j+(PBK_j)^T)\right.$$

$$+2H_i^T PH_i+2Pr_2^2]+\lambda\sum_{k=1}^{2}\{(2G_{i,k}^T PG_{i,k})+(G_{i,k}^T P)+(PG_{i,k})+P+3Pr_{3,k}^2\bigg\}$$

$$+\bar{Q}+K_j^T RK_j\}x(t)\}+2E\left\{\sum_{i=1}^{l}\sum_{i<j}^{l}h_i(z)h_j(z)x^T(t)\{[0.5(A_i^T P+PA_i+A_j^T P+PA_j)\right.$$

$$+0.5(PBK_i+(PBK_i)^T+PBK_j+(PBK_j)^T)+(H_i^T PH_i+H_j^T PH_j)$$

$$+\lambda\sum_{k=1}^{2}\{(G_{i,k}^T PG_{i,k}+G_{j,k}^T PG_{j,k})+0.5(G_{i,k}^T P+PG_{i,k}+G_{j,k}^T P+PG_{j,k})\}$$

$$+0.5(K_i^T RK_i+K_j^T RK_j)+(r_1^2 I+PP+2Pr_2^2+\lambda P+3\lambda Pr_{3,2}^2+\bar{Q})\}x(t)\bigg\}$$

$$<E\{V(x_0)\}+E\left\{\sum_{i=1}^{l}h_i^2(z)x^T(t)\Phi_{ii}^1 x(t)\right\}+2E\left\{\sum_{i=1}^{l}\sum_{i<j}^{l}h_i(z)h_j(z)x^T(t)\Phi_{ij}^1 x(t)\right\}$$

where

$$\Phi_{ij}^{(1)} = [0.5(A_i^T P + PA_i + A_j^T P + PA_j) + 0.5(PBK_i + (PBK_i)^T + PBK_j + (PBK_j)^T)$$

$$+ (H_i^T PH_i + H_j^T PH_j) + \lambda \sum_{k=1}^{2} \{(G_{i,k}^T PG_{i,k} + G_{j,k}^T PG_{j,k}) + 0.5(G_{i,k}^T P + PG_{i,k}$$

$$+ G_{j,k}^T P + PG_{j,k})\} + (r_1^2 I + PP + 2Pr_2^2 + \lambda P + 3\lambda Pr_{3,2}^2 + \bar{Q}) + 0.5(K_i^T RK_i$$

$$+ K_j^T RK_j)$$

And the initial inequality $EV(x_0) \leq \alpha$ holds; we have

$$J_1(u(t)) \leq E\{V(x_0)\} \leq \alpha$$

By (5.32)–(5.36), the HJBI in (5.29) can be replaced by

$$E\{V(x_0)\} + E\left\{ \sum_{i=1}^{l} \sum_{j=1}^{l} h_i(z) h_j(z) x^T(t) \{[(A_i^T P + PA_i) + r_1^2 I + PP + (PBK_j + (PBK_j)^T)] \right.$$

$$+ 2H_i^T PH_i + 2Pr_2^2] + \tfrac{1}{\beta} PDD^T P + \bar{Q} + K_j^T RK_j + \lambda \sum_{k=1}^{2} \{(2G_{i,k}^T PG_{i,k}) + (G_{i,k}^T P)$$

$$+ (PG_{i,k}) + P + 3Pr_{3,k}^2\}\} x(t) + \beta v^T(t) v(t)\Big\}$$

$$= E\{V(x_0)\} + E\left\{ \sum_{i=1}^{l} h_i^2(z) x^T(t) \{[(A_i^T P + PA_i) + r_1^2 I + PP + (PBK_j + (PBK_j)^T)] \right.$$

$$+ 2H_i^T PH_i + 2Pr_2^2] + \lambda \sum_{k=1}^{2} \{(2G_{i,k}^T PG_{i,k}) + (G_{i,k}^T P) + (PG_{i,k}) + P + 3Pr_{3,k}^2\} + \bar{Q}$$

$$+ K_j^T RK_j + \tfrac{1}{\beta} PDD^T P\} x(t)\} + 2E\left\{ \sum_{i=1}^{l} \sum_{i<j}^{l} h_i(z) h_j(z) x^T(t) \{[0.5(A_i^T P + PA_i + A_j^T P \right.$$

$$+ PA_j) + 0.5(PBK_i + (PBK_i)^T + PBK_j + (PBK_j)^T) + (H_i^T PH_i + H_j^T PH_j)$$

$$+ \lambda \sum_{k=1}^{2} \{(G_{i,k}^T PG_{i,k} + G_{j,k}^T PG_{j,k}) + 0.5(G_{i,k}^T P + PG_{i,k} + G_{j,k}^T P + PG_{j,k})\}$$

$$+ (r_1^2 I + PP + 2Pr_2^2 + \lambda P + 3\lambda Pr_{3,1}^2 + 3\lambda Pr_{3,2}^2 + 0.5(K_i^T RK_i + K_j^T RK_j) + \bar{Q}$$

$$+ \tfrac{1}{\beta} PDD^T P)\} x(t)\} + E\{\beta v^T(t) v(t)\}$$

$$< E\{V(x_0)\} + E\left\{ \sum_{i=1}^{l} h_i^2(z) x^T(t) \Phi_{ii}^2 x(t) \right\} + 2E\left\{ \sum_{i=1}^{l} \sum_{i<j}^{l} h_i(z) h_j(z) x^T(t) \Phi_{ij}^2 x(t) \right\}$$

$$+ E\{\beta v^T(t) v(t)\}$$

where

$$\Phi_{ij}^{(2)} = \Phi_{ij}^{(1)} + \tfrac{1}{\beta} PDD^T P$$

If the following two algebraic Riccati-like inequalities are satisfied,

$$\Phi_{ij}^{(1)} \leq 0 \tag{5.37}$$

$$\Phi_{ij}^{(2)} \leq 0, \forall i < j(i, j = 1, 2, ..., l) \tag{5.38}$$

then we have $J_1(u(t)) \leq \alpha$ and $J_2(u(t)) \leq \beta$, respectively. Let $\Lambda = P^{-1}$ and $Z_j = K_j \Lambda$; then the inequalities in (5.37) and (5.38) can be represented by the following two inequalities in (5.39) and (5.40), respectively

$$\Xi_{ij}^{(1)} \leq 0 \tag{5.39}$$

$$\Xi_{ij}^{(2)} \leq 0, \forall i < j(i, j = 1, 2, ..., l) \tag{5.40}$$

where

$$\begin{aligned}
\Xi_{ij}^{(1)} = [0.5(\Lambda A_i^T + A_i \Lambda + \Lambda A_j^T + A_j \Lambda) + 0.5((BZ_i + (BZ_i)^T) + (BZ_j + (BZ_j)^T)) \\
+ (\Lambda H_i^T \Lambda^{-1} H_i \Lambda + \Lambda H_j^T \Lambda^{-1} H_j \Lambda) + \lambda \sum_{k=1}^{2} \{ (\Lambda G_{i,k}^T \Lambda^{-1} G_{i,k} \Lambda + \Lambda G_{j,k}^T \Lambda^{-1} G_{j,k} \Lambda) \\
+ 0.5(\Lambda G_{i,k}^T + G_{i,k} \Lambda + \Lambda G_{j,k}^T + G_{j,k} \Lambda) \} + (\Lambda r_1^2 \Lambda + I + 2r_2^2 \Lambda + \lambda \Lambda + 3\lambda r_{3,1}^2 \Lambda \\
+ 3\lambda r_{3,2}^2 \Lambda + \Lambda \bar{Q} \Lambda) + 0.5(Z_i^T RZ_i + Z_j^T RZ_j)
\end{aligned}$$

where

$$\Xi_{ij}^{(2)} = \Xi_{ij}^{(1)} + \tfrac{1}{\beta} DD^T$$

By Schur complement, the quadratic inequalities in (5.39) and (5.40) are equivalent to the LMIs in (5.24) and (5.25), respectively.

6 Multi-Objective Control Design of Nonlinear Mean-Field Stochastic Jump-Diffusion Systems

6.1 INTRODUCTION

The mean-field stochastic system was first proposed by M. Kac [103] to study the stochastic model for the Vlasov equation of plasma. The mean-field theory was proposed to describe collective behaviors resulting from individuals' mutual interaction in various physical and sociological dynamic systems. With the development of stochastic control and filter theories [11, 15, 36, 34, 104–123], the mean-field stochastic system has become an active research field [105–108]. One main feature of the mean-field stochastic system is that the mean of the system state (the mean term) Ex_t appears in its system dynamics. Thus, designing a controller for the mean-field stochastic system is more difficult than doing so for the classical one. In the past ten years, most researchers have focused on the optimal H_2 control design and optimal H_∞ robust control design for linear mean-field stochastic systems [109–111]. However, in practical applications, H_2 and H_∞ control performances are usually in partial conflict with each other. To balance the conflict between H_2 and H_∞ control performance, the multi-objective H_2/H_∞ control design has emerged [15, 112–114]. Unlike the mixed H_2/H_∞ control design that only minimizes H_2 performance under the constraint of a prescribed H_∞ performance [11, 34, 36, 115–117], the multi-objective H_2/H_∞ control design is defined as how to design a controlled input u_t to concurrently minimize H_2 and H_∞ performance in the Pareto optimal sense for a dynamic system and try to completely describe the tradeoff between H_2 and H_∞ performance.

There are three main difficulties in solving the multi-objective H_2/H_∞ control problem for nonlinear mean-field stochastic jump diffusion (MFSJD) systems, as follows.

(I) In general, for a multi-objective nonlinear H_2/H_∞ control design, we always need to solve Hamilton-Jacobi inequalities. However, for the nonlinear MFSJD system, its HJIs are difficult to derive.

(II) Owing to the fact that both the mean term Ex_t and Poisson jump processes appear in the system dynamic equation, the stability criterion of the nonlinear MFSJD system needs to be derived.

(III) Multi-Objective H_2/H_∞ control design problems are difficult to solve directly and always need to achieve a solution via multi-objective evolutionary

DOI: 10.1201/9781003362142-8

algorithm search methods. However, the existing linear matrix inequality–constrained MOEAs always have higher computational loads for nonlinear MFSJD systems.

For issue (I), the Takagi-Sugeno fuzzy model is employed to overcome it. The T-S fuzzy interpolation method had been widely used in nonlinear control and filter design problems [104, 116, 118–121]. Based on it, one can employ a set of locally linear MFSJD systems to approximate a nonlinear MFSJD system. Thus, the HJIs of the nonlinear MFSJD systems can be replaced by a set of LMIs, and the corresponding multi-objective H_2/H_∞ control design problem is transformed into an LMI-constrained multi-objective optimization problem. Even though the T-S fuzzy model efficiently reflects the behavior of the nonlinear MFSJD system, we also consider the effects of approximation errors of the T-S model on the multi-objective H_2/H_∞ control design because they indeed affect the stability and control performance of the nonlinear MFSJD system.

For issue (II), we will propose sufficient conditions for the stability of the nonlinear fuzzy MFSJD systems in the mean square sense by decoupling the MFSJD system into two orthogonal subsystems: the mean subsystem of Ex_t and the variation subsystem of $\tilde{x}_t = x_t - Ex_t$. Then, we have the augmented system with system state $\overline{X}_t = [\tilde{x}_t^T, Ex_t^T]^T$. Because the augmented system is also a stochastic jump diffusion system, the Itô-Lévy formula can be applied to it. We can derive the stability criterion for the MFSJD system in the mean square sense.

For issue (III), an indirect method is proposed to solve the multi-objective H_2/H_∞ control design for MFSJD systems by simultaneously minimizing the upper bound vectors (α, β) of the H_2 and H_∞ performance in the Pareto optimal sense. By applying the proposed indirect method, the multi-objective H_2/H_∞ control problem can be transformed into an LMI-constrained MOP. Therefore, the multi-objective H_2/H_∞ control for MFSJD systems becomes an issue of how to specify a set of control gains to concurrently minimize (α, β) subject to a set of LMI constraints in the Pareto optimal sense. For solving the LMI-constrained MOP, we usually introduce the LMI-constrained MOEA search method in [6, 11, 122, 123]. The LMI-constrained MOEA searches for the Pareto front of the multi-objective H_2/H_∞ control design for the nonlinear MFSJD systems via the evolutionary algorithm, nondominated sorting, and elitist selection operator. By concurrently decreasing the upper bound vectors (α, β) in the nondominating sense, the Pareto optimal solution set of the multi-objective H_2/H_∞ control for the nonlinear MFSJD system can be approached indirectly. As long as the Pareto front is obtained, the multi-objective H_2/H_∞ control problem for the MFSJD system can be also solved because each upper bound vector (α, β) of the Pareto front corresponds to a group of control gains, guaranteeing that the H_2 and H_∞ control performance (α, β) is minimized simultaneously. However, the existing LMI-constrained MOEAs mean a higher computational load. To decrease the computational load and improve the computational efficiency, we proposed the front-squeezing LMI-constrained MOEA. The front-squeezing LMI-constrained MOEA concurrently employs the vice front and current parent front to narrow the search region down from two sides of the feasible and infeasible regions.

In this chapter, two main objectives are as follows: first, we achieve a multi-objective H_2/H_∞ fuzzy control design for nonlinear MFSJD systems. It can completely describe the tradeoff between the H_2 and H_∞ control performance. Thus, the optimal H_2 control design and optimal H_∞ robust control design for a nonlinear MFSJD system can be accomplished simultaneously, and second, a novel reverse-order LMI-constrained MOEA called a front-squeezing LMI-constrained MOEA is proposed to efficiently solve the MOP. Compared with the LMI-constrained MOEA in [11], it can reduce the computational load by more than half and has better performance in searching the Pareto front and Pareto optimal solutions of the multi-objective H_2/H_∞ control design for nonlinear MFSJD systems. Once the Pareto front is obtained, the designer can select a fuzzy controller from the Pareto optimal solution set according to their own preference.

For convenience, we adopt the following notation:

Notation: A^T: transpose of matrix A; $A \geq 0$ ($A > 0$): symmetric positive semidefinite (symmetric positive definite) matrix A; I: identity matrix; $\|x\|_2$: Euclidean norm for the given vector $x \in \mathbb{R}^n$; C^2: class of functions $V(x)$ twice continuously differential with respect to x; f_x: gradient column vector of n_x-dimensional twice continuously differentiable function $f(x)$ (i.e., $\frac{\partial^2 f(x)}{\partial x^2}$); f_{xx}: Hessian matrix with elements of second partial derivatives of n_x-dimensional twice continuously differentiable function $f(x)$, (i.e., $\frac{\partial^2 f(x)}{\partial x^2}$); $L_F^2(\mathbb{R}^+; \mathbb{R}^{n_y})$: space of nonanticipative stochastic process $y(t) \in \mathbb{R}^l$ with respect to an increasing σ-algebra F_t satisfying $\|y(t)\|_{L_F^2(\mathbb{R}^+;\mathbb{R}^{n_y})} \triangleq E\{\int_0^\infty y^T(t)y(t)dt\}^{\frac{1}{2}} < \infty$; $\|y(t)\|_{L_F^2(\mathbb{R}^+;\mathbb{R}^{n_y},Q)} \triangleq E\{\int_0^\infty y^T(t)Qy(t)dt\}^{\frac{1}{2}} < \infty$; $B(\Theta)$: Borel algebra generated by Θ; E: expectation operator; $P\{\cdot\}$: probability measure function; $\bar{\lambda}(M)$: maximum eigenvalue of real-value matrix M; and $S_{n_x}^+$: set of positive $n \times n$ symmetric real-valued matrices.

6.2   PRELIMINARIES

6.2.1   Nonlinear Fuzzy MFSJD Systems

Consider the following nonlinear MFSJD system driven by Wiener processes $\{W_t^{(r)}\}_{r=1}^m$ and the marked Poisson processes $\{N_p(t,\theta_q)\}_{r=1}^m$:

$$dx_t = [f(x_t, Ex_t) + \eta(x_t, Ex_t)u_t + g(x_t)v_t]dt$$
$$+ \sum_{r=1}^m \sigma_r(x_t, Ex_t)dW_t^{(r)}$$
$$+ \sum_{q=1}^n T(x_{t^-}, Ex_{t^-}, \theta_q)dN_p(t,\theta_q) \qquad (6.1)$$
$$x_0 = x(\omega,0) \text{ and } Ex_0 = E\{x(\omega,0)\}$$

where $f : \mathbb{R}^{n_x} \times \mathbb{R}^{n_x} \to \mathbb{R}^{n_x}$, $\eta : \mathbb{R}^{n_x} \times \mathbb{R}^{n_x} \to \mathbb{R}^{n_x} \times \mathbb{R}^{n_u}$, $\eta : \mathbb{R}^{n_x} \times \mathbb{R}^{n_x} \to \mathbb{R}^{n_x} \times \mathbb{R}^{n_u}$, $g : \mathbb{R}^{n_x} \to \mathbb{R}^{n_x} \times \mathbb{R}^{n_v}$, $\sigma_r : \mathbb{R}^{n_x} \times \mathbb{R}^{n_x} \to \mathbb{R}^{n_x}$, and $T_q : \mathbb{R}^{n_x} \times \mathbb{R}^{n_x} \times \Theta \to \mathbb{R}^{n_x}$ are nonlinear Borel measurable continuous functions, which are satisfied with local Lipschitz continuity with respect to $X_t \triangleq [x_t^T \quad Ex_t^T]^T$.

The càdlàg process $x_t \triangleq x(\omega,t) \in U_x \subset \mathbb{R}^{n_x}$ denotes the system state vector, where U_x is a compact region such that the behavior of system states can be governed by (6.1);

$x_0 = x(\omega,0)$ is the initial state vector; the term $u_t \triangleq u(\omega,t) \in L_F^2(\mathbb{R}^+;\mathbb{R}^{n_u})$ denotes the admissible control input with respect to $\{F_t\}_{t\geq0}$; $v_t \triangleq v(\omega,t) \in L_F^2(\mathbb{R}^+;\mathbb{R}^{n_v})$ denotes the unknown finite energy stochastic external disturbance; Ex_t denotes the mean of x_t; the term $\Theta = \{\theta_1,\theta_2,...,\theta_n\} - \{0\}$ denotes the Poisson mark space; and the term $N_p(t,\theta_q)$ denotes the classical Poisson process with respect to mark θ_q with $E\{N_p(t,\theta_q)\} = \lambda_q t$. Since the Poisson marks are mutually disjoint, the mark-driven Poisson processes $\{N_p(\cdot,\theta_q)\}_{q=1}^n$ are mutually independent. Furthermore, the Wiener processes $\{W_t^{(r)}\}_{r=1}^m$ are assumed to be the standard Wiener processes and mutually independent from the mark-driven Poisson processes $\{N_p(\cdot,\theta_q)\}_{q=1}^n$ for all $t \geq 0$.

Without loss of generality, the nonlinear MFSJD system in (6.1) has the following two properties: first, the system states of the nonlinear MFSJD system in (6.1) are available; second, the origin of $\mathbb{R}^{n_x}$ is an equilibrium point of interest in the given nonlinear MFSJD system. If the equilibrium point of interest in the nonlinear MFSJD system is not at the origin point, then the designer can shift it to the origin by changing variables.

The *ith* rule of the T-S fuzzy model for approximating the nonlinear MFSJD system in (6.1) is described by

$$\text{If } \xi_1(t) \text{ is } G_{i,1} \text{ and } \xi_2(t) \text{ is } G_{i,2},...,\xi_g(t) \text{ is } G_{i,1},$$

$$\begin{aligned}\text{then } dx_t &= (A_i x_t + A_i^\mu Ex_t + B_i u_t + G_i v_t)dt \\ &+ \sum_{r=1}^m [C_{i,r} x_t + C_{i,r}^\mu Ex_t]dW_t^{(r)} \\ &+ \sum_{q=1}^n [D_{i,q} x_t + C_{i,q}^\mu Ex_{t^-}]dN_p(t,\theta_q)\end{aligned} \tag{6.2}$$

where $G_{i,g}$ is a fuzzy set; $A_i, A_i^\mu, B_i, G_i, C_{i,r}, C_{i,r}^\mu, D_{i,q}$, and $D_{i,q}^\mu$ are deterministic real value matrices with appropriate dimensions; l is the number of fuzzy rules; and $\xi_1(t)$, $\xi_2(t), \ldots, \xi_g(t)$ are the fuzzy premise variables.

The overall nonlinear fuzzy MFSJD system in (6.2) can be inferred as follows:

$$\begin{aligned}dx_t &= \sum_{i=1}^l h_i(\xi_t)\Big\{(A_i x_t + A_i^\mu Ex_t + B_i u_t + G_i v_t)dt \\ &+ \sum_{r=1}^m [C_{i,r} x_t + C_{i,r}^\mu Ex_t]dW_t^{(r)} \\ &+ \sum_{q=1}^n [D_{i,q} x_{t^-} + D_{i,q}^\mu Ex_{t^-}]dN_p(t,\theta_q)\Big\}\end{aligned} \tag{6.3}$$

where $\xi_t = [\xi_1,...,\xi_g]^T$, $\mu_i(\xi_t) = \prod_{k=1}^g G_{i,k}(\xi_t) \geq 0$, $h_i(\xi_t) = [\mu_i(\xi_t)/\sum_{i=1}^l \mu_i(\xi_t)]$, and $\sum_{i=1}^l h_i(\xi_t) = 1$.

The physical meaning of the fuzzy model in (6.3) is that the locally linearized MFSJD systems in (6.2) at different operation points associated with a different fuzzy set $G_{i,k}$ are interpolated piecewise via the fuzzy weighting basis $\{h_i(\xi_t)\}_{i=1}^l$ to approximate the original nonlinear system in (6.1). Similarly, the control input u_t for the nonlinear MFSJD system can be approximated by the following fuzzy control input:

> Control Rule $i : i = 1, 2, ..., l$
>
> If $\xi_1(t)$ is $G_{i,1}$ and $\xi_2(t)$ is $G_{i,2}, ..., $ and $\xi_g(t)$ is $G_{i,1}$,   (6.4)
>
> then $u_t = K_i^1 x_t + (K_i^2 - K_i^1)Ex_t = K_i^1 \tilde{x}_t + K_i^2 Ex_t$

The overall fuzzy control input for the nonlinear MFSJD system is represented as

$$u_t = \sum_{i=1}^{l} h_i(\xi_t)[K_i^1 \tilde{x}_t + K_i^2 Ex_t] \tag{6.5}$$

Substituting (6.5) into the nonlinear fuzzy MFSJD system in (6.3) yields the closed-loop system of the following form:

$$
\begin{aligned}
dx_t = \sum_{i,j=1}^{l} h_i(\xi_t) h_j(\xi_t) &\Big\{ (A_i x_t + A_i^\mu Ex_t + B_i K_j^1 x_t \\
&+ B_i(K_j^2 - K_j^1)Ex_t + G_i v_t)dt + \sum_{r=1}^{m} [C_{i,r} x_t + C_{i,r}^\mu Ex_t] dW_t^{(r)} \\
&+ \sum_{q=1}^{n} [D_{i,q} x_{t^-} + D_{i,q}^\mu Ex_{t^-}] dN_p(t, \theta_q) \Big\}
\end{aligned}
\tag{6.6}
$$

Even though the T-S fuzzy model efficiently reflects the system behavior, we still need to consider the effects of approximation errors of the T-S model because they indeed affect the stability and control performance of the nonlinear MFSJD system.

After counting the effects of approximation errors of the T-S fuzzy interpolation method and setting the control input to be u_t in (6.5), we can represent (6.1) as follows:

$$
\begin{aligned}
dx_t = \sum_{i=1}^{l} \sum_{j=1}^{l} h_i(\xi_t) h_j(\xi_t) &\Big\{ ([A_i x_t + A_i^\mu Ex_t + \Delta f] \\
&+ (B_i + \Delta\eta)[K_j^1 x_t + (K_j^2 - K_j^1)Ex_t] + [G_i + \Delta g(x_t)]v_t)dt \\
&+ \sum_{r=1}^{m} [C_{i,r} x_t + C_{i,r}^\mu Ex_t + \Delta\sigma_r] dW_t^{(r)} \\
&+ \sum_{q=1}^{n} [D_{i,q} x_{t^-} + D_{i,q}^\mu Ex_{t^-} + \Delta T_q] dN_p(t, \theta_q) \Big\}
\end{aligned}
\tag{6.7}
$$

where the functions Δf, Δg, $\Delta\eta$, $\Delta\sigma_r$, and ΔT_q are used to denote the T-S fuzzy approximation errors in the terms $f, \eta, g, \sigma_r,$ and T_q, respectively.

Assumption 6.1: Assume the fuzzy approximation errors in (6.7) of the nonlinear MFSJD system in (6.1) are bounded in quadratic form for all $t > 0$, that is,

$$\Delta f^T \Delta f \leq \varepsilon_1 x_t^T x_t + \varepsilon_1^\mu (Ex_t)^T (Ex_t), \text{a.s.}$$

$$\Delta\eta^T\Delta\eta \le \varepsilon_2 I_{n_u} \text{ a.s. } \Delta g^T\Delta g \le \delta I_{n_x} \text{ a.s.}$$

$$\Delta\sigma_r^T\Delta\sigma_r \le \varepsilon_{3,r}x_t^T x_t + \varepsilon_{3,r}^\mu (Ex_t)^T(Ex_t), \text{ a.s.}$$

$$\Delta T_q^T\Delta T_q \le \varepsilon_{4,q}x_t^T x_t + \varepsilon_{4,q}^\mu (Ex_t)^T(Ex_t), \text{ a.s.}$$

$$\text{for } r = 1,...,m, \ q = 1,...,n.$$

Remark 6.1: One may ask that if there exist two T-S fuzzy models $S = \{A_i, A_i^\mu, B_i, G_i, C_{i,r}, C_{i,r}^\mu, D_{i,q}, D_{i,q}^\mu\}$ and $\hat{S} = \{\hat{A}_i, \hat{A}_i^\mu, \hat{B}_i,, \hat{G}_i,, \hat{C}_{i,r}, \hat{C}_{i,r}^\mu, \hat{D}_{i,q}, \hat{D}_{i,q}^\mu\}$ with the same approximation accuracy, which one would be better. To respond to the problem, we only need to examine their domination relation between Pareto fronts P_{front}^* of S and $\hat{P}_{front}^*$ of $\hat{S}$. More precisely, for a given region U defined on the objective space, if P_{front}^* completely dominates $\hat{P}_{front}^*$, we say model S is better than model $\hat{S}$ in the given region U. In this chapter, the prior concern is the Pareto front and Pareto optimal solutions of the LMI-constrained MOP in (6.22) with LMI constraints in (6.24)–(6.28) in the sequel, and they are highly dependent on the chosen T-S fuzzy model and its approximation error. Thus, the proposed method may be sensitive to the T-S fuzzy model. In addition, since the Pareto front and Pareto optimal solutions of multi-objective H_2/H_∞ control design problems are difficult to obtain through direct calculation, it is hard to give a theorem to predict which T-S fuzzy model would be better.

Remark 6.2: In recent years, the tensor product (TP) model has also been proposed for approximating nonlinear dynamic model and replacing HJIs with a set of LMIs [124–126]. In this chapter, the proposed multi-objective H_2/H_∞ control design method can be used not only in T-S fuzzy model but also in TP model.

6.2.2 H_2 AND H_∞ PERFORMANCE OF MFSJD SYSTEMS

To process the multi-objective H_2/H_∞ control design for MFSJD system, we give the following definitions.

The H_2 performance $J_2(u_t)$ indicates the value of the system quadratic cost function.

Definition 6.1([111]): For a given MFSJD system with the external disturbance $v(t) = 0$, its H_2 performance $J_2(u_t)$ is defined as follows:

$$J_2(u) \triangleq E\left\{\int_0^\infty [\bar{X}_t^T\bar{Q}_1\bar{X}_t + \bar{u}_t^T\bar{R}_1\bar{u}_t]dt\right\} \tag{6.8}$$

where $(Q_1 \times R_1 \times Q_1^\mu \times R_1^\mu) \in (S_{n_x}^+ \times S_{n_u}^+ \times S_{n_x}^+ \times S_{n_u}^+)$, $\bar{X}_t = [\tilde{x}_t^T Ex_t^T]^T$, $\bar{u}_t = [\tilde{u}_t^T Eu_t^T]^T$, $\bar{Q}_1 = diag(Q_1, Q_1^\mu)$, and $\bar{R}_1 = diag(R_1, R_1^\mu)$.

Definition 6.2 ([127, 128]): For a given mean-field stochastic system with $x_0 = 0$, the H_∞ performance $J_\infty(u_t)$ of the given mean-field stochastic system is defined as follows:

$$J_\infty(u_t) = \sup_{\substack{v(t)\in L^2(\mathbb{R}_+,\mathbb{R}^{n_v}),\\ v\ne 0, x_0=0}} \left(\frac{\|\bar{X}_t\|_{L_F^2(\mathbb{R}^+;\mathbb{R}^{n_u},\bar{Q}_2)}^2 + \|\bar{u}_t\|_{L_F^2(\mathbb{R}^+;\mathbb{R}^{n_u},\bar{R}_2)}^2}{\|v_t\|_{L_F^2(\mathbb{R}^+;\mathbb{R}^{n_v})}^2}\right) \tag{6.9}$$

where $(Q_2 \times R_2 \times Q_2^\mu \times R_2^\mu) \in (S_{n_x}^+ \times S_{n_u}^+ \times S_{n_x}^+ \times S_{n_u}^+)$, $M \in \mathbb{R}^{n_x \times n_x}$, $\bar{Q}_2 = diag(MQ_2M,$
$MQ_2^\mu M)$, and $\bar{R}_2 = diag(R_2, R_2^\mu)$.

The H_∞ performance $J_\infty(u_t)$ can be regarded as the worst-case effect from the exogenous disturbance signal $v_t \in L_F^2(\mathbb{R}_+; \mathbb{R}^{n_v})$ on $M\tilde{x}_t$, MEx_t, $\tilde{u}_t$, and Eu_t. We need to eliminate the effects of initial energy, too (i.e., the energy $V(\bar{X}_t)$, which is caused by initial state $\bar{X}_0$). Thus, if the effect of initial condition is considered, the definition of the H_∞ performance needs to be modified as follows:

$$
J_\infty(u_t) = \sup_{\substack{v(t) \in L^2(\mathbb{R}_+, \mathbb{R}^{n_v}), \\ v \neq 0, x_0 = 0}} \left(\frac{\|\bar{X}_t\|^2_{L_F^2(\mathbb{R}^+; \mathbb{R}^{n_u}, \bar{Q}_2)} + \|\bar{u}_t\|^2_{L_F^2(\mathbb{R}^+; \mathbb{R}^{n_u}, \bar{R}_2)} - E\{V(\bar{X}_0)\}}{\|\bar{v}_t\|^2_{L_F^2(\mathbb{R}^+; \mathbb{R}^{n_v})}} \right) \tag{6.10}
$$

where $V(\cdot) \geq 0 \in C^2$ can be thought of as the generalized energy function.

6.3 STABILITY ANALYSIS OF NONLINEAR FUZZY MFSJD SYSTEMS

Stability is the most important concept in studying the stochastic dynamic system. To treat the state feedback control problem for the nonlinear MFSJD system in (6.1), we need to define the stabilization for a stochastic system first.

Definition 6.3 ([44, 49]): Suppose $x_{eq} = 0$ is an equilibrium point of interest in the following nonlinear MFSJD system:

$$
dx_t = [f(x_t, Ex_t) + \eta(x_t, Ex_t)u_t]dt + \sum_{r=1}^m \sigma_r(x_t, Ex_t)dW_t^{(r)}
$$
$$
+ \sum_{q=1}^n T(x_{t^-}, Ex_{t^-}, \theta_q)dN_p(t, \theta_q) \tag{6.11}
$$

If there exist some positive constants C_1 and C_2 such that the following inequality holds:

$$
E\left\{\|x(t)\|_2^2\right\} \leq C_1 \exp(-C_2 t) \tag{6.12}
$$

then the nonlinear MFSJD system in (6.11) is said to be exponentially mean square stable (i.e., exponentially stable in the mean square sense).

To derive the sufficient condition for the stability of nonlinear fuzzy MFSJD systems in the mean square sense, we introduce the decoupled method that could decouple the MFSJD system into two orthogonal subsystems in the mean-square sense: a mean subsystem of Ex_t and a variation subsystem of $\tilde{x}_t = x_t - Ex_t$.

Remark 6.3: Since $x_t = (x_t - Ex_t) + Ex_t$ for all $t > 0$, we have $E\{\|x_t\|_2^2\} = E\{\|(x_t - Ex_t) + Ex_t\|_2^2\}$. Let $\tilde{x}_t \triangleq (x_t - Ex_t)$; then we obtain (i) $E\{\|x_t\|_2^2\} = E\{\tilde{x}_t^T E\tilde{x}_t + (Ex_t)^T (Ex_t)\}$ for all $t > 0$ and (ii) $E\{\tilde{x}_t^T Ex_t\} = E\{Ex_t^T \tilde{x}_t\} = 0$ for all $t > 0$. Based on (ii), we find that $\tilde{x}_t$ and Ex_t are orthogonal.

Taking the expectation to both sides of the nonlinear fuzzy MFSJD system in (6.7), we have the following fuzzy mean subsystem:

$$
\begin{aligned}
dEx_t = \sum_{i=1}^{l}\sum_{j=1}^{l} h_i(\xi_t)h_j(\xi_t)\Bigg\{&\Bigg[\hat{A}_i + \sum_{q=1}^{n}\lambda_q \hat{D}_{i,q} + B_i K_j^2\Bigg]E\tilde{x}_t \\
&+ E\{\Delta f\} + \sum_{q=1}^{n}\lambda_q E\{\Delta T_q\} + G_i Ev_t + E\{\Delta gv_t\} \\
&+ \{\Delta\eta\cdot[K_j^1\tilde{x}_t + K_j^2 Ex_t]\}\Bigg\} dt
\end{aligned}
\tag{6.13}
$$

where $Ev_t \triangleq E\{v_t\}$, $\hat{A}_i = (A_i + A_i^\mu)$, and $\hat{D}_{i,q} = (D_{i,q} + D_{i,q}^\mu)$.

The fuzzy variation subsystem for the nonlinear fuzzy MFSJD system in (6.7) is obtained as follows:

$$
\begin{aligned}
dx_t = \sum_{i=1}^{l}\sum_{j=1}^{l} h_i(\xi_t)h_j(\xi_t)\Bigg\{&\Bigg([A_i + B_i K_j^1]\tilde{x}_t - \sum_{q=1}^{n}\lambda_q \hat{D}_{i,q} Ex_{t^-}\\
&+ G_i\tilde{v}_t\Big)dt + \sum_{r=1}^{m}[C_{i,r}\tilde{x}_t + \hat{C}_{i,r} Ex_t]dW_t^{(r)} \\
&+ \sum_{q=1}^{n}[D_{i,q}\tilde{x}_{t^-} + \hat{D}_{i,q} Ex_{t^-}]dN_p(t,\theta_q)\\
&+ [(\Delta f - E\{\Delta f\}) + \Delta\eta(K_j^1\tilde{x}_t + K_j^2 Ex_t) - E\{\Delta\eta(K_j^1\tilde{x}_t \\
&+ K_j^2 Ex_t)\} + \Delta gv_t - E\{\Delta gv_t\}]dt + \sum_{r=1}^{m}\Delta\sigma_r dW_t^{(r)} \\
&+ \sum_{q=1}^{n}\Delta T_q dN_p(t,\theta_q) - \sum_{q=1}^{n}\lambda_q E\{\Delta T_q\}dt\Bigg\}
\end{aligned}
\tag{6.14}
$$

where $\hat{C}_{i,r} = C_{i,r} + C_{i,r}^\mu$ and $\hat{D}_{i,q} = D_{i,r} + D_{i,r}^\mu$.

Now, let $\bar{X}_t = [\tilde{x}_t^T, Ex_t^T]^T$ and $\bar{v}_t = [\tilde{v}_t^T, Ev_t^T]^T$; then the fuzzy augmented stochastic jump diffusion system for the nonlinear fuzzy MFSJD system in (6.7) is defined as follows:

$$
\begin{aligned}
d\bar{X}_t = \sum_{i=1}^{l}\sum_{j=1}^{l} h_i(\xi_t)h_j(\xi_t)\Bigg\{&\Bigg([\bar{F}_i + \bar{B}_i \bar{K}_j]\bar{X}_t + \Delta\bar{F} - \sum_{q=1}^{n}\lambda_q \Delta\bar{D}_q^\mu \\
&+ \bar{U}_j + \bar{G}_i\bar{v}_t + \Delta\bar{G}\Big)dt + \sum_{r=1}^{m}(\bar{C}_{i,r}\bar{X}_t + \Delta\bar{C}_r)dW_t^{(r)} \\
&+ \sum_{q=1}^{n}(\bar{D}_{i,q}\bar{X}_t + \Delta\bar{D}_q)dN_p(t,\theta_q)\Bigg\}
\end{aligned}
\tag{6.15}
$$

where

$$
\bar{F}_i = \begin{bmatrix} A_i & -\sum_{q=1}^{n}\lambda_q \hat{D}_{i,q} \\ 0 & \hat{A}_i + \sum_{q=1}^{n}\lambda_q \hat{D}_{i,q} \end{bmatrix}, \quad
\Delta\bar{F} = \begin{bmatrix} \Delta f - E\{\Delta f\} \\ E\{\Delta f\} \end{bmatrix}, \quad
\bar{B}_i = \begin{bmatrix} B_i & 0 \\ 0 & B_i \end{bmatrix},
$$

$$\bar{K}_j = \begin{bmatrix} K_j^1 & 0 \\ 0 & K_j^2 \end{bmatrix}, \quad \bar{G}_i = \begin{bmatrix} G_i & 0 \\ 0 & G_i \end{bmatrix}, \quad \bar{C}_{i,r} = \begin{bmatrix} C_{i,r} & \hat{C}_{i,r} \\ 0 & 0 \end{bmatrix}, \quad \bar{D}_{i,q} = \begin{bmatrix} D_{i,q} & \hat{D}_{i,q} \\ 0 & 0 \end{bmatrix},$$

$$\Delta \bar{U}_j = \begin{bmatrix} \Delta \eta (K_j^1 \tilde{x}_t + K_j^2 Ex_t) - E\{\Delta \eta (K_j^1 \tilde{x}_t + K_j^2 Ex_t)\} \\ E\{\Delta \eta (K_j^1 \tilde{x}_t + K_j^2 Ex_t)\} \end{bmatrix}, \quad \Delta C_t = \begin{bmatrix} \Delta \sigma_r \\ 0 \end{bmatrix},$$

$$\Delta \bar{G} = \begin{bmatrix} \Delta gv_t - E\{\Delta gv_t\} \\ E\{\Delta gv_t\} \end{bmatrix}, \quad \Delta \bar{D}_q^\mu = \begin{bmatrix} E\{\Delta T_q^T\} & 0 \end{bmatrix}^T, \text{ and } \Delta \bar{D}_q = \begin{bmatrix} \Delta T_q^T & 0 \end{bmatrix}^T.$$

Lemma 6.1 ([129, 130]): Let $V : \mathbb{R}^{n_x} \to \mathbb{R}$, $V(\cdot) \in C^2(\mathbb{R}^{n_x}) \geq 0$. For the following autonomous nonlinear stochastic jump diffusion system:

$$dX_t = (f(X_t) + g(X_t)u_t + v_t)dt + \sum_{r=1}^m \sigma_r(X_t)dW_t^{(r)}$$

$$+ \sum_{q=1}^n T(X_{t^-}, \theta_q)dN_p(t, \theta_q)$$

(6.16)

the Itô-Lévy formula of $V(X_t)$ for the nonlinear stochastic jump diffusion system in (6.16) is given as

$$dV(X_t) = \left[V_x^T (f(X_t) + g(X_t)u_t + v_t) \right.$$

$$+ \frac{1}{2} \sum_{r=1}^m \sigma_r^T(X_t)V_{XX}\sigma_r(X_t) \right] dt$$

$$+ \sum_{r=1}^m V_X^T \sigma_r(X_t)dW_t^{(r)} + \sum_{q=1}^n \{V(X_t + T(X_{t^-}, \theta_q))$$

$$- V(X_t)\}dN_p(t, \theta_q)$$

(6.17)

Theorem 6.1: Consider the nonlinear fuzzy MFSJD system in (6.7) with $v_t = 0$. Suppose $V : \mathbb{R}^{2n_x} \to \mathbb{R}^+$, $V(\cdot) \in C^2(\mathbb{R}^{2n_x}) \geq 0$ to be a scalar function with respect to the nonlinear fuzzy augmented stochastic jump diffusion system in (6.15). If (i) $LV(\bar{X}_t) \leq -m_1 \bar{X}_t^T \bar{X}_t$ with $m_1 > 0$, (ii) $m_2 \bar{X}_t^T \bar{X}_t \leq V(\bar{X}_t) \leq m_3 \bar{X}_t^T \bar{X}_t$ with $m_2, m_3 > 0$, where

$$LV(\bar{X}_t) = \sum_{i=1}^l \sum_{j=1}^l h_i(\xi_t)h_j(\xi_t)V_{\bar{X}}^T \left\{ \left[(\bar{F}_i + \bar{B}_i \bar{K}_j)\bar{X}_t + \Delta \bar{F} + \bar{U}_j \right. \right.$$

$$- \sum_{q=1}^n \lambda_q \Delta \bar{D}_q^\mu \right] + \frac{1}{2} \sum_{r=1}^m (\bar{C}_{i,r}\bar{X}_t + \Delta \bar{C}_r)^T V_{\bar{X}\bar{X}} (\bar{C}_{i,r}\bar{X}_t$$

(6.18)

$$+ \Delta \bar{C}_r) + \sum_{q=1}^n \lambda_q [V(\bar{X}_t + [\bar{D}_{i,q}\bar{X}_{t^-} + \Delta \bar{D}_q]) - V(\bar{X}_t)] \right\}$$

then the fuzzy control input u_t in (6.5) stabilizes the nonlinear fuzzy MFSJD system in (6.7) in the mean square sense.

Proof: Please refer to Appendix 6.8.1.

Remark 6.4: The Lyapunov stability criterion is widely used to analyze the stability properties of dynamic systems [44, 49]. Theorem 6.1 can be regarded as the Lyapunov stability criterion of a nonlinear MFSJD system in the mean-square sense. Suppose the nonlinear MFSJD system in (6.1) can be approximated by the nonlinear fuzzy MFSJD system in (6.7). Based on the fact that $E\{x_t^T x_t\} = E\{\tilde{x}_t^T \tilde{x}_t + (Ex_t)^T (Ex_t)\} = E\{\bar{X}_t^T \bar{X}_t\}$ for all $t > 0$, if the nonlinear fuzzy augmented stochastic jump diffusion system in (6.15) can be stabilized by u_t in (6.5), we also have $\lim_{t \to \infty} E\{x_t^T x_t\} = 0$, which means the nonlinear MFSJD system can be stabilized by u_t in the mean square sense.

6.4 MULTI-OBJECTIVE H_2/H_∞ CONTROL DESIGN FOR NONLINEAR FUZZY MFSJD SYSTEMS

The concurrent optimization problem with respect to H_2 and H_∞ performance is a dynamically constrained MOP defined as follows:

Definition 6.4: The multi-objective H_2/H_∞ control design for the nonlinear MFSJD system is defined as

$$\min_{u(t) \in U_F} (J_2(u_t), J_\infty(u_t))$$

$$\text{subject to (6.15)}$$

(6.19)

where U_F is the set of all the admissible control laws.

For a multi-objective H_2/H_∞ control design problem, there exists no unique solution u_t corresponding to both the minimum values of the H_2 and H_∞ performance. Rather, there exists a set of feasible optimal solutions for the multi-objective H_2/H_∞ control design problem in the Pareto optimal sense. The "optimal" control design of the MOP for a dynamic system can be seen as how to select a preferred solution from the Pareto front to obtain the Pareto optimal solution. In general, for an MOP, the domination concept is employed to decide which admissible control law is better (i.e., we use domination to define the "minimization" in the MOP in (6.19)). Thus, the solution of the MOP in (6.19) is a set of admissible control laws with a nondominated objective vector. The definition of domination is given as follows.

Definition 6.5 ([15]): For a multi-objective H_2/H_∞ control problem, an objective vector (α, β) of the control law u_t is said to dominate another objective vector $(\dot{\alpha}, \dot{\beta})$ of the control law $\dot{u}_t$ if and only if both the following conditions are true: (I) $\alpha \leq \dot{\alpha}$ and $\beta \leq \dot{\beta}$, and (II) at least one of the two inequalities in (I) with strict inequality. Moreover, the admissible control law u_t is regarded as a better solution than $\dot{u}_t$.

Unfortunately, there is no efficient method to directly solve the control u_t of the MOP in (6.19). Thus, we introduce an indirect method to help us solve the MOP in (6.19).

Theorem 6.2: Suppose α and β are the upper bounds of the H_2 and H_∞ performance, respectively; that is, $J_2(u_t) \leq \alpha$ and $J_\infty(u_t) \leq \beta$. The MOP in (6.19) is equivalent to the following MOP:

$$\min_{u(t) \in U_F} (\alpha, \beta) \tag{6.20}$$
$$\text{subject to } J_2(u_t) \leq \alpha \text{ and } J_\infty(u_t) \leq \beta$$

Proof: Please refer to Appendix 6.8.2.

To transform the MOP in (6.20) into an LMI-constrained MOP, we need the following lemma.

Lemma 6.2 ([40]): For any two real-valued vectors $A \in \mathbb{R}^{n_x}$ and $B \in \mathbb{R}^{n_x}$, we have

$$A^T B + B^T A \leq \gamma^2 A^T A + \frac{1}{\gamma^2} B^T B \tag{6.21}$$

where γ is any nonzero real number.

With the help of the T-S fuzzy model and Theorem 6.1, the MOP in (6.20) for the nonlinear MFSJD system in (6.1) can be transformed into an LMI-constrained MOP as follows.

Theorem 6.3: If the following LMI-constrained MOP can be solved, then the multi-objective H_2/H_∞ fuzzy control design problem in (6.19) for the nonlinear MFSJD systems in (6.1) can be solved

$$\min_{\zeta = \{W_1, \{Y_j^1\}_{j=1}^i\}, W_2, \{Y_j^2\}_{j=1}^i\}} \{\alpha, \beta\} \tag{6.22}$$
$$\text{subject to LMIs in } (6.23) - (6.27)$$

$$diag(W_1, W_2) \geq \alpha^{-1} Tr\{R_{\bar{x}_0}\} I_{2n_x} \tag{6.23}$$

$$\begin{bmatrix} \tilde{\Psi}_{H_2}^{i,j} & (Y_j^1)^T & W_1 & W_1 C_{i,1}^T & \cdots & W_1 C_{i,m}^T & W_1 D_{i,1}^T & \cdots & W_1 D_{i,n}^T \\ * & -\breve{R}_1^{-1} & 0 & 0 & \cdots & 0 & 0 & \cdots & 0 \\ * & * & -\breve{Q}_1^{-1} & 0 & \cdots & 0 & 0 & \cdots & 0 \\ * & * & * & \frac{-W_1}{2} & \cdots & 0 & 0 & \cdots & 0 \\ * & * & * & * & \ddots & 0 & 0 & \cdots & 0 \\ * & * & * & * & * & \frac{-W_1}{2} & 0 & \cdots & 0 \\ * & * & * & * & * & * & \frac{-W_1}{2\lambda_1} & \cdots & 0 \\ * & * & * & * & * & * & * & \ddots & 0 \\ * & * & * & * & * & * & * & * & \frac{-W_1}{2\lambda_n} \end{bmatrix} \leq 0 \tag{6.24}$$

$$
\begin{bmatrix}
\hat{\Psi}_{H_2}^{i,j} & (Y_j^2)^T & W_2 & W_2 \widehat{C}_{i,1}^T & \cdots & W_2 \widehat{C}_{i,m}^T & W_2 \widehat{D}_{i,1}^T & \cdots & W_2 \widehat{D}_{i,n}^T \\
* & -(\breve{R}_1^\mu)^{-1} & 0 & 0 & \cdots & 0 & 0 & \cdots & 0 \\
* & * & -(\breve{Q}_1^\mu)^{-1} & 0 & \cdots & 0 & 0 & \cdots & 0 \\
* & * & * & \frac{-W_1}{2} & \cdots & 0 & 0 & \cdots & 0 \\
* & * & * & * & \ddots & 0 & 0 & \cdots & 0 \\
* & * & * & * & * & \frac{-W_1}{2} & 0 & \cdots & 0 \\
* & * & * & * & * & * & \frac{-W_1}{2\lambda_1} & \cdots & 0 \\
* & * & * & * & * & * & * & \ddots & 0 \\
* & * & * & * & * & * & * & * & \frac{-W_1}{2\lambda_n}
\end{bmatrix} \leq 0
\tag{6.25}
$$

$$
\begin{bmatrix}
\breve{\Psi}_{H_\infty}^{i,j} & (Y_j^1)^T & W_1 & G_i & W_1 C_{i,1}^T & \cdots & W_1 C_{i,m}^T & W_1 D_{i,1}^T & \cdots & W_1 D_{i,n}^T \\
* & -(\breve{R}_2)^{-1} & 0 & 0 & 0 & \cdots & 0 & 0 & \cdots & 0 \\
* & * & -(\breve{Q}_2)^{-1} & 0 & 0 & \cdots & 0 & 0 & \cdots & 0 \\
* & * & * & \frac{-\beta}{(1+\delta)}I & 0 & \cdots & 0 & 0 & \cdots & 0 \\
* & * & * & * & \frac{-W_1}{2} & \cdots & 0 & 0 & \cdots & 0 \\
* & * & * & * & * & \ddots & 0 & 0 & \cdots & 0 \\
* & * & * & * & * & * & \frac{-W_1}{2} & 0 & \cdots & 0 \\
* & * & * & * & * & * & * & \frac{-W_1}{2\lambda_1} & \cdots & 0 \\
* & * & * & * & * & * & * & * & \ddots & 0 \\
* & * & * & * & * & * & * & * & * & \frac{-W_1}{2\lambda_n}
\end{bmatrix} \leq 0
\tag{6.26}
$$

$$
\begin{bmatrix}
\hat{\Psi}_{H_\infty}^{i,j} & (Y_j^2)^T & W_2 & G_i & W_2 \hat{C}_{i,1}^T & \cdots & W_2 \widehat{C}_{i,m}^T & W_2 \widehat{D}_{i,1}^T & \cdots & W_2 \widehat{D}_{i,n}^T \\
* & -(\breve{R}_2^\mu)^{-1} & 0 & 0 & 0 & \cdots & 0 & 0 & \cdots & 0 \\
* & * & -(\breve{Q}_2^\mu)^{-1} & 0 & 0 & \cdots & 0 & 0 & \cdots & 0 \\
* & * & * & \frac{-\beta}{(1+\delta)}I & 0 & \cdots & 0 & 0 & \cdots & 0 \\
* & * & * & * & \frac{-W_1}{2} & \cdots & 0 & 0 & \cdots & 0 \\
* & * & * & * & * & \ddots & 0 & 0 & \cdots & 0 \\
* & * & * & * & * & * & \frac{-W_1}{2} & 0 & \cdots & 0 \\
* & * & * & * & * & * & * & \frac{-W_1}{2\lambda_1} & \cdots & 0 \\
* & * & * & * & * & * & * & * & \ddots & 0 \\
* & * & * & * & * & * & * & * & * & \frac{-W_1}{2\lambda_n}
\end{bmatrix} \leq 0
\tag{6.27}
$$

where

$$\tilde{\Psi}_{H_2}^{i,j} \triangleq A_i W_1 + W_1 A_i^T + B_i Y_j^1 + (B_i Y_j^1)^T + \sum_{q=1}^{n} \lambda_q (D_{i,q} W_1 + W_1 D_{i,q}^T) + 2I_{n_x}$$
$$+ \sum_{q=1}^{n} \lambda_q W_1$$

$$\hat{\Psi}_{H_2}^{i,j} \triangleq \hat{A}_i W_2 + W_2 \hat{A}_i^T + B_i Y_j^2 + (B_i Y_j^2)^T + \sum_{q=1}^{n} \lambda_q (\hat{D}_{i,q} W_2 + W_2 \hat{D}_{i,q}^T) + 2I_{n_x}$$
$$+ \sum_{q=1}^{n} \lambda_q W_2$$

$$\tilde{\Psi}_{H_\infty}^{i,j} \triangleq \tilde{\Psi}_{H_2}^{i,j} + \frac{1}{\beta}(1+\delta)I_{n_x}, \quad \hat{\Psi}_{H_\infty}^{i,j} \triangleq \hat{\Psi}_{H_2}^{i,j} + \frac{1}{\beta}(1+\delta)I_{n_x}, \quad \breve{R}_1 \triangleq (R_1 + \varepsilon_2 I_{n_u}),$$

$$\breve{Q}_1 \triangleq (Q_1 + \tilde{\Delta} I_{n_x}), \quad \breve{R}_1^\mu \triangleq (R_1^\mu + \varepsilon_2 I_{n_x}), \quad \breve{Q}_1^\mu \triangleq (Q_1^\mu + \hat{\Delta} I_{n_x}), \quad R_{\bar{X}_0} \triangleq E\{\bar{X}_0 \bar{X}_0^T\},$$

$$\Gamma \triangleq \alpha TR\{R_{\bar{X}_0}\}^{-1}, \quad \hat{\Delta} \triangleq s(\varepsilon_1 + \varepsilon_1^\mu) + \sum_{q=1}^{n} 3\Gamma \lambda_q (\varepsilon_{4,q} + \varepsilon_{4,q}^\mu) + \sum_{r=1}^{m} 2\Gamma (\varepsilon_{3,r} + \varepsilon_{3,r}^\mu),$$

and $\tilde{\Delta} \triangleq \varepsilon_1 + \sum_{r=1}^{m} 2\Gamma \varepsilon_{3,r} + \sum_{q=1}^{n} 3\Gamma \lambda_q \varepsilon_{4,q}$.

Proof: Please refer to Appendix 6.8.3.

Definition 6.6 ([11, 131]): For the given LMI-constrained MOP in (6.22), the given vector (α, β) is a feasible objective vector if there exists a feasible solution

$$\zeta_{(\alpha,\beta)} = \{W_1, \{Y_j^1\}_{j=1}^l, W_2, \{Y_j^2\}_{j=1}^l\} \in \mho \qquad (6.28)$$

that is satisfied by the LMIs in (6.23)–(6.27) such that the corresponding fuzzy controlled input

$$u_t = \sum_{j=1}^{l} h_j(\xi_t)[K_j^1 x_t + (K_j^2 - K_j^1)Ex_t] \qquad (6.29)$$

with $K_j^1 = Y_j^1 W_1^{-1}$ and $K_j^2 = Y_j^2 W_2^{-1}$ can guarantee $(J_2(u_t), J_\infty(u_t)) \le (\alpha, \beta)$.

Theorem 6.4: For the nonlinear fuzzy MFSJD system in (6.7), if (i) the external noise $v_t = 0$ and (ii) $\zeta = \{W_1, \{Y_j^1\}_{j=1}^l, W_2, \{Y_j^2\}_{j=1}^l\}$ is a feasible solution of the MOP in (6.22), then the fuzzy control input $u_t = \sum_{j=1}^{l} h_j(\xi_t)[K_j^1 x_t + (K_j^2 - K_j^1)Ex_t]$ stabilizes the nonlinear MFSJD system in (6.1) exponentially in the mean square sense.

Proof: Please refer to Appendix 6.8.4.

Definition 6.7 ([131]): A feasible solution $\zeta_{(\alpha,\beta)} = \{W_1, \{Y_j^1\}_{j=1}^l, W_2, \{Y_j^2\}_{j=1}^l\}$ is said to be Pareto optimal with respect to the given LMI-constrained MOP in (6.22) if and only if there is no other feasible objective vector $(\dot{\alpha}, \dot{\beta})$ associated with a feasible solution $\zeta_{(\dot{\alpha},\dot{\beta})} = \{\dot{W}_1, \{\dot{Y}_j^1\}_{j=1}^l, \dot{W}_2, \{\dot{Y}_j^2\}_{j=1}^l\}$ such that (α, β) is dominated by $(\dot{\alpha}, \dot{\beta})$.

For the multi-objective H_2/H_∞ fuzzy control design problem, it is obvious that if a feasible objective vector (α, β) is said to dominate another feasible objective vector $(\dot{\alpha}, \dot{\beta})$, the feasible solution $\zeta_{(\alpha,\beta)} = \{W_1, \{Y_j^1\}_{j=1}^l, W_2, \{Y_j^2\}_{j=1}^l\}$ is better than $\zeta_{(\dot{\alpha},\dot{\beta})} = \{\dot{W}_1, \{\dot{Y}_j^1\}_{j=1}^l, \dot{W}_2, \{\dot{Y}_j^2\}_{j=1}^l\}$ in the Pareto optimal sense.

Definition 6.8 ([131]): For the given LMI-constrained MOP in (6.22), the Pareto optimal set P_{set}^* is defined as

$$P_{set}^* \triangleq \{\zeta_{(\alpha,\beta)} \mid \text{there does not exist another feasible solution } \zeta_{(\dot{\alpha},\dot{\beta})} \text{ such that}$$
$$(\dot{\alpha},\dot{\beta}) \text{ dominates } (\alpha,\beta)\}$$

Definition 6.9 ([131]): For the given LMI-constrained MOP in (6.22), the Pareto front P_{front}^* is defined as

$$P_{front}^* \triangleq \{(\alpha,\beta) \mid \zeta_{(\alpha,\beta)} \in P_{set}^*\}$$

Once the Pareto front is obtained, the corresponding "optimal" fuzzy control gains $\{K_j^1\}_{j=1}^l$ and $\{K_j^2\}_{j=1}^l$ of the nonlinear MFSJD systems can also be obtained.

(1) The optimal H_2 performance $J_2(u_t^*) = \alpha^*$ and the H_2 optimal control law $u_t^* = \sum_{j=1}^l h_j(\xi_t)[K_j^{1*}x_t + (K_j^{2*} - K_j^{1*})Ex_t]$ of the nonlinear MFSJD system in (6.1) can be obtained by solving the following LMI-constrained single optimization problem:

$$\min_{\{W_1>0,\{Y_j^1\}_{j=1}^l, W_2>0,\{Y_j^2\}_{j=1}^l\}} (\alpha)$$
$$\text{s.t } (6.23)-(6.25) \text{ for all } i,j=1,2,...,l \tag{6.30}$$

(2) Replacing the term $\Gamma \triangleq \alpha Tr\{R_{\bar{x}_0}\}$ in (6.26) and (6.27) by $\rho > 0$ with $\rho I_{2n_x} \geq P$ (i.e., the upper bound of P), the optimal H_∞ robust performance $J_\infty(u_t^\dagger) = \beta^\dagger$ and the optimal H_∞ robust fuzzy control law $u_t^\dagger = \sum_{j=1}^l h_j(\xi_t)[K_j^{1\dagger}x_t + (K_j^{2\dagger} - K_j^{1\dagger})Ex_t]$ of the nonlinear MFSJD system in (6.1) can be realized by solving the following LMI-constrained SOP:

$$\min_{\{W_1>0,\{Y_j^1\}_{j=1}^l, W_2>0,\{Y_j^2\}_{j=1}^l, \rho\}} (\beta)$$
$$\text{subject to } (6.26)-(6.27) \text{ with } diag(W_1,W_2) \geq \rho^{-1}I_{(2n_x)} \tag{6.31}$$
$$\text{for all } i,j=1,2,...,l$$

6.5   FRONT-SQUEEZING LMI-CONSTRAINED MOEA

An LMI-constrained MOEA has been proposed for the LMI-constrained MOP [79, 132, 133]. Unlike the traditional MOEA with algebraic functional constraints, as in Chapter 1 and [11, 123, 134], the search region of the LMI-constrained MOEA is set in an objective space. Thus, the LMI-constrained MOEA has a high computational load.

To decrease the computational load and improve the performance of the reverse-order LMI-constrained MOEA, we proposed a front-squeezing LMI-constrained MOEA to help us solve the LMI-constrained MOP in (6.22). The Pareto front can be regarded as the boundary between the feasible and infeasible region.

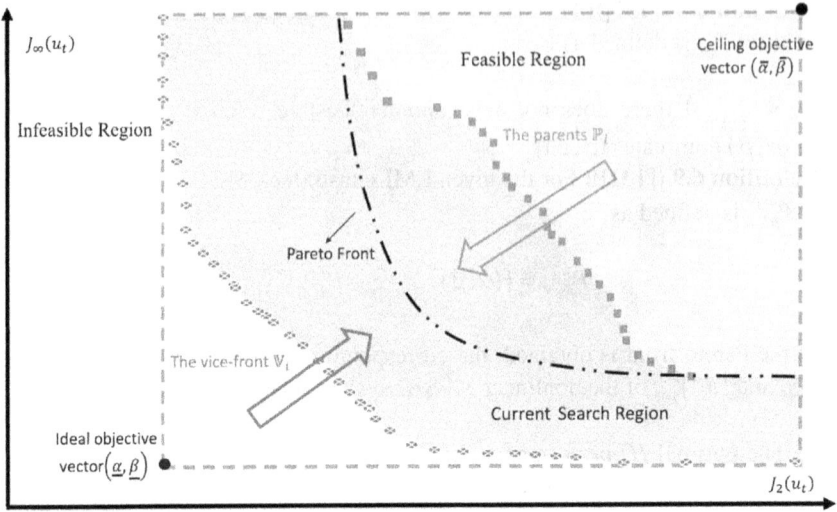

FIGURE 6.1   Front-squeezing MOEA concurrently searches for the Pareto front of an LMI-constrained MOP from the feasible and infeasible regions. Here, the dashed line is used to indicate the initial search region Γ.

The front-squeezing LMI-constrained MOEA concurrently employs the vice front V_i and the current parents P_i to approach the Pareto front from two sides of both regions shown in Figure 6.1, where the current parents P_i and the vice-front V_i are defined as the set of the parent objective vectors of the *ith* generation and the set of currently observed infeasible objective vectors that are the closest to the Pareto front in the Pareto optimal sense of the *ith* generation, respectively. As the iteration number goes up, the vice front and the current parent front will gradually move closer to each other. Eventually, they will overlap. Thus, the Pareto front can be obtained. To establish the vice front V_i, we need to introduce the concept of "inverse domination."

Definition 6.10: A given objective vector (α, β) is said to inversely dominate another objective vector $(\dot{\alpha}, \dot{\beta})$ if and only if both the following conditions are true: (I) $\alpha \geq \dot{\alpha}$ and $\beta \geq \dot{\beta}$, and (II) at least one of the two inequalities in (I) is a strict inequality.

Definition 6.11: For the multi-objective H_2/H_∞ control design of the nonlinear MFSJD system in (6.1), the ideal objective vector $(\underline{\alpha}, \underline{\beta})$ is defined as $(\underline{\alpha}, \underline{\beta}) \triangleq (\alpha^*, \beta^\dagger)$, where α^* and $\beta^\dagger$ can be obtained by solving the previous two SOPs in (6.30) and (6.31), respectively.

Design Procedure of the Multi-Objective H_2/H_∞ Control Design for the Non-linear MFSJD Systems: Based on the previous analysis, the detailed design procedure of the reverse-order LMI-constrained MOEA approach for the multi-objective H_2/H_∞ control design problem in (6.20) of the nonlinear fuzzy MFSJD system is given as follows:

Step 1: Initialization

Step 1.1 Set the initial search region $\Gamma = (\underline{\alpha}, \overline{\alpha}) \times (\underline{\beta}, \overline{\beta})$. For the multi-objective H_2/H_∞ control design problem, the ideal objective vector $(\underline{\alpha}, \underline{\beta})$ for the nonlinear fuzzy MFSJD system in (6.7) can be easily obtained by solving the two SOPs in (6.30) and (6.31), respectively.

Step 1.2 Set the maximum number of individuals N_{pop} in the EA, the iteration number N_{iter}, the crossover rate M_c, and the mutation rate M_r (In general, N_{pop} is set to be an even number; the crossover rate M_c and mutation rate M_r are set to be 0.9 and 0.1, respectively).

Step 1.3 Randomly select N_{pop} feasible objective vectors to be parent candidates z_1 from the search region Γ and set the vice front $V_1 = \{(\underline{\alpha}, \underline{\beta})\}$.

Step 2: Update: If the iteration index i with $1 \le i \le N_{iter}$.

Step 2.1 Select the $(N_{pop}/2)$ individuals from the parent candidates Z_i to be the parents P_i by using the crowded tournament selection operator.

Step 2.2 Make the parents P_i perform the crossover and the mutation operator to produce N_{pop} offspring that are bounded by vice front V_i and parents P_i are the candidate offspring Q_i (i.e., each candidate offspring should be dominated by vice front V_i but not dominated by its parents P_i)

Step 2.3 Classify the candidate offspring Q_i into feasible offspring $\hat{Q}_i$ and infeasible offspring $\tilde{Q}_i$ by examining whether the given candidate offspring (α, β) has a corresponding feasible solution $\zeta_{(\alpha,\beta)} = \{W_1, \{Y_j^1\}_{j=1}^l, W_2, \{Y_j^2\}_{j=1}^l\}$.

Step 2.4 Set the parent candidates $Z_{i+1} = Z_i \cup \hat{Q}_i$.

Step 2.5 Select at most $2N_{pop}$ elitist individuals from the set $\{V_i \cup \tilde{Q}_i\}$ to be the vice front V_{i+1} of the $(i+1)$th iteration by using the inversely dominant sort.

Step 2.6 Set the iteration index $i = i + 1$.

Step 3: Stop Criteria: Is the iteration index: $i > N_{iter}$?

Step 3.1 Once $i > N_{iter}$, stop the iteration procedure. Let current parents $P_{N_{iter}+1}$ be the Pareto front. Otherwise, go to Step 2.1.

Step 4: Select fuzzy control gains $\{K_j^1\}_{j=1}^l$ and $\{K_j^2\}_{j=1}^l$:

Step 4.1 Select a preferred-feasible objective individual $(\alpha^*, \beta^*) \in P_{front}^*$ according to the designer's preference. Once the preferred-feasible objective individual is selected, the corresponding $\zeta^* = \{W_1^*, \{Y_j^{1*}\}_{j=1}^l, W_2^*, \{Y_j^{2*}\}_{j=1}^l\} \in P_{set}^*$ is obtained. By using ζ^*, the proposed fuzzy controller $u_t^* = \sum_{j=1}^{l} h_j(\xi_t)[K_j^{1*}x_t + (K_j^{2*} - K_j^{1*})Ex_t]$ with $K_j^{1*} = Y_j^{1*}(W_1^*)^{-1}$ and $K_j^{2*} = Y_j^{2*}(W_2^*)^{-1}$ can be constructed, and the multi-objective H_2/H_∞ control problem can be solved with $J_2(u_t) = \alpha^*$ and $J_\infty(u_t) = \beta^*$ simultaneously.

Remark 6.5: The computation complexity of the proposed front-squeezing LMI-constrained MOEA is about $O(rn(n+1)MN^2)$, including $O(rn(n+1)/2)$ in solving LMIs and $O(2MN^2)$ in the MOEA, where n is the dimension of W_1 and W_2; N is the population number of MOEA; r is the number of fuzzy rules; and M is the generation number of MOEA.

6.6   SIMULATION EXAMPLE

In recent years, researchers have focused on system dynamic models to describe real economic and financial systems [11, 132, 135]. However, in practical cases, the financial dynamic system is a nonlinear stochastic system and may suffer from continuous and discontinuous parametric fluctuations due to national and international situation changes, oil price changes, the surplus between investment and savings, variable interest rates, false economy strategies, and so on. Thus, using a nonlinear stochastic dynamic model to describe a real financial system would be more appealing [136, 137]. For a financial system, the control input signal u_t can be seen as an intervention or investment strategy. With the rise of quantitative finance, the multi-objective H_2/H_∞ strategy has been introduced into stochastic financial systems. Here, we use a nonlinear stochastic MFSJD system to mimic a stock price system. Consider the following nonlinear MFSJD financial system driven by two Wiener processes $\{W_t^{(r)}\}_{r=1}^{2}$ and two marked Poisson processes $\{N_p(t,\theta_q)\}_{q=1}^{2}$:

$$dS_t = [f(S_t, ES_t) + \eta u_t + gv_t]dt + \sum_{r=1}^{2} \sigma_r(S_t, ES_t)dW_t^{(r)}$$

$$+ \sum_{q=1}^{2} T(S_{t^-}, ES_{t^-}, \theta_q)dN_p(t,\theta_q)$$

$$S_0 \in [0.04, 0.06] \times [0.045, 0.055] \times [11, 15], [8, 10] \quad\quad (6.32)$$

$$S_t = [S_1(t), S_2(t), S_3(t), S_4(t)]^T$$

$$ES_0 = [0.05, 0.05, 13, 9]^T$$

where S_t denotes the system state. The state $S_1(t)$ is the interest rate of bank A, $S_2(t)$ is the interest rate of bank B, $S_3(t)$ is the stock price of Company C, and $S_4(t)$ is the stock price of Company D. The mean term ES_t is used to denote the effects of the sample moving average. The terms $\sigma_1(S_t, ES_t)dW_t^1$ and $\sigma_2(S_t)dW_t^2$ can be regarded as two continuous intrinsic fluctuations dependent on system state S_t and its mean ES_t. Thus, $T(x_{t^-}, Ex_{t^-}, \theta_1)dN_p(t,\theta_1)$ and $T(x_{t^-}, Ex_{t^-}, \theta_2)dN_p(t,\theta_2)$ can be regarded as two discontinuous intrinsic fluctuations dependent on system states S_t and ES_t. The external disturbance vector $v_t = [0.01n_1(t), 0.01n_2(t), 0.2n_3(t), 0.2n_4(t)]^T$, where $n_1(t)$,

$n_2(t)$, $n_3(t)$, and $n_4(t)$ are random processes and used to denote external noise with mean zero variance 1 normal distribution.

Suppose that optimal H_2, optimal H_∞, and multi-objective H_2/H_∞ strategies are employed by the consortium or government to regulate the nonlinear MFSJD financial system in (6.32) to finally achieve a desired steady state S_d. To do this, the first step is to shift the origin of the nonlinear MFSJD financial system in (6.32) to S_d. Let $X_t = S_t - S_d$ and $S_d = [S_{d_1}, S_{d_2}, S_{d_3}, S_{d_4}]^T = [0, 0.025, 10, 7]^T$. Then we get the following nonlinear MFSJD financial regulation system:

$$
\begin{aligned}
dX_1 &= [-0.02X_1 + b_1u_1 + g_1v_1]dt + (0.01X_1 + 0.001X_3)dW_1 \\
&\quad + 0.05X_1dN_p(t, \theta_1) \\[4pt]
dX_2 &= [\tfrac{1}{c}(0.2 - (X_2))(X_4) - \tfrac{0.01}{c}(EX_2) + b_2u_2 + g_2v_2]dt \\
&\quad + 0.01(X_2)dW_1 + [0.001X_1 + 0.001X_4]dW_2 \\
&\quad + 0.05X_2dN_p(t, \theta_2) \\[4pt]
dX_3 &= [-5(X_2)(X_4) + 5\sin(X_3)X_1 + 0.15(X_3) \\
&\quad + 0.001(EX_3)(EX_4) + b_3u_3 + g_3v_3]dt \\
&\quad + [0.05\sin(X_3)(X_3) + 0.007EX_3(EX_4)]dW_1 \\
&\quad - 0.05[EX_3 + X_3]dN_p(t, \theta_1) \\[4pt]
dX_4 &= [-0.01(X_4)^2 + 0.5(X_3) - 2X_1 - 0.15(X_4) \\
&\quad + 0.002(EX_4)EX_4 + b_4u_4 + g_4v_4]dt + (0.025X_2 \\
&\quad + 0.03X_4)dW_2 + (0.0025EX_2 + 0.003EX_4)dW_2 \\
&\quad - 0.05[EX_4 + X_4]dN_p(t, \theta_2)
\end{aligned}
\tag{6.33}
$$

Now the consortium or government wants to regulate the nonlinear MFSJD financial system in (6.33) after Day 15 with $EX_{15} = [0.0415, 0.0404, -1.9984, -0.1658]^T$ and

$$
R_{\tilde{x}_{15}} = E\{\tilde{x}_{15}, \tilde{x}_{15}^T\} =
\begin{bmatrix}
0.0001 & 0.00001 & 0.0026 & 0.0048 \\
* & 0.000014 & -0.0011 & -0.0011 \\
* & * & 1.0588 & 1.5367 \\
* & * & * & 3.5242
\end{bmatrix}
$$

To reduce the design effort and computational complexity, the fuzzy rules of the fuzzy system are used as little as possible. Let the premise variables ξ_t be $\xi_t = [\xi_1, \xi_2, \xi_3]^T = [X_3, X_4, EX_4]^T$. We use 28 rules to approximate the nonlinear MFSJD financial system in (6.33). The fuzzy operation points for the nonlinear MFSJD financial regulation system in (6.33) are given as follows:

$$
\begin{aligned}
&X_3^{(1)} = -4, X_3^{(2)} = -2.67, X_3^{(3)} = -1.33, X_3^{(4)} = 0, \\
&X_3^{(5)} = 1.33, X_3^{(6)} = 2.67, X_3^{(7)} = 4, X_4^{(1)} = -8, \\
&X_4^{(2)} = 8, EX_4^{(1)} = -6, EX_4^{(2)} = 6
\end{aligned}
\tag{6.34}
$$

The *ith* rule of this T-S fuzzy model for the nonlinear financial MFSJD regulation system in (6.33) is described as

System Rule $i = 4(i_1 - 1) + 2(i_2 - 1) + i_3$, for $i_1 \in \{1,...,7\}, i_2, i_3 \in \{1,2\}$:

If $\xi_1 = X_3$ is $G_{i_1,1}$ and $\xi_2 = X_4$ is $G_{i_2,2}, ..., \xi_3 = X_4$ is $G_{i_3,3}$

then

$$dX_t = (A_i X_t + A_i^\mu EX_t + B_i u_t + G_i v_t) dt \qquad (6.35)$$

$$+ \sum_{r=1}^{m} [C_{i,r} X_t + C_{i,r}^\mu X_t] dW_t^{(r)}$$

$$+ \sum_{q=1}^{n} [D_{i,q} X_{t^-} + D_{i,q}^\mu X_{t^-}] dN_p(t, \theta_q)$$

where

$$A_i = \begin{bmatrix} -\frac{1}{50} & 0 & 0 & 0 \\ 0 & \frac{-X_4^{(i_2)}}{c} & 0 & \frac{0.2}{c} \\ 5\sin(X_3^{(i_1)}) & -5X_4^{(i_2)} & \frac{3}{20} & 0 \\ -2 & 0 & \frac{1}{2} & -(\frac{15+X_4^{(i_2)}}{100}) \end{bmatrix}, A_i^\mu = \begin{bmatrix} 0 & 0 & 0 & 0 \\ -\frac{0.01}{c} & 0 & 0 & 0 \\ 0 & 0 & \frac{EX_4^{(i_3)}}{200} & 0 \\ 0 & 0 & 0 & \frac{EX_4^{(i_3)}}{400} \end{bmatrix},$$

$$C_{i,1} = \begin{bmatrix} \frac{1}{100} & 0 & \frac{1}{1000} & 0 \\ 0 & \frac{1}{100} & 0 & 0 \\ 0 & 0 & \frac{\sin(X_3^{(i_1)})}{20} & 0 \\ 0 & 0 & 0 & 0 \end{bmatrix}, C_{i,1}^\mu = \begin{bmatrix} 0 & 0 & 0 & 0 \\ 0 & 0 & 0 & 0 \\ 0 & 0 & \frac{7EX_4^{(i_3)}}{1000} & 0 \\ 0 & 0 & 0 & 0 \end{bmatrix},$$

$B_i = diag(\frac{1}{200}, \frac{1}{100}, \frac{1}{2}, \frac{1}{2})$, $G_i = diag(0.01, 0.01, 0.2, 0.2)$,

$C_{i,2} = diag(0,0,0.025,0.03), C_{i,2}^\mu = diag(0,0,0.0025,0.003)$,

$D_{i,1} = diag(0.05,0,-0.05,0), D_{i,1}^\mu = diag(0,0,-0.05,0)$,

$D_{i,2} = diag(0,0.05,0,-0.05)$ and $D_{i,2}^\mu = diag(0,0,0,-0.05)$.

The *ith* rule control law and the overall fuzzy control input of the nonlinear financial regulation MFSJD system are described in (6.4) and (6.5), respectively.

The weighting matrices and upper bounds of the fuzzy approximation errors are given as follows:

$Q_1 = Q_2 = diag(50,100,0.001,0.001)$, $Q_1^\mu = Q_2^\mu = diag(500,500,0.05,0.05)$

$R_1 = R_2 = diag(0.05,0.1,0.01,0.01)$, $R_1^\mu = R_2^\mu = diag(0.078,0.078,0.082,0.082)$

$\varepsilon_1 = 0.3, \varepsilon_1^\mu = 0.2, \varepsilon_2 = 0, \delta = 0,\ \varepsilon_{3,r} = 0.05, \varepsilon_{3,r}^\mu = 0.01$ for $r = 1, 2, .., m$ and

$\varepsilon_{4,q} = 0.02, \varepsilon_{4,q}^\mu = 0.01$, for $q = 1, 2, .., n$

Now the multi-objective H_2/H_∞ strategy problem for the nonlinear MFSJD system in (6.33) can be formulated as follows:

$$\min_{u_t \in U_F}(J_2(u_t), J_\infty(u_t)) \text{ subject to (6.33)} \tag{6.36}$$

By using Lemma 6.2, we can transform the previous MOP into the LMI-constrained MOP in (6.22). The parameters of the proposed front-squeezing LMI-constrained MOEA in the design procedure are given as follows:

$$\Gamma \triangleq [109.9, 200] \times [0.5981, 1],\ N_{pop} = 100,\ N_{iter} = 45,\ M_c = 0.9,\ M_r = 0.1$$

where $J_2(u_t^*) = \underline{\alpha} = 109.9$ and $J_\infty(u_t^\dagger) = \underline{\beta} = 0.5981$.

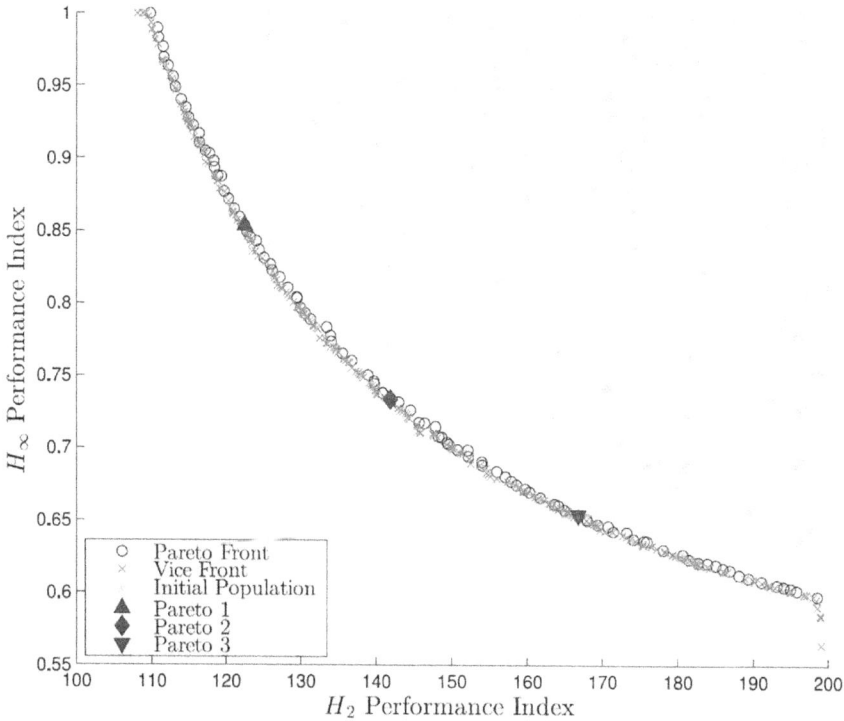

FIGURE 6.2 This figure shows the Pareto fronts P_{front}^* of the MOP in (6.36) by the proposed front-squeezing LMI-constrained MOEA. The Pareto front is denoted by a ○. The three chosen Pareto solutions are denoted by ▲, ◆, ▼, respectively. The initial population is denoted by *, and the vice front is denoted by ×.

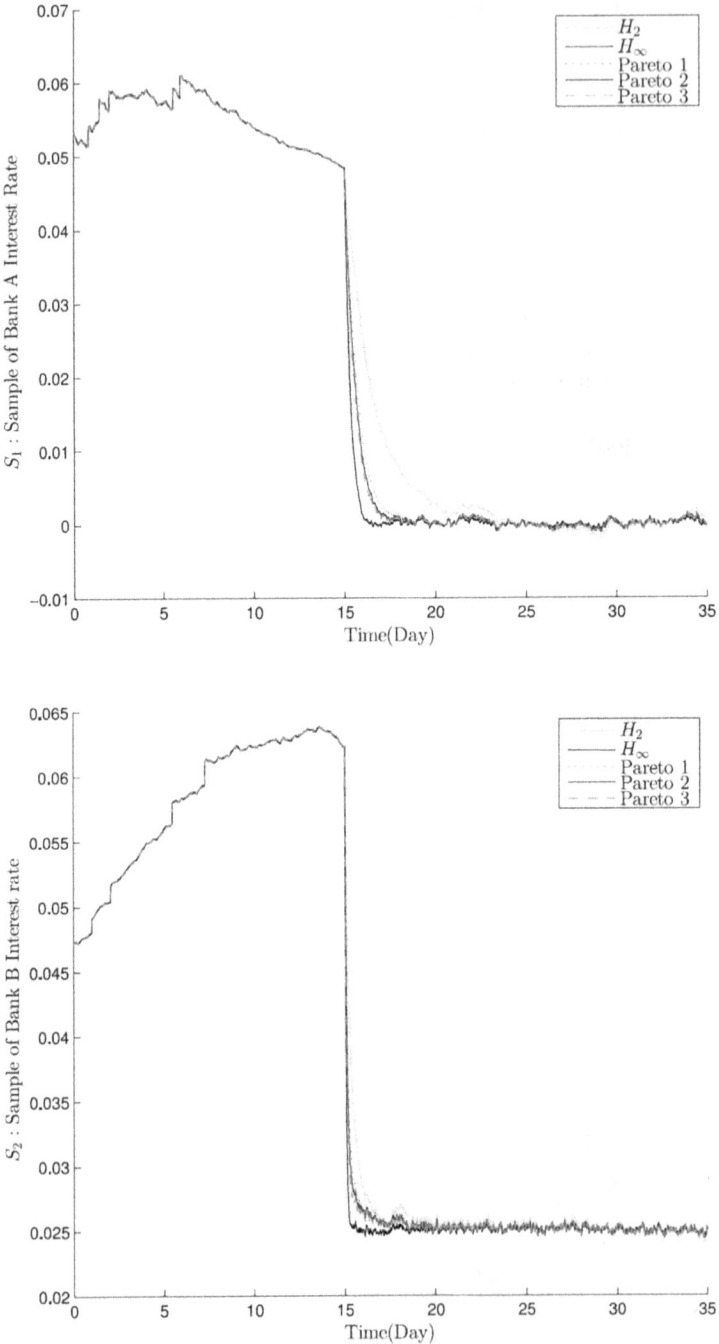

FIGURE 6.3 Trajectories of the system state in (6.33) for the three chosen Pareto solutions. (a) Trajectories of S_1 in (6.33). (b) Trajectories of S_2 in (6.33). (c) Trajectories of S_3 in (33). (d) Trajectories of S_4 in (6.33). (e) Trajectories of the marked Poisson processes in (6.33).

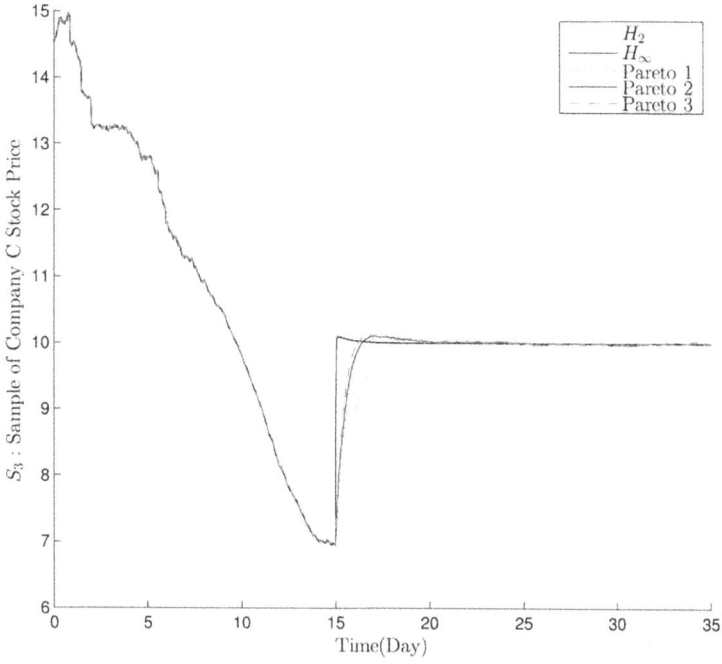

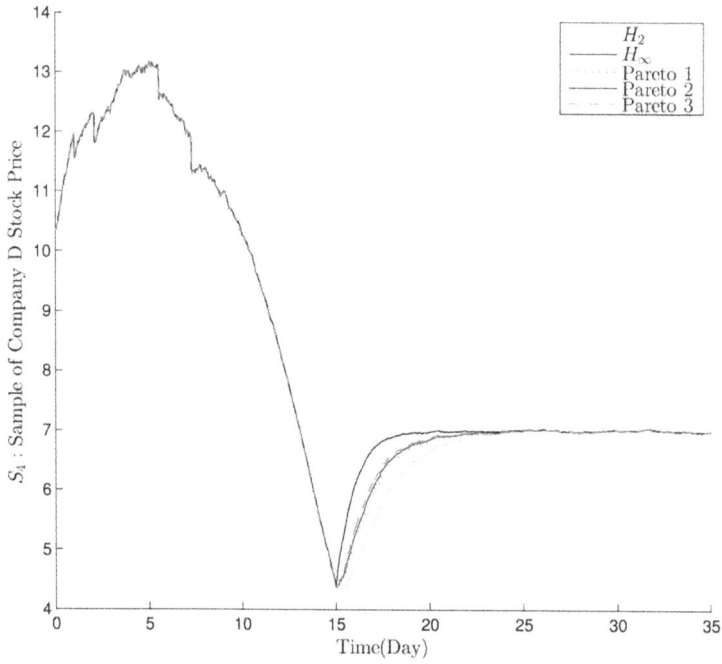

FIGURE 6.3   (Continued)

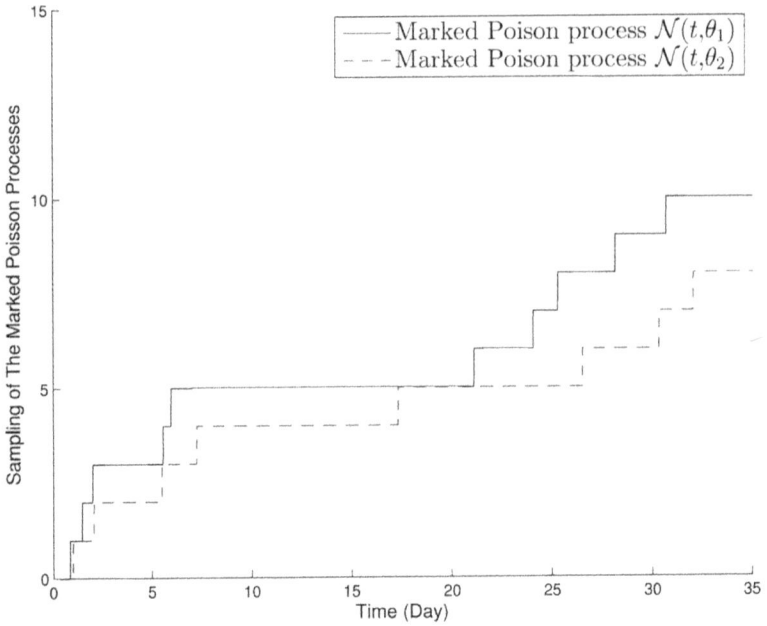

FIGURE 6.3 (Continued)

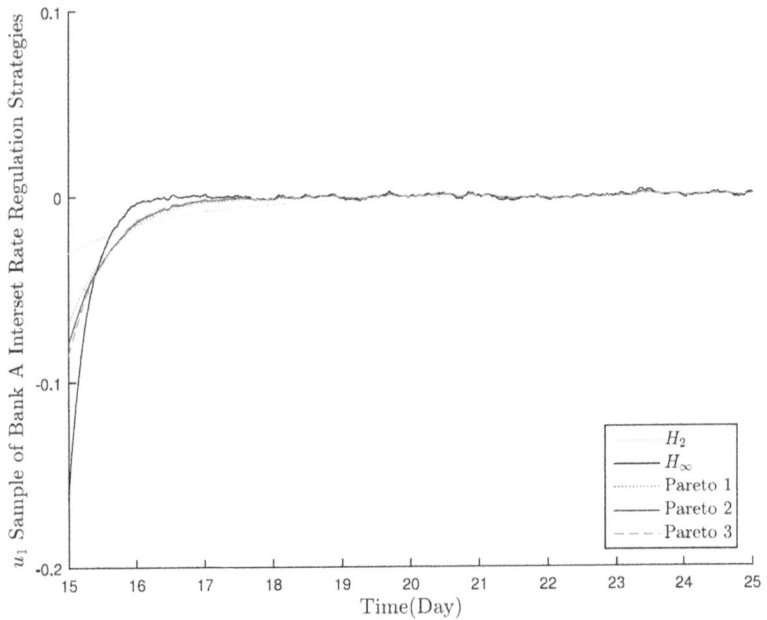

FIGURE 6.4 Profiles of fuzzy control input u_t for four chosen Pareto solutions. (a) Profile of fuzzy control input u_{1t}. (b) Profile of fuzzy control input u_{2t}. (c) Profile of fuzzy control input u_{3t}. (d) Profile of fuzzy control input u_{4t}.

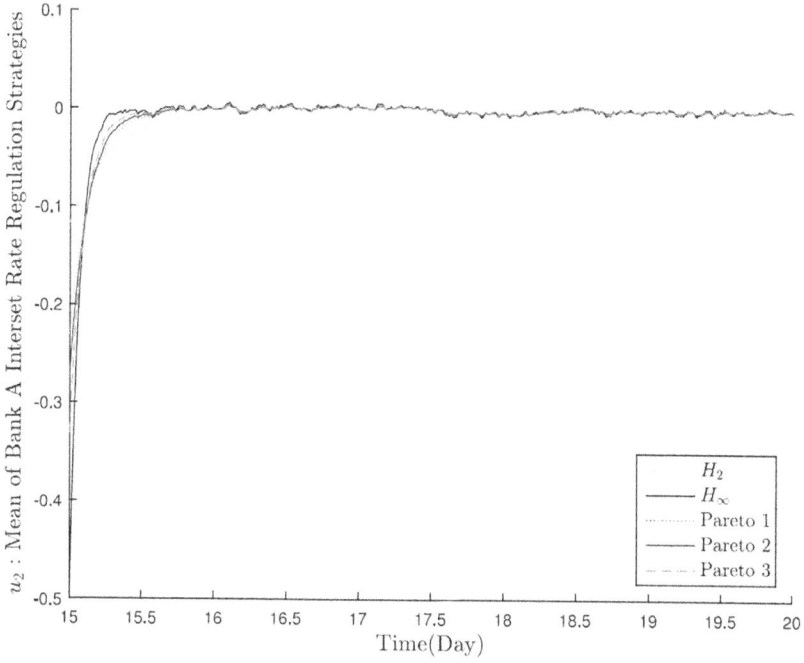

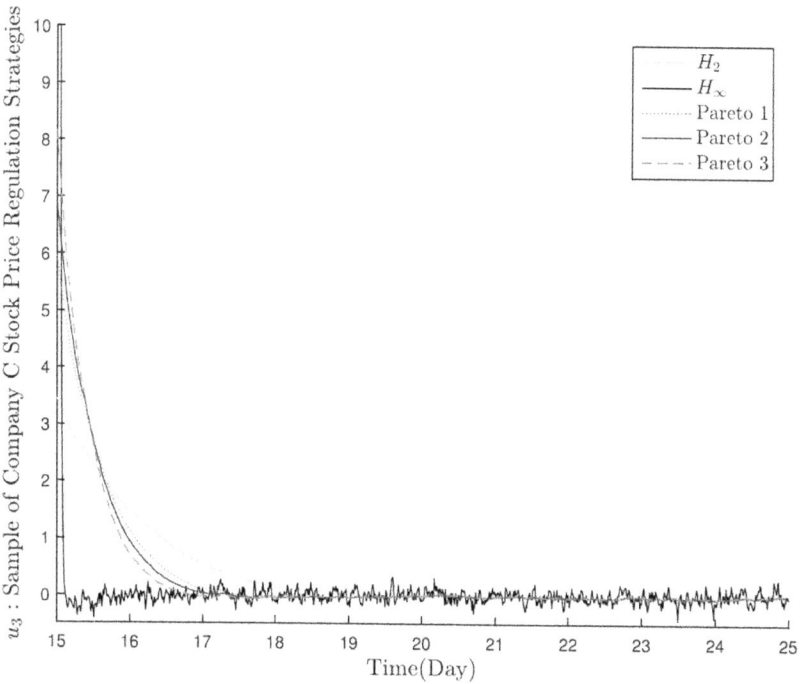

FIGURE 6.4   (Continued)

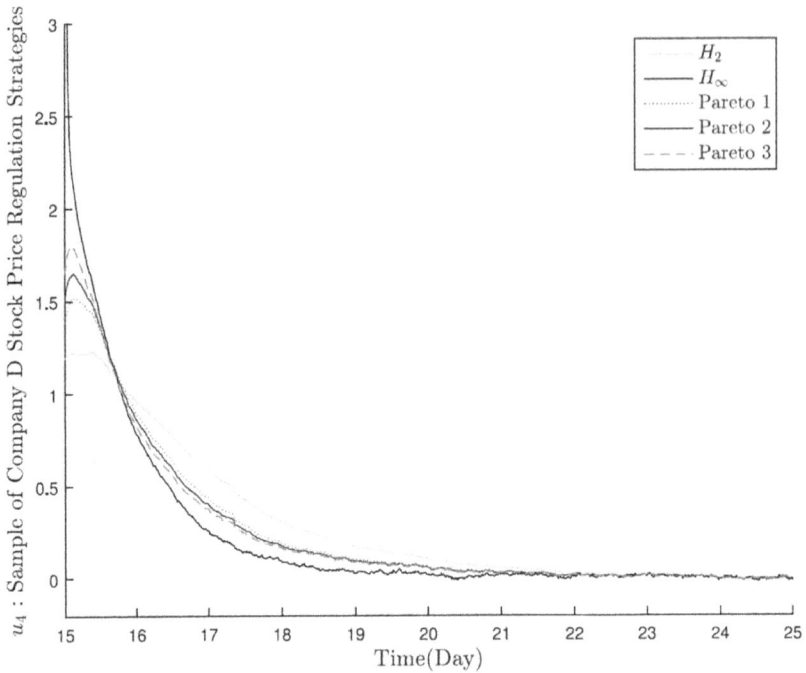

FIGURE 6.4 (Continued)

Once the iteration number N_{iter} is achieved, the Pareto front P^*_{front} of the MOP for the nonlinear MFSJD system in (6.33) can be obtained as shown in Figure 6.2.

We choose three Pareto optimal strategies from the Pareto front to compare their performance as follows:

Pareto Strategy 1	Pareto Strategy 2	Pareto Strategy 3
(122.5, 0.8535)	(141.8, 0.7336)	(166.7, 0.6541)

Detailed information about the three chosen Pareto strategies is given in Appendix 6.8.5. Figures 6.3 and 6.4 present the simulation results for the optimal H_2, optimal H_∞, and three multi-objective H_2/H_∞ fuzzy strategies of the nonlinear MFSJD financial system. It is clear that the optimal H_∞ robust intervention strategy has the smallest J_∞ performance but spends the greatest funds to regulate the stock price to the desired steady state with the minimal time to accomplish it. The optimal H_2 intervention strategy has the smallest J_2 performance and tends to regulate the stock price to the desired steady state via minimal funds. However, it cannot suffer a volatile change of the stock market. Consider the three chosen Pareto optimal strategies. Pareto optimal strategy 3 is lopsided on H_∞ performance J_∞. Pareto optimal strategy 1 is lopsided on H_2 performance J_2, and Pareto optimal strategy 2 is a relatively balanced intervention strategy with respect to J_2 and J_∞. For the three chosen strategies,

Pareto optimal strategy 3 is most robust; that is, this intervention strategy has the minimal vibration when the stock market has drastic intrinsic and external changes. On the other hand, Pareto optimal strategy 1 concurrently has minimal J_2 performance and the best J_∞ performance of the three chosen Pareto optimal strategies, so it may suffer slight intrinsic and extrinsic changes of the stock market. It is easy to observe that Pareto strategy 2 is the preferred solution of the three chosen Pareto strategies because Pareto optimal strategy 2 provides a compromise between J_2 and J_∞ to achieve the desired steady state robustly.

Remark 6.6: The computational complexity of the conventional LMI-constrained MOEA in [11] is $O(\frac{1}{2}rn(n+1)MN^2)$. Thus, the proposed front-squeezing LMI-constrained MOEA has a higher computation complexity (see Remark 6.5). However, it costs less time to approach the Pareto front by concurrently narrowing the search region down via the current parent front and vice front, as shown in Figure 6.1. The front-squeezing LMI-constrained MOEA spends 72 h and 53 min to obtain the Pareto front in Figure 6.2. However, the conventional LMI-constrained MOEA in [11] spends 162 h and 19 mins to obtain a similar results. It is clear that the front-squeezing LMI-constrained MOEA has better computational efficiency.

6.7   CONCLUSION

This chapter introduced the multi-objective H_2/H_∞ fuzzy control design problem of nonlinear MFSJD systems for concurrently minimizing H_2 and H_∞ performance of a nonlinear MFSJD system in the Pareto optimal sense. The multi-objective H_2/H_∞ control design problem for nonlinear MFSJD systems is an MOP constrained by a nonlinear MFSJD system. Based on the decoupling technique, we can decouple the nonlinear fuzzy MFSJD system into two orthogonal subsystems (i.e., a mean Ex_t subsystem and a variation $\tilde{x}_t$ subsystem) and obtain a corresponding fuzzy augmented stochastic jump diffusion system in (6.15). By introducing the Itô-Lévy formula, sufficient conditions for the stability of nonlinear fuzzy MFSJD systems can be derived. By using the T-S fuzzy model and the proposed indirect method, the dynamically constrained MOP in (6.19) can be replaced by the LMI-constrained MOP in (6.22). Now the LMI-constrained MOP can be efficiently solved by the proposed front-squeezing LMI-constrained MOEA. As the iteration number grows, the Pareto front of the nonlinear MFSJD system can be obtained by a rapid two-sided approach scheme from feasible and infeasible regions. When the Pareto front is obtained, the designer can select a preferred fuzzy controller from the set of Pareto optimal controllers according to their own preference and complete the multi-objective H_2/H_∞ fuzzy control design. Finally, we give a simulation example to illustrate the design procedure and confirm the results through a computer simulation.

6.8   APPENDIX

6.8.1   PROOF OF THEOREM 6.1

Based on Lemma 6.1, the Itô-Lévy formula of the augmented fuzzy MFSJD control system can be represented as

$$dV(\bar{X}_t) = \sum_{i=1}^{l}\sum_{j=1}^{l} h_i(\xi_t)h_j(\xi_t)\Big(\Big\{V_{\bar{X}}^T\Big[(\bar{F}_i + \bar{B}_i\bar{K}_j)\bar{X}_t$$

$$+\Delta\bar{F} + \Delta\bar{U}_j - \sum_{q=1}^{n}\lambda_q\Delta\bar{D}_q^\mu\Big] + \tfrac{1}{2}\sum_{r=1}^{m}(\bar{C}_{i,r}\bar{X}_t + \Delta\bar{C}_r)^T$$

$$\cdot V_{\bar{X}\bar{X}}(\bar{C}_{i,r}\bar{X}_t + \Delta\bar{C}_r)\Big\}dt + \sum_{r=1}^{m}(\bar{C}_{i,r}\bar{X}_t + \Delta\bar{C}_r)dW_t^{(r)}$$

$$+\sum_{q=1}^{n}\Big[V(\bar{X}_t + [\bar{D}_{i,q}\bar{X}_{t^-} + \Delta\bar{D}_q]) - V(\bar{X}_t)\Big]dN_p(t,\theta_q)\Big)$$

Condition (i) in Theorem 6.1 implies that

$$E\{dV(\bar{X}_t)\} = E\{LV(\bar{X}_t)\}dt \le -m_1 E\{\bar{X}_t^T \bar{X}_t\}dt < 0$$

By using condition (ii) in Theorem 6.1, we get

$$dE\{V(\bar{X}_t)\} \le \tfrac{-m_1}{m_3} E\{V(\bar{X}_t)\}dt < 0$$

Since $E\left\{\|\bar{X}_t\|_2^2\right\} = E\{\tilde{x}_t^T \tilde{x}_t + Ex_t^T x_t\} = E\{x_t^T x_t\}$, we have $\lim_{t\to\infty} E\left\{\|\bar{X}_t\|_2^2\right\} = 0$.

6.8.2 PROOF OF THEOREM 6.2

Please refer to [11].

6.8.3 PROOF OF THEOREM 6.3

Let $V(\bar{X}_t) = \tilde{x}_t^T P_1\tilde{x}_t + Ex_t^T P_2 Ex_t = \bar{X}_t^T P\bar{X}_t$ with $P = diag(P_1, P_2) > 0$ be the chosen Lyapunov function for the augmented fuzzy MFSJD control system in (6.15).

Suppose $E\{\bar{X}_0^T P\bar{X}_0\} \le \alpha$; we have

$$E\{\bar{X}_0^T P\bar{X}_0\} \le \bar{\sigma}(P)Tr\{R_{\bar{X}_0}\}$$

where $\Gamma \triangleq \alpha Tr\{R_{\bar{X}_0}\}^{-1}$ and $R_{\bar{X}_0} \triangleq E\{\bar{X}_0\bar{X}_0^T\}$.

For H_2 performance, we followed the following steps:

If

(i) $E\left\{\int_0^\infty (\bar{X}_t^T \bar{Q}_1\bar{X}_t + \bar{u}_t^T \bar{R}_1\bar{u}_t)dt + dV(\bar{X}_t)\right\} \le 0$

(ii) $E\{\bar{X}_0^T P\bar{X}_0\} \le \alpha$

(6.37)

can be held, then we have $J_2(u(t)) \le \alpha$.

Based on Lemma 6.2, the Itô-Lévy formula, and Assumption 6.1, it is clear that

$$E\{dV(\bar{X}_t)\} \leq \sum_{i=1}^{l}\sum_{j=1}^{l} h_i(\xi_t) h_j(\xi_t) E\left\{\bar{X}_t^T \left(P\bar{F}_i + \bar{F}_i^T P + P\bar{B}_i\bar{K}_j\right.\right.$$
$$+ \bar{K}_j^T \bar{B}_i^T P + 2PP + \bar{K}_j^T \Lambda_2 \bar{K}_j + \sum_{r=1}^{m} 2\bar{C}_{i,r}^T P\bar{C}_{j,r}$$
$$+ \sum_{q=1}^{n} 2\lambda_q \bar{D}_{i,q}^T P\bar{D}_{i,q} + \sum_{q=1}^{n} \lambda_q (P\bar{D}_{i,q} + \bar{D}_{i,q}^T P) + \sum_{q=1}^{n} \lambda_q P \tag{6.38}$$
$$+ \left.\left.\left[\Lambda_1 + \sum_{r=1}^{m} 2\Gamma\Lambda_{3,r} + \sum_{q=1}^{n} 3\Gamma\lambda_q\Lambda_{4,r}\right]\right)\bar{X}_t\right\}dt$$

where $\Lambda_{3,r} \triangleq diag(\varepsilon_{3,r} I_{n_x}, (\varepsilon_{3,r} + \varepsilon_{3,r}^{\mu})I_{n_x})$, for $r = 1,...,m$;
$\Lambda_{4,q} \triangleq diag(\varepsilon_{4,r} I_{n_x}, (\varepsilon_{4,r} + \varepsilon_{4,r}^{\mu})I_{n_x})$, for $q = 1,...,n$; $\Lambda_1 \triangleq diag(\varepsilon_1 I_{n_x}, (\varepsilon_1 + \varepsilon_1^{\mu})I_{n_x})$,
$\Lambda_2 \triangleq diag(\varepsilon_2 I_{n_x}, \varepsilon_2 I_{n_x})$.

Thus, we have

$$E\left\{\int_0^{\infty} (\bar{X}_t^T \bar{Q}_1 \bar{X}_t + \bar{u}_t^T \bar{R}_1 \bar{u}_t)dt + dV(\bar{X}_t)\right\}$$
$$\leq E\left\{\int_0^{\infty} \sum_{i=1}^{l}\sum_{j=1}^{l} \bar{X}_t^T h_i(\xi_t) h_j(\xi_t)\left(\bar{Q}_1 + \bar{K}_j^T (\bar{R}_1 + \Lambda_2)\bar{K}_j + P\bar{F}_i + \bar{F}_i^T P\right.\right.$$
$$+ P\bar{B}_i\bar{K}_j + \bar{K}_j^T \bar{B}_i^T P + 2PP + \sum_{r=1}^{m} 2\bar{C}_{i,r}^T P\bar{C}_{j,r} + \sum_{q=1}^{n} \lambda_q (P\bar{D}_{i,q} + \bar{D}_{i,q}^T P)$$
$$+ \sum_{q=1}^{n} \lambda_q \bar{D}_{i,q}^T P\bar{D}_{i,q} + \sum_{q=1}^{n} \lambda_q P + \left.\left.\left[\Lambda_1 + \sum_{r=1}^{m} 2\Gamma\Lambda_{3,r} + \sum_{q=1}^{n} 3\Gamma\lambda_q\Lambda_{4,r}\right]\right)\bar{X}_t\right\}dt$$

Since

$$E\left\{\bar{X}_t^T (P\bar{F}_i + \bar{F}_i^T P)\bar{X}_t\right\} = E\left\{\bar{X}_t^T \left[diag(P\bar{F}_i + \bar{F}_i^T P)\right]\bar{X}_t\right\}$$
$$E\left\{\bar{X}_t^T (\bar{C}_{i,r}^T P\bar{C}_{j,r})\bar{X}_t\right\} = E\left\{\bar{X}_t^T \left[diag(\bar{C}_{i,r}^T P\bar{C}_{j,r})\right]\bar{X}_t\right\}$$
$$E\left\{\bar{X}_t^T (\bar{D}_{i,q}^T P\bar{D}_{i,q})\bar{X}_t\right\} = E\left\{\bar{X}_t^T \left[diag(\bar{D}_{i,q}^T P\bar{D}_{i,q})\right]\bar{X}_t\right\}$$

we obtain

$$E\left\{\int_0^{\infty} (\bar{X}_t^T \bar{Q}_1 \bar{X}_t + \bar{u}_t^T \bar{R}_1 \bar{u}_t)dt + dV(\bar{X}_t)\right\}$$
$$\leq E\left\{\int_0^{\infty} \sum_{i=1}^{l}\sum_{j=1}^{l} \bar{X}_t^T h_i(\xi_t) h_j(\xi_t)\right. \tag{6.39}$$
$$\left.\cdot \left[diag\left(\tilde{M}_{H_2}^{i,j}(K_j^1, P_1), \tilde{M}_{H_2}^{i,j}(K_j^2, P_1, P_2)\right)\right]\bar{X}_t dt\right\}$$

where

$$\tilde{M}_{H_2}^{i,j}(K_j^1, P_1) \triangleq Q_1 + P_1 A_i + A_i^T P_1 + P_1 B_i K_j^1 + (P_1 B_i K_j^1)^T$$
$$+ (K_j^1)^T (R_1 + \varepsilon_2 I_{n_x}) K_j^1 + 2 P_1 P_1 + \sum_{r=1}^m 2 C_{i,r}^T P_1 C_{i,r}$$
$$+ \sum_{q=1}^n 2 \lambda_q D_{i,q}^T P_1 D_{i,q} + \sum_{q=1}^n \lambda_q (P_1 D_{i,q} + D_{i,q}^T P_1) \qquad (6.40)$$
$$+ \tilde{\Delta} I_{n_x} + \sum_{q=1}^n \lambda_q P_1$$

$$\hat{M}_{H_2}^{i,j}(K_j^2, P_1, P_2) \triangleq Q_1^\mu + P_2 \hat{A}_i + \hat{A}_i^T P_2 + P_2 B_i K_j^2 + (P_2 B_i K_j^2)^T$$
$$+ (K_j^2)^T (R_1^\mu + \varepsilon_2 I) K_j^2 + 2 P_2 P_2 + \sum_{r=1}^m 2 \hat{C}_{i,r}^T P_1 \hat{C}_{i,r}$$
$$+ \sum_{q=1}^n 2 \lambda_q \hat{D}_{i,q}^T P_1 \hat{D}_{i,q} + \sum_{q=1}^n \lambda_q (P_2 \hat{D}_{i,q} + \hat{D}_{i,q}^T P_2) \qquad (6.41)$$
$$+ \hat{\Delta} I_{n_x} + \sum_{q=1}^n \lambda_q P_2$$

with $\quad \tilde{\Delta} \triangleq \varepsilon_1 + \sum_{r=1}^m 2\Gamma \varepsilon_{3,r} + \sum_{q=1}^n 3\Gamma \lambda_q \varepsilon_{4,q} \quad$ and $\quad \hat{\Delta} \triangleq (\varepsilon_1 + \varepsilon_1^\mu) + \sum_{r=1}^m 2\Gamma(\varepsilon_{3,r} + \varepsilon_{3,r}^\mu)$
$+ \sum_{q=1}^n 3\Gamma \lambda_q (\varepsilon_{4,q} + \varepsilon_{4,q}^\mu).$

If the following BMIs can also be held:

$$diag\left(\tilde{M}_{H_2}^{i,j}(K_j^1, P_1), \widehat{M}_{H_2}^{i,j}(K_j^2, P_1, P_2) \right) \leq 0$$
$$\text{for all } i, j \in 1, 2, ..., l \qquad (6.42)$$

then we have $J_2(u(t)) \leq \alpha$.

Multiplying both sides of the BMIs in (6.40) and (6.41) by $W = P^{-1} = diag(W_1, W_2)$, and setting $Y_j^1 = K_j^1 W_1$, $Y_j^2 = K_j^2 W_2$ and then applying the Schur complement in [19], we have the LMIs in (6.24) and (6.25), respectively. When the LMIs in (6.24) and (6.25) can be held, the inequality in (6.39) is also held so that we finish the proof of (i).

For H_∞ performance, we followed the following:

Since

$$E\left\{ 2\bar{X}_t^T P(\bar{G}_i \bar{v}_t + \Delta \bar{G}) \right\} \leq E\left\{ \bar{X}_t^T P\left[\tfrac{(1+\delta)}{\beta}(\bar{G}_i \bar{G}_i^T + I_{2n_x}) \right] P\bar{X}_t \right\} + \beta E\{\bar{v}_t^T \bar{v}_t\}$$
$$= E\left\{ \bar{X}_t^T P\left[\tfrac{(1+\delta)}{\beta} diag([\bar{G}_i \bar{G}_i^T + I_{n_x}], [\bar{G}_i \bar{G}_i^T + I_{2n_x}]) \right] P\bar{X}_t \right\} + \beta E\{\bar{v}_t^T \bar{v}_t\}$$

and

$$E\left\{ \int_0^\infty (\bar{X}_t^T \bar{Q}_2 \bar{X}_t + \bar{u}_t^T \bar{R}_2 \bar{u}_t) dt \right\} - E\left\{ \bar{X}_0^T P \bar{X}_0 \right\}$$
$$\leq E\left\{ \int_0^\infty (\bar{X}_t^T diag\left(\tilde{M}_{H_\infty}^{i,j}(K_j^1, P_1), \tilde{M}_{H_\infty}^{i,j}(K_j^2, P_1, P_2) \right) \bar{X}_t + \beta E\{\bar{v}_t^T \bar{v}_t\} dt \right\}$$

we have

$$\tilde{M}_{H_\infty}^{i,j}(K_j^1,P_1)$$

$$\triangleq P_1A_i + A_i^TP_1 + P_1B_iK_j^1 + (P_1B_iK_j^1)^T$$

$$+P_1\left[(2+\tfrac{(1+\delta)}{\beta})I_{n_x}\right]P_1 + Q_2 + \sum_{q=1}^{n}\lambda_q(P_1D_{i,q} + D_{i,q}^TP_1) \tag{6.43}$$

$$+(K_j^1)^T(R_2 + \varepsilon_2I_{n_x})K_j^1 + \sum_{r=1}^{m}2C_{i,r}^TP_1C_{i,r} + \tilde{\Delta}I_{n_x}$$

$$+\sum_{q=1}^{n}2\lambda_qD_{i,q}^TP_1D_{i,q} + \sum_{q=1}^{n}\lambda_qP_1 + \tfrac{(1+\delta)}{\beta}P_1G_iG_i^TP_1$$

$$\hat{M}_{H_\infty}^{i,j}(K_j^2,P_1,P_2)$$

$$\triangleq P_2\hat{A}_i + \hat{A}_i^TP_2 + P_2B_iK_j^2 + (P_2B_iK_j^2)^T + Q_2^\mu$$

$$+P_2\left[(2+\tfrac{(1+\delta)}{\beta})I_{n_x}\right]P_2 + \sum_{q=1}^{n}\lambda_q(P_2\hat{D}_{i,q} + \hat{D}_{i,q}^TP_2) \tag{6.44}$$

$$+(K_j^2)^T(R_2^\mu + \varepsilon_2I_{n_x})K_j^2 + \sum_{r=1}^{m}2\hat{C}_{i,r}^TP_1\hat{C}_{i,r} + \hat{\Delta}I_{n_x}$$

$$+2\lambda_q\sum_{q=1}^{n}\hat{D}_{i,q}^TP_1\hat{D}_{i,q} + \sum_{q=1}^{n}\lambda_qP_2 + \tfrac{(1+\delta)}{\beta}P_1G_iG_i^TP_1$$

If the following BMIs can be held:

$$diag\left(\tilde{M}_{H_\infty}^{i,j}(K_j^1,P_1),\widehat{M}_{H_\infty}^{i,j}(K_j^2,P_1,P_2)\right)\leq 0$$

$$\text{for all } i,j \in 1,2,...,l \tag{6.45}$$

then we have $J_\infty(u(t)) \leq \beta$.

To transform the BMIs in (6.43) and (6.44) to LMIs, we multiply both sides of the BMIs in (6.43) and (6.44) by $W = P^{-1} = diag(W_1,W_2)$. Setting $Y_j^1 = K_j^1W_1$ and $Y_j^2 = K_j^2W_2$ and then applying the Schur complement, we have the LMIs in (6.26) and (6.27), respectively.

6.8.4 PROOF OF THEOREM 6.4

Suppose that for all $i,j = 1,2,...,l$, the matrix set $\{W_1\{Y_j^1\}_{j=1}^l, W_2\{Y_j^2\}_{j=1}^l\}$ is a feasible solution of the LMIs in (6.24)–(6.26), and we have

$$E\{LV(\bar{X}_t)\} \leq E\{-m_1\bar{X}_t^T\bar{X}_t\}dt < 0 \tag{6.46}$$

where m_1 is the smallest eigenvalue of the positive definite matrix $\bar{Q}_1$.

Thus, we have $\lim_{t\to\infty}E\{\|x\|_2^2\} = 0$.

6.8.5 DATA OF SIMULATION

The values of $P_1 = W_1^{-1}$ and $P_2 = W_2^{-1}$ for the optimal H_2 strategy, the optimal H_∞ robust strategy, and three chosen Pareto strategies as follows:

(1) The optimal H_2 strategy:

$$P_1^* = \begin{bmatrix} 483.5878 & 0.4891 & -0.0530 & -0.0346 \\ * & 764.6808 & 0.1217 & 0.4867 \\ * & * & 0.0290 & 0.0025 \\ * & * & * & 0.0127 \end{bmatrix}$$

$$P_2^* = \begin{bmatrix} 4.2032 & -0.0098 & 0.0005 & -0.0005 \\ * & 5.0570 & -0.0010 & 0.0048 \\ * & * & 0.0005 & 0.0001 \\ * & * & * & 0.0001 \end{bmatrix} \times 10^3$$

(2) The optimal $H\infty$ robust strategy:

$$P_1^\dagger = \begin{bmatrix} 495.4262 & -0.7817 & -0.0522 & -0.0241 \\ * & 768.6164 & 0.1818 & 0.4328 \\ * & * & 0.0301 & 0.0030 \\ * & * & * & 0.0121 \end{bmatrix}$$

$$P_2^\dagger = \begin{bmatrix} 5.9945 & -0.0125 & 0.0007 & -0.0007 \\ * & 8.1168 & -0.0011 & 0.0059 \\ * & * & 0.0008 & 0.0001 \\ * & * & * & 0.0002 \end{bmatrix} \times 10^3$$

(3) The Pareto optimal strategy 1:

$$P_1^\Delta = \begin{bmatrix} 485.6257 & 0.5489 & -0.0532 & -0.0358 \\ * & 770.1123 & 0.1323 & 0.4907 \\ * & * & 0.0289 & 0.0026 \\ * & * & * & 0.0122 \end{bmatrix}$$

$$P_2^\Delta = \begin{bmatrix} 4.4283 & -0.0101 & 0.0005 & -0.0006 \\ * & 5.3183 & -0.0011 & 0.0049 \\ * & * & 0.0006 & 0.0001 \\ * & * & * & 0.0001 \end{bmatrix} \times 10^3$$

(4) The Pareto optimal strategy 2:

$$P_1^{\Diamond} = \begin{bmatrix} 492.3809 & 0.3656 & -0.0551 & -0.0374 \\ * & 781.2651 & 0.1565 & 0.4839 \\ * & * & 0.0290 & 0.0027 \\ * & * & * & 0.0120 \end{bmatrix}$$

$$P_2^{\Diamond} = \begin{bmatrix} 5.1635 & -0.0114 & 0.0007 & -0.0006 \\ * & 6.2833 & -0.0011 & 0.0053 \\ * & * & 0.0007 & 0.0001 \\ * & * & * & 0.0002 \end{bmatrix} \times 10^3$$

(5) The Pareto optimal strategy 3:

$$P_1^{\nabla} = \begin{bmatrix} 492.5810 & -0.7460 & -0.0514 & -0.0230 \\ * & 765.2412 & 0.1712 & 0.4322 \\ * & * & 0.0301 & 0.0030 \\ * & * & * & 0.0124 \end{bmatrix}$$

$$P_2^{\nabla} = \begin{bmatrix} 5.5531 & -0.0120 & 0.0006 & -0.0007 \\ * & 7.5511 & -0.0011 & 0.0058 \\ * & * & 0.0008 & 0.0001 \\ * & * & * & 0.0002 \end{bmatrix} \times 10^3$$

7 Multi-Objective Fault-Tolerance Observer-Based Control Design of Stochastic Jump-Diffusion Systems

7.1 INTRODUCTION

Along with the development of modern industrial production, due to the fact that actuators and sensors in control systems have become much more vulnerable to fault signals, higher requirements of safety and reliability for control plants have been put forward. In order to ensure safety and reliability in the industrial process, the capability of fault tolerance during the control process has become an important issue and caught the attention of control engineers. In the case of passive fault-tolerant control (FTC) design, which considers a fault a specific external disturbance, it aims to achieve control performance with a prescribed fault tolerance level [138–140]. On the other hand, for active FTC schemes, fault signals are estimated by an observer, and estimated faults will be used to eliminate the effect of real faults. FTC techniques have been widely investigated and applied in several fields. To attenuate high-frequency variation effects from actuators, FTC is applied for attitude control of a satellite in [141]. In [142], FTC was used to stabilize a linear quantum system with the consideration of random voltage fluctuations. In [143], operator theory-based fault-tolerant control was applied to a multi-input multi-output microreactor. Also, in [144], fault-tolerant control was developed for a hypersonic flight vehicle with multiple sensor faults.

For almost the physical systems, there has strongly nonlinearity behavior in the physical system. Also, most fault signals are nonlinear functions, which are coupled with state variables and control inputs in physical systems. While considering the effect of faults in nonlinear systems, nonlinear fault functions will increase the difficulty of FTC analysis [143–145]. In recent years, the Takagi-Sugeno fuzzy model has been considered an efficient tool to describe nonlinear dynamic systems [83, 146]. By utilizing the T-S fuzzy approximation method, a nonlinear system can be approximated by interpolating a set of local linear systems, and control design for nonlinear systems can be simplified by a set of fuzzy controllers. The T-S fuzzy model–based control scheme has been widely investigated in several control issues [147–150]. Moreover, the fuzzy adaptive FTC scheme has been proposed for

DOI: 10.1201/9781003362142-9

fault-tolerant tracking control of a class of uncertain nonlinear systems with nonaffine nonlinear faults [151]. A back-stepping fuzzy control scheme is proposed to deal with a stabilization problem with consideration of actuator faults in [152]. In [153], event-triggered FTC design for network-based fuzzy systems is also discussed.

In general, environmental noise is inevitable during the control process. Thus, it is expected that FTC design can not only achieve great FTC performance while the system is affected by a fault but also achieve the prescribed robust control performance for disturbance attenuation. For this reason, several mixed FTC designs, such as mixed optimal FTC schemes [154, 155] and mixed robust FTC schemes [156, 157], have been developed. By extending the concept of a mixed control method [70, 158], multi-objective design has become a popular issue and is addressed in many research studies on control and estimation [6, 11, 122, 159–161]. Compared with conventional mixed H_2/H_∞ control design schemes [70, 158], which have a unique optimal control strategy, there exist several optimal control strategies with corresponding multi-objective optimal performance in a multi-objective control design. Based on the MO control design, engineers are free to choose the preferred control strategy from several optimal control strategies according to their own demands. At present, there have been very few studies to address the fuzzy MO FTC problem of nonlinear systems under sensor and actuator faults. Further, in conventional FTC designs, the descriptor observer is always employed for fault signal estimation [152–157]. Since the augmented descriptor observer system is singular, more effort is needed for robust FTC design. Therefore, conventional FTC schemes are not suitable for MO FTC. In this chapter, a smoothed dynamic model is proposed to efficiently describe actuator and sensor fault signals. Since the augmented system with fault dynamic model is nonsingular so that the simple conventional Luenberger observer could be employed to precisely estimate state variables and fault signals for the consequent H_2/H_∞ observer-based FTC design of T-S fuzzy system with actuator and sensor faults as well as external disturbance and measurement noise.

Despite the fact that the design concept of MO control is suitable for modern industry, it is not easy to solve the corresponding multi-objective optimization problems in general. Especially in fuzzy-based MOPs, computational complexity will be great according to the number of fuzzy if-then rules. Among wide ranges of different algorithms of nature-inspired optimization methods [112, 162–164], multi-objective evolution algorithms are a powerful nature-inspired optimization method to solve MOPs [1–3]. In general, conventional MOEAs update design parameters via an evolution algorithm to search the multi-objective vector for Pareto optimal solutions via nondominated sorting and crowded tournament selection schemes in the design parametric space [1–3, 162–164]. Since the fuzzy controller and observer of complex fuzzy systems consist of a large number of local controllers and observers, respectively, it is almost impossible to employ an EA to update these fuzzy controller and observer parameters to achieve an MOP of complex T-S fuzzy systems, such as multi-objective H_2/H_∞ observer-based FTC of T-S fuzzy systems with actuator and sensor faults. This is why conventional MOEAs have not been addressed in the MOPs of nonlinear T-S fuzzy systems, even though they are very powerful in MOPs of other, simpler designed systems. Besides, for conventional MOEAs, the corresponding

objective vector of the child population randomly generated by crossover and mutation operations may exceed the real Pareto front if the current population is close to
the real Pareto optimality. Clearly, if the generated child population is infeasible, it
will be directly discarded. In this situation, it is very difficult to generate a feasible
child population while the current population is close to the real Pareto optimality
[7, 165]. Hence, it is more appealing to override the bottleneck of the conventional
MOEA to solve a more complex MOP such as multi-objective H_2/H_∞ observer-based
FTC of T-S fuzzy systems.

In this chapter, a simple Luenberger observer-based FTC is designed for the T-S
fuzzy system with actuator and sensor faults as well as external disturbance and
measurement noise. In order to achieve the optimal robust attenuation of external
disturbances on the FTC performance of the controller and observer, the optimal H_∞
observer-based FTC design is proposed to minimize the effect of the external disturbance and measurement noise on the FTC performance of control and observation.
While the effects of external disturbance and measurement noise have been optimally
attenuated by the H_∞ observer-based control strategy, the H_2 observer-based control
strategy is also proposed to achieve the optimal quadratic control and observation
simultaneously, that is, to achieve the MOP of multi-objective H_2/H_∞ observer-based
FTC of T-S fuzzy systems.

On the other hand, a reverse-order MOEA is proposed to simplify the design
procedure of the complex MO H2/H∞ fuzzy observer-based FTC design problem.
Instead of searching controller and observer parameters directly, we search the
H_2/H_∞ objective vector (α, β) directly by EA algorithm, nondominated sorting, and
crowded tournament selection scheme and then find the corresponding controller
and observer parameters using the LMI toolbox in MATLAB via the convex optimization algorithm indirectly. Based on the reverse-order MOEA and LMI toolbox in
MATLAB, we can efficiently solve the complex MOP of the multi-objective H_2/H_∞
observer-based FTC of T-S fuzzy system. In the future, the proposed reverse-order
MOEA could be applied to other complex MOPs of nonlinear control and estimation
design problems. An observer-based 3-D missile guidance system with actuator and
sensor faults due to sudden cheating side-step maneuvering and hostile jamming,
respectively, is given to illustrate the design procedure and then validate the performance of the proposed multi-objective H_2/H_∞ optimal fault-tolerant guidance control
design. Also, a multi-objective H_2/H_∞ FTC design for an inverted pendulum system
is carried out in comparison with the state-of-the-art FTC method.

The main purposes of this chapter are as follows:

(1) A novel nonsingular smoothed dynamic model is introduced to efficiently
describe the actuator and sensor fault signals. Thus, instead of constructing a conventional singular descriptor estimator for the fault signal, the
simple Luenberger observer could be employed to precisely estimate state
variables and actuator and sensor fault signals for the H_2/H_∞ observer-
based FTC design. Further, compared to the descriptor-based FTC design,
which has to solve a set of algebraic constraints, the proposed FTC design
can be transformed into an equivalent LMI-based constrained optimization problem, and it is easier to solve for practical application.

(2) Instead of using a conventional MOEA to search the design parameters of fuzzy controller and observer gains for the multi-objective $H2/H\infty$ observer-based FTC design problem, a reverse-order LMI-constrained MOEA is proposed to directly search the optimal multi-objective vector, and then the corresponding design variables of the controller and observer can be easily obtained by using the MATLAB LMI toolbox. Further, the proposed reverse-order MOEA could be applied to efficiently solve other complex MOPs of control and estimation in nonlinear or T-S fuzzy systems.

(3) To improve the rate of convergence of MOEA, a mechanism is included to deal with the problem of an infeasible population generated by the muta-tion and crossover operators in MOEA. By embedding this additional mechanism in the proposed reverse-order LMI-constrained MOEA, the infeasible population is replaced by the mean of itself and the closest feasi-ble population in the previous iteration. In this situation, the information of the infeasible population can be utilized to find the closest feasible popula-tion for each iteration. Hence, the convergence of the proposed MOEA can be effectively improved over conventional MOEAs.

This chapter is divided into five sections, as follows: the T-S fuzzy system model with actuator and sensor faults is introduced in Section 7.2. The virtual fault dynamic models are also given in Section 7.2. In Section 7.3, we develop the multi-objective optimal H_2/H_∞ observer-based FTC design for the T-S fuzzy system with sensor and actuator faults. A reverse-order MOEA for the multi-objective observer-based FTC design is proposed in Section 7.4. The simulation of a fault-tolerant missile guid-ance control design of a 3-D tactical missile system and FTC design for inverted pendulum system are proposed to verify the effectiveness of the proposed method in Section 7.5. Conclusions are given in Section 7.6.

Notation: A^T: transpose of matrix A; $A \geq 0$ $(A > 0)$; symmetric positive semi-definite (symmetric positive definite) matrix A; $\|x\|$: Euclidean norm for the given vector $x(t) \in \mathbb{R}^n$; $L_2(\mathbb{R}^+; \mathbb{R}^n) = \{v(t): \mathbb{R}^+ \to \mathbb{R}^n \mid (\int_0^\infty v^T(t)v(t)dt)^{1/2} < \infty\}$; $\lambda_{max}(P)$: maximum eigenvalue of real-value symmetric matrix P; I_a: identity matrix with demission $a \times a$; $0_{a \times b}$: zero matrix with dimension $a \times b$; $eig(A)$: set that collects the eigenvalues of matrix A; S: set of one-dimensional complex numbers; and $col[D]$: column space of matrix D.

7.2 SYSTEM DESCRIPTION

We consider a continuous-time nonlinear system with actuator and sensor faults, which could be described by the T-S fuzzy model. The ith fuzzy if-then rule of the nonlinear system can be represented as follows [83, 146]:

If $z_1(t)$ is F_{i1} and ... and $z_g(t)$ is F_{ig},

Then

$$\dot{x}(t) = A_i x(t) + B_{u,i} u(t) + B_{a,i} f_a(t) + B_{w,i} w(t)$$ (7.1)

$$y(t) = C_i x(t) + D_i f_s(t) + n(t)$$

for $i = 1, ..., I$

where $x(t) \in \mathbb{R}^n$ is the state vector, $u(t) \in \mathbb{R}^{n_u}$ is the input vector, $y(t) \in \mathbb{R}^m$ denotes the measurement output by sensors, and $f_a(t) \in \mathbb{R}^{n_a}$ and $f_s(t) \in \mathbb{R}^{n_s}$ are the fault signals on actuators and sensors, respectively. $w(t) \in \mathbb{R}^{n_w}$ denotes the external disturbance, and $n(t) \in \mathbb{R}^m$ denotes the measurement noise at the sensor. $z_1(t),...,z_g(t)$ are the premise variables, F_{ij} is the ith fuzzy set of the jth premise variable for $i = 1,...,I$ and $j = 1,...,g$, where I is the number of fuzzy rules. The matrices $A_i \in \mathbb{R}^{n \times n}$, $B_{u,i} \in \mathbb{R}^{n \times n_u}$, $B_{a,i} \in \mathbb{R}^{n \times n_a}$, $B_{w,i} \in \mathbb{R}^{n \times n_w}$, $C_i \in \mathbb{R}^{m \times n}$, and $D_i \in \mathbb{R}^{n \times n_s}$, for $i = 1,...,I$. Thus, the overall T-S fuzzy system in (7.1) is inferred as follows [146]:

$$\dot{x}(t) = \sum_{i=1}^{I} h_i(z(t))(A_i x(t) + B_{u,i} u(t) + B_{a,i} f_a(t) + B_{w,i} w(t))$$
$$y(t) = \sum_{i=1}^{I} h_i(z(t))(C_i x(t) + D_i f_s(t) + n(t))$$
(7.2)

where $z(t) = [z_1(t),...,z_g(t)]$, $q_i(z(t)) = \prod_{j=1}^{g} F_{ij}(z_j(t))$, and $h_i(z(t)) = \dfrac{q_i(z(t))}{\sum_{j=1}^{I} q_i(z(t))}$, which satisfies $1 \geq h_i(z(t)) \geq 0$ and $\sum_{i=1}^{I} h_i(z(t)) = 1$.

Assumption 7.1: The T-S fuzzy system in (7.2) is controllable and observable; that is, the pair $(A_i, B_{u,i})$ is controllable and the pair (A_i, C_i) is observable for $i = 1,...,I$.

In this chapter, in order to efficiently estimate fault signals $f_a(t)$ and $f_s(t)$ by the conventional Luenberger observer for the FTC design in the sequel, a novel dynamic smoothed model is proposed for fault signals $f_a(t)$ and $f_s(t)$. To begin with, based on the derivative definition of $\dot{f}_a(t) = \lim_{h \to 0} \dfrac{f_a(t+h) - f_a(t)}{h}$, the smoothed model of $f_a(t)$ is given as follows:

$$\dot{f}_a(t) = \frac{1}{h}(f_a(t+h) - f_a(t)) + \varepsilon_{1,a}(t)$$
$$\dot{f}_a(t-h) = \frac{1}{h}(f_a(t) - f_a(t-h)) + \varepsilon_{2,a}(t)$$
$$\vdots$$
$$\dot{f}_a(t-kh) = \frac{1}{h}(f_a(t-(k-1)h) - f_a(t-kh)) + \varepsilon_{k,a}(t)$$
(7.3)

where $\varepsilon_{1,a}(t),...,\varepsilon_{k,a}(t)$ denote the corresponding approximation errors of the derivative at different smoothed time points for actuator fault $f_a(t)$. The constant h is a small enough time interval. Further, in order to reduce the effect of future fault signal $f_a(t+h)$ on the dynamic smoothed model of fault signal $f_a(t)$, $f_a(t+h)$ could be also represented by extrapolation (e.g., Lagrange extrapolation [166]) as follows:

$$f_a(t+h) = \sum_{i=0}^{k} a_i f_a(t-ih) + \delta_a(t)$$
(7.4)

where a_i, $i = 0,...,k$ are the extrapolation coefficients, and $\delta_a(t)$ indicates the extrapolation error of $f_a(t+h)$. Then, we could obtain the following dynamic smoothed model of actuator fault signal $f_a(t)$.

$$\dot{F}_a(t) = A_{f_a} F_a(t) + \varepsilon_a(t)$$
(7.5)

where $F_a(t) = [f_a(t)^T, f_a(t-h)^T, ..., f_a(t-kh)^T]^T$, the smoothed model error of actuator $\varepsilon_a(t) = [(\varepsilon_{1,a}(t) + \delta_a(t)/h)^T, \varepsilon_{2,a}^T(t), ..., \varepsilon_{k,a}^T(t)]^T$, and

$$
A_{f_a} = \begin{bmatrix}
\frac{\bar{a}_0}{h} I_{n_a} & \frac{a_1}{h} I_{n_a} & \frac{a_2}{h} I_{n_a} & \cdots & \frac{a_k}{h} I_{n_a} \\
\frac{1}{h} I_{n_a} & -\frac{1}{h} I_{n_a} & 0 & \cdots & 0 \\
0 & \frac{1}{h} I_{n_a} & -\frac{1}{h} I_{n_a} & \cdots & 0 \\
\vdots & \ddots & \ddots & \ddots & \vdots \\
0 & \cdots & 0 & \frac{1}{h} I_{n_a} & -\frac{1}{h} I_{n_a}
\end{bmatrix}
$$

where $\bar{a}_0 = -1 + a_0$. Similarly, the dynamic smoothed model for the sensor fault signal $f_s(t)$ is similar to the virtual signal model in (7.5) as follows:

$$
\dot{F}_s(t) = A_{f_s} F_s(t) + \varepsilon_s(t) \tag{7.6}
$$

where $F_s(t) = [f_s(t)^T, f_s(t-h)^T, ..., f_s(t-kh)^T]^T$, the smoothed model error of sensor $\varepsilon_s(t) = [(\varepsilon_{1,s}(t) + \delta_s(t)/h)^T, \varepsilon_{2,s}^T(t), ..., \varepsilon_{k,s}^T(t)]^T$, and

$$
A_{f_s} = \begin{bmatrix}
\frac{\bar{b}_0}{h} I_{n_s} & \frac{b_1}{h} I_{n_s} & \frac{b_2}{h} I_{n_s} & \cdots & \frac{b_k}{h} I_{n_a} \\
\frac{1}{h} I_{n_s} & -\frac{1}{h} I_{n_s} & 0 & \cdots & 0 \\
0 & \frac{1}{h} I_{n_s} & -\frac{1}{h} I_{n_s} & \cdots & 0 \\
\vdots & \ddots & \ddots & \ddots & \vdots \\
0 & \cdots & 0 & \frac{1}{h} I_{n_s} & -\frac{1}{h} I_{n_s}
\end{bmatrix}
$$

where $\bar{b}_0 = -1 + b_0$, b_i, $i = 0, ..., k$ are the extrapolation coefficients.

Remark 7.1: For the T-S fuzzy system in (7.2) with the signal faults on actuators and sensors, it is difficult to construct a suitable fault estimation scheme due to the unknown prior knowledge of the actuator faults and the sensor faults. In the traditional augmented descriptor observer design, the unknown input is not easily estimated (i.e., the fault signal decoupling condition cannot always be satisfied [140]). Therefore, more efforts are needed to efficiently estimate the state and fault information. In this chapter, unlike the conventional descriptor model [167], the nonsingular fault dynamic models of $f_a(t)$ and $f_s(t)$ in (7.5) and (7.6) are to be embedded in the augmented system with the T-S fuzzy system in (7.2). In this situation, the conventional Luenberger observer could be employed to precisely estimate the state variables and actuator and sensor fault signals to efficiently compensate the effect of fault signals and external disturbance for the FTC design.

For the convenience of estimating $x(t)$ and $f_a(t)$, $f_s(t)$ simultaneously, the dynamic smoothed model of fault signals in (7.5) and (7.6) could be embedded as an internal model of the T-S fuzzy system in (7.2) as the following augmented system:

$$\dot{\overline{x}}(t) = \sum_{i=1}^{I} h_i(z(t))(\overline{A}_i \overline{x}(t) + \overline{B}_{u,i} u(t) + \overline{B}_{w,i} \overline{w}(t))$$

$$y(t) = \sum_{i=1}^{I} h_i(z(t))(\overline{C}_i \overline{x}(t) + \overline{D}_i \overline{w}(t))$$

(7.7)

with the augmented state $\overline{x}(t) = [F_a^T(t), F_s^T(t), x^T(t)]^T$; the mapping matrix $C_{f_a} = [I_{n_a}, 0_{n_a \times n_a}, ..., 0_{n_a \times n_a}]$; the vector $\overline{w}(t) = [\varepsilon_a^T(t), \varepsilon_s^T(t), w^T(t), n^T(t)]^T$; the mapping matrix $C_{f_s} = [I_{n_s}, 0_{n_s \times n_s}, ..., 0_{n_s \times n_s}]$; and $\overline{B}_{u,i} = \begin{bmatrix} 0 \\ 0 \\ B_{u,i} \end{bmatrix}$, $\overline{B}_{w,i} = \begin{bmatrix} I & 0 & 0 & 0 \\ 0 & I & 0 & 0 \\ 0 & 0 & B_{w,i} & 0 \end{bmatrix}$,

$\overline{C}_i = \begin{bmatrix} 0 & D_i C_{f_s} & C_i \end{bmatrix}$, and $\overline{D}_i = \begin{bmatrix} 0 & 0 & 0 & I \end{bmatrix}$. Before the estimating the state and actuator and sensor signals for fault-tolerant control of the fuzzy system in (7.1), we need to guarantee the augmented state $\overline{x}(t)$ in (7.7) is observable from $y(t)$.

Theorem 7.1: For the T-S fuzzy system in (7.1), if the local matrices (A_i, C_i) for $i = 1,...,I$ are observable, that is,

$$rank \begin{bmatrix} sI_n - A_i \\ C_i \end{bmatrix} = n \text{ for } s \in S$$

(7.8)

and the following conditions hold

$$eig(A_i) \cap eig(A_{f_a}) = \varnothing$$
$$eig(A_i) \cap eig(A_{f_s}) = \varnothing \quad (7.9)$$
$$eig(A_{f_s}) \cap eig(A_{f_a}) = \varnothing$$

$$col \begin{bmatrix} -B_{a,i} C_{f_a} \\ 0 \end{bmatrix} \cap col \begin{bmatrix} sI_n - A_i \\ C_i \end{bmatrix} = \varnothing \text{ for } s \in A_{f_a} \quad (7.10)$$

$$rank \begin{bmatrix} sI_{n_a(k+1)} - A_{f_a} \\ -B_{a,i} C_{f_a} \end{bmatrix} = n_a(k+1) \, s \in S \quad (7.11)$$

$$rank \begin{bmatrix} sI_{n_s(k+1)} - A_{f_s} \\ D_i C_{f_s} \end{bmatrix} = n_s(k+1) \, s \in S \quad (7.12)$$

then the ith augmented fuzzy system $(\overline{A}_i, \overline{C}_i)$ in (7.7) is observable for $i = 1,...,I$.

Proof: Please refer to Appendix 7.7.1.

Remark 7.2: The physical meaning of the conditions in (7.11) and (7.12) is that the fault state $f_a(t)$ of the dynamic smoothed model in (7.5) and fault state $f_s(t)$ of the dynamic smoothed model in (7.6) are all observable in the augmented system in (7.7).

Suppose the following conventional fuzzy Luenberger observer is proposed to deal with the estimation of the state variables and actuator and sensor fault signals of the nonlinear system in (7.2) or the state of the augmented system in (7.7):

Observer Rule i:

If $z_1(t)$ is F_{i1} and ... and $z_g(t)$ is F_{ig},

Then

$$\dot{\hat{\bar{x}}}(t) = \bar{A}_i\hat{\bar{x}}(t) + \bar{B}_{u,i}u(t) + L_i(y(t) - \hat{y}(t))$$

$$\hat{y}(t) = \bar{C}_i\hat{\bar{x}}(t)$$

(7.13)

where $L_i \in \mathbb{R}^{((n_a+n_s)(k+1)+n)\times m}$ is the observer parameters for $i = 1,...,I$. The vectors $\hat{\bar{x}}(t) \in \mathbb{R}^{(n_a+n_s)(k+1)+n}$ and $\hat{y}(t) \in \mathbb{R}^m$ are the estimated state and measurement output for the T-S fuzzy system in (7.7), respectively. Then the overall fuzzy observer can be designed as follows:

$$\dot{\hat{\bar{x}}}(t) = \sum_{i=1}^{I} h_i(z(t))(\bar{A}_i\hat{\bar{x}}(t) + \bar{B}_{u,i}u(t) + L_i(y(t) - \hat{y}(t)))$$
$$= \sum_{i=1}^{I}\sum_{j=1}^{I} h_i(z(t))h_j(z(t))(\bar{A}_i\hat{\bar{x}}(t) + \bar{B}_{u,i}u(t)$$
$$+ L_i(\bar{C}_j(\bar{x}(t) - \hat{\bar{x}}(t)) + \bar{D}_j\bar{w}(t)))$$
$$\hat{y}(t) = \sum_{i=1}^{I} h_i(z(t))\bar{C}_i\hat{\bar{x}}(t)$$

(7.14)

Remark 7.3: In general, the state variables $\bar{x}(t)$ in (7.7) are not accessible. Thus, the estimated state can be specified as the premise variables in the T-S fuzzy model (7.14), that is, $z(t) = \hat{\bar{x}}(t)$.

In this chapter, we employ the T-S fuzzy observer-based controller to deal with the fault-tolerant controller design of the T-S fuzzy system in (7.14). Thus, the jth fuzzy control rule is given as follows:

Control Rule j:

If $z_1(t)$ is F_{i1} and ... and $z_g(t)$ is F_{ig},

Then $u(t) = K_j\hat{\bar{x}}(t)$

(7.15)

for $j = 1,...,I$. Hence, the overall fuzzy controller can be represented as:

$$u(t) = \sum_{j=1}^{I} h_j(z(t)) K_j\hat{\bar{x}}(t)$$

(7.16)

where $K_j \in \mathbb{R}^{n_u \times ((n_a + n_s)(k+1)+n)}$ denotes the jth fuzzy controller gain to be designed for $j = 1,...,I$. Let us denote the estimation error as $e(t) = \bar{x}(t) - \hat{\bar{x}}(t)$; then we can formulate the augmented fuzzy observer-based fault-tolerant control systems as follows:

$$
\begin{bmatrix} \dot{\bar{x}}(t) \\ \dot{e}(t) \end{bmatrix}
$$

$$
= \sum_{i=1}^{I} \sum_{j=1}^{I} h_i(z(t)) h_j(z(t)) \left(\begin{bmatrix} \bar{A}_i + \bar{B}_{u,i} K_j & -\bar{B}_{u,i} K_j \\ 0 & \bar{A}_i - L_i \bar{C}_j \end{bmatrix} \begin{bmatrix} \bar{x}(t) \\ e(t) \end{bmatrix} \right. \tag{7.17}
$$

$$
\left. + \begin{bmatrix} \bar{B}_{w,i} \\ \bar{B}_{w,i} - L_i \bar{D}_j \end{bmatrix} \bar{w}(t) \right)
$$

Let us denote $\tilde{x}(t) = \begin{bmatrix} \bar{x}^T(t) & e^T(t) \end{bmatrix}^T$ and

$$
\tilde{A}_{ij} = \begin{bmatrix} \bar{A}_i + \bar{B}_{u,i} K_j & -\bar{B}_{u,i} K_j \\ 0 & \bar{A}_i - L_i \bar{C}_j \end{bmatrix}
$$

$$
\tilde{D}_{ij} = \begin{bmatrix} \bar{B}_{w,i} \\ \bar{B}_{w,i} - L_i \bar{D}_j \end{bmatrix}
$$

then the augmented system in (7.17) could be expressed as follows:

$$
\dot{\tilde{x}} = \sum_{i=1}^{I} \sum_{j=1}^{I} h_i(z(t)) h_j(z(t)) (\tilde{A}_{ij} \tilde{x}(t) + \tilde{D}_{ij} \bar{w}(t)) \tag{7.18}
$$

which is the observer-based output feedback T-S fuzzy system. Since the augmented disturbance $\bar{w}(t)$ due to external disturbance, modeling errors of fault signals and measurement noises in (7.18) will significantly influence the state estimation and control performance, the following MO observer-based control is designed to H_∞ optimally attenuate the effect of $\bar{w}(t)$ on the observer-based fault-tolerant control and H_2 optimally achieve the observer-based fault-tolerant quadratic control performance simultaneously.

7.3 MULTI-OBJECTIVE OPTIMAL H_2/H_∞ OBSERVER-BASED FAULT-TOLERANT CONTROL FOR T-S FUZZY SYSTEM WITH ACTUATOR AND SENSOR FAULTS

In the closed-loop observer-based output feedback T-S fuzzy control system in (7.18), the effects of the smoothed model error of sensor and actuator faults, external disturbance and measurement noise in $\bar{w}(t)$ will deteriorate the control and estimation performance of the observer-based FTC system and even lead to the instability of the fuzzy observer-based FTC system. In this situation, how to eliminate the effect of the smoothed model error of sensor and actuator faults, the external disturbance and the measurement noise in $\overline{w}(t)$ to guarantee robust control performance will be an important design goal for the fuzzy observer-based fault-tolerant control system. Since H_∞ control is the most important robust control design to efficiently eliminate the effect of uncertain $\overline{w}(t)$ on the FTC system, it will be employed to deal with the robust observer-based FTC design problem for the T-S fuzzy system in (7.18). Let us consider the following H_∞ observer-based fault-tolerant control performance of (7.18):

$$H_\infty(\{L_i, K_j\}_{i,j=1}^l)$$

$$= \sup_{\overline{w}(t) \in L_2(\mathbb{R}^+; \mathbb{R}^{n_{\overline{w}}})} \frac{\int_0^{t_f} \tilde{x}^T(t)\tilde{Q}_1\tilde{x}(t)dt - \tilde{x}^T(0)P\tilde{x}(0)}{\int_0^{t_f} \overline{w}^T(t)\overline{w}(t)dt} \qquad (7.19)$$

where t_f is the terminal time of control, and the matrix $\tilde{Q}_1 = diag\{0_{n_a(k+1) \times n_a(k+1)},$ $0_{n_s(k+1) \times n_s(k+1)}, \tilde{Q}_{1,x}, \tilde{Q}_{1,e}\}$ is specified beforehand according to the design purpose with the weighting matrix $\tilde{Q}_{1,x} \geq 0$ on the system state $x(t)$ and the weighting matrix $\tilde{Q}_{1,e} \geq 0$ on the estimation error $e(t)$. $n_{\overline{w}} = (n_a + n_s)(k+1) + n_w + m$ is the dimension of $\overline{w}(t)$. The term $\tilde{x}^T(0)P\tilde{x}(0)$ for some $P = P^T > 0$ in the numerator of (7.19) is used to deduct the effect of initial condition $\tilde{x}(0)$ on the H_∞ control and estimation performance. The main purpose of the H_∞ control and estimation performance is to eliminate the effect of external disturbance and measurement noise on the control and estimation performance of the observer-based T-S fuzzy FTC system in (7.18).

Most of the time, the nonlinear system is always in the normal condition without the occurrence of external disturbance and measurement noise. The consideration of H_∞ FTC design is not enough, and it may lead to a very conservative design. Hence, the following H_2 optimal control and estimation design, without the consideration of disturbance signal $\overline{w}(t)$, is more appealing for control engineers to achieve optimal quadratic observer-based fault-tolerant control via a proper choice of weighting matrices:

$$H_2(\{L_i, K_j\}_{i,j=1}^l) = \int_0^{t_f} \tilde{x}^T(t)\tilde{Q}_2\tilde{x}(t) + u^T(t)\tilde{R}_2u(t)dt \qquad (7.20)$$

where $\tilde{R}_2$ is the weighting matrix of control effort, and $\tilde{Q}_2 = diag\{0_{n_a(k+1) \times n_a(k+1)},$ $0_{n_s(k+1) \times n_s(k+1)}, \tilde{Q}_{2,x}, \tilde{Q}_{2,e}\}$ with the weighting matrix $\tilde{Q}_{2,x} \geq 0$ on the system state $x(t)$ and the weighting matrix $\tilde{Q}_{2,e} \geq 0$ on the estimation error $e(t)$. The weighting matrices $\tilde{Q}_2 \geq 0$ and $\tilde{R}_2 > 0$ are specified beforehand according to the design purpose of the FTC. Thus, the multi-objective optimal H_2/H_∞ observer-based FTC design problem for the fuzzy observer-based control system in (7.18) is given as follows:

$$\min_{\substack{L_i, K_j \\ j,i=1,\dots,l}} (H_2(\{L_i, K_j\}_{i,j=1}^l), H_\infty(\{L_i, K_j\}_{i,j=1}^l))$$

$$\text{subject to (7.18)} \qquad (7.21)$$

In this MOP, there exists a set of Pareto optimal solutions for the multi-objective optimal H_2/H_∞ observer-based FTC design problem in the Pareto optimal sense. In general, it is not easy to solve the MOP in (7.21) directly. Thus, in this study, an indirect method is proposed in the following to solve the MOP in (7.21) for multi-objective optimal H_2/H_∞ FTC design from the suboptimal perspective:

$$\min_{\substack{\{L_i,K_j\}_{i,j=1}^I \\ \alpha>0,\beta>0}} (\alpha,\beta)$$
$$\text{subject to } H_2(\{L_i,K_j\}_{i,j=1}^I) \le \alpha, H_\infty(\{L_i,K_j\}_{i,j=1}^I) \le \beta \qquad (7.22)$$

Since the MOP needs to minimize the multiple objective functions simultaneously, we use Pareto domination instead of conventional minimization. The solution of the MOP in (7.22) is a set of fuzzy control and observer parameters $\{L_i^*,K_j^*\}_{i,j=1}^I$ with the corresponding nondominated objective vector (α^*,β^*). Before further analysis, the following fundamental definitions are provided:

Definition 7.1 (Pareto Dominance [2]): For the MOP in (7.22), a solution $\{L_i^1,K_j^1\}_{i,j=1}^I$ with feasible objective vector (α^1,β^1) is said to dominate another solution $\{L_i^2,K_j^2\}_{i,j=1}^I$ with feasible objective vector (α^2,β^2) if and if only $\alpha^1 \le \alpha^2$ and $\beta^1 \le \beta^2$ and at least one of two inequalities with strict inequality.

Definition 7.2 (Pareto Optimal Solution [2]): For the MOP in (7.22), the solution $\{L_i^*,K_j^*\}_{i,j=1}^I$ with the corresponding objective vector (α^*,β^*) is a Pareto optimal solution if and only if it could not be dominated by any other solution.

Remark 7.4: Under the concept of Pareto optimal solution [2], the Pareto optimal solution of the MOP in (7.21) is not unique; that is, there exists a set of Pareto optimal solutions for the MOP in (7.21). Since there exist multiple solutions, it is not easy to solve the MOP in (7.21) by conventional optimization techniques.

Theorem 7.2: The MOP in (7.22) is equivalent to the MOP in (7.21) if the Pareto optimal solutions are achieved.

Proof: The proof is simple if we can prove that two inequalities in the constraints of (7.22) become equalities when the optimization objective vector (α^*,β^*) is achieved. We will prove this by contradiction. Suppose the inequality constraints in (7.22) are strictly held for a Pareto optimal solution (α',β'); that is, there exists a Pareto optimal solution (α',β') such that $H_2\left(\{L_i',K_j'\}_{i,j=1}^I\right) < \alpha'$ and $H_\infty(\{L_i',K_j'\}_{i,j=1}^I) < \beta'$ are strictly held. However, we can find (α^*,β^*) such that $H_2\left(\{L_i',K_j'\}_{i,j=1}^I\right) = \alpha^*$ and $H_\infty\left(\{L_i',K_j'\}_{i,j=1}^I\right) = \beta^*$. This immediately shows that the solution (α',β') is dominated by (α^*,β^*), and (α',β') is not the Pareto optimal solution in (7.22), which completes the inference of contradiction.

Based on Theorem 7.2, the MOP for the multi-objective H_2/H_∞ observer-based FTC problem becomes the MOP in (7.22). Before we solve the MOP in (7.22), the following lemma is given:

Lemma 7.1 [90]: For any two vectors $x \in \mathbb{R}^n$, $y \in \mathbb{R}^m$ and positive matrix P, we have

$$x^T Py + y^T Px \leq \frac{1}{\rho^2} x^T PPx + \rho^2 y^T y$$

where ρ is any nonzero real number.

Theorem 7.3: The MOP in (7.22) for the multi-objective optimal H_2/H_∞ fault-tolerant control design can be transformed into the following MOP:

$$\min_{\substack{P>0,\alpha>0,\beta>0 \\ (L_i,K_j)_{i,j=1}^I}} (\alpha,\beta) \tag{7.23}$$

$$\text{subject to } \tilde{A}_{ij}^T P + P\tilde{A}_{ij} + \tilde{Q}_2 + \bar{K}_j^T \tilde{R}_2 \bar{K}_j \leq 0 \tag{7.24}$$

$$\tilde{x}^T(0)P\tilde{x}(0) \leq \alpha \tag{7.25}$$

$$P\tilde{A}_{ij}^T + \tilde{A}_{ij}^T P + \tilde{Q}_1 + \frac{1}{\beta} P\tilde{D}_{ij}\tilde{D}_{ij}^T P \leq 0 \tag{7.26}$$

$$\text{where } \bar{K}_j = [K_j, -K_j], \forall i,j = 1,...,I$$

Proof: Please refer to Appendix 7.7.2.

By choosing the general Lyapunov function $V(\tilde{x}(t)) = \tilde{x}^T(t)P\tilde{x}(t)$ with positive definite matrix $P > 0$, the multi-objective optimal H_2/H_∞ FTC design can be transformed into the MOP in (7.23)–(7.26). However, since the constraints in (7.24)–(7.26) are bilinear constraints, it cannot be solved directly by any current optimization technique. Thus, by choosing the Lyapunov function for the augmented fuzzy observer-based control system in (7.18) as the sums of Lyapunov functions of two sub-systems, that is, $V(\tilde{x}(t)) = \bar{x}^T(t)P_1\bar{x}(t) + e^T(t)P_2 e(t)$ with $P_1 > 0$ and $P_2 > 0$, the design problem can be transformed into the MOP in (7.28)–(7.31), and it can be easily solved by MATLAB LMI toolbox with the proposed two-step design procedure and reverse-order MOEA in the sequel. To begin with, some decoupling techniques are employed for the weighting matrices $\tilde{Q}_1$ of H_∞ control performance and $\tilde{Q}_2$ of H_2 control performance as follows:

$$\tilde{Q}_1 = \begin{bmatrix} \tilde{Q}_{1,\tilde{x}} & 0 \\ 0 & \tilde{Q}_{1,e} \end{bmatrix}, \tilde{Q}_2 = \begin{bmatrix} \tilde{Q}_{2,\tilde{x}} & 0 \\ 0 & \tilde{Q}_{2,e} \end{bmatrix} \tag{7.27}$$

where the matrices $\tilde{Q}_{1,\tilde{x}} = diag\{0_{n_a(k+1)\times n_a(k+1)}, 0_{n_s(k+1)\times n_s(k+1)}, \tilde{Q}_{1,x}\}$, $\tilde{Q}_{2,\tilde{x}} = diag$ $\{0_{n_a(k+1)\times n_a(k+1)}, 0_{n_s(k+1)\times n_s(k+1)}, \tilde{Q}_{2,x}\}$. Thus, we can get the following result:

Theorem 7.4: The MOP in (7.23) can be transformed into the following BMI-constrained MOP:

$$(\alpha^*, \beta^*) = \min_{\substack{W_1 > 0, P_2 > 0 \\ (L_i, K_j)_{i,j=1}^I, \alpha > 0, \beta > 0}} (\alpha, \beta) \tag{7.28}$$

subject to

$$\begin{bmatrix} \Delta_{2,1}'' & -\bar{B}_{u,i} K_j & W_1 K_j^T & W_1 (\tilde{Q}_{2,\tilde{x}})^{1/2} \\ * & \Delta_{2,2}'' & -K_j^T & 0 \\ * & * & -\tilde{R}_2^{-1} & 0 \\ * & * & * & -I \end{bmatrix} \le 0 \tag{7.29}$$

$$\begin{bmatrix} e^T(0) P_2 e(0) - \alpha & \tilde{x}^T(0) \\ \tilde{x}(0) & -W_1 \end{bmatrix} \le 0 \tag{7.30}$$

$$\begin{bmatrix} \Delta_{\infty,1}'' & -\bar{B}_{u,i} K_j & \bar{B}_{w,i} & W_1 (\tilde{Q}_{1,\tilde{x}})^{1/2} \\ * & \Delta_{\infty,2}'' & P_2 \bar{B}_{w,i} - P_2 L_i \bar{D}_j & 0 \\ * & * & -\beta I & 0 \\ * & * & * & -I \end{bmatrix} \le 0 \tag{7.31}$$

for $i, j = 1, ..., I$, where $\Delta_{2,1}'' = \bar{A}_i W_1 + \bar{B}_{u,i} K_j W_1 + W_1 \bar{A}_i^T + W_1 K_j^T \bar{B}_{u,i}^T$, $\Delta_{2,2}'' = P_2 \bar{A}_i - P_2 L_i \bar{C}_j$ $+ \bar{A}_i^T P_2 - \bar{C}_j^T L_i^T P_2 + \tilde{Q}_{2,e}$, $\Delta_{\infty,1}'' = \bar{A}_i W_1 + \bar{B}_{u,i} K_j W_1 + W_1 \bar{A}_i^T + W_1 K_j^T \bar{B}_{u,i}^T$, and $\Delta_{\infty,2}'' = P_2 \bar{A}_i$ $- P_2 L_i \bar{C}_j + \bar{A}_i^T P_2 - \bar{C}_j^T L_i^T P_2 + \tilde{Q}_{1,e}$.

Proof: Please refer to Appendix 7.7.3.

If we could solve (α^*, β^*) of the MOP in (7.28)–(7.31) with the corresponding Pareto optimal solution $\{W_1^*, P_2^*, (L_i, K_j)_{i,j=1}^I\}$, then we could obtain the fuzzy controller parameters K_j^* in (7.16) and fuzzy observer parameters L_i^* in (7.14) for the following fuzzy observer-based FTC scheme:

$$\dot{\hat{x}}(t) = \sum_{i=1}^I h_i(z(t)) \bar{A}_i \hat{x}(t) + \bar{B}_{u,i} u(t) + L_i^*(y(t) - \hat{y}(t))$$
$$u(t) = \sum_{i=1}^I h_j(z(t)) K_j^* \hat{\bar{x}}(t), \text{ for } i, j = 1, ..., I \tag{7.32}$$

Remark 7.5: (i) If we only consider the optimal H_2 observer-based FTC of T-S fuzzy system, then the MOP in (7.28) is reduced to the following single-objective optimization problem:

$$\alpha^+ = \min_{\substack{W_1 > 0, P_2 > 0 \\ (L_i, K_j)_{i,j=1}^I, \alpha > 0}} \alpha \tag{7.33}$$

subject to the matrix constraints in (7.29), (7.30)

(ii) If we only consider the optimal H_∞ observer-based FTC of the T-S fuzzy system, then the MOP in (7.28) is reduced to the following SOP:

$$\beta^+ = \min_{\substack{W_i > 0, P_2 > 0 \\ (L_i, K_j)_{i,j=1}^l, \beta > 0}} \beta \tag{7.34}$$

subject to the matrix constraints in (7.31)

Remark 7.6: Since the constraints in (7.29) and (7.31) are BMIs, it cannot be efficiently solved via current optimization techniques. As a result, a two-step design procedure is developed to solve the MOP in (7.28)–(7.31):

(i) If the inequality constraints in (7.29) and (7.31) hold, then the sufficient and necessary conditions are that the diagonal terms of (7.29) and (7.31) must be negative. Thus, by letting $Z_j = K_j W_1$, we first solve the linear constraints $\Delta''_{2,1} < 0$ in (7.29) and $\Delta''_{\infty,1} < 0$ in (7.31) to obtain the solution W_1^* and $\{Z_i^*\}_{i=1}^l$ by using the MATLAB convex optimization toolbox. Once the solutions W_1^* and $\{Z_i^*\}_{i=1}^l$ are obtained, the corresponding fuzzy controller can be constructed as $K_j^* = Z_i^* (W_1^*)^{-1}$.

(ii) By letting $Y_i = P_2 L_i$ with the fixed matrices W_2^* and $\{K_j^*\}_{j=1}^l$ obtained from step 1 in (7.29) and (7.31), the constraints in (7.29) and (7.31) become linear matrix inequalities, which are solvable by utilizing the MATLAB convex optimization toolbox. Thus, we can solve (7.29)–(7.31) efficiently to obtain the optimal solutions P_2^* and $\{Y_i^*\}_{i=1}^l$. Once the solutions P_2^* and $\{Y_i^*\}_{i=1}^l$ are obtained, the corresponding fuzzy observer can be constructed as $L_i^* = (P_2^*)^{-1} Y_i^*$.

Remark 7.7: If the robust H_∞ performance in (7.19) or (7.52) is satisfied with positive definite weighting matrix $\tilde{Q}_1$ and a prescribed disturbance attenuation level $\beta > 0$ and external disturbance $\bar{W}(t) \in L_2(R^+, R^n)$, the right-hand side in (7.52) is finite as $t_f \to \infty$. As a result, it reveals the state variables will converge to zero by the proposed controller; that is, asymptotic stability for the fuzzy observer-based FTC system in (7.18) is guaranteed.

7.4 REVERSE-ORDER LMI-CONSTRAINED MOEA FOR MULTI-OBJECTIVE OPTIMAL H_2/H_∞ OBSERVER-BASED FAULT-TOLERANT DESIGN OF T-S FUZZY SYSTEMS

In the conventional MOEA, the algorithm employs an evolution algorithm to search the parameter of $\{W_1, P_2, K_i, L_i\}_{i=1}^L$ to solve the MOP in (7.28)–(7.31) directly. However, there are too many fuzzy controller and observer parameters to be searched to achieve the optimal objective vector (α^*, β^*). Obviously, it is almost impossible to employ the conventional MOEA to the MOP in (7.28)–(7.31) to solve the multi-objective H_2/H_∞ observer-based FTC design problem. Different from the conventional multi-objective evolution algorithm that directly searches the solutions $\{\alpha, \beta, W_1, P_2, K_i, L_i\}_{i=1}^L$ of the multi-objective optimal H_2/H_∞ observer-based FTC design in (7.28), the proposed reverse-order MOEA only searches the feasible objective vectors (α, β). Then, by substituting the objective vectors (α, β) into (7.29)–(7.31), the corresponding solutions $\{K_i, L_j\}_{i,j=1}^I$ can be easily obtained by using MATLAB LMI toolbox. As a result, the multi-objective optimal H_2/H_∞ observer-based FTC design in (7.28) can be effectively solved by the proposed reverse-order LMI-constrained MOEA and MATLAB LMI toolbox. Before further analysis of the LMI-constrained MOP in (7.28)–(7.31), some fundamental definitions of MOPs are given in the following:

Definition 7.3 ([2]): For the LMI-constrained MOP in (7.28)–(7.31) for multi-objective H_2/H_∞ observer-based FTC design in the T-S fuzzy system with actuator and sensor faults in (7.1), the ideal objective vector is defined as (α^+, β^+), where α^+ and α^+ can be obtained by solving the SOP in (7.33) and (7.34), respectively.

Definition 7.4 ([2]): For the LMI-constrained MOP in (7.28)–(7.31), the solution $\{W_1^*, P_2^*, \{K_j^*, L_i^*\}_{i,j=1}^I\}$ with the corresponding objective vector (α^*, β^*) is called the Pareto optimal solution if there does not exist any feasible solution $\{W_1, P_2, \{K_j, L_i\}_{i,j=1}^I\}$ with the corresponding objective vector (α, β) that dominates the solution $\{W_1^*, P_2^*, \{K_j^*, L_i^*\}_{i,j=1}^I\}$.

Definition 7.5 ([2]): For the LMI-constrained MOP in (7.28)–(7.31), the Pareto optimal solution set is defined as follows:

$\Omega_{Opt} = \{\{W_1^*, P_2^*, \{K_j^*, L_i^*\}_{i,j=1}^I\} \|$ The feasible solutions in (7.28)–(7.31) are not dominated by other feasible solutions$\}$,

that is, the Pareto optimal solution set collects the Pareto optimal solutions for the MOP in (7.28)–(7.31).

Definition 7.6 ([2]): For the LMI-constrained MOP in (7.28)–(7.31), the Pareto front is defined as follows:

$\Omega_{Front} = \{(\alpha^*, \beta^*) \|$ The objective vector (α^*, β^*) of the solution $\{W_1^*, P_2^*, \{K_j^*, L_i^*\}_{i,j=1}^I\}$ $\in \Omega_{Opt}\};$

that is, the Pareto front Ω_{Front} collects the objective vectors of the Pareto optimal solutions in Ω_{Opt}.

Once the Pareto front Ω_{Front} is obtained, the corresponding fuzzy controller gains $\{K_j^*\}_{j=1}^I$ and fuzzy observer gain $\{L_i^*\}_{i=1}^I$ can be constructed for multi-objective H_2/H_∞ fault-tolerant control of the T-S fuzzy system with actuator and sensor failure. Based on the analysis, a design procedure based on the proposed reverse-order LMI-constrained MOEA method for the multi-objective optimal H_2/H_∞ observer-based FTC design of the T-S fuzzy system with actuator and sensor faults is given in detail:

Design procedure of the multi-objective H_2/H_∞ observer-based FTC for the T-S fuzzy system

Step 1: Choose the search region $R = [\alpha_{low}, \alpha_{up}] \times [\beta_{low}, \beta_{up}]$. The objective vector $[\alpha_{low}, \beta_{low}]$ at the low bound can be obtained by solving the two SOPs in (7.33) and (7.34); that is, $\alpha_{low} = \alpha^+$, $\beta_{low} = \beta^+$. Assume the population number N_p, iteration number N_i, crossover rate C_r, and mutation rate C_m of the EA in the proposed reverse-order LMI-constrained MOEA. Set the iteration index as $j = 1$.

Step 2: Randomly generate the feasible objective vector $\{(\alpha_i, \beta_i)\}_{i=1}^{N_p}$ from the search region R by examining whether $\{(\alpha_i, \beta_i)\}_{i=1}^{N_p}$ are feasible in the LMI constraints in (7.29)–(7.31). Define the parent set as $P_{Parent}^{j} = \{(\alpha_i, \beta_i)\}_{i=1}^{N_p}$.

Step 3: Employ the crossover and mutation operators for the parent set P_{Parent}^{j} and produce N_p child population, that is, $\{(\tilde{\alpha}_i, \tilde{\beta}_i)\}_{i=1}^{N_p}$. Define the child set as $P_{Parent}^{j} = \{(\tilde{\alpha}_i, \tilde{\beta}_i)\}_{i=1}^{N_p}$. If there exists a child population $(\tilde{\alpha}, \tilde{\beta})$ that is not feasible for the constraints in (7.29)–(7.31), then a population (α, β) from the parent set that is closest to $(\tilde{\alpha}, \tilde{\beta})$ is selected, and the child population is replaced by $\frac{1}{2}(\alpha + \tilde{\alpha}, \beta + \tilde{\beta})$. The mechanism will be repeatedly executed until the fixed child population is feasible for the constraints in (7.29)–(7.31).

Step 4: Apply the nondominating sorting operator to the set $P_{Parent}^{j} \cup P_{Child}^{j}$ to obtain the corresponding nondominated front $F^{j} = \{F_1^{j}, F_2^{j}, ...\}$.

Step 5: Apply the crowded comparison assignment operator to the sets F_i^{j} and generate the corresponding crowding distance of each element in F_i^{j} for $i = 1, ...$. Based on their crowding distance, the sets $\{F_i^{j}\}_{i \in N}$ can be sorted in descending order.

Step 6: Let $t_j \in N$ be the minimum positive integer such that $\sum_{i=1}^{t_j} |F_i^{j}| > N_p$, where $|S|$ denotes the cardinality of the set S; that is, $t_j = \arg\min_{t_j \in N} \sum_{i=1}^{t_j} |F_i^{j}| > N_p$. Update the iteration index $j = j+1$ and let the parent set $P_{Parent}^{j} = \{F_i^{j}\}_{i=1}^{t_j-1} \cup \tilde{F}_i^{t_j}$ for the next iteration where $\tilde{F}_i^{t_j}$ is the set containing the first G^j elements in F_i^{j} and $G^j = N_p - \sum_{i=1}^{t_j} |F_i^{j}|$.

Step 7: Repeat Steps 3 to 6 until the iteration index $j = N_i$, and set the final population $P_{Parent}^{N_i} = \Omega_{Front}$ as the Pareto front.

Step 8: Choose a desired Pareto optimal objective vector $(\alpha^*, \beta^*) \in \Omega_{Front}$ according to the one's preference with the corresponding Pareto optimal solution $\{W_1^*, P_2^*, \{K_j^*, L_i^*\}_{i,j=1}^{I}\}$ of the MOP in (7.28)–(7.31). Once the preferable optimal solution is selected, the multi-objective optimal fuzzy observer-based fault-tolerant controller in (7.32) can be constructed as $\{K_j^*\}_{j=1}^{I}$ and $\{L_i^*\}_{i=1}^{I}$ for the multi-objective H_2/H_∞ observer-based FTC in (7.32) of the T-S fuzzy system in (7.2).

The detailed crossover operator, mutation operator, nondominating sorting operator, and crowded comparison operator are defined as follows:

Crossover Operator:

(I) Define two populations $(\alpha_1, \beta_1), (\alpha_2, \beta_2) \in \mathbb{R}^2$ and randomly generate a number $r \in [0, 1]$.

(II)   The crossover operator $f_C((\alpha_1,\beta_1),(\alpha_2,\beta_2))$ is defined as follows:

$$f_C((\alpha_1,\beta_1),(\alpha_2,\beta_2)) = \begin{cases} (\alpha_1,\beta_2) \text{ if } r<C_r \\ \frac{1}{2}(\alpha_1+\alpha_2,\beta_1+\beta_2), o.w. \end{cases}$$

where $C_r \in [0,1]$ is the crossover rate.

Mutation Operator:

(I)   Define one population $(\alpha,\beta) \in \mathbb{R}^2$ and randomly generate two numbers $r_1, r_2 \in [0,1]$.
(II)   The mutation operator $f_M((\alpha,\beta))$ is defined as follows:

$$f_M((\alpha,\beta)) = \begin{cases} (\alpha,\beta) & \text{if } r_1 \geq C_r, r_2 \geq C_m \\ (\alpha+d_1,\beta) & \text{if } r_1 \leq C_r, r_2 \geq C_m \\ (\alpha,\beta+d_2) & \text{if } r_1 \geq C_r, r_2 \leq C_m \\ (\alpha+d_1,\beta+d_2) & \text{if } r_1 \leq C_r, r_2 \leq C_m \end{cases}$$

where $C_m \in [0,1]$ denotes mutation rate, $d_1 = 1-(2(1-r_1))^{1/5}$, and $d_2 = 1-(2(1-r_2))^{1/5}$.

Crowded Comparison Assignment Operator:

(I)   Define a finite set I with cardinality $l = |I|$ and set $\{I_i = 0\}_{i=1}^{l}$.
(II)   Set the objective index $j = 1$.
(III) Sort I in descending order according to the jth objective value.
(IV)   Assign values to the first element I_1 and last element I_l in the sorted set I as follows:

$$I_1 = \infty, I_l = \infty$$

(V)   Assign values to the elements $\{I_i\}_{i=2}^{l-1}$ in the sorted set I:

$$I_i = I_i + (f_j^{i+1} - f_j^{i-1})/(f_j^{max} - f_j^{min})$$
$$i = 2,...,l-1$$

where f_j^{i+1} and f_j^{i-1} are the jth objective value of the $i+1$th and $i-1$th elements in the sorted set I, respectively, and f_j^{max} and f_j^{min} are the maximum and minimum jth objective values in the sorted set I, respectively.
(VI)   Update the objective index $j = j+1$ and repeat Steps (III)–(V).

Nondominating Sorting Operator:

(I)   Given a set P, for each element $p \in P$, generate the corresponding counter index n_p and domination set S_p as follows:

$$S_p = \{q \| q \in P, p < q\}$$
$$n_p = \{q \| q \in P, q \leq p\}$$

(II)   Define the first domination front F_1 as

$$F_1 = \{p \| p \in P, n_p = 0\}$$

and the jth domination front F_j as

$$F_j = \{q \| p \in F_{j-1}, q \in S_p, n_p = \sum_{i=1}^{j-1} |F_i|\}, \text{for } j \in \mathbb{N}$$

To sum up, the multi-objective optimal H_2/H_∞ FTC design problem is transformed into the BMI-constrained multi-objective problem. Consequently, the BMI-constrained MOP can be solved by utilizing the proposed reverse-order MOEA and the MATLAB LMI toolbox.

Remark 7.8: The computational complexity of the proposed reverse-order MOEA for the LMI-constrained MOP in (7.28)–(7.31) is about $O(n^2 I N_i N_p^2)$, which includes $O(n^2 I)$ in solving two-step LMIs in (7.29)–(7.31) and $O(N_i N_p^2)$ in the reverse-order MOEA, where I is the number of fuzzy rules, n is the dimension of positive-definite matrix P, N_p is the population number, and N_i is the iteration number.

Remark 7.9: Since the augmented system in (7.18) includes the smoothed model of the actuator and sensor signals, the dimensions of the augmented system will be enlarged if we use more points to extrapolate the actuator fault $f_a(t+h)$ and sensor fault $f_s(t+h)$. As a result, the computing complexity directly depends on the dimensions of the smoothed model of actuator and sensor signals.

Remark 7.10: In the proposed reverse-order MOEA for multi-objective H_2/H_∞ observer-based FTC design, the designer should give the following parameters for the MOEA:

(I)   Search region $R = [\alpha_{low}, \alpha_{up}] \times [\beta_{low}, \beta_{up}] \in \mathbb{R}^2$ with $\alpha_{low} \in \mathbb{R}_{\geq 0}$, $\alpha_{up} \in \mathbb{R}_{\geq 0}$, $\beta_{low} \in \mathbb{R}_{\geq 0}$, and $\beta_{up} \in \mathbb{R}_{\geq 0}$
(II)   Population number $N_p \in \mathbb{N}$
(III) Iteration number $N_i \in \mathbb{N}$
(IV)   Crossover rate $C_r \in [0,1]$
(V)   Mutation rate $C_m \in [0,1]$

On the other hand, at the start of the algorithm, N_p populations are randomly generated from the search region R and are set as the initial population.

Remark 7.11: Recently, with the advance of interpolation methods, the nonlinear inequality constraint can be equivalently transformed into a combination of local linearized matrix inequalities with corresponding interpolation functions, such as the fuzzy interpolation method and global linearization method. As a result, by utilizing these interpolation methods, the proposed reverse-order MOEA can be applied to general MOPs with nonlinear inequality constraints, such as multi-objective optimal controller design for nonlinear systems.

Remark 7.12: In the proposed reverse-order MOEA, the initial populations are randomly generated from the predefined search region. Clearly, if the initial populations are close to the real Pareto front, the MOEA will take less time to approach the real Pareto front, that is, global optimality. As a result, the convergence of the proposed reverse-order MOEA is related to the choice of search region. In this study, the search region is defined as $R = [\alpha_{low}, \alpha_{up}] \times [\beta_{low}, \beta_{up}]$, where α_{low} and β_{low} can be obtained by solving the two SOPs in (7.33) and (7.34). On the other hand, α_{up} and β_{up} can be chosen as several times of α_{low} and β_{low}, respectively, such as $\alpha_{up} = \theta\alpha_{low}$, for some $\theta \in \mathbb{R}_+$. In this situation, the randomly generated initial populations will fall into the feasible region of the constraints in (7.24)–(7.26).

7.5   SIMULATION EXAMPLE

To illustrate the design procedure and validate the performance of proposed MO H_2/H_∞ observer-based FTC scheme, two simulation examples are provided. A tactical missile guidance system suffering from actuator and sensor fault signals is given. Based on the proposed smoothed model, the fault signals are embedded in an augmented missile system. Then the proposed MO H_2/H_∞ observer-based FTC problem is given to the tactical missile guidance system. Finally, the MO optimal H_2/H_∞ observer-based FTC design problem is transformed into the BMI-constrained multi-objective problem, which can easily be solved by utilizing the proposed reverse-order MOEA and MATLAB LMI toolbox. On the other hand, the FTC design problem of balancing and swing-up of an inverted pendulum on a cart is considered with a comparison between the proposed method and conventional FTC design in [168].

Example 7.1: Consider the 3-D missile guidance system in the spherical coordinate (r, ψ, θ) with the origin fixed at the missile. The pursuit-evasion geometry between a missile such as a Patriot at the origin and a target such as an incoming ballistic missile is shown in Figure 7.1. Let $(\vec{e}_r, \vec{e}_\psi, \vec{e}_\theta)$ be the unit vector along the coordinate axis. The 3-D relative velocity is obtained through the differentiation of the relative distance vector $\vec{r}$ along with the line of sight as follows [92, 100]:

$$\dot{\vec{r}} = \dot{r}\vec{e}_r + r\dot{\psi}\cos(\theta)\vec{e}_\psi + r\dot{\theta}\vec{e}_\theta \qquad (7.35)$$

Hence, the relative acceleration at the direction of $\vec{e}_r$, $\vec{e}_\psi$, and $\vec{e}_\theta$ can be obtained by differentiating the previous equation in the following:

$$\ddot{r} - r\dot{\theta} - r\dot{\psi}^2 \cos^2 \theta = w_r$$
$$r\ddot{\psi}\cos\theta + 2\dot{r}\dot{\psi}\cos\theta - 2r\dot{\theta}\dot{\psi}\sin\theta = w_\psi - u_\psi \qquad (7.36)$$
$$r\ddot{\theta} + 2\dot{r}\dot{\theta} + r\dot{\psi}^2 \cos\theta \sin\theta = w_\theta - u_\theta$$

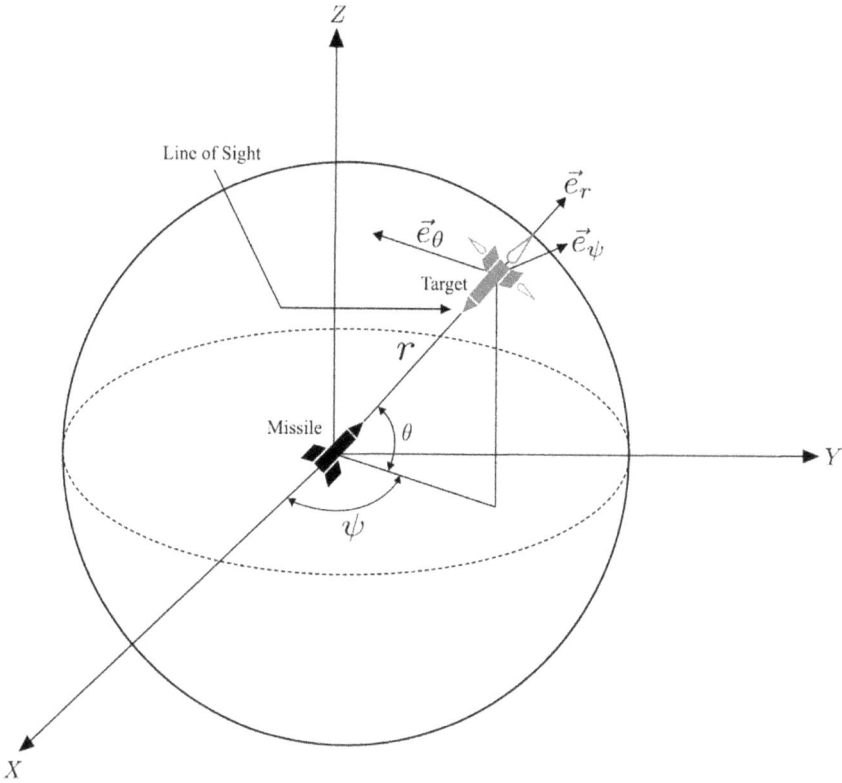

FIGURE 7.1 3-D pursuit-evasion geometry in the missile guidance system.

where u_ψ and u_θ are the control inputs, and w_r, w_ψ, and w_θ are the target acceleration vectors. Therefore, the kinematic between the missile and target in (7.36) can be represented by the following state space system:

$$\dot{x}(t) = F(x(t)) + B_u u(t) + B_w w(t)$$
$$y(t) = Cx(t) + n(t)$$

(7.37)

where $x(t) = [r(t), \psi(t), \theta(t), v_r(t), v_\psi(t), v_\theta(t)]^T$ denotes the state vector; $F(x(t))$ denotes the vector field; $u(t) = [u_\psi(t), u_\theta(t)]^T$ denotes the missile acceleration vector due to the guidance control; and $w(t) = [w_r(t), w_\psi(t), w_\theta(t)]^T$ denotes the target acceleration vector, which is unavailable and is considered the external disturbance to the missile guidance system. $y(t)$ denotes the measurement output of the missile by laser sensor with measurement noise $n(t)$. The detailed information of these matrices is as follows:

$$F(x(t)) = \begin{bmatrix} v_r \\ \dfrac{v_\psi}{r\cos(\theta)} \\ \dfrac{v_\theta}{r} \\ \dfrac{v_\psi^2 + v_\theta^2}{r} \\ \dfrac{v_\psi(v_\theta \tan\theta - v_r)}{r} \\ \dfrac{-(v_r v_\theta + v_\psi^2 \tan\theta)}{r} \end{bmatrix}, B_u = \begin{bmatrix} 0_{4\times2} \\ -I_2 \end{bmatrix}, B_w = \begin{bmatrix} 0_{3\times3} \\ I_3 \end{bmatrix}, C = I_3$$

To avoid the attack of a tactical missile such as the Patriot, the target will generate a jamming signal to interfere with the laser sensor through the wireless channel, which will lead to an equivalent sensor fault signal. On the other hand, the target will perform sudden side-step maneuvering through its two-side jets, which will lead to an equivalent actuator fault in the missile system. By considering the sensor and actuator faults by hostile jamming and sudden side-step maneuvering, respectively, the missile guidance system in (7.37) with actuator and sensor faults should be revised as:

$$\begin{aligned} \dot{x}(t) &= F(x(t)) + B_u u(t) + B_w w(t) + B_a f_a(t) \\ y(t) &= Cx(t) + n(t) + Df_s(t) \end{aligned} \tag{7.38}$$

where $f_a(t) \in \mathbb{R}$ denotes the actuator fault due to sudden side-step maneuvering through two-side jets in the target, and $f_s(t) \in \mathbb{R}$ denotes the sensor fault of the missile due to hostile jamming from the target. The corresponding fault matrices are given as follows:

$$B_a = [0,0,0,0,1,1]^T, D = [0,0,0,1,1,1]^T$$

In this simulation example, the following fuzzy premise variables are chosen, $z_2(t) = \theta(t)$, $z_3(t) = v_\psi(t)$, and $z_4(t) = v_\theta(t)$ with the corresponding fuzzy operation points:

$$r_i = 599.7 \text{ for } i = 1 - 24$$

$$r_i = 2560 \text{ for } i = 25 - 48$$

$$\theta_i = -0.6452 \text{ for } i = 1-12, 25-36$$

$$\theta_i = 1.2872 \text{ for } i = 13-24, 37-48$$

$$v_{\psi,i} = -50 \text{ for } i = 1-4, 13-16, 25-38, 37-40$$

$$v_{\psi,i} = 75.6 \text{ for } i = 5-8, 17-20, 29-32, 41-44$$

$$v_{\psi,i} = 551.1 \text{ for } i = 9-12, 20-24, 33-36, 45-48$$

$$v_{\theta,i} = -121 \text{ for } i = 1+4s$$

$$v_{\theta,i} = 0 \text{ for } i = 2+4s$$

$$v_{\theta,i} = 135.3 \text{ for } i = 3+4s$$

$$v_{\theta,i} = 310.5 \text{ for } i = 4+4s$$

where $s = 0-11$

Also, the ith if-then rule of the T-S fuzzy system for the nonlinear system in (7.38) is given as follows:

If $r(t)$ is r_i and $\theta(t)$ is θ_i
and $v_\psi(t)$ is $v_{\psi,i}$ and $v_\theta(t)$ is $v_{\theta,i}$
Then $\hspace{4cm}$ (7.39)
$$\dot{x}(t) = A_i x(t) + B_{u,i} u(t) + B_{a,i} f_a(t) + B_{w,i} w(t)$$
$$y(t) = C_i x(t) + D_i f_s(t) + n(t)$$

and the detailed local linearized matrices $\{A_i\}_{i=1}^{48}$ are given in [169].

To model the actuator and sensor faults, the third-order smoothed model in (7.5) is employed for the actuator fault signal, and the third-order smoothed model in (7.6) is employed for the sensor fault signal as follows:

$$A_{f_a} = \begin{bmatrix} \dfrac{\bar{a}_0}{h} & \dfrac{a_1}{h} & \dfrac{a_2}{h} & \dfrac{a_3}{h} \\ \dfrac{1}{h} & -\dfrac{1}{h} & 0 & 0 \\ 0 & \dfrac{1}{h} & -\dfrac{1}{h} & 0 \\ 0 & 0 & \dfrac{1}{h} & -\dfrac{1}{h} \end{bmatrix}, A_{f_s} = \begin{bmatrix} \dfrac{\bar{b}_0}{h} & \dfrac{b_1}{h} & \dfrac{b_2}{h} & \dfrac{b_3}{h} \\ \dfrac{1}{h} & -\dfrac{1}{h} & 0 & 0 \\ 0 & \dfrac{1}{h} & -\dfrac{1}{h} & 0 \\ 0 & 0 & \dfrac{1}{h} & -\dfrac{1}{h} \end{bmatrix}$$

with the specified extrapolation parameters $h = 10^{-3}$, $\bar{a}_0 = -1 + a_0$, $a_0 = 0.5$, $a_1 = 0.3$, $a_2 = 0.1$, $a_3 = 0.1$, $\bar{b}_0 = -1 + b_0$, $b_0 = 0.6$, $b_1 = 0.2$, $b_2 = 0.1$ and $b_3 = 0.1$. Hence, the overall fuzzy observer-based FTC system can be constructed as follows:

$$\dot{\tilde{x}}(t) = \sum_{i=1}^{I}\sum_{j=1}^{I} h_i(z(t))h_j(z(t))(\tilde{A}_{ij}\tilde{x}(t) + \tilde{D}_{ij}\bar{w}(t)) \tag{7.40}$$

In the missile guidance control, when $v_\psi(t) \to 0$ and $v_\theta(t) \to 0$, it means the missile and target in are the head-on condition [100]. Among three relative velocities, only the relative velocity $v_r(t)$ along with line of sight decreases the distance between the missile and the target; that is, $v_r(t)$ could decrease the relative distance between missile. Therefore, $v_r(t)$ could not be zero in the missile guidance process; that is, $v_r(t) \neq 0$. In this situation, to ensure the head-on condition, the controlled output $\eta(t) = [v_\psi(t), v_\theta(t)]^T$ to be controlled as small as possible for the missile guidance system in (7.38) can be obtained as:

$$\eta(t) = Ex(t)$$

where

$$E = \begin{bmatrix} 0 & 0 & 0 & 0 & 1 & 0 \\ 0 & 0 & 0 & 0 & 0 & 1 \end{bmatrix}$$

Therefore, to meet the head-on condition for MO H_2/H_∞ observer-based fault-tolerant control, the weighting matrices $\tilde{Q}_1$ and $\tilde{Q}_2$ in (7.19) and (7.20) should be specified as:

$$\begin{aligned}
\tilde{Q}_1 &= diag\{0_{8\times8}, \tilde{Q}_{1,x}, \tilde{Q}_{1,e}\} \\
\tilde{Q}_2 &= diag\{0_{8\times8}, \tilde{Q}_{2,x}, \tilde{Q}_{2,e}\} \\
\tilde{Q}_{1,x} &= E^T Q_1^\eta E, \tilde{Q}_{2,x} = E^T \tilde{Q}_{2,x} E \\
Q_1^\eta &= diag(1,1), Q_2^\eta = diag(2,1) \\
\tilde{Q}_{1,e} &= I_{14\times14}, \tilde{Q}_{2,e} = I_{14\times14}, \tilde{R}_2 = 0.01 I_{2\times2}
\end{aligned} \tag{7.41}$$

Remark 7.13: To ensure MO H_2/H_∞ observer-based FTC performance and the head-on condition, the weighting matrices $\tilde{Q}_{1,x}$ and $\tilde{Q}_{2,x}$ in $\tilde{Q}_1$ and $\tilde{Q}_2$ should focus on the control performance of the controlled output $\eta(t) = [v_\psi(t), v_\theta(t)]^T$, respectively. On the other hand, to ensure the estimation performance on state variables and actuator and sensor fault signals in the missile guidance system, the weighting matrices $\tilde{Q}_{1,e}$ and $\tilde{Q}_{2,e}$ in $\tilde{Q}_1$ and $\tilde{Q}_2$ are chosen as the full rank matrix.

To apply the MOEA for the MO H_2/H_∞ observer-based FTC in (7.28), the design parameters of MOEA are given as:

Iteration Number N_i: 100
Search Region R: $[10^6, 10^{10}] \times [0.6, 10]$

Crossover Rate C_r: 0.85
Mutation Rate C_m: 0.15
Population Number N_p: 300

Once the iteration number is achieved in the MOEA, the Pareto front of the MO H_2/H_∞ fault-tolerant observer-based control design is plotted, as shown in Figure 7.2. In the Pareto front, each red point represents a Pareto optimal solution with the corresponding optimal H_2 observer-based FTC performance and optimal H_∞ observer-based FTC performance. Since the objective vectors of obtained Pareto optimalities cannot be decreased anymore in Figure 7.2, the obtained Pareto front is nearly equivalent to the real Pareto front. According to their requirements, the designer is free to choose the Pareto optimal solution. In this study, the marked point $(8.9 \times 10^6, 0.61)$ in Figure 7.2, which is the so-called knee-solution, is chosen as the solution. In general, the knee solution has the benefit of balanced performance between H_2 and H_∞ performance. From the knee point, the fuzzy controllers and observers can be constructed as $K_j^* = Z_j(W_j^*)^{-1}$ and $L_i^* = (P_2^*)^{-1} Y_i^*$ for $i, j = 1, ..., 48$. The detailed parameters of $\{K_j^*, L_i^*\}_{i,j=1}^{48}$ can be found in [169].

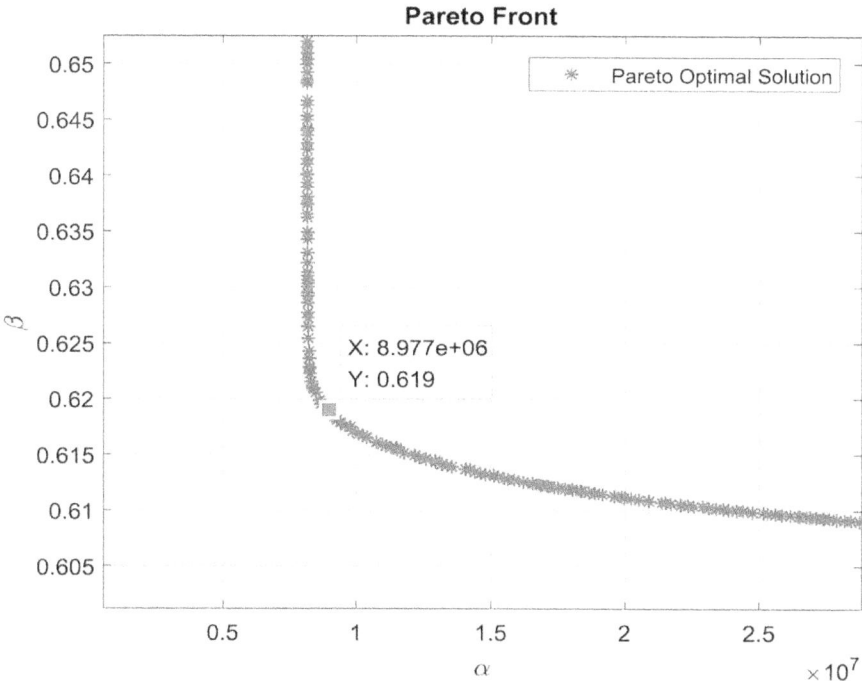

FIGURE 7.2 The Pareto front of the MO H_2/H_∞ fault-tolerant observer-based control design. In this figure, each red point represents the Pareto optimal solution in the MO H_2/H_∞ fault-tolerant observer-based control design. The marked point denotes the knee solution.

In this simulation, the ramp strategy of the target is chosen as the target acceleration vectors [100]:

$$w_r(t) = \lambda_T r(t), w_\psi(t) = \lambda_T \frac{-\dot{\theta}(t)}{\sqrt{\dot{\theta}^2(t) + \dot{\psi}(t)\cos^2\theta(t)}} \psi(t)$$

$$w_\theta(t) = \lambda_T \frac{\dot{\psi}(t)\cos\theta(t)}{\sqrt{\dot{\theta}^2(t) + \dot{\psi}(t)\cos^2\theta(t)}} \theta(t)$$

where λ_T denotes the target's navigation random gain from 0 to 2G. The measurement noise and the initial condition of the missile guidance system in (38) are given as:

$$n(t) = 0.5 * \cos(0.05 * t) * [1,1,1,1,1,1]^T$$

$$x_0 = [4900, pi/3, pi/3, -1000, 500, 500]^T$$

In general, to avoid the attack of the missile, the target can perform sudden side-step maneuvering with its two-side jets and transmit a jamming signal to interfere with the missile, which could lead to equivalent actuator and sensor faults, respectively. In this simulation, the equivalent actuator fault signal due to the sudden side-step maneuvering by two-side jets and sensor fault signal are shown as the square and cosine-type signals in Figure 7.3, respectively. Especially for the cosine-type sensor fault, the target aims to interfere with the estimation of relative velocity, yaw angular velocity, and pitch angular velocity with a large oscillation effect.

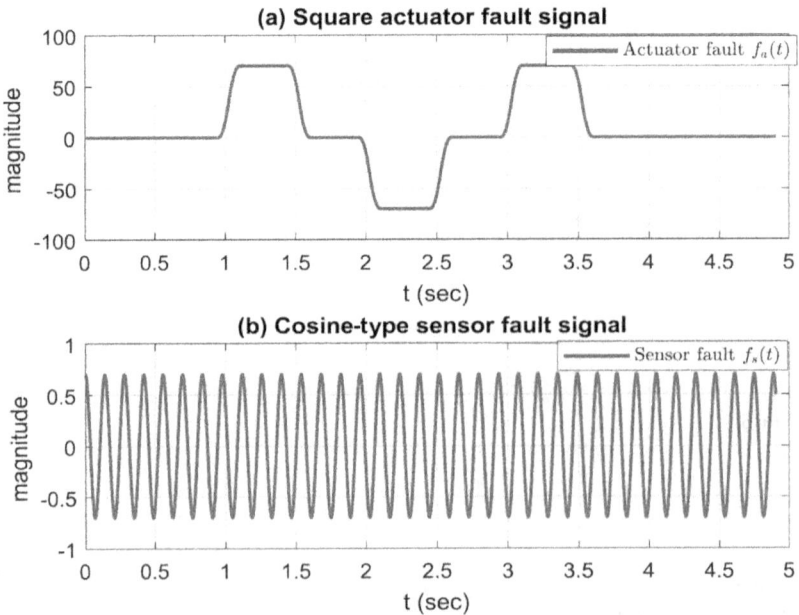

FIGURE 7.3 (a) The square actuator fault and (b) the cosine-type sensor fault signal.

In this simulation, the states of the missile guidance system and the corresponding estimated states are plotted in Figures 7.4–7.6. From the results in Figures 7.4–7.6, the Luenberger observer-based controller could successfully track the system states at $t = 0.06s$; that is, the Luenberger observer completely estimates the missile guidance system from $t = 0.06s$. From $t = 1s$, the target begins to perform sudden side-step maneuvering. Since the proposed Luenberger observer-based controller can quickly estimate the actuator fault, the effect of the actuator fault can be cancelled out by the proposed MO H_2/H_∞ observer-based fault-tolerant controller; that is, the actuator fault is cancelled by the estimated actuator fault. For the cosine-type sensor fault signal, Figures 7.5–7.6 reveal that the relative angular velocities of the yaw and pitch angles slightly fluctuate around the real states in the control process. With the proposed MO H_2/H_∞ observer-based fault-tolerant controller, the missile can successfully hit the target at $t = 4.9st$ with the head-on condition; that is, the relative distance, yaw angular velocity, and pitch angular velocity of missile guidance system approach 0 at $t = 4.9s$. Once the target is hit, the actuator and sensor fault signals vanish after $t = 4.9s$.

The simulation results in Figure 7.7 show the actuator and sensor faults and their estimation. In the beginning, due to the large estimation error (large initial condition) on the missile guidance system state, there have large but short estimated actuator and sensor fault signals. After that, the Luenberger observer can precisely estimate the actuator and sensor fault signals. The control strategies $u_\psi(t)$ and $u_\theta(t)$ are shown

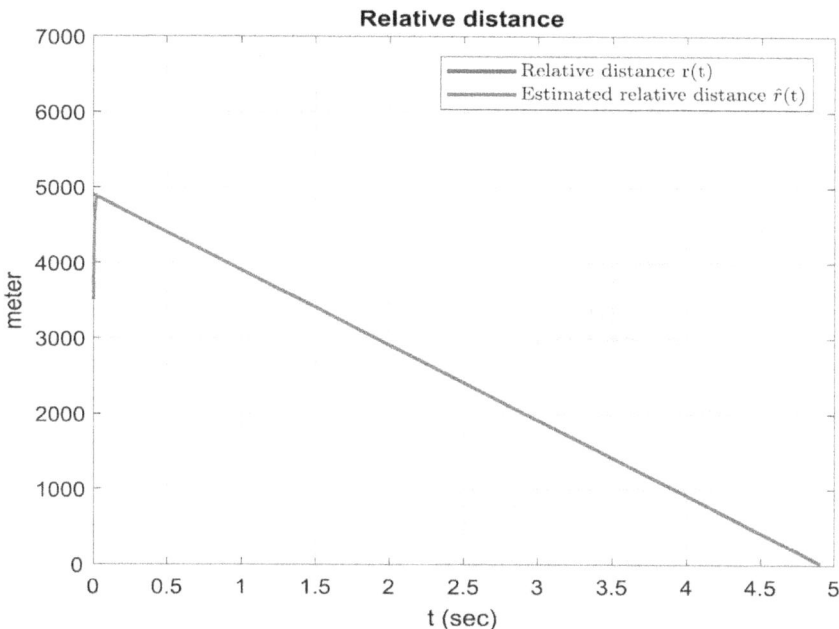

FIGURE 7.4 The relative distance and its estimation of the missile guidance system by the proposed MO H_2/H_∞ observer-based fault-tolerant control under the effect of the square actuator and cosine-type sensor fault signals.

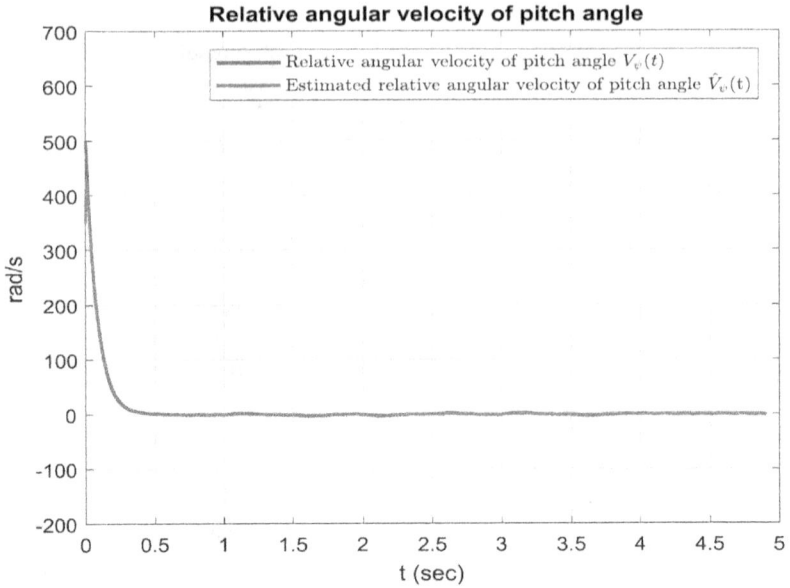

FIGURE 7.5 The relative pitch angular velocity and its estimation of missile guidance system by the proposed MO H_2/H_∞ observer-based fault-tolerant control under the effect of the square actuator and cosine-type sensor fault signals.

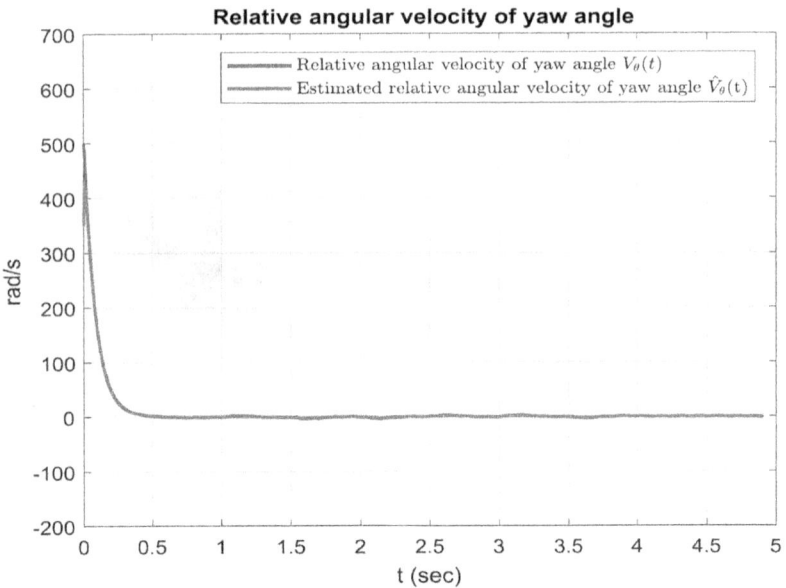

FIGURE 7.6 The relative yaw angular velocity and its estimation of missile guidance system by the proposed MO H2/H∞ observer-based fault-tolerant control under the effect of the square actuator and cosine-type sensor fault signals.

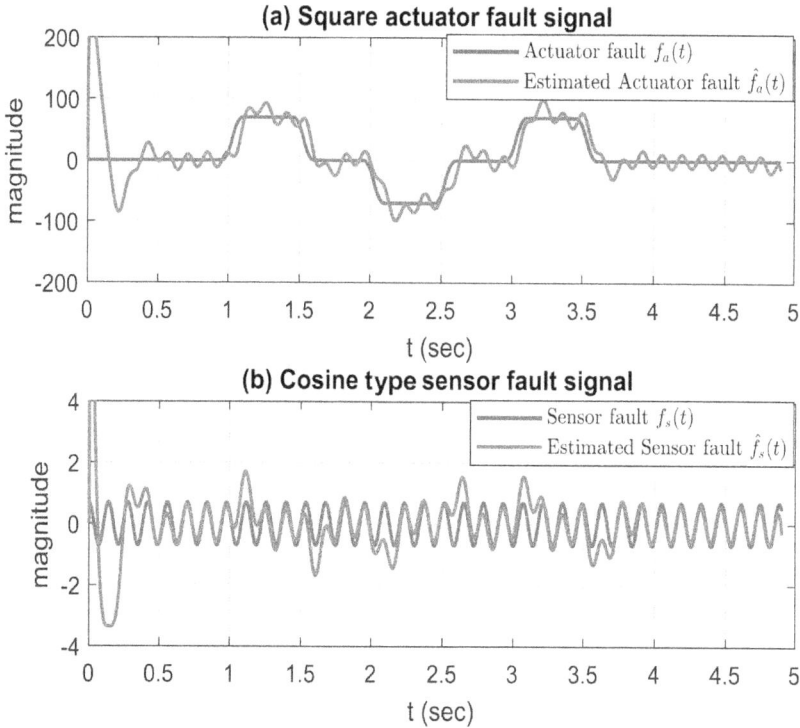

FIGURE 7.7 (a) The square actuator fault signal and its estimation. (b) The cosine-type sensor fault signal and its estimation.

in Figure 7.8, and it is seen that the proposed observer-based control strategies are effective to achieve the missile guidance task. Moreover, once the actuator fault signal is estimated, it can be seen that the control strategies will use the estimated actuator fault signal to eliminate the real actuator fault signal. In general, due to the external disturbance, the actuator and sensor fault signal from the target, it is very difficult for the missile to achieve the head-on condition; that is, the control period is increased by the actuator and sensor fault signal. Nevertheless, with the proposed MO H_2/H_∞ observer-based fault-tolerant controller, the effect of external disturbance on the missile guidance system is greatly attenuated, and the effect of the actuator and sensor fault signals is eliminated. Hence, as seen from the 3-D graph in Figure 7.9, the missile can maintain the head-on condition and successfully hit the target. The code file is included in [170] for performance validation.

Example 7.2: In this simulation example, the FTC design problem of balancing and swing-up of an inverted pendulum on a cart is considered. The pendulum motion can be explicitly expressed by the following fuzzy system with two if-then rules [168]:

$$\dot{x}(t) = \sum_{i=1}^{2} h_i(x(t))(A_i x(t) + B_i u(t) + B_{a,i} f_a(t) + B_{w,i} w(t)$$

$$y(t) = Cx(t) + n(t), x(0) = [3 \quad 1]^T$$

(7.42)

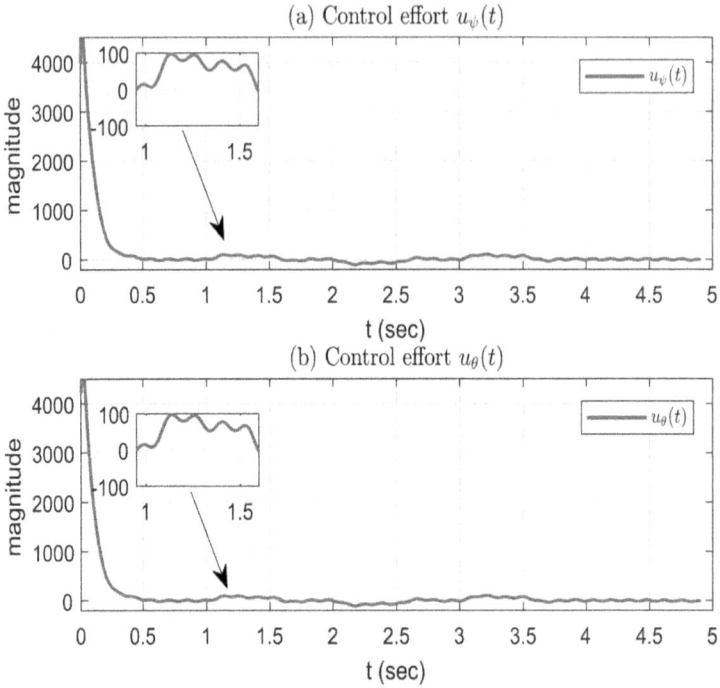

FIGURE 7.8   (a) The control signal of $u_\psi(t)$. (b) The control signal of $u_\theta(t)$. Once the actuator fault signal appears in the system, the control inputs will employ the estimated actuator fault signal to eliminate the effect of the real actuator fault signals.

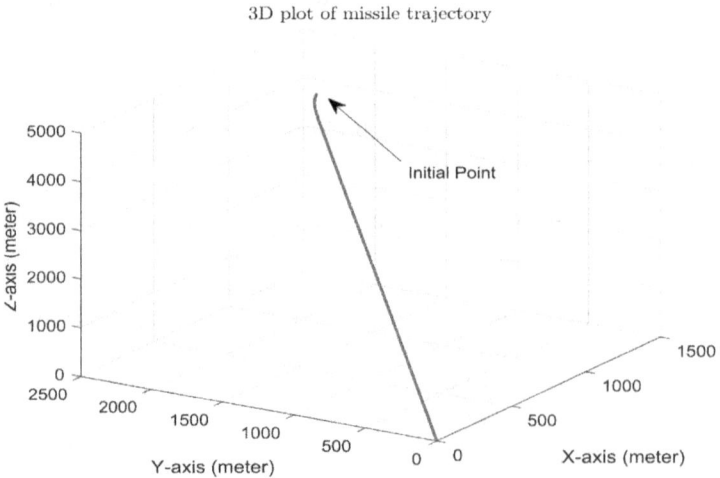

FIGURE 7.9   The relative yaw angular velocity and its estimation of missile guidance system by the proposed MO H_2/H_∞ observer-based fault-tolerant control under the effect of the square actuator and cosine-type sensor fault signals.

where $x(t) = [x_1(t) \quad x_2(t)]^T \in \mathbb{R}^2$ denotes the state vector with the angle of the pendulum $x_2(t)$ (rad/s), $u(t)$ is the force (N) to the cart, $f_a(t)$ is the actuator fault signal, $w(t)$ is the external disturbance to the cart, $y(t)$ represents the measurement output, and $n(t)$ is the measurement noise. The matrices in (7.42) are given as

$$A_1 = \begin{bmatrix} 0 & 1 \\ \dfrac{g}{4l/3 - aml} & 0 \end{bmatrix}, A_2 = \begin{bmatrix} 0 & 1 \\ \dfrac{2g}{\pi(4l/3 - aml\beta^2)} & 0 \end{bmatrix}, B_{a,1} = B_1 = \begin{bmatrix} 0 - \dfrac{g}{4l/3 - aml} \end{bmatrix}^T,$$

$$C = [1 \quad 0], \text{ and } B_{a,2} = B_2 = \begin{bmatrix} 0 - \dfrac{\alpha\beta}{4l/3 - aml\beta^2} \end{bmatrix}^T,$$

with the gravity constant $g = 9.8 m / s^2$, mass of the car $M = 2kg$, mass of the pendulum $m = 2kg$, half length of the pendulum $l = 0.5m$, interpolation functions $h_1(x_1(t)) = 1/(1 + \exp(x_1(t) + 0.5))$ and $h_2(x_1(t)) = 1 - h_1(x_1(t))$, and $a = 1/(M + m)$.

To efficiently estimate the actuator fault signal $f_a(t)$ in (7.42), the third-order smoothed signal model is chosen with the extrapolation coefficients $\alpha_0 = 0.6$, $\alpha_1 = 0.2$, $\alpha_2 = 0.1$, and $\alpha_3 = 0.1$ of A_{f_a} in (7.5). Also, the third-order smoothed signal model for the sensor fault signal is chosen with the extrapolation coefficients $\beta_0 = 0.6$, $\beta_1 = 0.2$, $\beta_2 = 0.15$, and $\beta_3 = 0.05$ of A_{f_s} in (7.6). Since the pendulum system is free of sensor fault signals, the weighting matrix in (7.27) is chosen as:

$$\tilde{Q}_1 = diag\{0_{8\times8}, I_2, I_4, 0_{4\times4}, I_2\}$$
$$\tilde{Q}_2 = diag\{0_{8\times8}, 2I_2, I_4, 0_{4\times4}, 2I_2\}$$
$$\tilde{R}_2 = 1$$

that is, the weighting of sensor fault estimation in $\tilde{Q}_1$ and $\tilde{Q}_2$ is zero. In this simulation example, the detailed parameters of the reverse MOEA are selected as search region $R = [10.5, 50] \times [0.9, 10]$, population number $N_p = 120$, iteration number $N_i = 100$, crossover rate $C_r = 0.8$, and mutation rate $C_m = 0.1$. Then, by solving the MOP in (7.28), the optimal control performance $(\alpha^*, \beta^*) = (23.21, 1.31)$ is selected, and the corresponding design variables can refer to [169].

In this simulation, the external disturbance and measurement noise are set as $0.1N(0,0.2)$ and $0.1N(0,0.2)$, respectively. The actuator fault signal $f_a(t) = 10$ for 0–20 s is set to represent an unknown constant fault in the actuator. The simulation results are shown in Figures 7.10–7.15. From Figures 7.10–7.11, the pendulum system can be effectively stabilized to zero with a high disturbance attenuation level. Furthermore, since the actuator fault signal $f_a(t)$ is precisely estimated in Figure 7.12, its effect can be effectively eliminated by the proposed FTC design. The FTC design in [168] is carried out for comparison with our design. From Figures 7.13–7.14, it can be clearly seen that the state variables severely oscillate during

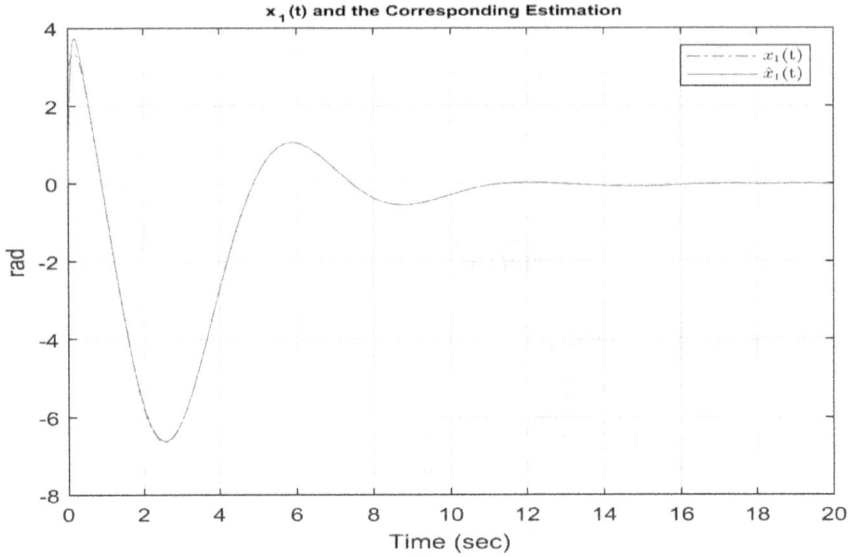

FIGURE 7.10   The trajectory of state $x_1(t)$ and its estimation.

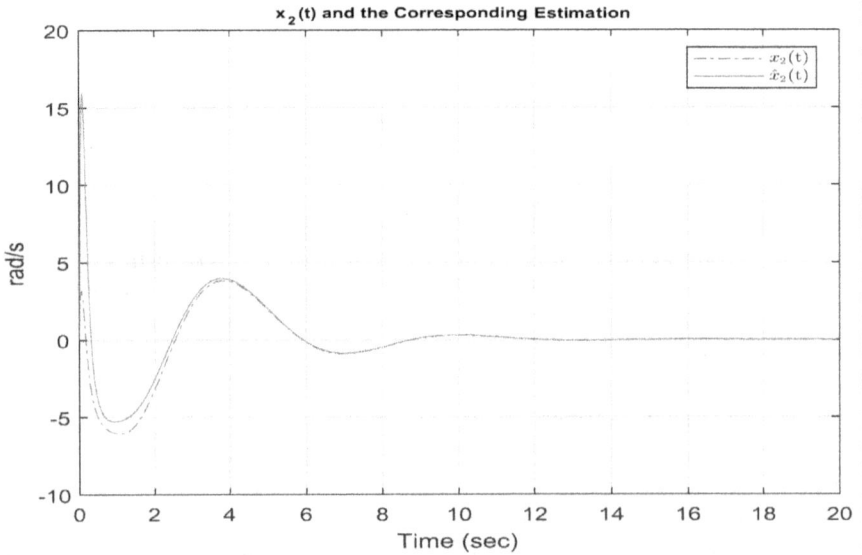

FIGURE 7.11   The trajectory of state $x_2(t)$ and its estimation.

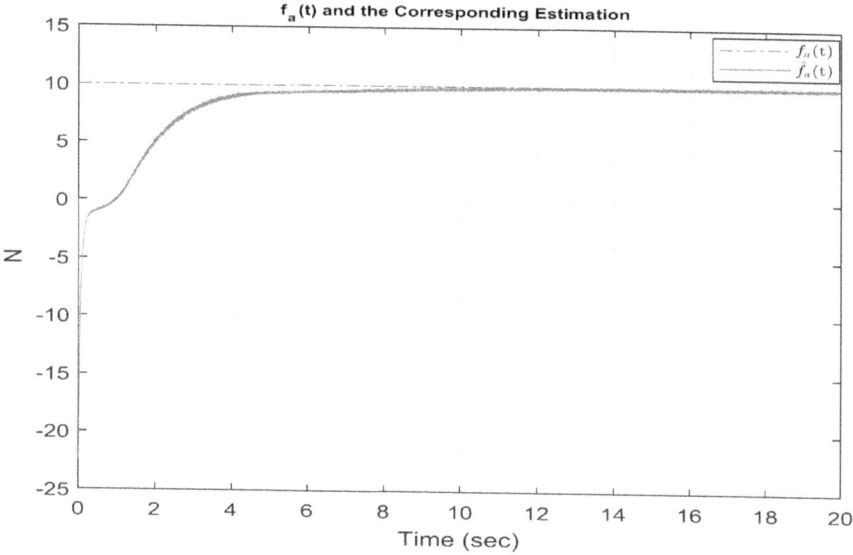

FIGURE 7.12   The fault signal $f_a(t)$ and its estimation.

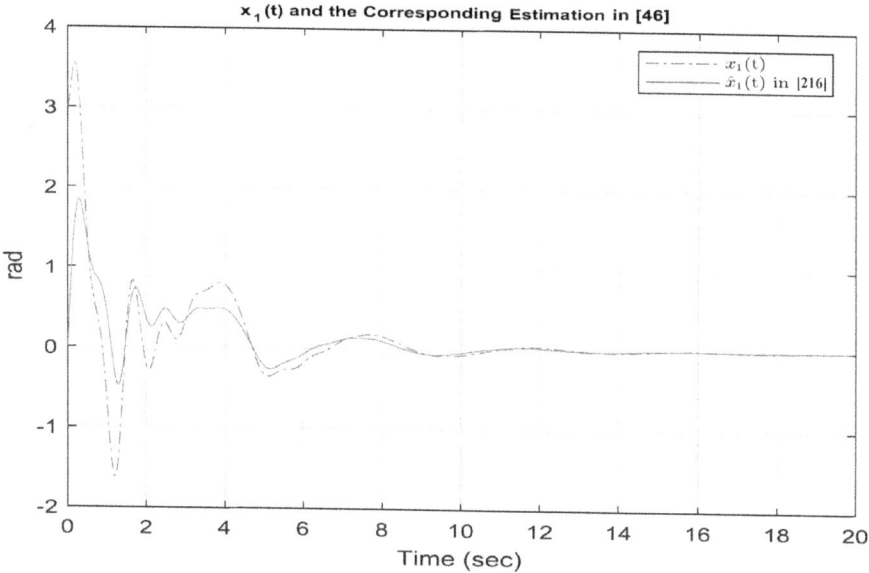

FIGURE 7.13   The trajectory of state $x_1(t)$ and its estimation by the method in [168].

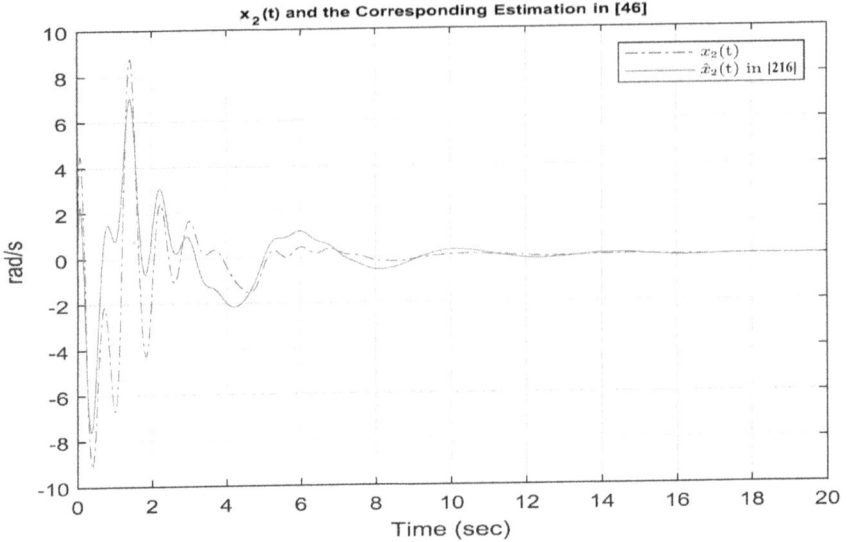

FIGURE 7.14 The trajectory of state $x_2(t)$ and its estimation by the method in [168].

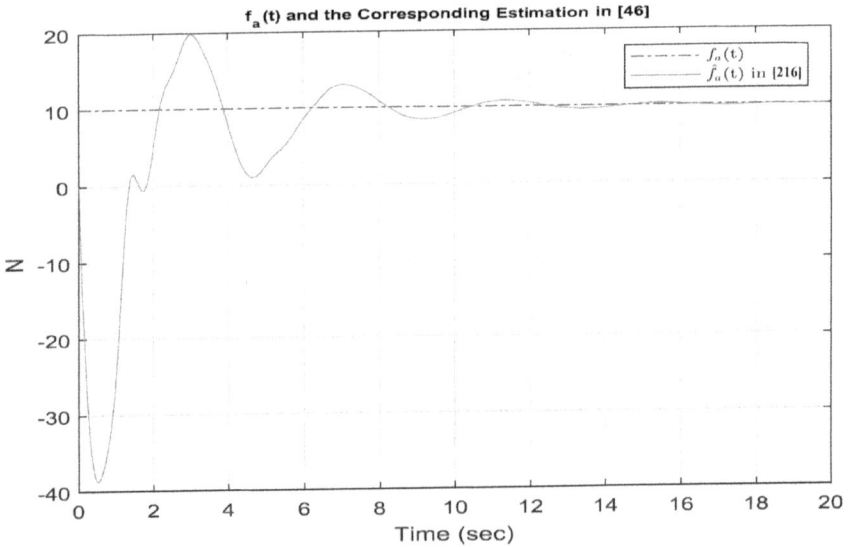

FIGURE 7.15 The fault signal $f_a(t)$ and its estimation by the method in [168].

the control process. On the other hand, from Figure 7.15, the transient state response for fault estimation is large, with some oscillation. As a result, due to the feedback of the imprecise fault signal estimation, the system may cause undesired control performance. The code file is included in [170] for performance validation.

7.6 CONCLUSION

Based on the proposed dynamic smoothed signal model, a novel reverse-order LMI-constrained MOEA was proposed to solve a complex multi-objective observer-based FTC of a T-S fuzzy system, which could not be solved by the conventional MOEA due to large fuzzy control and observer parameters to be selected by evolution algorithm for multi-objective optimization. Through the proposed indirect suboptimal method, the multi-objective observer-based FTC design problem is reduced to an LMI-constrained MOP, which could be efficiently solved by the proposed reverse-order MOEA with the help of the LMI toolbox in MATLAB via convex optimization algorithms such as the interior point method. Finally, the proposed reverse-order MOEA is applied to efficiently solve the complex multi-objective H_2/H_∞ observer-based 3-D missile fault-tolerant guidance control problem with sensor and actuator faults due to sudden cheating jet maneuvering and hostile interference. Also, the proposed method is applied to FTC design for an inverted pendulum system in comparison with the state-of-the-art FTC method. Further, the proposed reverse-order MOEA could be an efficient scheme to solve different MOP controls and estimation of complex T-S fuzzy systems in both the application and research fields. Also, along with the rapid development of the Internet of Things (IoT), information can be quickly exchanged in networked systems via wireless communication techniques to achieve more challenging tasks. Meanwhile, there may be malicious signals from attackers who try to damage the networked system and interfere with transmitted signals. Since the requirements of safety and reliability for networked systems are important issues, the proposed FTC method will be applied to the networked-based control system to achieve multi-objective optimal design with consideration of malicious attack signals in the future.

7.7 APPENDIX

7.7.1 PROOF OF THEOREM 7.1

By the rank test in [171], the ith augmented fuzzy system in (7.7) is observable if the following rank condition holds:

$$
rank\begin{bmatrix} sI_{n+(k+1)(n_a+n_s)} - \bar{A}_i \\ \bar{C}_i \end{bmatrix}
$$

$$
= rank\begin{bmatrix} sI_{(k+1)n_a} - A_{f_a} & 0 & 0 \\ 0 & sI_{(k+1)n_a} - A_{f_s} & 0 \\ -B_{a,i}C_{f_a} & 0 & sI_n - A_i \\ 0 & D_iC_{f_s} & C_i \end{bmatrix} \tag{7.43}
$$

$$
= n + n_a(k+1) + n_s(k+1), \forall s \in S
$$

where S denotes the set of complex s-domain. In the following, the proof is separated into two cases, with (i) $s \in S \setminus (eig\{A_i\} \cup eig\{A_{f_s}\} \cup eig\{A_{f_a}\})$ and (ii) $s \in eig\{A_i\} \cup eig\{A_{f_s}\} \cup eig\{A_{f_a}\}$.

In case (i), we immediately have following rank condition:

$$rank[sI_{(k+1)n_a} - A_{f_a}] = n_a(k+1)$$
$$rank[sI_{(k+1)n_a} - A_{f_s}] = n_s(k+1)$$
$$rank[sI_n - A_i] = n \tag{7.44}$$
$$s \in S \setminus (eig\{A_i\} \cup eig\{A_{f_s}\} \cup eig\{A_{f_a}\})$$

As a result, by (7.44), (7.43) is satisfied for $s \in S \setminus (eig\{A_i\} \cup eig\{A_{f_s}\} \cup eig\{A_{f_a}\})$, that is,

$$rank \begin{bmatrix} sI_{(k+1)n_a} - A_{f_a} & 0 & 0 \\ 0 & sI_{(k+1)n_a} - A_{f_s} & 0 \\ -B_{a,i}C_{f_a} & 0 & sI_n - A_i \\ 0 & D_iC_{f_s} & C_i \end{bmatrix} \tag{7.45}$$
$$= n + n_a(k+1) + n_s(k+1)$$
$$s \in S \setminus (eig\{A_i\} \cup eig\{A_{f_s}\} \cup eig\{A_{f_a}\})$$

In case (ii), by the assumption in (7.9) that the eigenvalues of (A_i, A_{f_s}, A_{f_a}) are mutually independent and (7.10), we can decouple the rank condition in (7.43) as the sum of three rank conditions

$$rank \begin{bmatrix} sI_{(k+1)n_a} - A_{f_a} & 0 & 0 \\ 0 & sI_{(k+1)n_a} - A_{f_s} & 0 \\ -B_{a,i}C_{f_a} & 0 & sI_n - A_i \\ 0 & D_iC_{f_s} & C_i \end{bmatrix}$$
$$= rank \begin{bmatrix} sI_{(k+1)n_a} - A_{f_a} \\ -B_{a,i}C_{f_a} \end{bmatrix} + rank \begin{bmatrix} sI_n - A_i \\ C_i \end{bmatrix} \tag{7.46}$$
$$+ rank \begin{bmatrix} sI_{(k+1)n_a} - A_{f_s} \\ D_iC_{f_s} \end{bmatrix}$$
$$s \in eig\{A_i\} \cup eig\{A_{f_s}\} \cup eig\{A_{f_a}\}$$

By applying the rank conditions in (7.8), (7.11), and (7.12), the rank condition in (7.46) can be written as:

$$rank \begin{bmatrix} sI_{(k+1)n_a} - A_{f_a} \\ -B_{a,i}C_{f_a} \end{bmatrix} + rank \begin{bmatrix} sI_{(k+1)n_a} - A_{f_s} \\ D_i C_{f_s} \end{bmatrix}$$

$$+ rank \begin{bmatrix} sI_n - A_i \\ C_i \end{bmatrix} \tag{7.47}$$

$$= n + n_a(k+1) + n_s(k+1)$$

$$s \in eig\{A_i\} \cup eig\{A_{f_s}\} \cup eig\{A_{f_a}\}$$

Thus, the observable condition for the ith augmented fuzzy system in (7.7) is guaranteed.

7.7.2 Proof of Theorem 7.2

Consider the H_2 observer-based fault-tolerant control in (7.20) and the Lyapunov function $V(\tilde{x}(t)) = \tilde{x}^T(t)P\tilde{x}(t)$ with the case $\bar{w}(t) = 0$ in (7.18); we have

$$H_2(\{L_i, K_j\}_{i,j=1}^l)$$

$$= \int_0^{t_f} \tilde{x}^T(t)\tilde{Q}_2\tilde{x}(t) + u^T(t)\tilde{R}_2 u(t)dt$$

$$= \int_0^{t_f} \tilde{x}^T(t)\tilde{Q}_2\tilde{x}(t) + u^T(t)\tilde{R}_2 u(t)dt + dV(\tilde{x}(t))$$

$$+ \tilde{x}^T(0)P\tilde{x}(0) - \tilde{x}^T(t_f)P\tilde{x}(t_f) \tag{7.48}$$

$$= \sum_{i=1}^l \sum_{j=1}^l h_i(z(t))h_j(z(t)) \int_0^{t_f} \left(P\tilde{A}_{ij} + \tilde{A}_{ij}^T P + \tilde{Q}_2 \right.$$

$$\left. + \bar{K}_j^T \tilde{R}_2 \bar{K}_j \right) \tilde{x}(t)dt + \tilde{x}^T(0)P\tilde{x}(0) - \tilde{x}^T(t_f)P\tilde{x}(t_f)$$

where $\bar{K}_j = [K_j, -K_j]$ with the positive-definite matrix $P > 0$. Clearly, if the matrix inequality constraints in (7.24) and (7.25) hold, we immediately have the following result:

$$H_2(\{K_i, L_j\}_{i,j=1}^l) \le \alpha$$

which satisfies the optimal H_2 constraint in (7.22).

On the other hand, for the robust H_∞ FTC performance in (7.19), we have

$$\int_0^{t_f} [\tilde{x}^T(t)\tilde{Q}_1\tilde{x}(t)]dt$$

$$= \int_0^{t_f} ([\tilde{x}^T(t)\tilde{Q}_1\tilde{x}(t)] + dV(\tilde{x}(t)) + \tilde{x}^T(0)P\tilde{x}(0)$$

$$- \tilde{x}^T(t_f)P\tilde{x}(t_f)$$

$$= \sum_{i,j=1}^l h_i(t)h_j(t) \int_0^{t_f} [\tilde{x}^T(t)(P\tilde{A}_{ij} + \tilde{A}_{ij}^T P + \tilde{Q}_1)\tilde{x}(t) \tag{7.49}$$

$$+ \tilde{x}^T(t)P\tilde{D}_{ij}\bar{w}(t) + \bar{w}^T(t)\tilde{D}_{ij}^T P\tilde{x}(t)]dt$$

$$+ \tilde{x}^T(0)P\tilde{x}(0) - \tilde{x}^T(t_f)P\tilde{x}(t_f)$$

By utilizing Lemma 7.1, with some $\beta > 0$, we have

$$\tilde{x}^T(t)P\tilde{D}_{ij}\overline{w}(t) + \overline{w}^T(t)\tilde{D}_{ij}^T P\tilde{x}(t)$$

$$\leq \frac{1}{\beta}\tilde{x}^T(t)P\tilde{D}_{ij}\tilde{D}_{ij}^T P\tilde{x}(t) + \beta\overline{w}^T(t)\overline{w}(t) \tag{7.50}$$

$$\forall i,j = 1,...,I$$

Substituting (7.50) into (7.49), we have:

$$\int_0^{t_f} \tilde{x}^T(t)\tilde{Q}_1\tilde{x}(t)dt$$

$$\leq \sum_{i,j=1}^{I} \int_0^{t_f} [\tilde{x}^T(t)(P\tilde{A}_{ij} + \tilde{A}_{ij}^T P + \tilde{Q}_1 \tag{7.51}$$

$$\frac{1}{\beta} P\tilde{D}_{ij}\tilde{D}_{ij}^T P)\tilde{x}(t) + \beta\overline{w}^T(t)\overline{w}(t) + \tilde{x}^T(0)P\tilde{x}(0)$$

If the matrix constraints in (7.26) hold, we immediately have:

$$\int_0^{t_f} \tilde{x}^T(t)\tilde{Q}_1\tilde{x}(t)dt$$

$$\leq \int_0^{t_f} \beta\overline{w}^T(t)\overline{w}(t) + \tilde{x}^T(0)P\tilde{x}(0) \tag{7.52}$$

$$\forall\overline{w}(t) \in L_2(\mathbb{R}^+;\mathbb{R}^{n_\overline{w}})$$

and $H_\infty(\{L_i, K_j\}_{i,j=1}^I) \leq \beta$. Hence, the MOP in (7.22) is transformed into the matrix inequality–constrained MOP in (7.23).

7.7.3 Proof of Theorem 7.3

By using the Schur complement [90], the inequalities in (7.24) are equivalent to:
Suppose $E\{\overline{X}_0^T P\overline{X}_0\} \leq \alpha$; we have

$$\begin{bmatrix} P\tilde{A}_{ij} + \tilde{A}_{ij}^T P + \tilde{Q}_2 & \overline{K}_j^T \\ * & -\tilde{R}_2^{-1} \end{bmatrix} \leq 0 \text{ for } i,j = 1,...,I \tag{7.53}$$

From (7.16) and (7.17), (7.53) can be rewritten as:

$$\begin{bmatrix} \Delta_{2,1} & -P_1\overline{B}_{u,i}K_j & K_j^T \\ * & \Delta_{2,2} & -K_j^T \\ * & * & -\tilde{R}_2^{-1} \end{bmatrix} \leq 0 \text{ for } i,j = 1,...,I \tag{7.54}$$

where $\Delta_{2,1} = P_1\overline{A}_i + P_1\overline{B}_{u,i}K_j + \overline{A}_i^T P_1 + K_j^T \overline{B}_{u,i}^T P_1 + \tilde{Q}_{2,\overline{x}}$ and $\Delta_{2,2} = P_2\overline{A}_i - P_2L_i\overline{C}_j + \overline{A}_i^T P_2$ $- \overline{C}_j^T L_i^T P_2 + \tilde{Q}_{2,e}$.

By multiplying $diag\{W_1, I, I\}$ by both sides of (7.54) with $W_1 = P_1^{-1}$, we have:

$$\begin{bmatrix} \Delta'_{2,1} & -\bar{B}_{u,i}K_j & W_1 K_j^T \\ * & \Delta'_{2,2} & -K_j^T \\ * & * & -\tilde{R}_2^{-1} \end{bmatrix} \leq 0 \text{ for } i, j = 1, ..., I \qquad (7.55)$$

where $\Delta'_{2,1} = \bar{A}_i W_1 + \bar{B}_{u,i} K_j W_1 + W_1 \bar{A}_i^T + W_1 K_j^T \bar{B}_{u,i}^T + W_1 \tilde{Q}_{2,\tilde{x}} W_1$ and $\Delta'_{2,2} = P_2 \bar{A}_i - P_2 L_i \bar{C}_j + \bar{A}_i^T P_2 - \bar{C}_j^T L_i^T P_2 + \tilde{Q}_{2,e}$. By using the Schur complement again, (7.55) can be rewritten as:

$$\begin{bmatrix} \Delta''_{2,1} & -\bar{B}_{u,i}K_j & W_1 K_j^T & W_1(\tilde{Q}_{2,\tilde{x}})^{1/2} \\ * & \Delta''_{2,2} & -K_j^T & 0 \\ * & * & -\tilde{R}_2^{-1} & 0 \\ * & * & * & -I \end{bmatrix} \leq 0 \text{ for } i, j = 1, ..., I \qquad (7.56)$$

where $\Delta''_{2,1} = \bar{A}_i W_1 + \bar{B}_{u,i} K_j W_1 + W_1 \bar{A}_i^T + W_1 K_j^T \bar{B}_{u,i}^T$ and $\Delta''_{2,2} = P_2 \bar{A}_i - P_2 L_i \bar{C}_j + \bar{A}_i^T P_2 - \bar{C}_j^T L_i^T P_2 + \tilde{Q}_{2,e}$.

For the constraint in (7.25), by using the Schur complement [90], the inequality in (7.25) can be transformed into the matrix inequality as follows:

$$\tilde{x}^T(0) P \tilde{x}(0) \leq \alpha$$
$$\Rightarrow e^T(0) P_2 e(0) + \bar{x}^T(0) P_1 \bar{x}(0) - \alpha \leq 0$$
$$\Rightarrow \begin{bmatrix} e^T(0) P_2 e(0) - \alpha & \bar{x}^T(0) \\ \bar{x}(0) & -W_1 \end{bmatrix} \leq 0$$

At last, we consider the H_∞ matrix constraints in (7.26). By using the Schur complement [90], the inequalities in (7.26) can be rewritten as:

$$\begin{bmatrix} P\tilde{A}_{ij} + \tilde{A}_{ij}^T P + \tilde{Q}_1 & P\tilde{D}_{ij} \\ * & -\beta I \end{bmatrix} \leq 0 \text{ for } i, j = 1, ..., I \qquad (7.57)$$

From (7.16) and (7.18), (7.57) can be rewritten as:

$$\begin{bmatrix} \Delta_{\infty,1} & -P_1 \bar{B}_{u,i} K_j & P_1 \bar{B}_{w,i} \\ * & \Delta_{\infty,2} & P_2 \bar{B}_{w,i} - P_2 L_i \bar{D}_j \\ * & * & -\beta I \end{bmatrix} \leq 0 \text{ for } i, j = 1, ..., I \qquad (7.58)$$

where $\Delta_{\infty,1} = P_1\bar{A}_i + P_1\bar{B}_{u,i}K_j + \bar{A}_i^T P_1 + K_j^T \bar{B}_{u,i}^T P_1 + \tilde{Q}_{1,\bar{x}}$ and $\Delta_{\infty,2} = P_2\bar{A}_i - P_2L_i\bar{C}_j + \bar{A}_i^T P_2$
$-\bar{C}_j^T L_i^T P_2 + \tilde{Q}_{1,e}$.

By multiplying $diag\{W_1, I, I\}$ by both sides of (7.58) with $W_1 = P_1^{-1}$, we have:

$$\begin{bmatrix} \Delta'_{\infty,1} & -\bar{B}_{u,i}K_j & \bar{B}_{w,i} \\ * & \Delta'_{\infty,2} & P_2\bar{B}_{w,i} - P_2L_i\bar{D}_j \\ * & * & -\beta I \end{bmatrix} \leq 0 \text{ for } i,j = 1,\ldots,I \tag{7.59}$$

where $\Delta'_{\infty,1} = \bar{A}_i W_1 + \bar{B}_{u,i}K_j W_1 + W_1\bar{A}_i^T + W_1 K_j^T \bar{B}_{u,i}^T + W_1\tilde{Q}_{1,\bar{x}}W_1$ and $\Delta'_{\infty,2} = P_2\bar{A}_i - P_2L_i\bar{C}_j$
$+ \bar{A}_i^T P_2 - \bar{C}_j^T L_i^T P_2 + \tilde{Q}_{1,e}$. By applying the Schur complement again, (7.59) can be rewritten as:

$$\begin{bmatrix} \Delta''_{\infty,1} & -\bar{B}_{u,i}K_j & \bar{B}_{w,i} & W_1(\tilde{Q}_{1,\bar{x}})^{1/2} \\ * & \Delta''_{\infty,2} & P_2\bar{B}_{w,i} - P_2L_i\bar{D}_j & 0 \\ * & * & -\beta I & 0 \\ * & * & * & -I \end{bmatrix} \leq 0 \text{ for } i,j = 1,\ldots,I \tag{7.60}$$

where $\Delta''_{\infty,1} = \bar{A}_i W_1 + \bar{B}_{u,i}K_j W_1 + W_1\bar{A}_i^T + W_1 K_j^T \bar{B}_{u,i}^T$ and $\Delta''_{\infty,2} = P_2\bar{A}_i - P_2L_i\bar{C}_j + \bar{A}_i^T P_2$
$-\bar{C}_j^T L_i^T P_2 + \tilde{Q}_{1,e}$. It is obvious that the BMIs in (7.60) are the constraints in (7.31).

Part III

Multi-Objective Optimization
Designs in Signal Processing
and Systems Communication

8 Multi-Objective H2/H∞ Optimal Filter Design of Nonlinear Stochastic Signal Processing Systems

8.1 INTRODUCTION

This filtering approach has been used to address many engineering problems, and, in general, it is difficult to design an efficient filter to estimate the state of non-linear systems in control and signal processing. The extended Kalman filter (EKF) is regarded as an adaptive linearized filter based on the current state estimate of nonlinear systems [172–174]. The design of the EKF relies on exact knowledge of nonlinear dynamics and the second-order statistical properties of external distur-bance and measurement noise. Some filtering problems have been solved for linear uncertain systems [175, 176]. The H_∞ filter was recently proposed for robust state estimation of nonlinear stochastic systems [177–189]. However, this filter requires the solution of a complex Hamilton-Jacobi equation, which cannot easily be solved except in the case of simple nonlinear systems. In recent years, there has been a growing interest in the Takagi-Sugeno fuzzy model for nonlinear systems since it is a powerful solution that bridges the gap between linear control and complex nonlinear systems [182, 184, 185, 190–197]. The important advantage of the T-S fuzzy model is its universal approximation of any smooth nonlinear function by interpolating certain local linear systems.

Since it requires exact knowledge of the noise covariance values, the H_2 filter is an optimal filter in the least-mean-squares error sense [172, 173]. However, in the real-world scenario, H_2 filtering performance will deteriorate if the noise statistics are not exact. Recently, as a means to remedying the shortcomings of the H_2 filter, the H_∞ robust filter has received considerable attention for its robust properties against external disturbances from the point of view of worst-case attenuation [175–189, 198, 199]. The advantage of using the H_∞ filter in comparison with the H_2 filter in terms of performance is that it may lead to a more robust but conservative result. More recently, the mixed H_2/H_∞ filter was designed so that H_2 optimal filtering per-formance may be obtained with a prescribed H_∞ filtering performance to attenuate the effect of uncertain noise on the filtering error; that is, the conventional H_2/H_∞ filter design is an H_2 optimal filter design problem under a prescribed robust H_∞ filtering constraint and is essentially only a single objective problem in filter design [9, 34, 71, 74, 198, 200–204]. Although most existing studies specify filter design characteristics in the form of an SOP, multiple objectives for filter design arise

DOI: 10.1201/9781003362142-11

naturally in real-life scenarios. These objectives normally conflict with each other such that they cannot be maximized or minimized simultaneously. As an alternative to finding the global optimal solution in an SOP, Pareto optimal solutions have been obtained for MOPs [2, 77, 205], and recently MOPs have become an active research area [77, 206]. However, most MOPs only focus on algebraic systems and have still not been applied to the design problem of dynamic systems. Therefore, further efforts are needed to extend these results to the multi-objective H_2/H_∞ filter design problem of nonlinear dynamic systems.

In this chapter, the proposed multi-objective fuzzy filter design is able to simultaneously achieve optimal H_2 and H_∞ filtering performance for nonlinear signal processing; that is, the multi-objective H_2/H_∞ fuzzy filter design is an MOP for nonlinear signal processing. In classical optimization methods, the MOP is converted to an SOP by emphasizing one particular Pareto optimal solution at a time using the weighted-sum method [207–209]. As this approach is to be used for discovering multiple solutions, different weightings have to be applied several times, hopefully finding a solution in each simulation run. Recently, a number of MOEAs have been proposed to find multiple Pareto optimal solutions of algebraic systems efficiently in one single simulation run [2, 77, 205, 210, 211]. Although the MOEAs have been widely employed to solve the MOPs of certain algebraic functional systems, further efforts are required to apply them to the MO fuzzy filter design problem for nonlinear stochastic fuzzy dynamic systems. Generally, the MOP for a fuzzy filter design in a nonlinear stochastic fuzzy dynamic system is more difficult than specifying a variable to minimize the multi-objective function in conventional MOPs for algebraic systems.

In this chapter, we propose a multi-objective H_2/H_∞ filter design to simultaneously achieve optimal H_2 and H_∞ filtering performance for nonlinear stochastic systems. In this situation, H_2 filtering performance requires minimization for least-mean-squares filtering performance, and H_∞ filtering performance needs to be minimized for the worst-case effect of uncertain disturbance and measurement noise on the filtering error; that is, the H_∞ filter is designed as a robust filter for all possible external disturbance $w(k)$ and measurement noise $n(k)$ by minimizing their worst-case effect on filtering performance. Therefore, the MO fuzzy filter design problem translates to minimizing the H_2 and H_∞ filtering performance simultaneously for nonlinear fuzzy signal systems with uncertain disturbance and measurement noise. For convenience of design, the T-S fuzzy model is first employed to interpolate several local stochastic systems to approximate a nonlinear stochastic signal system. We assume that the fuzzy signal system is corrupted with uncertain disturbance and measurement noise. However, it is not easy to solve the multi-objective H_2 and H_∞ filtering design problems of nonlinear stochastic signal systems directly. However, employing an indirect method, the multi-objective fuzzy filter design can be reduced to an MOP of how to minimize the two upper bounds of H_2 and H_∞ filtering performance under the constraint of a set of LMIs; that is, the MOP of multi-objective H_2/H_∞ filtering design by the fuzzy filter can be converted into an LMI-constrained MOP. Due to the convexity of LMIs [212], the LMI-constrained MOP of multi-objective H_2/H_∞ filter design can be solved efficiently by a set of Pareto optimal solutions using the convex optimization method via an LMI-based MOEA search method with the help of the LMI toolbox in MATLAB. Unlike conventional EA search algorithms in SOPs, in order to minimize H_2 and H_∞ filtering performance simultaneously, the MOEA search is used for simultaneous optimization through a nondominated sorting scheme to avoid one dominating another in the reverse-order LMI-based MOEA search process.

For a wider perspective, a reverse-order LMI-based MOEA method is developed to search for the set of Pareto optimal solutions for fuzzy filter gains via a number of genetic operators, such as selection, mutation, crossover, and certain processes. Next, one of the set of Pareto optimal solutions is selected for the fuzzy filter design, allowing the designer to achieve multi-objective H_2/H_∞ filter performance according to their own preference. Furthermore, for comparison, we suggest another multi-objective H_2/H_∞ fuzzy filter design method based on the weighted sum scheme. The main purposes of this chapter are as follows: (1) A broader perspective to designing the MOP H_2/H_∞ fuzzy filter is provided for nonlinear stochastic T-S fuzzy systems, and it can actually be easily applied to other multi-objective filter design problems; (2) an indirect method is proposed to transform the complex MO H_2/H_∞ filter design problem to a solvable LMI-constrained MOP; and (3) a reverse-order LMI-based MOEA through a nondominated sorting scheme is developed to efficiently solve the MO H_2/H_∞ filtering problem with the help of the MATLAB LMI toolbox.

This chapter is divided into five sections as follows. In Section 8.2, the T-S fuzzy stochastic system is introduced to interpolate several local stochastic linear systems to approximate nonlinear stochastic signal processing, and the MO H_2/H_∞ fuzzy filter design problem is formulated as an MOP. In Section 8.3, the MOP of multi-objective H_2/H_∞ fuzzy filter design is indirectly formulated as an LMI-constrained MOP. In Section 8.4, the MOP of the multi-objective H_2/H_∞ fuzzy filter design is solved using the proposed LMI-based MOEA. Finally, a simulation example regarding state estimation of a trajectory estimation of reentry vehicles by radar is given to illustrate the design procedure and confirm the performance of the proposed MO H_2/H_∞ fuzzy filter design in nonlinear systems.

Notation: Throughout the chapter, $x(k)$, $\hat{x}(k)$, and $\tilde{x}(k) \in R^n$ denote the state vector, state estimation, and estimation error, respectively. $\min(\alpha^*, \beta^*)$ is used to denote the simultaneous minimization of α and β, and its simultaneous minimum solutions (α^*, β^*) are not always unique and are called Pareto optimal solutions. MOEAs denote multi-objective evolution algorithms to search for the Pareto optimal solutions by the nondominating sorting method.

8.2   SIGNAL SYSTEM DESCRIPTION AND PROBLEM FORMULATION

The filtering problem in nonlinear discrete time signal processing systems with noisy measurements is depicted in Figure 8.1. The physical nonlinear signal processing system is represented using a T-S fuzzy model, and the output measurement $y(k)$ is corrupted by uncertain measurement noise $n(k)$. We model the whole signal processing problem mathematically as follows.

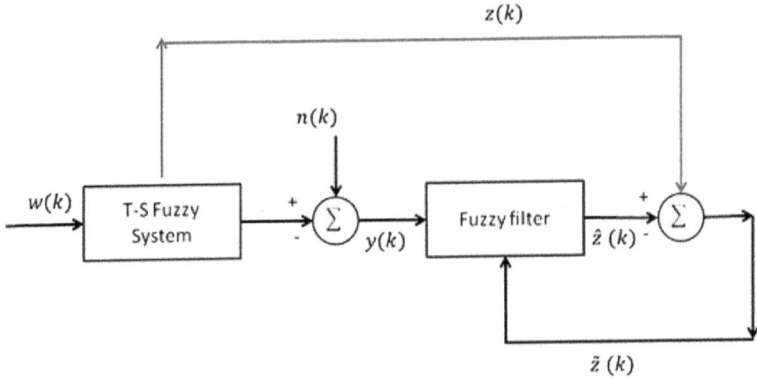

FIGURE 8.1 The T-S fuzzy system represents a nonlinear discrete-time stochastic signal processing system with unavailable state variables, and a fuzzy filter is designed to estimate the state vector of the T-S fuzzy system.

8.2.1 Physical Signal Processing System

The physical signal system in Figure 8.1 is a nonlinear discrete-time signal processing system that is represented by the stochastic T-S fuzzy model as follows [184, 185, 188, 189, 194–197, 213]:

System Rule i: If $\theta_1(k)$ is M_{i1} and $\theta_2(k)$ is M_{i2}, ... and $\theta_p(k)$ is M_{ip}, then

$$x(k+1) = A_i x(k) + B_i w$$
$$y(k) = C_i x(k) + D_i n(t) \tag{8.1}$$
$$z(k) = L_i x(k), \qquad\qquad i = 1,...,r$$

where M_{ij} are fuzzy sets; $x(k) \in R^n$ is the state vector; $z(k) \in R^q$ is the signal to be estimated; $y(k) \in R^m$ is the measurement output; A_i, B_i, C_i, D_i, and L_i are known constant matrices with appropriate dimensions; r is the number of if-then rules; $\theta(k) = [\theta_1(k), \theta_2(k),...,\theta_p(k)]$ is the measurable premise variable vector; and the driving noise $w(k) \in R^n$ and measurement noise $n(k) \in R^m$ are all assumed to be an arbitrary signal in $l_2[0,\infty)$. The fuzzy basic functions are given by

$$h_i(\theta(k)) = \prod_{j=1}^{p} M_{ij}(\theta_j(k)) \left/ \sum_{i=1}^{r} \prod_{j=1}^{p} M_{ij}(\theta_j(k)) \right.$$

where $M_{ij}(\theta_j(k))$ represents the grade of membership of $\theta_j(k)$ in the fuzzy set M_{ij}, and it is assumed that $\prod_{j=1}^{p} M_{ij}(\theta_j(k)) \geq 0$ for all k. In the following, for the sake of brevity, we will drop the argument of $h_i(\theta(k))$. Therefore, we have $h_i(\theta(k)) \geq 0$ for $i = 1, 2,...,r$ and $\sum_{i=1}^{r} h_i(\theta(k)) = 1$.

A more compact presentation of the T-S discrete-time fuzzy model in (8.1) is given by [184, 185] as

$$x(k+1) = A(h)x(k) + B(h)w(k)$$
$$y(k) = C(h)x(k) + D(h)n(k) \qquad (8.2)$$
$$x(k) = L(h)x(k)$$

where

$$A(h) = \sum_{i=1}^{r} h_i(\theta(k))A_i, B(h) = \sum_{i=1}^{r} h_i(\theta(k))B_i, \ C(h) = \sum_{i=1}^{r} h_i(\theta(k))C_i,$$

$$D(h) = \sum_{i=1}^{r} h_i(\theta(k))D_i, \ L(h) = \sum_{i=1}^{r} h_i(\theta(k))L_i, \ h = (h_1, h_2, ..., h_r)$$

The physical explanation of the T-S fuzzy system in (8.1) or (8.2) is that r local linear discrete-time stochastic signal processing systems are interpolated by choosing fuzzy basic functions with fuzzy rules to approximate a nonlinear stochastic signal system. Fuzzy parameters A_i, B_i, C_i and D_i of r local linear systems can be easily identified by the Linear Model Identification tool in MATLAB.

8.2.2 FUZZY FILTER FOR STATE ESTIMATION

In this study, we consider the following fuzzy filter to estimate $z(k)$ in (8.1) or (8.2):

(1) Filter rule i: If $\theta_1(k)$ is M_{i1}, $\theta_2(k)$ is M_{i2} ... and $\theta_p(k)$ is M_{ip}, then

$$\hat{x}(k+1) = A_i\hat{x}(k) + K_i(y(k) - C_i\hat{x}(k))$$
$$\hat{z}(k) = L_i\hat{x}(k), \qquad i = 1, 2, ..., r \qquad (8.3)$$

where $\hat{x}(k) \in R^n$ is the state estimation, $\hat{z}(k) \in R^q$ is the signal estimation, and K_i are filter gains to be determined.

Thus, the fuzzy filter can be represented in the following form:

$$\hat{x}(k+1) = A(h)\hat{x}(k) + K(h)(y(k) - C(h)\hat{x}(k))$$
$$\hat{z}(k) = L(h)\hat{x}(k) \qquad (8.4)$$

where $K(h) = \sum_{i=1}^{r} h_i(\theta(k))K_i$.

(2) Filter error system: Based on the fuzzy dynamic system in (8.1) and (8.2) and the fuzzy filter system in (8.3) and (8.4), we can obtain the filtering estimation error system by subtracting (8.2) from (8.4) as follows:

$$\tilde{x}(k+1) = [A(h) - K(h)C(h)]\tilde{x}(k) + [B(h)$$
$$- K(h)D(h)]\begin{bmatrix} w(k) \\ n(k) \end{bmatrix} \qquad (8.5)$$
$$\tilde{z}(k) = L(h)\tilde{x}(k)$$

where the state estimation error $\tilde{x}(k) = x(k) - \hat{x}(k)$, and the signal estimation error $\tilde{z}(k) = z(k) - \hat{z}(k)$.

The problem in filter design is to locate the filter gains such that the MO H_2/H_∞ filtering performance on signal estimation error $\tilde{z}(k)$ is guaranteed to achieve the optimal mean-squares error and optimal filter robustness simultaneously.

8.2.3 MULTI-OBJECTIVE H_2/H_∞ FUZZY FILTER DESIGN

Based on the filtering error system in (8.5), the design purpose in this chapter is described as follows. We wish to design fuzzy filter gains K_i, $i = 1,...,r$ in (8.3) and (8.4) such that the resulting H_2 and H_∞ filtering performance is minimized together, that is, denoting R_1, R_2 as the weighting factors: (1) The H_2 filtering error covariance $E\{\tilde{z}^T(k+1)R_2\tilde{z}(k+1)\}$ should be minimized to achieve the minimum mean square estimation error, and (2) the robust H_∞ filtering performance

$$E\left\{\sum_{k=0}^{\infty}\tilde{z}^T(k)R_1\tilde{z}(k)\right\}\Bigg/ E\left\{\sum_{k=0}^{\infty}[w^T(k)w(k)]+n^T(k)n(k)]\right\}$$

should be simultaneously minimized to achieve optimal filtering robustness. Consequently, the MO H_2/H_∞ fuzzy filter design problem can be formulated as the following MOP:

$$\min_{k}(J_1(K),J_2(K)) \tag{8.6}$$

where

$$J_1(K) = \frac{E\left\{\sum_{k=0}^{\infty}\tilde{z}^T(k)R_1\tilde{z}(k)\right\}}{E\left\{\sum_{k=0}^{\infty}[w^T(k)w(k)]+n^T(k)n(k)]\right\}}$$

$$J_2(K) = E\left\{\tilde{z}^T(k+1)R_2\tilde{z}(k+1)\right\}$$
$$= Tr\left(E\left\{\tilde{z}(k+1)R_2\tilde{z}^T(k+1)\right\}\right)$$

and $K = [K_1,K_2,...,K_r]$ denotes the set of fuzzy filter gains K_i in (8.3) or (8.4). $\min_K (J_1(K),J_2(K))$ means to specify K to minimize $J_1(K)$ and $J_2(K)$ at the same time.

Remark 8.1:

(1) At present, most MOPs focus on static systems [2, 77, 205, 210, 211], and they are not easily applied to solving the MOP in (8.6) directly for the nonlinear stochastic systems in (8.5). In the following section, an indirect approach is proposed to simplify the solving process of the MOP in (8.6) for multi-objective H_2/H_∞ fuzzy filter design.

(2) The conventional H_2/H_∞ filter designs [9, 34, 71, 200–202, 208] focused on the minimization of $J_2(K)$ under a prescribed H_∞ filtering performance $J_1(K) \leq \rho$ for some given ρ, that is, the so-called mixed H_2/H_∞ filter design [9, 201, 202], which is actually an SOP with a given H_∞ filtering performance constraint. For the multi-objective H_2/H_∞ filtering in (6), the H_∞ filtering performance $J_1(K)$ and H_2 filtering performance $J_2(K)$ must be minimized simultaneously. Therefore, it is an MOP and different from the constrained SOP of the mixed H_2/H_∞ filter design.

8.3 MULTI-OBJECTIVE H_2/H_∞ FUZZY FILTER DESIGN

Based on the previous analysis, the multi-objective H_2/H_∞ fuzzy filter design problem requires solving the MOP in (8.6). Since it is very difficult to solve the MOP in (8.6) directly for the nonlinear fuzzy stochastic filter error system in (8.5), an indirect method is proposed to minimize the upper bounds of these filter design objectives. In other words, we consider the following upper bounds to each objective:

$$J_1(K) = \frac{E\left\{\sum_{k=0}^{\infty}\left\{\tilde{z}^T(k)R_1\tilde{z}(k)\right\}\right\}}{E\left\{\sum_{k=0}^{\infty}\left\{w^T(k)w(k)\right\} + n^T(k)n(k)\right\}\right\}} \leq \alpha$$

$$J_2(K) = E\left\{\tilde{z}^T(k+1)R_2\tilde{z}(k+1)\right\} \leq \beta \tag{8.7}$$

for some positive scalar values α and β, respectively. Based on (8.6) and (8.7), the MOP for the multi-objective H_2/H_∞ fuzzy filter design in (8.6) can be reformulated in the following solvable manner:

$$\min_K (\alpha, \beta)$$
$$\text{subject to } J_1(K) \leq \alpha \text{ and } J_2(K) \leq \beta \tag{8.8}$$

And we have the following lemma.

Lemma 8.1: The MOP in (8.6) is equivalent to the MOP in (8.8).

Proof: The proof of Lemma 8.1 is straightforward. One only needs to prove that both inequality constraints in problem (8.8) become equality for Pareto optimal solutions. We will show this by contradiction. Given a 3-tuple Pareto optimal solution (K^*, α^*, β^*) of the MOP in (8.8), we assume that either inequality in (8.8) remains a strict inequality in the Pareto optimal solution. Without loss of generality, suppose that $J_1(K^*) < \alpha^*$. As a result, there exists α_1 such that $\alpha_1 < \alpha^*$ and $J_1(K^*) = \alpha_1$. Now, for the same K^*, the solution (α_1, β^*) dominates the Pareto optimal solution (α^*, β^*), leading to a contradiction. This implies that both inequality constraints in MOP (8.8) indeed become equality for Pareto optimal solutions. The MOP in (8.8) is hence equivalent to the MOP described in (8.6).

From the upper bound α of H_∞ filtering performance in (8.7), we get

$$E\left\{\sum_{k=0}^{\infty} \tilde{x}^T(k)L^T(h)R_1L(h)\tilde{x}(k)\right\}$$
$$\leq \alpha E\left\{\sum_{k=0}^{\infty} w^T(k)w(k)+n^T(k)n(k)\right\} \tag{8.9}$$

If the effect of the initial condition $\tilde{x}(0)$ is considered for the case of the previous H_∞ filtering performance, then the expression should be modified as [178–180]

$$E\left\{\sum_{k=0}^{\infty} \tilde{x}^T(k)L^T(h)R_1L(h)\tilde{x}(k)\right\}$$
$$\leq E\left\{\tilde{x}^T(0)P\tilde{x}(0)\right\}+\alpha E\left\{\sum_{k=0}^{\infty} w^T(k)w(k)+n^T(k)n(k)\right\} \tag{8.10}$$

for some positive definite matrix $P > 0$.

Similarly, from (8.7), we get

$$J_2(K) = E\left\{\tilde{x}^T(k+1)L^T(h)R_2L(h)\tilde{x}(k+1)\right\}$$
$$= Tr\left(E\left\{L(h)\left[\tilde{x}(k+1)R_2\tilde{x}^T(k+1)\right]L^T(h)\right\}\right) \leq \beta \tag{8.11}$$

Since we assume the statistics of $w(k)$ and $n(k)$ are uncertain, it is not easy to obtain the covariance $E\left\{\tilde{x}^T(k+1)R_2\tilde{x}(k+1)\right\}$ in (8.11) for the H_2 filtering performance. In this situation, the covariance of the state estimation error is calculated based on unitary noise power (i.e., $E\{w(k)w^T(k)\} = I$ and $E\{n(k)n^T(k)\} = I$) [9, 34, 200–202]. If the nominal noise powers are available, we could consider the case of $E\{w(k)w^T(k)\} = \Lambda_1$ and $E\{n(k)n^T(k)\} = \Lambda_2$ for some nominal Λ_1 and Λ_2 instead of for unitary noise power. Therefore, the design of fuzzy filter gain $K = [K_1, K_2,..., K_r]$ will be solved for the MOP in (8.8) such that the two upper bounds α and β will be minimized simultaneously under the constraints in (8.10) and (8.11). Then, we obtain the following MOP for the multi-objective H_2/H_∞ fuzzy filter design.

Theorem 8.1: The MOP in (8.8) for multi-objective H_2/H_∞ filter design can be transformed into the following MOP in (8.12)–(8.15), where M is the dimension of $L(h)$.

$$\min_{P>0,K}(\alpha,\beta) \tag{8.12}$$

subject to

$$\begin{bmatrix} P-L^T(h)R_1L(h) & 0 & 0 & (A(h)-K(h)C(h))^T P \\ 0 & \alpha I & 0 & B^T(h)P \\ 0 & 0 & \alpha I & -(K(h)D(h))^T P \\ P(A(h)-K(h)C(h)) & PB(h) & -(K(h)D(h)P) & P \end{bmatrix} > 0 \tag{8.13}$$

$$
\begin{bmatrix}
P & PK(h)D(h)\sqrt{R_2} & PB(h)\sqrt{R_2} & P(A(h)-K(h)C(h))\sqrt{R_2} \\
(PK(h)D(h)\sqrt{R_2})^T & I & 0 & 0 \\
(PB(h)\sqrt{R_2})^T & 0 & I & 0 \\
(A(h)-K(h)C(h))^T P\sqrt{R_2} & 0 & 0 & P
\end{bmatrix} > 0 \quad (8.14)
$$

$$
\begin{bmatrix}
\beta I_{M\times M} & \sqrt{M}L \\
\sqrt{M}L^T(h) & P
\end{bmatrix} > 0 \quad\quad (8.15)
$$

Proof: Please refer to Appendix 8.7.1.

Since the MOP in (8.12)–(8.15) for MO H_2/H_∞ fuzzy filter design still contains fuzzy system matrices $A(h)$, $B(h)$, $C(h)$, $D(h)$, $L(h)$, and $K(h)$ in (8.2), it is still not easy to solve for filter gain K. By substituting these fuzzy matrices into the MOP in (8.12)–(8.15), we get the following main result for MO H_2/H_∞ fuzzy filter design.

Theorem 8.2: The multi-objective H_2/H_∞ fuzzy filter design becomes how to solve the following MOP in (8.16)–(8.19), for $i,j = 1,2,...,r$.

$$
\min_{P>0, K=[K_1,...,K_r]} (\alpha, \beta) \quad\quad (8.16)
$$

subject to

$$
\begin{bmatrix}
P - L_i^T R_1 L_j & 0 & 0 & (A_i - K_j C_i)^T P \\
0 & \alpha I & 0 & B_i^T P \\
0 & 0 & \alpha I & -(K_j D_i)^T P \\
P(A_i - K_j C_i) & PB_i & -PK_j D_i & P
\end{bmatrix} > 0 \quad (8.17)
$$

$$
\begin{bmatrix}
P & PK_j D_i\sqrt{R_2} & PB_i\sqrt{R_2} & P(A_i - K_j C_i)\sqrt{R_2} \\
(PK_j D_i\sqrt{R_2})^T & I & 0 & 0 \\
(PB_i\sqrt{R_2})^T & 0 & I & 0 \\
(A_i - K_j C_i)^T P\sqrt{R_2} & 0 & 0 & P
\end{bmatrix} > 0 \quad (8.18)
$$

$$
\begin{bmatrix}
\beta I_{M\times M} & \sqrt{M}L_i \\
\sqrt{M}L_i^T & P
\end{bmatrix} > 0 \quad\quad (8.19)
$$

Proof: Please refer to Appendix 8.7.2.

In general, it is still not easy to solve the MOP using bilinear matrix inequality constraints with the multiplication of P and K_j to be solved in (8.17) and (8.18). Fortunately, the BMIs in (8.17) and (8.18) can be reformulated as equivalent LMIs, which are easily solved using the LMI toolbox in MATLAB.

Let $Z_j = PK_j$ or $K_j = P^{-1}Z_j$; then, the BMIs in (8.17) become the following LMIs of P and Z_j:

$$
\begin{bmatrix}
P - L_i^T R_1 L_j & 0 & 0 & A_i^T P - C_i^T Z_j^T \\
0 & \alpha I & 0 & B_i^T P \\
0 & 0 & \alpha I & -D_i^T Z_j^T \\
PA_i - Z_j C_i & PB_i & -Z_j D_i & P
\end{bmatrix} > 0 \tag{8.20}
$$

Similarly, the BMIs in (8.18) become the following LMIs of P and Z_j (8.21).

$$
\begin{bmatrix}
P & Z_j D_i \sqrt{R_2} & PB_i \sqrt{R_2} & (PA_i - Z_j C_i)\sqrt{R_2} \\
D_i^T Z_j^T \sqrt{R_2} & I & 0 & 0 \\
B_i^T P \sqrt{R_2} & 0 & I & 0 \\
(A_i^T P - C_i^T Z_j^T)\sqrt{R_2} & 0 & 0 & P
\end{bmatrix} > 0 \tag{8.21}
$$

Therefore, the MOP with BMI constraints in (8.16)–(8.19) for MO H_2/H_∞ fuzzy filter design is equivalent to the following MOP with LMI constraints:

$$
(\alpha^*, \beta^*) = \min_{P>0, Z=[Z_1,...,Z_r]} (\alpha, \beta) \tag{8.22}
$$

subject to LMIs (8.19) – (8.21)

where (α^*, β^*) denotes the minimum solutions of (α, β) in the previous MOP, that is, the so-called Pareto optimal solutions. In general, Pareto optimal solutions are not unique. Therefore, there may exist many optimal solutions for (α^*, β^*).

After solving the MOP in (8.22), the corresponding filter gains are obtained as $K_j^* = P^{*-1}Z_j^*$ and $j = 1, 2, ..., r$, which correspond to the optimal solution (α^*, β^*).

Remark 8.2:

(1) If we solely consider H_∞ filtering performance, then the MOP in (8.22) is reduced to the following SOP for H_∞ fuzzy filter design [179–185]

$$
\alpha_0 = \min_{P>0, Z=[Z_1,...,Z_r]} \alpha \tag{8.23}
$$

subject to (8.21)

(2) If we solely consider $H2$ filtering performance, then the MOP in (8.22) can be reduced to the following SOP for $H2$ fuzzy filter design [173]:

$$\beta_0 = \min_{P>0, Z=[Z_1,...,Z_r]} \beta$$

(8.24)

subject to (8.19) and (8.20)

(3) If conventional mixed H_2/H_∞ filter design is considered, then we need to solve the following constrained SOP for some prescribed noise attenuation level a of H_∞ filtering [9, 34, 74, 198, 200–202, 194–197, 213]:

$$\beta_* = \min_{P>0, Z=[Z_1,...,Z_r]} \beta$$

(8.25)

subject to (8.19) − (8.21)

Remark 8.3: The LMI constraints of the MOP in (8.22) for MO H_2/H_∞ filter design are the same as the SOP in (8.25) for mixed H_2/H_∞ filter design. However, the optimal solution β_* in (8.25) can be obtained more easily by decreasing β until no positive solution $P > 0$ is obtained for the LMIs in (8.19)–(8.21), while more time is needed to search for the simultaneous minimization (α^*, β^*) in (8.22) via the MOEA through a nondominated sorting scheme [2, 77, 205]. More detailed discussion about computational complexity will be given as follows.

The SOP problem in (8.23) and (8.24) is termed an eigenvalue problem and can be easily solved by decreasing α or β with the help of the LMI optimization toolbox in MATLAB, as long as the no-solution case for $P > 0$ exists. In contrast to the SOPs in (8.23)–(8.25), the global optimal solution for the MOP in (8.22) may not exist. In other words, there exists no unique solution $Z_j^* = PK_j^*$ such that α and β are minimized simultaneously. Therefore, more effort is needed for the MOP in (8.22) to seek a set of Pareto optimal solutions [2, 77, 205]. Before we further analyze the MOP in (8.22), we introduce the weighted sum method for the MOP with some given weightings w_1 or w_2. The weighted sum method has been applied to solve the MO problem. This technique converts the MOP to an SOP by emphasizing one optimal solution of the weighted sum, that is, how to minimize $w_1\alpha + w_2\beta$ for some weightings w_1 and w_2. Based on the weighted sum method, the MOP for multi-objective H_2/H_∞ fuzzy filter design in (8.22) can be transformed into an SOP by choosing fuzzy filter gain K_j to minimize $w_1\alpha + w_2\beta$ with $w_1 + w_2 = 1$. We then obtain the following result:

Theorem 8.3: The multi-objective H_2/H_∞ fuzzy filter for fuzzy signal systems in (8.1) can be solved by the following weighted method for all $w_1 \geq 0$, $w_2 \geq 0$, and $w_1 + w_2 = 1$

$$\min_{P>0, Z=[Z_1,...,Z_r]} w_1\alpha + w_2\beta$$

(8.26)

subject to LMIs (8.19), (8.20) *and* (8.21)

Proof: Using the weighted sum $w_1\alpha + w_2\beta$ to replace (α,β) and as in the proof of the indirect method in Appendix 8.7.1, we can easily prove Theorem 8.3 by following a similar argument to that used in proving Theorems 8.1 and 8.2. We omit the detailed procedure here.

The MOP in (8.26) is referred to as the weighted-sum SOP for MO H_2/H_∞ fuzzy filter design. It is different from finding Pareto optimal solutions for the MOP in (8.22) in one single run. To locate different Pareto optimal solutions, we need to apply different weightings to the SOP in (8.26) several times for different weighted sum solutions. How to choose the weightings w_1 and w_2, however, is another design problem, especially when α and β in (8.26) are in different scales.

Based on the previous analysis, we will combine the LMI technique from the LMI toolbox with the MOEA [212] to solve the LMI-constrained MOP in (8.22) or the weighted sum method in (8.26). In the following section, we first introduce some key notions for solving the LMI-constrained MOP in (8.22), such as dominance and Pareto fronts [2, 77, 205].

Remark 8.4: To find the feasible set for α and β, additional notation is used: $\bar{\alpha}$, $\bar{\beta}$ are denoted as the upper bounds on α and β, respectively, and α_0 in (8.23) and β_0 in (8.24) are denoted as the lower bounds on α and β, respectively (i.e., $\alpha_0 \le \alpha \le \bar{\alpha}$ and $\beta_0 \le \beta \le \bar{\beta}$).

8.4 MULTI-OBJECTIVE H_2/H_∞ FUZZY FILTER DESIGN VIA THE LINEAR MATRIX INEQUALITY–BASED MULTI-OBJECTIVE EVOLUTION ALGORITHM

The Pareto optimal solutions we want to find in the MOP are a set of solutions that are nondominated with respect to each other [77]. Before solving the MOP in (8.22) for the Pareto optimal solutions of the multi-objective H_2/H_∞ fuzzy filter, some key notions for the MOP are introduced as follows.

8.4.1 PARETO DOMINANCE RELATION IN THE MULTI-OBJECTIVE OPTIMIZATION PROBLEM

For simplicity of the ensuing derivation, the LMI-constrained MOP in (8.22) is abstractly formulated as

$$\min_{\{P,Z\}\in\Omega} (\alpha,\beta)$$

$$\text{subject to LMIs (8.19), (8.20) and (8.21)} \tag{8.27}$$

where Ω represents the feasible set in which the element of Ω must be satisfied with the LMIs in (8.19), (8.20), and (8.21). Since we want to search for some P and Z to minimize α and β simultaneously, three definitions for the solutions of the MOP in (8.27) are provided to guarantee the domination of candidate solutions (α,β) in the search process as follows:

Pareto dominance: For two solutions (P_1,Z_1) and (P_2,Z_2) in Ω for two objective values (α_1,β_1) and (α_2,β_2) subject to the LMIs in (8.19), (8.20), and

(8.21), respectively, (P_1, Z_1) is said to dominate (P_2, Z_2) if $\alpha_1 \leq \alpha_2$ and $\beta_1 \leq \beta_2$ for at least one inequality being a strict inequality.

Pareto optimality: A solution (P^*, Z^*) with objective value (α^*, β^*) is said to be a Pareto optimal solution of (8.27) if there does not exist another feasible solution (P_1, Z_1) with objective value (α_1, β_1) such that the objective value (α_1, β_1) dominates (α^*, β^*).

Pareto front: For the MOP in (8.27), the Pareto front PF^* defined as $PF^* = \{(\alpha^*, \beta^*) | (P^*, Z^*)\}$ is a Pareto optimal solution, and (α^*, β^*) is generated by (P^*, Z^*) subject to the LMIs in (8.19)–(8.21).

Based on these key notions, a reverse-order LMI-based MOEA is proposed to solve the MO H_2/H_∞ fuzzy filter design problem in (8.27). Since the LMI technique can transform the filter design problem into a convex optimization problem, it has been widely applied to solve the SOP of filter design. But for the MOP in (8.27), the LMI technique cannot solve the MOP directly. Hence, the LMI technique is combined with the MOEA to deal with the MOP in (8.27) for the H_2/H_∞ fuzzy filter design problem.

8.4.2 LINEAR MATRIX INEQUALITY–BASED MULTI-OBJECTIVE EVOLUTION ALGORITHM APPROACH FOR MULTI-OBJECTIVE FUZZY FILTER DESIGN

An MOEA is a stochastic search method through nondominated sorting scheme based on the mechanism of natural selection [2, 77, 205, 214]. It is also particularly suitable to multi-objective optimization due to its population-based nature, which allows a set of Pareto optimal solutions to be obtained in a single run. At present, a number of MOEAs have been proposed to solve MOPs with simple algebraic constraints [2, 77, 205]. In the case of MO H_2/H_∞ fuzzy filter design, we propose an LMI-based MOEA to solve the MOP in (8.27). Some important steps for LMI-based MOEAs and some characteristics of Pareto optimal solutions of MOPs are introduced as follows.

Traditional MOEAs randomly select candidate chromosomes (solutions) from the feasible region as the initial population. Therefore, the LMI constraints in (8.19)–(8.21) should be involved in the traditional MOEA. First, a number of chromosomes will be randomly generated from the feasible region. Next, in order to ensure some of these chromosomes (solutions) are feasible for the MOP in (8.22), each feasible chromosome will be checked by the existence of a positive definite solution of P per the LMIs in (8.19)–(8.21) via the LMI toolbox in MATLAB. If certain candidate solutions (or chromosomes) cannot guarantee positively defined solutions, that is, $P > 0$, they should be deleted from these candidate chromosomes. After this process, the remaining chromosomes can be regarded as the initial population for the MOP (8.22). Note that the three LMIs in (8.19)–(8.21) set more constraints than traditional methods on the evolutionary search for feasible initial chromosomes. Moreover, the LMI toolbox in MATLAB can help efficiently check whether these chromosomes satisfy the LMIs in (8.19)–(8.21) to accelerate the selection speed of the initial population.

Based on the previous analysis, the detailed design procedure of reverse-order LMI-based MOEA for the multi-objective H_2/H_∞ fuzzy filter problem in (8.22) is proposed as follows:

8.4.3 DESIGN PROCEDURE

Step 1: Select the feasible range for prescribed attenuation level α, β, the maximum number of generations, the total population size, the crossover ratio, and the mutation ratio in the MOEA.

Step 2: Initialize a number of chromosomes (solutions) and consider those chromosomes satisfied under the LMI constraints in (8.19)–(8.21) as the initial population.

Step 3: Evaluate objectives α_i and β_i of the current population. Additionally sort the current population into different fronts and assign a fitness value and then a rank of nondomination to each individual (α_i, β_i).

Step 4: Select half of the current population through nondominated sorting method as the parent population based on the rank and fitness value of each individual (α_i, β_i).

Step 5: Perform crossover and mutation operations of the parent population, and check the three LMIs in (8.19)–(8.21) to construct feasible offspring for the MOP in (8.22).

Step 6: Combine the parent population with offspring and perform step 4 to select individuals according to their rank and fitness value until the population size reaches the total population size. A new generation is thus created for the Pareto optimal solution search.

Step 7: Repeat steps 4, 5, 6, and 7 until the maximum number of generations is reached.

Step 8: Obtain $Z = [Z_1, Z_2, ..., Z_r]$ by solving $P > 0$ and Z_j, $j = 1, ..., r$ from the LMIs in (8.19), (8.20), and (8.21) for each chromosome of the last populations at the maximum number of generations and obtain the corresponding fuzzy filter gain $K_j = P^{-1}Z_j$ for all j in (8.3) and $K(h) = \sum_{i-1}^{r} h_i(\theta(k))K_i$ in (8.4). After finishing this LMI-based MOEA procedure, the Pareto optimal solutions will be obtained for the MO H_2/H_∞ fuzzy filter design problem in (8.16)–(8.19).

Remark 8.5:

(1) In the previous LMI-based MOEA design procedure, individuals with a high rank of nondomination have more copies than other individuals, which causes the elitist characteristic in the MOEA and allows better convergence to the Pareto optimal solutions of (8.27).

(2) The computation complexities of the proposed LMI-based MOEA are about $O(rn(n+1)MN^2)$, including $O(rn(n+1)/2)$ in solving LMIs, and $O(2MN^2)$ in the MOEA, where n is the dimension of P, N is the population number of the MOEA, r is the number of fuzzy rules, and M is the generation number of the MOEA.

8.5   SIMULATION EXAMPLES

To illustrate the design process and demonstrate the performance of the MO H_2/H_∞ fuzzy filter, we provide the simulation example. We use the proposed evolutionary search algorithm to search a set of Pareto optimal solutions (α^*, β^*) to solve the MOP in (8.22) for the multi-objective H_2/H_∞ fuzzy filter design problem. We choose a nondominated sorting genetic algorithm (NSGA-II) [2] as an MOEA search due to its popularity and combine it with the LMI toolbox in MATLAB to locate Pareto optimal solutions for the MOP in (8.22). The simulation settings for the evolution search algorithm in the design procedure are as follows: population size $N = 50$, generation number $M = 50$, crossover ratio $P_c = 0.9$, and mutation ratio $P_m = 0.01$. They were chosen for the modified NSGA-II. The corresponding Pareto optimal solutions P^* and Z_j^* for $j = 1, 2, ..., 16$ are also solved by the LMIs in (19)–(21). Next, the corresponding filter gains in (8.3) are also obtained as $K_j^* = P^{*-1} Z_j^*$ for $j = 1, 2, ..., 16$.

Example: The six-degrees-of-freedom reentry vehicle flies for several hundred kilometers along a ballistic trajectory above the rotating earth. The estimation of a reentry vehicle trajectory plays an important role in target tracking systems. A trajectory estimation of reentry vehicles (RVs) by radar is depicted in Figure 8.2. The RV model in radar coordinates (O_R, X_R, Y_R, Z_R) centered at the radar site can be expressed as follows [215, 216]:

$$\dot{v}_x = -\{\rho(v_x^2 + v_y^2 + v_z^2)/2\delta\}g \cos r_1 \sin r_2 + d_4$$
$$\dot{v}_y = -\{\rho(v_x^2 + v_y^2 + v_z^2)/2\delta\}g \cos r_1 \cos r_2 + d_5$$
$$\dot{v}_z = \{\rho(v_x^2 + v_y^2 + v_z^2)/2\delta\}g \sin r_1 - g + d_6$$

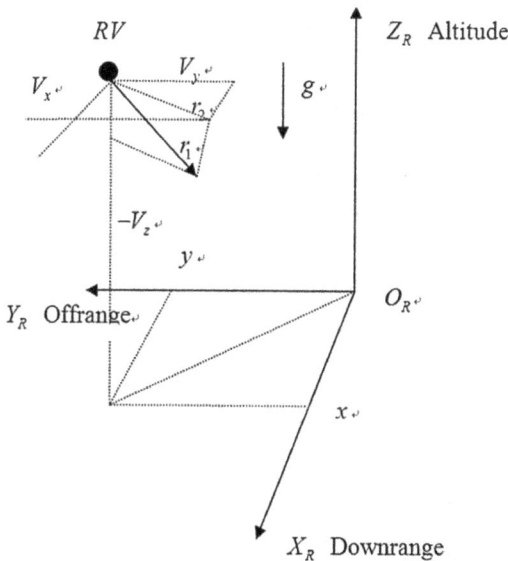

FIGURE 8.2   Tactical ballistic missile geometry in the example.

where v_x, v_y, and v_z are velocity components (m/s) along X_R, Y_R, and Z_R, respectively; d_4, d_5, and d_6 are external disturbances; ρ and δ are air density and ballistic coefficients, respectively; and $g = 9.8m / s^2$ is the gravity force, where r_1 and r_2 are defined as follows:

$$\delta = \frac{W}{SC_{DO}}, r_1 = \tan^{-1}\left(\frac{-v_z}{\sqrt{v_x^2 + v_y^2}}\right), r_2 = \tan^{-1}\left(\frac{v_x}{v_y}\right)$$

In this example, $\delta = 2240 Kg / m^2$ is assumed to be a constant. The air density is a function of altitude and described as $\rho = 0.00237 e^{-z/10000}$.

Let us denote $x = [x_1 \quad x_2 \quad x_3 \quad x_4 \quad x_5 \quad x_6]^T = [x \quad y \quad z \quad v_x \quad v_y \quad v_z]^T$, where x, y, and z denote the RV position along X_R, Y_R, and Z_R, respectively. The RV is measured by a precision radar with a sampling rate of 4 Hz (sampling time $T = 0.25s$). The nonlinear discrete time state equation can be written as follows:

$$x_1(k+1) = x_1(k) + Tx_4(k) + w_1(k)$$
$$x_2(k+1) = x_2(k) + Tx_5(k) + w_2(k)$$
$$x_3(k+1) = x_3(k) + Tx_6(k) + w_3(k)$$
$$x_4(k+1) = x_4(k) - T\{[\rho(x_4^2(k) + x_5^2(k) + x_6^2(k))]/2\delta\}g\cos r_1 \sin r_2 + w_4(k)$$
$$x_5(k+1) = x_5(k) - T\{[\rho(x_4^2(k) + x_5^2(k) + x_6^2(k))]/2\delta\}g\cos r \cos r_2 + w_5(k)$$
$$x_6(k+1) = x_6(k) - T\{[\rho(x_4^2(k) + x_5^2(k) + x_6^2(k))]/2\delta\}g\sin r_1 + w_6(k)$$

where

$$r_1 = \tan^{-1}\left(\frac{-x_6}{\sqrt{x_4^2 + x_5^2}}\right), r_2 = \tan^{-1}\left(\frac{x_4}{x_5}\right), \rho = 0.00237 e^{-z/10000}, w_1(k) = T\zeta_1, w_2(k) = T\zeta_2,$$

$w_3(k) = T\zeta_3$, $w_4(k) = T(d_4 + \zeta_4)$, $w_5(k) = T(d_5 + \zeta_5)$, and $w_6(k) = -T(g - d_6 - \zeta_6)$

and ζ_i, where $i = 1,...,6$, stand for the process noises [215].

For conventional air defense missile systems, ground radar is the major instrument for detecting the RV. The radar measurement equation is

$$y(k) = Cx(k) + n(k)$$

where $C = I$, and $n(k)$ denotes the measurement noise and jamming. The fuzzy filter design problem is to robustly estimate the trajectory of $x(k)$ of a reentry vehicle system from the radar output $y(k)$. It is assumed that $x_3 \in [0, 30000]$, x_4 and $x_5 \in [800, 0]$, and $x_6 \in [-2000, 0]$ [215]. The T-S fuzzy model for the RV system is identified with triangle membership functions as follows:

Rule 1: If x_3 is about 0, x_4 is about −800, x_5 is about −800, and x_6 is about −2000, then

$$x(k+1) = A_1 x(k) + Bw(k), y(k) = C_1 x(k) + Dn(k)$$

Rule 2: If x_3 is about 0, x_4 is about -800, x_5 is about -800, and x_6 is about 0, then

$$x(k+1) = A_2 x(k) + Bw(k), y(k) = C_2 x(k) + Dn(k).$$

Rule 3: If x_3 is about 0, x_4 is about -800, x_5 is about 0, and x_6 is about -2000, then

$$x(k+1) = A_3 x(k) + Bw(k), y(k) = C_3 x(k) + Dn(k).$$

Rule 4: If x_3 is about 0, x_4 is about -800, x_5 is about 0, and x_6 is about 0, then

$$x(k+1) = A_4 x(k) + Bw(k), y(k) = C_4 x(k) + Dn(k).$$

Rule 5: If x_3 is about 0, x_4 is about 0, x_5 is about -800, and x_6 is about -2000, then

$$x(k+1) = A_5 x(k) + Bw(k), y(k) = C_5 x(k) + Dn(k).$$

Rule 6: If x_3 is about 0, x_4 is about 0, x_5 is about -800, and x_6 is about 0, then

$$x(k+1) = A_6 x(k) + Bw(k), y(k) = C_6 x(k) + Dn(k).$$

Rule 7: If x_3 is about 0, x_4 is about 0, x_5 is about 0, and x_6 is about -2000, then

$$x(k+1) = A_7 x(k) + Bw(k), y(k) = C_7 x(k) + Dn(k).$$

Rule 8: If x_3 is about 0, x_4 is about 0, x_5 is about 0, and x_6 is about 0, then

$$x(k+1) = A_8 x(k) + Bw(k), y(k) = C_8 x(k) + Dn(k).$$

Rule 9: If x_3 is about 30000, x_4 is about -800, x_5 is about -800, and x_6 is about -2000, then

$$x(k+1) = A_9 x(k) + Bw(k), y(k) = C_9 x(k) + Dn(k).$$

Rule 10: If x_3 is about 30000, x_4 is about -800, x_5 is about -800, and x_6 is about 0, then

$$x(k+1) = A_{10} x(k) + Bw(k), y(k) = C_{10} x(k) + Dn(k).$$

Rule 11: If x_3 is about 30000, x_4 is about -800, x_5 is about 0, and x_6 is about -2000, then

$$x(k+1) = A_{11} x(k) + Bw(k), y(k) = C_{11} x(k) + Dn(k).$$

Rule 12: If x_3 is about 30000, x_4 is about -800, x_5 is about 0, and x_6 is about 0, then

$$x(k+1) = A_{12}x(k) + Bw(k), y(k) = C_{12}x(k) + Dn(k).$$

Rule 13: If x_3 is about 30000, x_4 is about 0, x_5 is about -800, and x_6 is about -2000, then

$$x(k+1) = A_{13}x(k) + Bw(k), y(k) = C_{13}x(k) + Dn(k).$$

Rule 14: If x_3 is about 30000, x_4 is about 0, x_5 is about -800, and x_6 is about 0, then

$$x(k+1) = A_{14}x(k) + Bw(k), y(k) = C_{14}x(k) + Dn(k).$$

Rule 15: If x_3 is about 30000, x_4 is about 0, x_5 is about 0, and x_6 is about -2000, then

$$x(k+1) = A_{15}x(k) + Bw(k), y(k) = C_{15}x(k) + Dn(k).$$

Rule 16: If x_3 is about 30000, x_4 is about 0, x_5 is about 0, and x_6 is about 0, then

$$x(k+1) = A_{16}x(k) + Bw(k), y(k) = C_{16}x(k) + Dn(k).$$

where $B = I, D = I, A_i$, and C_i for $i = 1, 2, ..., 16$ are identified by the LMI toolbox in MAT-LAB $w(k) = [w_1(k) \quad w_2(k) \quad \cdots \quad w_6(k)]^T$ and $n(k) = [n_1(k) \quad n_2(k) \quad \cdots \quad n_6(k)]^T$. The disturbance d_i $(i = 4, 5, 6)$, process noises ζ_i $(i = 1, ..., 6)$, and measurement noises n_i $(i = 1, 2, ..., 6)$ are assumed as follows: $d_i = N(0, 10)$ for $i = 4, 5, 6$; $\zeta_i = N(0, 1)$; $n_i = N(0, 10)$ for $i = 1, 2, ..., 6$ where $N(m, \sigma^2)$ represents the normal distribution with mean m and variance σ^2. Triangular membership functions are used in Rules 1 to 16. It is observed that the state variables corresponding to the nonlinear terms are $x_3, x_4, x_5,$ and x_6. It is assumed that $x_3 \in [0, 30000]$, x_4 and $x_5 \in [800, 0]$, and $x_6 \in [-2000, 0]$.

Based on the proposed design procedure, we use a population size of 50 and run the LMI-based MOEA for 50 generations. Fifty Pareto optimal solutions belonging to the final population are shown as the Pareto front in Figure 8.3, of which three are selected for comparison in Table 8.1. We can see that Pareto solution 1 sacrifices more H_2 filtering performance than H_∞ filtering performance; Pareto solution 2 makes a compromise between the optimal H_2 solution and the optimal H_∞ solution, and Pareto solution 3 sacrifices more H_∞ filtering performance than H_2 filtering performance. According to the Monte Carlo simulation results of 100 runs for trajectory estimation of reentry vehicles by radar in Table 8.2, the MO filtering performances of three Pareto optimal solutions are not much different with respect to the squared estimation error in the example. They all also show good performance when compared with the EKF [172] and the mixed H_2/H_∞ filter in (8.25) with a given H_∞ performance $\alpha = 0.31$. The computation time to solve the MO H_2/H_∞ filter for the trajectory estimation of the reentry vehicle with 100 runs of Monte Carlo simulation is about 215 min. The method we have proposed will help application of trajectory estimation

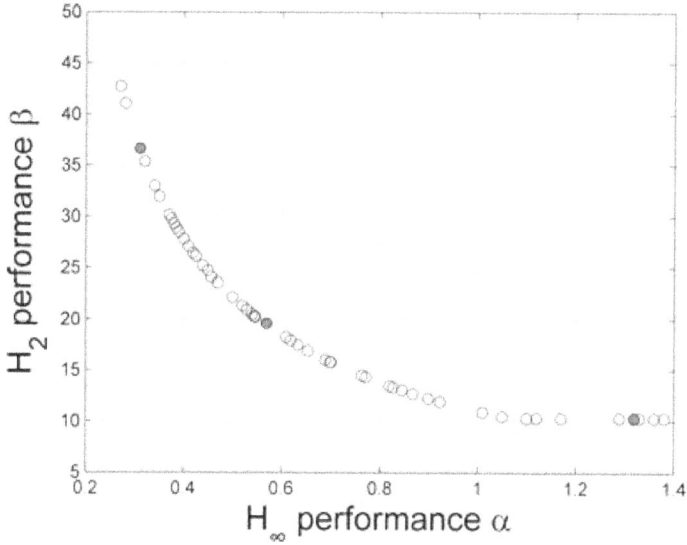

FIGURE 8.3   Pareto front for 50 Pareto optimal solutions by the proposed method in the example, in which three Pareto optimal solutions are chosen for comparison in Table 8.2.

TABLE 8.1
Three Pareto Optimal Solutions of the Example for Comparison

	Pareto Solution 1	Pareto Solution 2	Pareto Solution 3
(α^*, β^*)	(0.31, 36.63)	(0.57, 19.58)	(1.32, 10.35)

TABLE 8.2
Ratios of the Estimation Errors to State Variables in the Example

$\dfrac{\|x_i - \hat{x}_i\|_2}{\|x_i\|_2}$	Pareto Solution 1	Pareto Solution 2	Pareto Solution 3	Extended Kalman Filter	Mixed H2/ H∞ Filter
$i = 1$	6.16×10^{-4}	7.07×10^{-4}	8.96×10^{-4}	8.11×10^{-4}	7.71×10^{-4}
$i = 2$	1.97×10^{-3}	1.84×10^{-3}	1.30×10^{-3}	1.18×10^{-3}	1.30×10^{-3}
$i = 3$	1.23×10^{-4}	1.55×10^{-4}	1.83×10^{-4}	1.62×10^{-4}	1.38×10^{-4}
$i = 4$	0.16	0.28	0.23	0.24	0.31
$i = 5$	0.41	0.30	0.33	0.34	0.4
$i = 6$	0.057	0.057	0.066	0.069	0.06

for reentry vehicles. In the example, the operator or designer is therefore able to select any Pareto optimal solution for MO fuzzy filter design according to their own preference.

8.6 CONCLUSION

In this chapter, we introduced an MO fuzzy H_2/H_∞ filter design for the state estimation of fuzzy systems for stochastic nonlinear systems. The proposed multi-objective filter design takes advantage of the compromise between the optimal H_2 and H_∞ filtering design. Due to the conflicting objectives, the Pareto front representation for MOP H_2/H_∞ optimal solutions can be considered compromise solutions of the optimal H_2/H_∞ filtering design in the objective function space based on the MO optimization method for nonlinear fuzzy stochastic systems. An indirect method is introduced for solving the MOP of the multi-objective H_2/H_∞ filter under a set of BMI constraints in the sense of minimizing the upper bounds of both the H_∞ and H_2 filter performance simultaneously. To deal with the BMI-constrained MOP, a simple LMI-constrained MOP is proposed to replace the BMI-constrained MOP to simplify the MO H_2/H_∞ filter design as an LMI-constrained MOP. Because of the convex property of the LMI-based MOEA, the MOP for multi-objective fuzzy H_2/H_∞ filter design can be efficiently solved. Unlike conventional MOPs for algebraic systems, the proposed design method for MO H_2/H_∞ filter design for stochastic nonlinear dynamic systems is novel and suitable for practical applications. Furthermore, the proposed MO H_2/H_∞ filter design method can be easily extended to other MO H_2/H_∞ filter design problems, such as MO l_2/H_∞ filter design. Simulation examples demonstrate that multi-objective optimal filtering performance can be achieved using the proposed method.

8.7 APPENDIX

8.7.1 PROOF OF THEOREM 8.1

The following lemma is useful for the proof of following theorem.

Lemma 8.2: Schur complements [212]:

The following LMI

$$\begin{pmatrix} R_1(x) & S(x) \\ S^T(x) & R_2(x) \end{pmatrix} > 0$$

where $R_1(x) = R_1^T(x)$, $R_2(x) = R_2^T(x)$, and $S(x)$ depends on x, is equivalent to $R_2(x) > 0$, $R_1(x) - S(x)R_2^{-1}(x)S(x)^T > 0$. Then we get the following result.

The proof is divided into these steps:

In the first step, we show that H_∞ performance in (8.10) can be converted into the inequality in (8.13); that is, if the inequality in (8.13) holds for α, then H_∞ performance has an upper bound α. The second step is then to show that H_2 performance in (8.11) has an upper bound β if (8.14) and (8.15) both hold. The third step is to show that the multi-objective problem in (8.8) is equivalent to solving the MOP in (8.12)–(8.15).

For the H_∞ filtering part: Consider the H_∞ filtering performance in (8.9). By substituting $\tilde{x}(k+1)$ into (8.5) into the following equation:

$$\sum_{k=0}^{\infty} E[\tilde{x}^T(k)L^T(h)R_1L(h)\tilde{x}(k)]$$

$$= E\tilde{x}^T(0)P\tilde{x}(0) - E\tilde{x}^T(\infty)P\tilde{x}(\infty)$$

$$+ E\sum_{k=0}^{\infty}[\tilde{x}^T(k)L^T(h)R_1L(h)\tilde{x}(k) + \tilde{x}^T(k+1)P\tilde{x}(k+1)$$

$$- \tilde{x}^T(k)P\tilde{x}(k)]$$

$$\leq E\left\{\tilde{x}^T(0)P\tilde{x}(0) + \alpha\sum_{k=0}^{\infty}[w^T(k)w(k) + n^T(k)n(k)]\right\} \tag{8.28}$$

$$+ E\sum_{k=0}^{\infty}\{\tilde{x}^T(k)L^T(h)R_1L(h)\tilde{x}(k) + (\tilde{x}^T(k)(A^T(h)$$

$$- C^T(h)K^T(h)) + w^T(k)B^T(h)$$

$$- n^T(k)D^T(h)K^T(k))P((A(h) - K(h)C(h))\tilde{x}(k)$$

$$+ B(h)w(k) - K(h)D(h)n(k))$$

$$- \tilde{x}^T(k)P\tilde{x}(k) - \alpha w^T(k)w(k) - \alpha n^T(k)n(k)\}$$

If the following inequality holds:

$$\tilde{x}^T(k)(A^T(h) - C^T(h)K^T(h)) + w^T(k)B^T(h)$$
$$-n^T(k)D^T(h)K^T(k))P((A(h)$$
$$-K(h)C(h))\tilde{x}(k) + B(h)w(k) - K(h)D(h)n(k))$$
$$-\tilde{x}^T(k)P\tilde{x}(k) - \alpha w^T(k)w(k) - \alpha n^T(k)n(k)$$
$$\tilde{x}^T(k)L^T(h)R_1L(h)\tilde{x}(k) < 0$$

then the inequality in (8.28) is reduced to the H_∞ filtering performance in (8.10); that is, if the following inequality holds:

$$\begin{bmatrix} \tilde{x}^T(k) \\ w(k) \\ n(k) \end{bmatrix}^T \left\{ \begin{bmatrix} (A(h)-K(h)C(h))^T \\ (B(h))^T \\ (-K(h)D(h))^T \end{bmatrix} P \begin{bmatrix} (A(h)-K(h)C(h))^T \\ (B(h))^T \\ (-K(h)D(h))^T \end{bmatrix}^T \right.$$

$$- \left. \begin{bmatrix} P-L^T(h)R_1L(h) & 0 & 0 \\ 0 & \alpha I & 0 \\ 0 & 0 & \alpha I \end{bmatrix} \right\} \begin{bmatrix} \tilde{x}^T(k) \\ w(k) \\ n(k) \end{bmatrix} < 0 \tag{8.29}$$

then the H_∞ filtering performance in (8.10) is guaranteed. Then the inequality in (8.29) holds for any $\tilde{x}(k)$, $w(k)$, and $n(k)$ if the following inequality holds:

$$
\begin{bmatrix} (A(h)-K(h)C(h))^T \\ (B(h))^T \\ (-K(h)D(h))^T \end{bmatrix} P \begin{bmatrix} (A(h)-K(h)C(h))^T \\ (B(h))^T \\ (-K(h)D(h))^T \end{bmatrix}^T
$$
$$
- \begin{bmatrix} P-L^T(h)R_1L(h) & 0 & 0 \\ 0 & \alpha I & 0 \\ 0 & 0 & \alpha I \end{bmatrix} < 0
\tag{8.30}
$$

After applying Schur complement Lemma 8.1 [212] to (8.30), we get the equivalent inequality in (8.13); that is, if the inequality (8.13) holds, then the H_∞ filter performance has an upper bound α in (8.7) or (8.9).

For the H_2 filtering part: Consider H_2 filtering performance in (8.11). By substituting $\tilde{x}(k+1)$ in (8.5) into (8.11), we get

$$
\begin{aligned}
J_2(K) &= Tr(L(h)E[\tilde{x}(k+1)R_2\tilde{x}^T(k+1)]L^T(h)) \\
&= Tr(L(h)E\{[A(h)-K(h)C(h)\tilde{x}(k)+B(h)w(k) \\
&\quad - K(h)D(h)n(k)]R_2[A(h)-K(h)C(h)\tilde{x}(k) \\
&\quad + B(h)w(k)-K(h)D(h)n(k)]^T L^T(h)\}) \le \beta
\end{aligned}
\tag{8.31}
$$

In H_2 optimal filtering design, we assume $\tilde{x}(k)$, $n(k)$ and $w(k)$ are mutually orthogonal. By the unitary power of $w(k)$ and $n(k)$ in the H_2 filtering design case, we get

$$
\begin{aligned}
J_2(K) &= Tr\{L(h)[(A(h)-K(h)C(h))E(\tilde{x}(k)\tilde{x}^T(k))R_2(A(h) \\
&\quad - K(h)C(h))^T + B(h)R_2B^T(h) \\
&\quad + K(h)D(h)R_2D^T(h)K^T(h)]L^T(h)\}
\end{aligned}
\tag{8.32}
$$

Denote $Q = E[\tilde{x}(k)\tilde{x}^T(k)]$; then we get

$$
\begin{aligned}
J_2(K) &= Tr\{L(h)[(A(h)-K(h)C(h))QR_2(A(h) \\
&\quad - K(h)C(h))^T - Q + B(h)R_2B^T \\
&\quad + K(h)D(h)R_2D^T(h)K^T(h)]L^T(h)\} \\
&\quad + TrL(h)QL^T(h) < \beta
\end{aligned}
\tag{8.33}
$$

If the following inequality holds:

$$
\begin{aligned}
&(A(h)-K(h)C(h))QR_2(A(h)-K(h)C(h))^T \\
&- Q + B(h)R_2B^T + K(h)D(h)R_2D^T(h)K^T(h) < 0
\end{aligned}
\tag{8.34}
$$

then the H_2 performance has an upper bound β

$$J_2(K) \le TrL(h)QL^T(h) < \beta \qquad (8.35)$$

Let $P = Q^{-1}$ and multiply by P on the left and right sides of (8.34); then we get

$$
\begin{aligned}
&P(A(h) - K(h)C(h))P^{-1}R_2(A(h) - K(h)C(h))^T P \\
&-P + PB(h)R_2B^T P + PK(h)D(h)R_2D^T(h)K^T(h)P < 0
\end{aligned}
\qquad (8.36)
$$

Applying the Schur complement [212] to (8.36), we obtain the LMI in (8.14). The upper bound of the H_2 filter performance in (8.35) is satisfied.

For the multi-objective H_2/H_∞ part: Based on steps 1 and 2 in the previous proof, the MOP in (8.8) can be solved if the MOP in (8.12)–(8.15) holds.

8.7.2 Proof of Theorem 8.2

Substituting $A(h)$, $B(h)$, $C(h)$, $D(h)$, $L(h)$, and $K(h)$ into (8.13), we get,

$$
\sum_{i=1}^{r}\sum_{j=1}^{r} h_i(\theta(k))h_j(\theta(k))
\begin{bmatrix}
P - L_i^T R_1 L_j & 0 & 0 & (A_i - K_j C_i)^T P \\
0 & \alpha I & 0 & B_i^T P \\
0 & 0 & \alpha I & -(K_j D_i)^T P \\
P(A_i - K_j C_i) & PB_i & -PK_j D_i & P
\end{bmatrix} > 0 \quad (8.37)
$$

From (8.37), it is seen that if the inequalities in (8.17) hold for all i, j, then the inequality (8.14) also holds. Similarly, if the inequalities in (8.19) hold, then the inequality in (8.15) also holds.

9 Security-Enhanced Filter Design for Stochastic Systems under Malicious Attack via Multi-Objective Estimation Method

9.1 INTRODUCTION

In recent years, robust H_∞ filter techniques have played an important role in many application fields such as image processing, industry engineering, and signal processing [217, 218]. In real situations, external disturbance and measurement noise are unavailable for designers, and these effects are inevitable during the industrial process. Hence, the robust H_∞ filter is employed to detect and estimate the signal when the transmitted signal is influenced by external disturbance or measurement noise [217, 218]. In contrast to other filter design schemes, such as the particle filter (PF) [219, 220] or H_2 filter [221, 222], one of the advantages of the robust H_∞ filter is that it is not necessary to know the exact statistical characteristics of the external disturbance and measurement noise. Thus, the robust H_∞ filter is suitable to deal with the estimation of transmitted signals that suffer from uncertain external disturbance and measurement noise. The robust H_∞ filtering problem has been investigated for a variety of dynamic systems such as nonlinear network systems [223], Markov jumping linear systems [224], uncertain nonlinear systems [225], and fuzzy model–based nonlinear systems [226].

In order to make filter design for signal transmission systems more practical, robust filter designs for stochastic nonlinear systems have been discussed in recent years [10, 223, 227, 228]. Some practical applications have considered the stochastic effects on system models in many fields, such as engineering, biology, finance, and economics [11, 29, 34, 188, 229–231]. Further, the discontinuous (jump) Poisson process has also gained increasing attention in recent years [129, 130, 232, 233]. The Poisson process could be captured as event-driven uncertainties, such as corporate defaults, operational failures, or insured events [10, 129, 130, 224, 231–234]. Therefore, a stochastic system with continuous Wiener fluctuation and discontinuous Poisson fluctuation has been utilized to describe intrinsic continuous and discontinuous random fluctuation of signal processing in the transmission system [10, 217, 223, 227, 228, 235–237]. By using the Itô-Lévy stochastic differential equations, the stochastic properties of continuous and discontinuous random fluctuation in stochastic systems could be explicitly expressed for further design consideration.

DOI: 10.1201/9781003362142-12

Recently, due to the rapid development of communication networks and smart devices, the Internet is widely used for many services, such as information retrieval, video streaming, file sharing, online shopping, banking, and social networking. The Internet of Things framework connects many devices that can communicate and share information to achieve common goals in many fields. Although the IoT has profoundly changed our lives, it generates many issues to be further considered, such as energy challenges [238–240], effectiveness [241], reliability [242], security, and privacy [243–245]. From the security perspective, the IoT may suffer from active attacks when signals or data are transmitted. The topic of security attacks has been investigated during the past decade [246–248].

Owing to the complicated network of the IoT, which persistently transmits signals or data through the Internet, robust filter design only considering the effect of external disturbances and measurement noise is insufficient. Active malicious attack signals usually target the potential weak points of the system and disrupt transmitted information so that fluent transmission is interrupted. There are several kinds of the active malicious attacks, such as the well-known denial of service (DoS), jamming, and false data injection (FDI) [249]. The attackers aim to interrupt the transmitted signal and make systems vulnerable to malicious attacks. As a result, the topic of malicious attack signals is an urgent issue for future transmission systems. Recently, descriptor filters based on the singular descriptor model have been widely utilized to deal with the estimation of malicious attacks [247, 248, 250–253]. In general, by using the descriptor filter method, the singularity may have serious effects on the descriptor filter of a transmission system like instability or breakdown if strict algebraic conditions are not satisfied during the design.

In general, most transmission systems have strongly nonlinear behavior. Thus, the filter design problem for nonlinear systems will be transformed into a Hamilton-Jacobi inequality–constrained optimization problem. It is well known that the HJI is a partial differential inequality, and it will increase the difficulty in the analysis and design of nonlinear filters. In contrast to solving the linear matrix inequality–constrained optimization problem for filter design of the linear stochastic system case, the HJI-constrained optimization problem for filter design of nonlinear stochastic systems is still difficult to analyze. Therefore, global linearization technology was employed to efficiently treat the filter design of nonlinear systems. By using global linearization technology, nonlinear systems could be interpolated by several local linearized systems at the vertices of the polytope of global linearization systems [254, 255]. In this situation, the HJI-constrained optimization problem for filter design of the nonlinear system could be transformed into an LMI-constrained optimization problem.

Since the H_2 optimal filter cannot attenuate the effect of external disturbance, for practical applications, the robust H_∞ filter design problem is also an important issue to be addressed by researchers, for example, a robust H_∞ filter for a class of nonlinear stochastic systems [10, 29, 217, 223, 226, 230, 234]. Nevertheless, designers intend filter design to be not only robust but also optimal. By integrating the advantages of both H_∞ and H_2 filters, a mixed H_2/H_∞ filter design has been proposed in [7, 34, 255]. The concept of mixed H_2/H_∞ filter design aims to minimize the H_2 filter performance index under a prescribed H_∞ filter performance level, that is, a single-objective optimization problem. By extending the mixed H_2/H_∞ filter design, the MO concept has been utilized for filter design. MO H_2/H_∞ filter design could achieve both optimal H_∞ robust filter performance and optimal H_2 filter performance simultaneously.

By the previous analysis, the security-enhanced filter (SEF) is designed for linear and nonlinear stochastic jump-diffusion systems with external disturbances as well as malicious attacks on the system and sensors that may corrupt the system state and output measurement simultaneously. A novel smoothed signal model is proposed to efficiently describe malicious attack signals. To simplify the design, the smoothed signal model of malicious attack signals is embedded in the stochastic system model as an augmented stochastic system. Unlike conventional complicated descriptor-based filters to deal with the estimation of malicious attack signals by isolating the singularity of the descriptor system, we can efficiently estimate the system state and malicious signals directly from the state estimation of the augmented stochastic system by the traditional Luenberger-type filter. In this situation, malicious signals are embedded in the augmented system and will not corrupt the state estimation of the augmented stochastic system again. The SEF design based on the proposed stochastic MO H_2/H_∞ filtering performance is introduced for the augmented stochastic system with efficient attenuation on the effect of external disturbance and measurement noise to achieve optimal stochastic H_2 and H_∞ state and malicious attack signal estimation simultaneously. Since the stochastic MO H_2/H_∞ SEF design problem cannot be solved directly, a suboptimal method is proposed to transform the stochastic MO H_2/H_∞ SEF design problem into an LMI-constrained MOP. Further, in the case of nonlinear stochastic systems, global linearization technology [19, 254] is employed to interpolate the HJI-constrained MOP for MO H_2/H_∞ SEF by a set of LMI-constrained MOPs at the vertices of the polytope of the global linearization systems. A reverse-order LMI-constrained MOEA is also proposed to efficiently solve the LMI-constrained MOP for MO H_2/H_∞ SEF design of stochastic jump-diffusion systems under malicious attack signals. Two simulation examples are provided to validate the effectiveness of the stochastic MO H_2/H_∞ SEF method. A radar system for trajectory estimation of a nonlinear stochastic incoming ballistic missile with external disturbance, measurement noise, and malicious attack signals is given in the first example. In the second example, a network-based mass-spring system is provided to compare the effectiveness between the proposed security-enhanced filter design, our method, and other fault estimation filter design methods.

The purposes of this chapter are as follows:

(1) To remedy the conventional designs of complex descriptor filter based on a singular descriptor model [246–248, 250–253], a novel smoothed signal model is proposed to efficiently describe the dynamic of malicious attack signals on system and sensor. Then a simple traditional Luenberger-type filter can be employed to efficiently estimate the system states and malicious attack signals to achieve the MO H_2/H_∞ SEF design for the stochastic jump-diffusion system. Hence, the proposed MO H_2/H_∞ SEF design can achieve optimal H_2 security filtering and H_∞ performance on attenuating the effect of external disturbance and measurement noise simultaneously.

(2) Since the HJI-constrained MOP for nonlinear MO $H2/H\infty$ SEF is not easy to solve directly by the existing analytical or numerical methods, the

global linearization method is employed to transform the HJI-constrained MOP to an LMI-constrained MOP. Thus, the stochastic SEF design can be simply implemented for practical applications.

(3) In general, the LMI-constrained MOP is not easy to solve by conventional MOEAs directly. An LMI-constrained MOEA is also proposed to efficiently solve the LMI-constrained MOP for SEF design of the stochastic jump-diffusion system under malicious attack signals with the help of the LMI toolbox.

This chapter is divided into six sections as follows: The linear stochastic jump-diffusion system and the signal smoothed model are introduced in Section 9.2. In the Section 9.3, the linear MO H_2/H_∞ SEF design problem is formulated as an MOP, and a suboptimal method is proposed to transform the MO H_2/H_∞ SEF design problem to an LMI-constrained MOP. In Section 9.4, the nonlinear MO H_2/H_∞ SEF design problem is formulated as an HJI-constrained MOP and then transformed into an LMI-constrained MOP by using the global linearization scheme. Two simulation examples, the estimation of incoming ballistic missiles and a network-based mass-spring system, are provided in Section 9.5 to illustrate the design procedure and validate the performance of the proposed method. Finally, this is followed by concluding remarks in Section 9.6.

Notation: $\mathbb{R}_{\geq 0} := [0,\infty)$; M^T: transpose of the matrix M; $M > 0$: positive definite matrix; $E[\cdot]$: expectation operation; C^2: class of function $V(x)$, which is twice continuously differentiable with respect to x; $L_2^{F_t}(\mathbb{R}^+;\mathbb{R}^n) = \{v(t): \mathbb{R}^+ \to \mathbb{R}^n \| E[\int_0^\infty v^T(t)v(t)dt)^{\frac{1}{2}}] < \infty\}$; I_a: identity matrix with dimension $a \times a$; $0_{a \times b}$: zero matrix with dimension $a \times b$; $eig(A)$: set of eigenvalues of A; and $col(A)$: column space of A.

9.2   SYSTEM DESCRIPTION AND PRELIMINARIES

9.2.1   STOCHASTIC JUMP DIFFUSION SYSTEM AND SMOOTHED ATTACK SIGNAL MODEL

Consider the following continuous-time linear stochastic jump diffusion system with external disturbance and malicious attack signals on system components and sensors:

$$dx(t) = (Ax(t) + B_d v(t) + B_a f_a(t))dt + Gx(t)dw(t)$$
$$+ Jx(t)dp(t) \tag{9.1}$$
$$y(t) = Cx(t) + B_n n(t) + B_s f_s(t)$$

where $x(t) \in \mathbb{R}^n$ is the state vector; $y(t) \in \mathbb{R}^m$ is the measurement output; $v(t) \in \mathbb{R}^{n_n}$ is the measurement noise; and $f_a(t) \in \mathbb{R}^a$ and $f_s(t) \in \mathbb{R}^s$ are the malicious attack signals on the system and sensors, respectively. They could corrupt the system state and sensor measurement output, respectively, and are to be detected. $w(t) \in \mathbb{R}$ is a standard 1-D Wiener process, and $p(t) \in \mathbb{R}$ is a Poisson counting process with intensity λ. The system matrices $A, B_d, B_a, G, J, C, B_n$, and B_s in (9.1) have the appropriate dimensions.

Assumption 9.1: The continuous-time linear stochastic jump diffusion system in (9.1) is observable; that is, the pair (A, C) is observable.

Assumption 9.2: The malicious attack signals $f_a(t)$ and $f_s(t)$ in the stochastic jump diffusion system in (9.1) are differentiable; that is, $\dot{f}_a(t)$ and $\dot{f}_s(t)$ exist.

In the linear stochastic jump-diffusion system in (9.1), the diffusion term $Gx(t)dw(t)$ denotes continuous but nondifferentiable random fluctuation, and the jumping term $Jx(t)dp(t)$ denotes discontinuous random fluctuation. The Wiener process $w(t)$ and Poisson counting process $p(t)$ are defined on the complete filtered probability space $(\Omega, F, \{F_t\}_{t \in \mathbb{R}_{\geq 0}}, P)$, where Ω denotes the sample space, σ-field F_t is generated by the Wiener process $w(s)$ and Poisson counting process $p(s)$ for $s < t$, $F = \cup_{t \geq 0} F_t$, and P is the probability measure. The Wiener process $w(t)$ and Poisson counting process $p(t)$ are assumed to be mutually independent. Some important properties of Wiener process and Poisson counting process are given as follows [10]:

$$
\begin{aligned}
&1) E[w(t)] = E[dw(t)] = 0 \\
&2) E[dw^2(t)] = dt \\
&3) E[dp(t)] = \lambda dt \\
&4) E[dw(t)dp(t)] = 0
\end{aligned}
\tag{9.2}
$$

Remark 9.1: In future cyber-physical systems, the system integrates computational computers, a communication network, and physical components. This fact implies the system will not only suffer from the effects of external disturbance on physical components but also receive malicious attack signals from the communication network. Among the several types of malicious attack signals, common malicious attack signals include a large block function and high-frequency periodic function. In particular, the large block function aims to break down the system and lead to undesired operation. On the other hand, the high-frequency periodic function in a communication network is used to attack the estimation of the sensor.

Since the system dynamic and the sensor output suffer from the malicious attack signals $f_a(t)$ and $f_s(t)$, respectively, it may lead to system performance degradation and even system breakdown. In order to solve this security filtering problem, a novel smoothed signal model is proposed to help filter and estimate the malicious attack signal efficiently. Since the derivative and right derivative are equivalent for a differentiable function, by Assumption 9.2, a smoothed signal model of the malicious attack signal can be constructed based on the right derivative of $\dot{f}_a(t) = \lim_{h \to 0^+} \frac{f_a(t+h) - f_a(t)}{h}$ as follows:

$$
\begin{aligned}
\dot{f}_a(t) &= \tfrac{1}{h}(f_a(t+h) - f_a(t)) + \xi_{1,a}(t) \\
\dot{f}_a(t-h) &= \tfrac{1}{h}(f_a(t) - f_a(t-h)) + \xi_{2,a}(t) \\
&\qquad \vdots \\
\dot{f}_a(t-kh) &= \tfrac{1}{h}(f_a(t-(k-1)h) - f_a(t-kh)) + \xi_{k+1,a}(t)
\end{aligned}
\tag{9.3}
$$

where $h > 0$ is a small enough time interval, $\xi_{1,a}(t),...,\xi_{k+1,a}(t)$ denotes the corresponding approximation errors of derivative at different smoothed time points, and $f_a(t) = x_a(t)$ for $-kh \leq t \leq 0$ with unknown continuous function $x_a(t)$. Further, the malicious attack signal $f_a(t+h)$ could be represented by extrapolation (for example, Lagrange extrapolation [256]) as follows:

$$f_a(t+h) = \sum_{i=0}^{k} \alpha_i f_a(t-ih) + \tau_a(t) \tag{9.4}$$

where $\{\alpha_i\}_{i=0}^{k}$ are the extrapolation coefficients, and $\tau_a(t)$ represents the extrapolation error of $f_a(t+h)$. Then,= the smoothed signal model of the malicious attack signal $f_a(t)$ in (9.1) could be written as follows:

$$\dot{F}_a(t) = A_a F_a(t) + \xi_a(t) \tag{9.5}$$

where $F_a(t) = [f_a^T(t), f_a^T(t-h),..., f_a^T(t-kh)]^T$, the error vector $\xi_a(t) = [(\xi_{1,a}(t) + \tau_a(t))^T, \xi_{2,a}^T(t),..., \xi_{k+1,a}^T(t)]^T$, and the smoothed matrix

$$A_a = \begin{bmatrix} \frac{\bar{\alpha}_0}{h} I_a & \frac{\alpha_1}{h} I_a & \frac{\alpha_2}{h} I_a & \cdots & \frac{\alpha_k}{h} I_a \\ \frac{1}{h} I_a & -\frac{1}{h} I_a & 0 & \cdots & 0 \\ 0 & \frac{1}{h} I_a & -\frac{1}{h} I_a & \ddots & \vdots \\ \vdots & \ddots & \ddots & \ddots & 0 \\ 0 & \cdots & 0 & \frac{1}{h} I_a & -\frac{1}{h} I_a \end{bmatrix}$$

where $\bar{\alpha}_0 = \alpha_0 - 1$.

Remark 9.2: Since the malicious attack signal is unavailable for the design, it is not easy to find a set of fixed parameters $\{\alpha_i\}_{i=0}^{k}$ to exactly extrapolate the malicious function at a future sample point by the conventional Lagrange extrapolation. In general, the conventional Lagrange extrapolation needs time-varying extrapolation coefficients $\{\alpha_i(t)\}_{i=0}^{k}$ for the extrapolation, and it will make the filter design more complex. In this study, a modified Lagrange extrapolation method, that is, the proposed smoothed signal model with fixed extrapolation coefficients $\{\alpha_i\}_{i=0}^{k}$, is proposed to simplify the design procedure. Since the extrapolation error is considered in (9.4), the smoothed signal model in (9.4) is able to extrapolate the value of the malicious function at a future sample point by the current and previous values of malicious functions as possible. In general, due to the continuity of the malicious signal, it is expected that the value of the malicious function at a future sample point is closer to the value of the current sample point under a small sample period $h > 0$. Thus, the extrapolation parameters $\{\alpha_i\}_{i=0}^{k}$ should be chosen as a positive decreasing sequence, and the parameter α_0 is close but not equal to 1; that is, $\alpha_i \geq \alpha_{i+1} \geq 0$ for $i = 0,...,k-1$ and $\sum_{i=0}^{k} \alpha_i = 1$.

Similarly, the smoothed signal model of the malicious attack signal $f_s(t)$ can be constructed as:

$$\dot{f}_s(t) = \tfrac{1}{h}(f_s(t+h) - f_s(t)) + \xi_{1,s}(t)$$
$$\dot{f}_s(t-h) = \tfrac{1}{h}(f_s(t) - f_s(t-h)) + \xi_{2,s}(t)$$
$$\vdots$$
$$\dot{f}_s(t-kh) = \tfrac{1}{h}(f_s(t-(k-1)h) - f_s(t-kh)) + \xi_{k+1,s}(t) \tag{9.6}$$

where $\xi_{1,s}(t),...,\xi_{k+1,s}(t)$ denote the approximation errors of derivative at different smoothed time points, and $f_s(t) = x_s(t)$ for $-kh \le t \le 0$, with unknown continuous function $x_s(t)$. Further, the malicious attack signal $f_s(t+h)$ can be represented by extrapolation as follows:

$$f_s(t+h) = \sum_{i=0}^{k} \beta_i f_s(t-ih) + \tau_s(t) \tag{9.7}$$

where $\{\beta_i\}_{i=0}^{k}$ are the extrapolation coefficients, and $\tau_s(t)$ represents the extrapolation error of $f_s(t+h)$.

Then, the smoothed signal model of the malicious attack signal on the sensor in (9.1) is given as follows:

$$\dot{F}_s(t) = A_s F_s(t) + \xi_s(t) \tag{9.8}$$

where $F_s(t) = [f_s^T(t), f_s^T(t-h),..., f_s^T(t-kh)]^T$, the error vector $\xi_s(t) = [(\xi_{1,s}(t) + \tau_s(t))^T, \xi_{2,s}^T(t),..., \xi_{k+1,s}^T(t)]^T$, and the smoothed matrix

$$A_s = \begin{bmatrix} \frac{\bar{\beta}_0}{h}I_s & \frac{\beta_1}{h}I_s & \frac{\beta_2}{h}I_s & \cdots & \frac{\beta_k}{h}I_s \\ \frac{1}{h}I_s & -\frac{1}{h}I_s & 0 & \cdots & 0 \\ 0 & \frac{1}{h}I_s & -\frac{1}{h}I_s & \ddots & \vdots \\ \vdots & \ddots & \ddots & \ddots & 0 \\ 0 & \cdots & 0 & \frac{1}{h}I_s & -\frac{1}{h}I_s \end{bmatrix}$$

where $\bar{\beta}_0 = \beta_0 - 1$. Then, the augmented system is proposed to include the smoothed signal models (9.5) and (9.8) of malicious attack signals on the system and sensor in (9.1):

$$d\bar{x}(t) = (\bar{A}\bar{x}(t) + \bar{B}\bar{v}(t))dt + \bar{G}\bar{x}(t)dw(t) + \bar{J}\bar{x}(t)dp(t)$$
$$y(t) = \bar{C}\bar{x}(t) + \bar{D}\bar{v}(t) \tag{9.9}$$

where the augmented state $\bar{x}(t) = [x^T(t), F_a^T(t), F_s^T(t)]^T$; $\bar{v}(t) = [v^T(t), n^T(t), \xi_a^T(t),$

$\xi_s^T(t)]^T$; the mapping matrices $M_a = [I_a, 0, ..., 0]$ and $M_s = [I_s, 0, ..., 0]$, and $\bar{A} = \begin{bmatrix} A \\ 0 \\ 0 \end{bmatrix}$,

$\bar{B} = \begin{bmatrix} B_d & 0 & 0 & 0 \\ 0 & 0 & I & 0 \\ 0 & 0 & 0 & I \end{bmatrix}$, $\bar{G} = diag\{G, 0, 0\}$, $\bar{J} = diag\{J, 0, 0\}$, $\bar{C} = [C \quad 0 \quad B_s \quad M_s]$, and

$\bar{D} = [0 \quad B_n \quad 0 \quad 0]$. Because the augmented state includes the malicious attack signal on the system and sensor for the filter design, we will estimate $\bar{x}(t)$ directly, and the augmented state $\bar{x}(t)$ needs to be observable. Hence, the following theorem is proposed to guarantee the observability of the augmented state in (9.9).

Theorem 9.1: For the continuous-time linear stochastic jump diffusion system in (9.10), if the system (A, C) is observable; that is,

$$rank\begin{bmatrix} sI_n - A \\ C \end{bmatrix} = n, \forall s \in C \tag{9.10}$$

and the following conditions hold

$$eig(A) \cap eig(A_a) = \varnothing$$
$$eig(A) \cap eig(A_s) = \varnothing \tag{9.11}$$
$$eig(A_s) \cap eig(A_a) = \varnothing$$

$$rank\begin{bmatrix} sI_{s(k+1)} - A_s \\ B_s M_s \end{bmatrix} = s(k+1), \forall s \in C \tag{9.12}$$

$$rank\begin{bmatrix} -B_a M_a \\ sI_{(k+1)a} - A_a \end{bmatrix} = a(k+1), \forall s \in C \tag{9.13}$$

$$col\begin{bmatrix} sI_n - A \\ C \end{bmatrix} \cap col\begin{bmatrix} -B_a M_a \\ 0 \end{bmatrix} = \varnothing, \forall s \in eig(A_a) \tag{9.14}$$

where C is the set of one dimensional complex number, then $(\bar{A}, \bar{C})$ in the augmented system in (9.9) is observable.

Proof: Please refer to Appendix 9.7.1.

After the observability of the augmented system in (9.9) is guaranteed, the following linear Luenberger-type filter is employed to estimate the system state in (9.9), which includes state variables and malicious attack signals:

$$d\hat{\bar{x}}(t) = (\bar{A}\hat{\bar{x}}(t) + L(y(t) - \hat{y}(t)))dt$$
$$\hat{y}(t) = \bar{C}\hat{\bar{x}}(t) \tag{9.15}$$

where L is the filter gain to be designed, and $\hat{\bar{x}}(t)$ and $\hat{y}(t) \in$ are the estimated state and the estimated measurement output, respectively.

Remark 9.3: If we estimate the state $x(t)$ in (9.1) by Luenberger-type filter directly, the estimated $\hat{x}(t)$ could be corrupted by attacked signals $f_a(t)$ and $f_s(t)$. However, for the augmented system in (9.9), $f_a(t)$ and $f_s(t)$ are embedded in $\bar{x}(t)$, and we could estimate state and malicious attack signals by $\bar{x}(t)$; that is, $\bar{x}(t)$ is only with the corruption of $\bar{v}(t)$ and without the corruption of malicious attack signals $f_a(t)$ and $f_s(t)$ again. Therefore, we could design an SEF by estimating the state of the augmented system in (9.9) by the Luenberger-type filter in (9.15).

9.2.2 PROBLEM FORMULATION

In this study, the estimation of $x(t)$, $f_a(t)$ and $f_s(t)$, that is, $\bar{x}(t)$ in (9.9), by the Luenberger-type filter in (9.15) still needs to efficiently attenuate the effect of $\bar{v}(t)$ by H_∞ filtering performance and achieve optimal filtering performance under random fluctuation due to $w(t)$ and $p(t)$ by H_2 filtering performance. In order to optimally estimate the state and malicious attack signals in (9.10), based on the augmented system in (9.9) without considering $\bar{v}(t)$, the following the stochastic H_2 filtering performance of the Luenberger-type filter in (9.15) is introduced [221, 222]:

$$H_2(L) = E \int_0^{t_f} (\bar{x}(t) - \hat{\bar{x}}(t))^T Q_2 (\bar{x}(t) - \hat{\bar{x}}(t)) dt \qquad (9.16)$$

where $Q_2 \geq 0$ denotes the weighting matrix on estimation error, and t_f denotes the terminal time. In (9.16), the design objective is to find a specific filter gain L to achieve the optimal H_2 security-enhanced filtering performance.

On the other hand, since the augmented disturbance $\bar{v}(t)$ is unpredictable for the stochastic jump diffusion system (9.9), the stochastic robust H_∞ filtering performance is proposed to reduce the effect of the augmented disturbance $\bar{v}(t)$ on the estimation error as follows [223, 224]:

$$H_\infty(L) = \sup_{\bar{v}(t) \in L_2^{F_t}(\mathbb{R}^+; \mathbb{R}^{n_{\bar{v}}})} \frac{E\left[\int_0^{t_f} (\bar{x}(t) - \hat{\bar{x}}(t))^T Q_\infty (\bar{x}(t) - \hat{\bar{x}}(t)) dt - E(\tilde{x}^T(0)P\tilde{x}(0)) \right]}{E \int_0^{t_f} \bar{v}^T(t)\bar{v}(t) dt]} \qquad (9.17)$$

where $Q_\infty \geq 0$ denotes the weighting matrix on the estimation error, $n_{\bar{v}} = n_v + n_n + (k+1)(a+s)$, and the term $-E[\tilde{x}^T(0)P\tilde{x}(0)]$ deletes the effect of the initial condition on the robust H_∞ filtering performance with positive matrix $P > 0$ and $\tilde{x}(0) = x(0) - \overset{\Delta}{\bar{x}}(0)$. In (9.17), we aim to find a filter gain L in (9.15) to attenuate the worst-case effect of the augmented disturbance $\bar{v}(t)$ in (9.9) on the estimation error to achieve stochastic robust H_∞ filtering performance.

Remark 9.4: Since the extrapolation errors $\tau_a(t)$ and $\tau_s(t)$ in (9.4) and (9.7) are unavailable and it will influence on the estimation performance, the worst-case effect of extrapolation errors $\tau_a(t)$ and $\tau_s(t)$ in (17) should be attenuated to be as small as possible. From the theoretical point of view, the extrapolation errors $\tau_a(t)$ and $\tau_s(t)$ can be reduced by (i) increasing the sampling point k, (ii) decreasing the sampling period h, and (iii) selecting appropriate parameters $\{\alpha_i\}_{i=0}^k$. Further, the worst-case effect of extrapolation errors is considered in the robust H_∞ filter design in (9.17). Thus, the proposed MO H_2/H_∞ SEF can efficiently attenuate the effect of extrapolation errors during the filtering process.

In order to achieve stochastic optimal H_∞ robust filtering performance and H_2 filtering performance simultaneously, the MO H_2/H_∞ SEF design of a stochastic system with external disturbance and malicious attack signals in (9.10) is formulated as follows:

$$\min_L (H_2(L), H_\infty(L)) \text{ s.t. } (9.9),\ (9.15) \tag{9.17}$$

9.3 STOCHASTIC MO H_2/H_∞ SEF DESIGN

In this section, we first construct an augmented system of (9.9) and (9.15) in the following for convenience of design:

$$d\tilde{x}(t) = (\tilde{A}\tilde{x}(t) + \tilde{B}\bar{v}(t))dt + \tilde{G}\tilde{x}(t)dw(t) + \tilde{J}\tilde{x}(t)dp(t)$$
$$y(t) = \tilde{C}\tilde{x}(t) + \tilde{D}\bar{v}(t) \tag{9.19}$$

where $\tilde{x}(0) = \left[\bar{x}^T(0) \hat{\bar{x}}^T(0) \right]^T$ and $\tilde{A} = \begin{bmatrix} \bar{A} & 0 \\ L\bar{C} & \bar{A} - L\bar{C} \end{bmatrix}$, $\tilde{B} = \begin{bmatrix} \bar{B} \\ L\bar{D} \end{bmatrix}$, $\tilde{G} = diag\{\bar{G}, 0\}$,

$\tilde{J} = diag\{\bar{J}, 0\}$, $\tilde{C} = [\bar{C}, 0]$, and $\tilde{D} = \bar{D}$. Based on the augmented system in (9.19), the stochastic H_2 filtering performance in (9.16) and the stochastic robust H_∞ filtering performance in (9.17) can be reformulated as follows:

$$H_2(L) = E \int_0^{t_f} \tilde{x}^T(t) \tilde{Q}_2 \tilde{x}(t) dt \tag{9.20}$$

and

$$H_\infty(L) = \sup_{\bar{v}(t) \in L_2^{F_l}(\mathbb{R}^+; \mathbb{R}^{m_p})} \frac{E[\int_0^{t_f} \tilde{x}^T(t) \tilde{Q}_\infty \tilde{x}(t) dt - (\tilde{x}^T(t) P \tilde{x}(t))]}{E \int_0^{t_f} \bar{v}^T(t) \bar{v}(t) dt} \tag{9.21}$$

respectively, where

$$\tilde{Q}_2 = \begin{bmatrix} Q_2 & -Q_2 \\ -Q_2 & Q_2 \end{bmatrix}, \tilde{Q}_\infty = \begin{bmatrix} Q_\infty & -Q_\infty \\ -Q_\infty & Q_\infty \end{bmatrix}$$

Therefore, the MO H_2/H_∞ SEF design problem in (9.18) for the stochastic augmented jump-diffusion system in (9.19) is reformulated as follows:

$$\min_L (H_2(L), H_\infty(L)) \text{ s.t. } (9.19) \tag{9.22}$$

In general, it is not easy to solve this MOP directly. Thus, a suboptimal method is proposed to solve the MOP indirectly by minimizing the upper bounds as follows:

$$(\alpha^*, \beta^*) = \min_L (\alpha, \beta) \tag{9.23}$$

$$\text{s.t. } (9.19), \ H_2(L) \le \alpha \text{ and } H_\infty(L) \le \beta \tag{9.24}$$

Before solving MOP in (9.23) and (9.24), some fundamental properties of the MOP are introduced as follows:

Property 9.1 (Pareto Dominance [7, 11]): For two feasible solutions L^1 and L^2 of filter gain in (9.15) with the corresponding objective values (α^1, β^1) and (α^2, β^2) of the MOP in (9.23), respectively, the solution L^1 is said to dominate L^2 if and only if $\alpha^1 \le \alpha^2$ and $\beta^1 \le \beta^2$ and at least one of the inequalities is a strict inequality.

Property 9.2 (Pareto Optimality [7, 11]): The feasible solution L^* with the objective vector (α^*, β^*) of the MOP in (9.23) is said the Pareto optimality with respect to the feasible solution set if and only if there does not exist another feasible solution that dominates it.

Property 3 (Pareto Optimal set [7, 11]): For the MOP in (9.23) with feasible solution L and the corresponding objective vector (α, β), the Pareto optimal solution set σ^* is defined as $\sigma^* \triangleq \{L^* | L^*$ and is of Pareto optimality in (9.23)$\}$.

Property 4 (Pareto Front [7, 11]): For the MOP in (9.23) and the Pareto optimal solution set σ^*, the Pareto front is defined as $P_F \triangleq \{(\alpha^*, \beta^* | L^* \in \sigma^*$ with corresponding objective vector $(\alpha^*, \beta^*)\}$. Since the solution of the MOP is not unique, all solutions (α^*, β^*) are in P_F.

Then we get the following result of the suboptimal method in (9.23) for the MOP of the SEF design in (9.22).

Theorem 9.2: The suboptimal MOP in (9.23) is equivalent to the MOP in (9.22) for the design of MO H_2/H_∞ SEF.

Proof: Please refer to Appendix 9.7.2.

From Theorem 9.2, the MO H_2/H_∞ SEF design becomes how to specify filter gain L in (9.15) to solve the MOP in (9.23). The following lemmas are useful for solving the MOP in (9.23) for the SEF design of the stochastic augmented system in (9.19).

Lemma 9.1: (Itô-Lévy lemma of linear stochastic system [130, 234]) For the linear stochastic jump-diffusion system in (19), the Itô-Lévy formula for the Lyapunov function $V(\tilde{x}(t)) = \tilde{x}^T(t)P\tilde{x}(t)$, $V(0) = 0$, $P > 0$ is given as follows:

$$dV(\tilde{x}(t))$$

$$= \left(\frac{\partial V(\tilde{x}(t))}{\partial \tilde{x}(t)}\right)^T (\tilde{A}\tilde{x}(t) + \tilde{B}\bar{v}(t))dt + \frac{1}{2}\tilde{x}^T(t)\tilde{G}^T \frac{\partial^2 V(\tilde{x}(t))}{\partial \tilde{x}^2(t)}\tilde{G}\tilde{x}(t)dt \qquad (9.25)$$

$$+ \left(\frac{\partial V(\tilde{x}(t))}{\partial \tilde{x}(t)}\right)^T \tilde{G}\tilde{x}(t)dw(t) + [V(\tilde{x}(t) + \tilde{J}\tilde{x}(t)) - V(\tilde{x}(t))]dp(t)$$

Lemma 9.2: For any two vectors x and y with appropriate dimensions and a positive definite matrix P, we have the following inequality [19]

$$x^T Py + y^T Px \le \frac{1}{\rho} x^T PPx + \rho y^T y$$

where ρ is a nonzero real number.

Based on Theorem 9.2 and Lemmas 9.1 and 9.2, we get the following result.

Theorem 9.3: The MOP in (9.23) for the design of the MO H_2/H_∞ SEF in (9.15) of the linear stochastic jump-diffusion system in (9.9) could be transformed into the following BMI-constrained MOP:

$$(\alpha^*, \beta^*) = \min_{L,P>0,\alpha>0,\beta>0} (\alpha, \beta) \qquad (9.26)$$

$$\text{s.t. } \tilde{Q}_2 + P\tilde{A} + \tilde{A}^T P + \tilde{G}^T P\tilde{G} + \lambda(\tilde{J}^T P + P\tilde{J} + \tilde{J}^T P\tilde{J}) < 0 \qquad (9.27)$$

$$P \le \frac{\alpha}{n_p} E[R_0^{-1}] \qquad (9.28)$$

$$\begin{bmatrix} \tilde{Q}_\infty + P\tilde{A} + \tilde{A}^T P + \tilde{G}^T P\tilde{G} & P\tilde{B} \\ +\lambda(\tilde{J}^T P + P\tilde{J} + \tilde{J}^T P\tilde{J}) & \\ * & -\beta I \end{bmatrix} < 0 \qquad (9.29)$$

where $R_0 = \tilde{x}^T(0)\tilde{x}(0)$, $n_p = 2(n + (k+1)(a+s))$.

Proof: Please refer to Appendix 9.7.3.

Remark 9.5: If the Lyapunov energy function $V(\tilde{x})$ of the augmented system (9.19) is the sum of two subsystems in (9.9) and (9.15), that is,

$$V(\tilde{x}(t)) = \bar{x}^T(t)P_1\bar{x}(t) + \hat{\tilde{x}}^T(t)P_2\hat{\tilde{x}}(t) = [\bar{x}^T(t), \hat{\tilde{x}}^T(t)]\begin{bmatrix} P_1 & 0 \\ 0 & P_2 \end{bmatrix}\begin{bmatrix} \bar{x}(t) \\ \hat{\tilde{x}}(t) \end{bmatrix}, \text{ we have:}$$

$$P\tilde{A} = \begin{bmatrix} P_1 & 0 \\ 0 & P_2 \end{bmatrix}\begin{bmatrix} \bar{A} & 0 \\ L\bar{C} & \bar{A} - L\bar{C} \end{bmatrix} = \begin{bmatrix} P_1\bar{A} & 0 \\ P_2 L\bar{C} & P_2\bar{A} - P_2 L\bar{C} \end{bmatrix}$$

$$P\tilde{B} = \begin{bmatrix} P_1 & 0 \\ 0 & P_2 \end{bmatrix}\begin{bmatrix} \bar{B} \\ L\bar{D} \end{bmatrix} = \begin{bmatrix} P_1\bar{B} \\ P_2 L\bar{D} \end{bmatrix}$$

Let $W = P_2 L$; then the BMI-constrained MOP in (9.26) becomes the following LMI-constrained MOP:

$$(\alpha^*, \beta^*) = \min_{P_1 > 0, P_2 > 0, W, \alpha > 0, \beta > 0} (\alpha, \beta)$$

$$\text{s.t.} (9.27)-(9.29)$$

(9.30)

After solving the LMI-constrained MOP in (9.30), the filter gain L^* for the SEF design is given by $L^* = (P_2^*)^{-1} W^*$.

9.4 MO H_2/H_∞ SEF DESIGN FOR NONLINEAR STOCHASTIC JUMP DIFFUSION SYSTEMS

Consider the following nonlinear stochastic jump-diffusion system with malicious attack signals on system components and sensors:

$$dx(t) = \{f(x(t)) + b_d(x(t))v(t) + b_a(x(t))f_a(t)\}dt$$
$$+ g(x(t))dw(t) + j(x(t))dp(t)$$

$$y(t) = c(x(t)) + b_n(x(t))n(t) + b_s(x(t))f_s(t)$$

(9.31)

where the nonlinear functions $f(x(t))$, $b_d(x(t))$, $b_a(x(t))$, $g(x(t))$, $j(x(t))$, $c(x(t))$, $b_n(x(t))$, and $b_s(x(t))$ are satisfied with the Lipschitz condition such that the nonlinear stochastic equation in (9.31) has an unique solution.

In order to avoid interference from the malicious attack signals $f_a(t)$ and $f_s(t)$, the smoothed signal models in (9.5) and (9.8) are embedded in the nonlinear stochastic jump-diffusion system in (9.31) as follows:

$$d\overline{x}(t) = \{\overline{f}(\overline{x}(t)) + \overline{b}(\overline{x}(t))\overline{v}(t)\}dt + \overline{g}(\overline{x}(t))dw(t)$$
$$+ \overline{j}(\overline{x}(t))dp(t)$$

$$y(t) = \overline{c}(\overline{x}(t)) + \overline{d}(\overline{x}(t))\overline{v}(t)$$

(9.32)

where the augmented state $\overline{x}(t) = [x^T(t), F_a^T(t), F_s^T(t)]^T$ and $\overline{v}(t) = [v(t)^T, n(t)^T, \xi_a(t)^T, \xi_s(t)^T]^T$; the mapping matrices $M_a = [I_a, 0_a, ..., 0_a]$ and $M_s = [I_s, 0_s, ..., 0_s]$;

and $\overline{f}(\overline{x}(t)) = \begin{bmatrix} f(x(t)) + b_a(x(t))M_a F_a(t) \\ A_a F_a(t) \\ A_s F_s(t) \end{bmatrix}, \overline{b}(\overline{x}(t)) = \begin{bmatrix} b_d(x(t)) & 0 & 0 & 0 \\ 0 & 0 & I & 0 \\ 0 & 0 & 0 & I \end{bmatrix}, \overline{g}(\overline{x}(t)) =$

$[g^T(x(t)) \quad 0 \quad 0]^T, \overline{j}(\overline{x}(t)) = [j^T(x(t)) \quad 0 \quad 0]^T, \overline{c}(\overline{x}(t)) = c(x(t)) + b_s(x(t))M_s F_s(t),$

and $\overline{d}(\overline{x}(t)) = [0 \quad b_n(x(t)) \quad 0 \quad 0]$.

In general, the observability criterion of the augmented nonlinear stochastic jump-diffusion systems in (9.32) cannot be expressed as a simple explicit condition (e.g., rank condition or matrix equality). The following lemma from [257] is provided to test the observability of the augmented nonlinear stochastic jump-diffusion systems in (9.32):

Lemma 3 ([257]): Define the augmented vector $z(t) = \left[y^T(t) \quad dy^T(t)/\mathrm{dt} \quad \cdots \right.$
$\left. d^{N_x} y^T(t)/d^N(t) \right]$ and the corresponding Jacobian matrix $Y_z(\bar{x}(t)) = \dfrac{\partial^2 z(t)}{\partial x^2(t)}$. The augmented nonlinear stochastic jump-diffusion system in (9.32) is observable at the equilibrium point $\bar{x}(t) = 0$ if there exist a constant $\varepsilon > 0$ and a constant matrix T such that the absolute values of the leading principal minors $\Delta_1((\bar{x}(t)), ..., \Delta_{N_x}(\bar{x}(t))$ of $TY_z(\bar{x}(t))$ satisfy the following ratio conditions:

$$\Delta_1((\bar{x}(t)) \geq \varepsilon$$
$$\vdots$$
$$\Delta_{N_x}(\bar{x}(t)) \geq \varepsilon, \forall \bar{x}(t)$$

where $N_x = (a+s)(k+1) + n$ is the dimension index of $\bar{x}(t)$, and the principal minor $\Delta_i(\bar{x}(t))$ is the determinant of the matrix by deleting the last $N_x - i$ columns and rows of Jacobian $TY_z(\bar{x}(t))$.

To simplify the filter design, the following Luenberger-type filter is employed to estimate for the stochastic nonlinear augmented systems in (9.32):

$$d\hat{\bar{x}}(t) = \{\bar{f}(\hat{\bar{x}}(t)) + l(\hat{\bar{x}}(t))(y(t) - \hat{y}(t))\}dt$$
$$\hat{y}(t) = \bar{c}(\hat{\bar{x}}(t))$$
(9.33)

where $l(\hat{x}(t))$ is the filter gain to be designed and $\hat{\bar{x}}(t)$ and $\hat{y}(t)$ are the estimated system state and the estimated measurement output, respectively.

The stochastic H_2 filtering performance for the Luenberger-type filter is proposed as follows:

$$H_2(l(\hat{\bar{x}}(t))) = E \int_0^{t_f} (\bar{x}(t) - \hat{\bar{x}}(t))^T Q_2 (\bar{x}(t) - \hat{\bar{x}}(t))dt$$
(9.34)

where $Q_2 \geq 0$ is the weighting matrix of the state estimation error in the H_2 filtering performance, and $\tilde{x}(t) = \left[\bar{x}^T(t), \hat{\bar{x}}^T(t) \right]^T$.

Since the augmented disturbance $\bar{v}(t)$ will influence the filtering performance, the stochastic robust H_∞ filtering performance is proposed to attenuate the effect of the augmented disturbance $\bar{v}(t)$ on the estimation error as follows:

$$H_\infty(l(\hat{\bar{x}}(t))) = \sup_{\bar{v}(t) \in L_2^{F_i}(\mathbb{R}^+; \mathbb{R}^{n_v})} \frac{E\left[\begin{array}{c} \int_0^{t_f} (\bar{x}(t) - \hat{\bar{x}}(t))^T Q_{\Psi}(\bar{x}(t) - \hat{\bar{x}}(t))dt \\ -V(\tilde{x}(0))] \end{array}\right]}{E\left[\int_0^{t_f} \bar{v}^T(t)\bar{v}(t)dt\right]} \tag{9.35}$$

where $Q_\infty \geq 0$ is the weighting matrix of state estimation error in H_∞ filtering performance, and $V(\tilde{x}(0))$ is the positive function to represent the effect of initial condition.

The MO H_2/H_∞ SEF design problem for the nonlinear stochastic jump-diffusion system is formulated as follows:

$$\min_{l(\hat{\bar{x}}(t))}(H_2(l(\hat{\bar{x}}(t))), H_\infty(l(\hat{\bar{x}}(t)))) \text{ s.t.} (9.32)-(9.33) \tag{9.36}$$

In (9.36), the H_2 and H_∞ filtering performance are simultaneously minimized from the MO optimization perspective. For convenience of the MO H_2/H_∞ SEF design in (9.36), the augmented system of the nonlinear stochastic system in (9.32) and its filter in (9.33) is constructed as follows:

$$d\tilde{x}(t) = \{\tilde{f}(\tilde{x}(t)) + \tilde{b}(\tilde{x}(t))\bar{v}(t)\}dt + \tilde{g}(\tilde{x}(t))dw(t)$$
$$+\tilde{j}(\tilde{x}(t))dp(t) \tag{9.37}$$
$$y(t) = \tilde{c}(\tilde{x}(t)) + \tilde{d}(\tilde{x}(t))\bar{v}(t)$$

where

$$\tilde{x}(t) = \left[\bar{x}^T(t), \hat{\bar{x}}^T(t)\right]^T, \tilde{f}(\tilde{x}(t)) = \left[\bar{f}^T(\bar{x}^T(t)), (\bar{f}(\hat{\bar{x}}(t)) + l(\hat{\bar{x}}(t))(\bar{c}(\bar{x}(t)) - \bar{c}(\hat{\bar{x}}(t)))^T\right]^T,$$

$$\tilde{b}(\tilde{x}(t)) = \left[\bar{b}^T(\bar{x}(t)), (l(\hat{\bar{x}}(t))\bar{d}(\bar{x}(t)))^T\right]^T, \tilde{g}(\tilde{x}(t)) = [\bar{g}^T(\bar{x}(t)), 0]^T, \tilde{j}(\tilde{x}(t)) = [\bar{j}^T(\bar{x}(t)), 0]^T,$$

$\tilde{c}(\tilde{x}(t)) = \bar{c}(\bar{x}(t))$, and $\tilde{d}(\tilde{x}(t)) = \bar{d}(\bar{x}(t))$. Then the H_2 and H_∞ SEF design criteria could be reformulated as follows:

$$H_2(l(\hat{\bar{x}}(t))) = E\int_0^{t_f} \tilde{x}^T(t)\tilde{Q}_2\tilde{x}(t)dt \tag{9.38}$$

$$H_\infty(l(\hat{\bar{x}}(t))) = \sup_{\bar{v}(t) \in L_2^{F_i}(\mathbb{R}^+; \mathbb{R}^{n_v})} \frac{E[\int_0^{t_f} \tilde{x}^T(t)\tilde{Q}_\infty\tilde{x}(t)dt - V(\tilde{x}(0))]}{E\int_0^{t_f} \bar{v}^T(t)\bar{v}(t)dt} \tag{9.39}$$

where $\tilde{Q}_2$ and $\tilde{Q}_\infty$ are the same as (9.20) and (9.21). Hence, the MO H_2/H_∞ SEF design problem for the nonlinear stochastic jump-diffusion augmented system in (9.37) could be reformulated as follows:

$$\min_{l(\hat{\bar{x}}(t))}(H_2(l(\hat{\bar{x}}(t))), H_\infty(l(\hat{\bar{x}}(t)))) \text{ s.t.} \tag{9.37)-(9.40}$$

Due to the difficulty in directly solving the MOP in (9.40), the suboptimal method is proposed to indirectly solve the MOP in (9.40) as follows:

$$(\alpha^*, \beta^*) = \min_{l(\hat{\tilde{x}}(t))} (\alpha, \beta) \tag{9.41}$$

$$\text{s.t. } H_2(l(\hat{\tilde{x}}(t))) \leq \alpha \text{ and } H_\infty(l(\hat{\tilde{x}}(t))) \leq \beta \tag{9.42}$$

Theorem 9.4: The suboptimal MOP in (9.41)–(9.42) is equivalent to the MOP in (9.40).

Proof: The proof is similar to Theorem 9.2.

Lemma 9.4: (Itô-Lévy formula of nonlinear stochastic system [130, 234]) For the nonlinear stochastic system in (9.33), the Itô-Lévy formula for a Lyapunov function $V(\tilde{x}(t)) \in C^2$, $V(\tilde{x}(t)) \geq 0$

$$
\begin{aligned}
&dV(\tilde{x}(t)) \\
&= \left(\frac{\partial V(\tilde{x}(t))}{\partial \tilde{x}(t)}\right)^T (\tilde{f}(\tilde{x}(t)) + \tilde{b}(\tilde{x}(t))\bar{v}(t))dt \\
&+ \frac{1}{2}\tilde{g}(\tilde{x}(t))^T \frac{\partial^2 V(\tilde{x}(t))}{\partial \tilde{x}^2(t)} \tilde{g}(\tilde{x}(t))dt + \left(\frac{\partial V(\tilde{x}(t))}{\partial \tilde{x}(t)}\right)^T \tilde{g}(\tilde{x}(t))dw(t) \\
&+ \left(V(\tilde{x}(t) + \tilde{j}(\tilde{x}(t))) - V(\tilde{x}(t))\right)dp(t)
\end{aligned}
\tag{9.43}
$$

Based on Theorem 9.4 and Lemma 9.3, we get the result:

Theorem 9.5: The MOP in (9.41) and (9.42) for the MO H_2/H_∞ SEF design of the nonlinear stochastic jump-diffusion system in (9.37) could be transformed into the following HJI-constrained MOP:

$$(\alpha^*, \beta^*) = \min_{l(\hat{\tilde{x}}(t)), V(\tilde{x}(t))} (\alpha, \beta) \tag{9.44}$$

$$
\begin{aligned}
&\left(\frac{\partial V(\tilde{x}(t))}{\partial \tilde{x}(t)}\right)^T \tilde{f}(\tilde{x}(t)) + \frac{1}{2}\tilde{g}(\tilde{x}(t))^T \frac{\partial^2 V(\tilde{x}(t))}{\partial \tilde{x}^2(t)} \tilde{g}(\tilde{x}(t)) \\
&+ \tilde{x}^T(t)\tilde{Q}_2\tilde{x}(t) + \lambda\left(V(\tilde{x}(t) + \tilde{j}(\tilde{x}(t))) - V(\tilde{x}(t))\right) \leq 0
\end{aligned}
\tag{9.45}
$$

$$E[V(\tilde{x}(t))] \leq \alpha \tag{9.46}$$

$$
\begin{aligned}
&\left(\frac{\partial V(\tilde{x}(t))}{\partial \tilde{x}(t)}\right)^T \tilde{f}(\tilde{x}(t))\bar{v}(t)dt + \frac{1}{2}\tilde{g}(\tilde{x}(t))^T \frac{\partial^2 V(\tilde{x}(t))}{\partial \tilde{x}^2(t)} \tilde{g}(\tilde{x}(t)) \\
&+ \tilde{x}^T(t)\tilde{Q}_\infty\tilde{x}(t) + \frac{1}{4\beta}\left(\frac{\partial V(\tilde{x}(t))}{\partial \tilde{x}(t)}\right)^T \tilde{b}(\tilde{x}(t))\tilde{b}^T(\tilde{x}(t))\frac{\partial V(\tilde{x}(t))}{\partial \tilde{x}(t)} \\
&+ \lambda\left(V(\tilde{x}(t) + \tilde{j}(\tilde{x}(t))) - V(\tilde{x}(t))\right) \leq 0
\end{aligned}
\tag{9.47}
$$

Proof: Please refer to Appendix 9.7.5.

In general, it is very difficult to solve the HJIs in (9.45)–(9.47) of the MOP in (9.44) for the MO H_2/H_∞ SEF design problem of the nonlinear stochastic system in (9.37). In this design, based on the following global linearization technique [254,

255], the nonlinear stochastic system in (9.32) and nonlinear Luenberger-type filter in (9.33) could be interpolated by several local linearized systems at the N vertices of a polytope of [254, 255]:

$$
\begin{bmatrix} \frac{\partial \bar{f}(\bar{x})}{\partial \bar{x}} \\ \frac{\partial \bar{b}(\bar{x})}{\partial \bar{x}} \\ \frac{\partial \bar{g}(\bar{x})}{\partial \bar{x}} \\ \frac{\partial \bar{j}(\bar{x})}{\partial \bar{x}} \\ \frac{\partial \bar{c}(\bar{x})}{\partial \bar{x}} \\ \frac{\partial \bar{d}(\bar{x})}{\partial \bar{x}} \\ \frac{\partial \bar{l}(\bar{x})}{\partial \bar{x}} \end{bmatrix} \in C_0 \left\{ \begin{bmatrix} \bar{A}_1 \\ \bar{B}_1 \\ \bar{G}_1 \\ \bar{J}_1 \\ \bar{C}_1 \\ \bar{D}_1 \\ \bar{L}_1 \end{bmatrix} \cdots \begin{bmatrix} \bar{A}_i \\ \bar{B}_i \\ \bar{G}_i \\ \bar{J}_i \\ \bar{C}_i \\ \bar{D}_i \\ \bar{L}_i \end{bmatrix} \cdots \begin{bmatrix} \bar{A}_N \\ \bar{B}_N \\ \bar{G}_N \\ \bar{J}_N \\ \bar{C}_N \\ \bar{D}_N \\ \bar{L}_N \end{bmatrix} \right\}, \forall \bar{x}(t) \tag{9.48}
$$

where C_0 denotes the convex hull of the polytope with N vertices. In other words, if the total local linearized systems at all $\bar{x}(t)$ are inside the convex hull C_0, then the trajectory $\bar{x}(t)$ of the nonlinear stochastic jump-diffusion system in (9.32) will be the convex combination of the trajectories of the following N local linearized stochastic jump-diffusion systems at the vertices of the polytope in (9.48)

$$
d\bar{x}(t) = (\bar{A}_i\bar{x}(t) + \bar{B}_i\bar{v}(t))dt + \bar{G}_i\bar{x}(t)dw(t) + \bar{J}_i\bar{x}(t)dp(t)
$$
$$
y(t) = \bar{C}_i\bar{x}(t) + \bar{D}_i\bar{v}(t), i = 1,...,N \tag{9.49}
$$

Based on the global linearization theory [254, 255], the trajectory of the nonlinear stochastic jump-diffusion system in (9.32) could be represented by the convex combination of N local linearized stochastic jump-diffusion systems as follows:

$$
d\bar{x}(t) = \sum_{i=1}^{N} \varpi_i(\bar{x}(t))[(\bar{A}_i\bar{x}(t) + \bar{B}_i\bar{v}(t))dt + \bar{G}_i\bar{x}(t)dw(t)
$$
$$
+ \bar{J}_i\bar{x}(t)dp(t)] \tag{9.50}
$$
$$
y(t) = \sum_{i=1}^{N} \varpi_i(\bar{x}(t))[\bar{C}_i\bar{x}(t) + \bar{D}_i\bar{v}(t)]
$$

where $\varpi_i(\bar{x}(t)), i = 1,...,N$ are interpolation functions with $0 \le \varpi_i(\bar{x}(t)) \le 1$ and $\sum_{i=1}^{N} \varpi_i(\bar{x}(t)) = 1$.

Remark 9.6: Even the nonlinear stochastic system in (9.32) is interpolated by a set of local linearized systems, the interpolated system in (9.50) it is still a nonlinear system and its observability can not be directly examined. However, by using Theorem 1, the observability of the local linearized system can be directly examined.

Similarly, the nonlinear Luenberger-type filter in (9.33) could be represented by the global linearization form as follows:

$$d\hat{\tilde{x}}(t) = \sum_{i=1}^{N} \varpi_i(\hat{\tilde{x}}(t))[(\bar{A}_i\hat{\tilde{x}}(t) + \bar{L}_i(y(t) - \hat{y}(t))]dt$$

$$\hat{y}(t) = \sum_{i=1}^{N} \varpi_i(\hat{\tilde{x}}(t))\bar{C}_i\hat{\tilde{x}}(t)$$

(9.51)

Thus, the nonlinear stochastic augmented system in (9.37) can be written as:

$$d\tilde{x}(t) = \sum_{i=1}^{N}\sum_{j=1}^{N} \varpi_i(\bar{x}(t))\,\varpi_j(\hat{\tilde{x}}(t))[(\tilde{A}_i\tilde{x}(t) + \tilde{B}_{ij}\bar{v}(t))dt$$

$$+\tilde{G}_i\tilde{x}(t)dw(t) + \tilde{J}_i\tilde{x}(t)dp(t)]$$

(9.52)

$$y(t) = \sum_{i=1}^{N} \varpi_i(\bar{x}(t))[\bar{C}_i\bar{x}(t) + \bar{D}_i\bar{v}(t)]$$

where $\tilde{A}_{ij} = \begin{bmatrix} \bar{A}_i & 0 \\ L_j\bar{C}_i & \bar{A}_j - L_j\bar{C}_j \end{bmatrix}$, $\tilde{B}_{ij} = \begin{bmatrix} \bar{B}_i \\ L_j\bar{D}_i \end{bmatrix}$, $\tilde{G}_i = diag\{\bar{G}_i, 0\}$, and $\tilde{J}_i = diag\{\bar{J}_i, 0\}$

for $i, j = 1, ..., N$.

Lemma 9.5 ([254, 255]): For any matrix S_i with appropriate dimension and inter-polation functions $\varpi_i(\bar{x}(t))$ with $0 \le \varpi_i(\bar{x}(t)) \le 1$ and $\sum_{i=1}^{N} \varpi_i(\bar{x}(t)) = 1$, we have:

$$\left(\sum_{i=1}^{N} \varpi_i(\bar{x}(t))S_i\right)^T P\left(\sum_{j=1}^{N} \varpi_j(\bar{x}(t))S_j\right) \le \sum_{i=1}^{N} \varpi_i(\bar{x}(t))S_i^T PS_i$$

(9.53)

Based on these results, we get the following theorem:

Theorem 9.6: Based on the global linearization systems in (9.50) and (9.51) and Lemma 9.4, the HJI-constrained MOP in (9.44)–(9.47) for the MO H_2/H_∞ SEF design of the nonlinear stochastic jump-diffusion system in (9.37) can be transformed into the following BMI-constrained MOP

$$(\alpha^*, \beta^*) = \min_{\bar{L}_j, P_1 > 0, P_2 > 0, \alpha > 0, \beta > 0} (\alpha, \beta)$$

(9.54)

s.t. $\tilde{Q}_2 + P\tilde{A}_{ij} + \tilde{A}_{ij}^T P + \tilde{G}_i^T P\tilde{G}_i + \lambda(\tilde{J}_i^T P + P\tilde{J}_i + \tilde{J}_i^T P\tilde{J}_i) < 0$ (9.55)

$$P \le \frac{\alpha}{n_p} E[R_0^{-1}]$$

(9.56)

$$\begin{bmatrix} \tilde{Q}_\infty + P\tilde{A}_{ij} + \tilde{A}_{ij}^T P + \tilde{G}_i^T P\tilde{G}_i & P\tilde{B}_{ij} \\ +\lambda(\tilde{J}_i^T P + P\tilde{J}_i + \tilde{J}_i^T P\tilde{J}_i) & \\ * & -\beta I \end{bmatrix} < 0$$

(9.57)

for $i, j = 1,..., N$, $n_p = 2(n + (k+1)(a+s))$ and $R_0 = E\tilde{x}(0)\tilde{x}^T(0)$.

Proof: Please refer to Appendix 9.7.5.

Remark 9.7: Since the terms in the BMIs in (9.55)–(9.57) are still coupling, the coupling term could be solved as follows:

Choose the Lyapunov energy function

$$V(\tilde{x}(t)) = \tfrac{1}{2}\bar{x}^T(t)P_1\bar{x}(t) + \tfrac{1}{2}\hat{\tilde{x}}^T(t)P_2\hat{\tilde{x}}(t) = \tfrac{1}{2}\tilde{x}^T(t)\begin{bmatrix} P_1 & 0 \\ 0 & P_2 \end{bmatrix}\tilde{x}(t) = \tfrac{1}{2}\tilde{x}^T(t)P\tilde{x}(t)$$

that is, the positive matrix $P = \begin{bmatrix} P_1 & 0 \\ 0 & P_2 \end{bmatrix}$. Let $W_j = P_2\bar{L}_j$ for $j = 1,...,N$, and we get

$$P\tilde{A}_{ij} = \begin{bmatrix} P_1 & 0 \\ 0 & P_2 \end{bmatrix}\begin{bmatrix} \bar{A}_i & 0 \\ \bar{L}_j\bar{C}_i & \bar{A}_j - \bar{L}_j\bar{C}_j \end{bmatrix} = \begin{bmatrix} P_1\bar{A}_i & 0 \\ W_j\bar{C}_i & P_2\bar{A}_j - W_j\bar{C}_j \end{bmatrix}$$

$$P\tilde{B}_{ij} = \begin{bmatrix} P_1 & 0 \\ 0 & P_2 \end{bmatrix}\begin{bmatrix} \bar{B} \\ \bar{L}_j\bar{D}_j \end{bmatrix} = \begin{bmatrix} P_1\bar{B}_i \\ W_j\bar{D}_j \end{bmatrix}$$

By substituting the previous $P\tilde{A}_{ij}$ and $P\tilde{B}_{ij}$ into (9.55)–(9.57), then we have the LMI-constrained MOP for (9.54) as follows:

$$(\alpha^*,\beta^*) = \min_{P_1>0,P_2>0,\{W_j\}_{j=1}^N,\alpha>0,\beta>0} (\alpha,\beta) \tag{9.58}$$
$$\text{s.t. } (9.55)-(9.57)$$

Remark 9.8: In this design, the state-independent weighting matrices Q_2 and Q_∞ are chosen in the H_2 and H_∞ filtering performance in (9.34) and (9.35), respectively. If the state-dependent weighting matrices $Q_2(\bar{x}(t))$ and $Q_\infty(\bar{x}(t))$ are chosen in these two performance indices, the design procedure is similar to the case of state-independent weighting matrices Q_2 and Q_∞. In fact, by using the global linearization method, $Q_2(\bar{x}(t))$ and $Q_\infty(\bar{x}(t))$ can be represented by convex interpolation of N local weighting matrices $Q_{\infty,i}$ and $Q_{2,i}$ at N vertices of polytope in (9.48) as $Q_2(\bar{x}(t)) = \sum_{i=1}^N \varpi_i(\bar{x}(t))Q_{2,i}$ and $Q_\infty(\bar{x}(t)) = \sum_{i=1}^N \varpi_i(\bar{x}(t))Q_{\infty,i}$, where $\{Q_{\infty,i}, Q_{2,i}\}_{i=1}^N$ are local linearized weighting matrices. Then the design can be achieved by replacing the terms $\tilde{Q}_2$ and $\tilde{Q}_\infty$ in (9.55) and (9.57) with $\tilde{Q}_{\infty,i}$ and $\tilde{Q}_{2,i}$, respectively.

After solving the LMI-constrained MOP in (9.58) for P_1^*, P_2^*, and W_j^*, $j = 1,...,N$, the filter gains for the MO H_2/H_∞ SEF design problem in (9.51) are given by $\bar{L}_j^* = (P_2^*)^{-1}W_j^*$, and then we can obtain the nonlinear MO H_2/H_∞ SEF in (9.51).

According to the previous analysis, a reverse-order LMI-constrained MOEA to solve the MOP in (9.58) for the MO H_2/H_∞ SEF in (9.41) of the nonlinear stochastic jump-diffusion system with malicious attack signals in (9.31) is proposed as follows:

Step 1: Select the search range $S = [\alpha^L, \alpha^U] \times [\beta^L, \beta^U]$ for the feasible objective vector (α, β) and set the iteration number N_i, population number N_p, crossover rate R_c, and mutation rate R_m in the LMI-constrained MOEA. Set iteration number $i = 1$.

Step 2: Select N_p feasible individuals satisfying the LMIs in (9.55)–(9.57) as the initial population P_1 from S.

Step 3: Perform the EA operator with crossover rate R_c and mutation rate R_m to generate P_1 feasible individuals by examining whether their corresponding objective vectors (α, β) are feasible vectors satisfying the LMIs in (9.55)–(9.57).

Step 4: Set the iteration index $i = i + 1$ and select N_p elite individuals from $2N_p$ feasible individuals in Step 3 through the nondominated sorting method and crowded comparison operation. Set the selected elite individuals as the population P_{i+1}.

Step 5: Repeat Steps 3 and 4 until the iteration number $i = N_i$ is satisfied; then set the final population $P_{N_i} = P_F$ as the Pareto front.

Step 6: Select a preferable feasible objective individual $(\alpha^*, \beta^*) \in P_F$ according to the designer's own preference with the corresponding optimal filter gains $\bar{L}_j^* = (P_2^*)^{-1} W_j^*$ and $j = 1, 2, ..., N$ of the MOP in (9.58). Once the preferable solution is selected. Then we can obtain the SEF from (9.51) for the nonlinear stochastic jump-diffusion system in (9.32).

Remark 9.9: (i) The time complexity of the proposed LMI-constrained MOEA is approximately estimated as $O(\varepsilon^6 N N_p^2 m)$ [19, 258], including $O(\varepsilon^6 N)$ for solving LMIs and $m N_p^2$ for MOEA, where $\varepsilon = 2(n + (a + s)(k + 1))$ is the order of the Lyapunov function matrix P, N is the number of local linearized models in (9.49) of the nonlinear stochastic system in (9.32), $m = 2$ is the number of objectives, and N_p is the population number of the LMI-constrained MOEA. On the other hand, the space (memory) complexity of the proposed LMI-constrained MOEA is approximately estimated as $O((mN_p + N_p^2) N_p N \varepsilon^4)$ [259, 260] with the memory $O(mN_p + N_p^2)$ in the MOEA and the memory $O(N_p N \varepsilon^4)$ in the LMI problem. (ii) In the conventional MOEAs [7, 11], the search filter gains $P_1 > 0$, $P_2 > 0$, W_j, and $j = 1, ..., N$ for the optimal objective vector (α^*, β^*) for the MOP in (9.58) directly. When N is large, it becomes very difficult to employ the conventional MOEA to solve the MOP in (9.58).

Remark 9.10: In the proposed LMI-constrained MOEA, the convergence speed greatly depends on the population number. Instinctively, if a large population number is adopted in MOEA, the number of initial populations that are close to the real Pareto front will increase. As a result, it will increase the success probability for reaching the real Pareto front. However, in such a case, due to the growth of the population number, the time complexity will be increased with a polynomial increase.

9.5 SIMULATION RESULTS

In this section, two simulation examples are provided to illustrate the effectiveness of the proposed MO SEF design. In the first example, an estimation problem for a stochastic nonlinear radar system with the influence of malicious attack signals on actuator and sensor, external disturbance, and measurement noise is carried out. In the second example, a networked-based mass-spring system is provided to compare the effectiveness between our method and the fault estimation filter design method in [261].

9.5.1   SEF Design for Stochastic Nonlinear Radar System

In this section, the MO H_2/H_∞ SEF is employed for a radar system to estimate the position and velocity (i.e., the trajectory) of an incoming ballistic missile, as shown in Figure 9.1. Suppose the radar sensor is interfered with by malicious attack signals from the missile and the missile also performs staggering side drift via sudden side-step maneuvering via the side jets. By using classical kinematic analysis, the dynamic equation of the motions of the incoming ballistic missile could be expressed as follows [10, 262]:

$$\ddot{x}(t) = \frac{-\rho(z(t))g\sqrt{\dot{x}(t)^2 + \dot{y}(t)^2 + \dot{z}(t)^2}}{2\beta}\dot{x}(t)$$

$$\ddot{y}(t) = \frac{-\rho(z(t))g\sqrt{\dot{x}(t)^2 + \dot{y}(t)^2 + \dot{z}(t)^2}}{2\beta}\dot{y}(t) \tag{9.59}$$

$$\ddot{z}(t) = \frac{-\rho(z(t))g\sqrt{\dot{x}(t)^2 + \dot{y}(t)^2 + \dot{z}(t)^2}}{2\beta}\dot{z}(t) - g$$

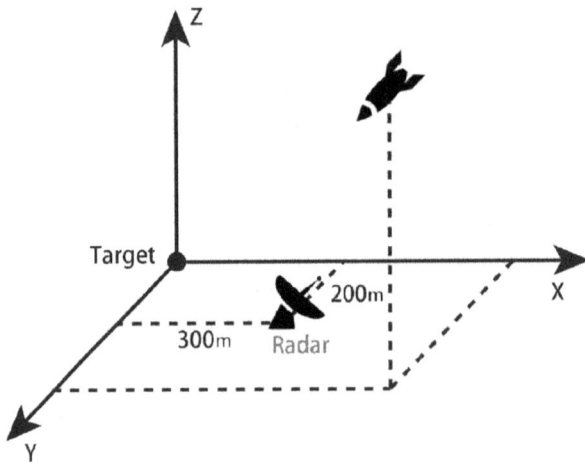

FIGURE 9.1   The incoming ballistic missile 3-D geometry in [10, 262], where the x axis denotes the downrange of the missile, y axis denotes the offrange of the missile, and z axis denotes the altitude of the missile. The target of the missile is set at the origin point of the Cartesian coordinate, and the radar is located at $(x, y, z) = (300, 200, 0)$.

where $x(t), y(t), z(t)$ are the target-centered Cartesian coordinates of the incoming ballistic missile on the ground; g is the gravity constant; β is the ballistic coefficient; and ρ is the density of the atmosphere at the incoming ballistic missile position, which is defined as follows [10, 262]:

$$\rho(z(t)) = \begin{cases} \rho_h e^{-\alpha_h z(t)}, \rho_h = 1.745, \alpha_h = 1.49 \times 10^{-4} \\ \quad \text{if } z(t) \geq 9144 \text{ meters} \\ \rho_l e^{-\alpha_l z(t)}, \rho_l = 1.227, \alpha_l = 1.093 \times 10^{-4} \\ \quad \text{if } z(t) < 9144 \text{ meters} \end{cases} \tag{9.60}$$

By using the dynamic motion of the incoming ballistic missile in (9.59), the non-linear incoming ballistic missile could be written as follows [10, 262]:

$$\dot{X}(t) = \left[\dot{x}(t), \dot{y}(t), \dot{z}(t), \tfrac{-\rho(z(t))g\omega(X)}{2\beta} \dot{x}(t), \tfrac{-\rho(z(t))g\omega(X)}{2\beta} \dot{y}(t), \right.$$
$$\left. \tfrac{-\rho(z(t))g\omega(X)}{2\beta} \dot{z}(t) - g \right]^T = f(X(t)) \tag{9.61}$$

where

$$X(t) = [x(t), y(t), z(t), \dot{x}(t), \dot{y}(t), \dot{z}(t)]^T$$

and

$$\omega(X(t)) = \sqrt{\dot{x}(t)^2 + \dot{y}(t)^2 + \dot{z}(t)^2}$$

In order to simplify the notations, we use

$$X(t) = [x_1(t), x_2(t), x_3(t), x_4(t), x_5(t), x_6(t)]^T$$

Then, suppose the incoming ballistic missile suffers from continuous random noise $g(X)dw(t)$ due to the stochastic parameter fluctuation and Poisson jump discontinuous noise $j(X)dp(t)$ due to corporate defaults, component failures, or insured events, respectively. Then $v(t)$ is the external disturbance, and $n(t)$ denotes the measurement noise of radar sensor. Both of them are assumed to be zero mean white noise with unit variance. Hence, the nonlinear incoming ballistic missile under the detection of the radar system can be represented by the following nonlinear stochastic jump-diffusion system:

$$dX(t) = [f(X) + Bu(t) + b_d(X)v(t)]dt + g(X)dw(t)$$
$$+ j(X)dp(t) \tag{9.62}$$
$$Y(t) = c(X) + b_n(X)n(t)$$

where $B = I_{6\times6}$, $u(t) = [u_1(t), u_2(t), u_3(t), u_4(t), u_5(t), u_6(t)]^T$ denotes the guidance law to guide the missile to the attack target, $Y(t)$ denotes the sensor measurement of the radar system, $c(X) = [x_1 - 300, x_2 - 200, x_3, x_4, x_5, x_6]^T$ is a shifted state to describe the distance from the radar to the target of the missile, $n(t)$ denotes the measurement noise of the radar sensor, and $v(t)$ denotes the environmental disturbance on the missile.

In order to avoid the attack of a ground anti-missile such as the Patriot, the incoming ballistic missile will perform sudden side-step maneuvering through its two-side jets, which are considered the defense and evasive signal $f_a(t)$ of the missile system or equivalent malicious attack signals on the radar detection system in (9.62). On the other hand, the sensor of the ground radar station will be interfered with by the incoming missile to cause deterioration of the accuracy of the detection capability of radar, which will be regarded as malicious attack signal $f_s(t)$ on the sensor of the radar detection system. The guidance law $u(t) = KX = [-0.1x_1 - x_4, -0.1x_2 - x_5, -0.08x_3 - x_6, -0.01x_4, -0.01x_5, -0.01x_6]^T$ is employed by the missile. Therefore, the nonlinear radar detection system of the incoming ballistic missile could be rewritten as follows:

$$dX(t) = [f(X) + BKX + b_d(X)v(t) + b_a(X)f_a(t)]dt$$
$$+ g(X)dw(t) + j(X)dp(t) \tag{9.63}$$

$$Y(t) = c(X) + b_n(X)n(t) + b_s(X)f_s(t)$$

where

$$f(X) + BKX = \begin{bmatrix} -0.1x_1 \\ -0.1x_2 \\ -0.08x_3 \\ \frac{-\rho g\omega(X)}{2\beta}x_4 - 0.01x_4 \\ \frac{-\rho g\omega(X)}{2\beta}x_5 - 0.01x_5 \\ (\frac{-\rho g\omega(X)}{2\beta}x_6 - 0.01x_6 \\ +(2\times10^{-4})x_3 - g) \end{bmatrix} = F(X)$$

$$b_a(X) = b_s(X) = [0_{1\times3}, 1, 1, 1]^T$$
$$g(X) = 0.03\times[0_{1\times3}^T, x_4, x_5, x_6]^T \tag{9.64}$$
$$j(X) = 0.01\times[0_{1\times3}^T, x_4, x_5, x_6]^T$$
$$b_d(X) = b_n(X) = diag\{0_{3\times3}, I_3\}$$

By using third-order signal-smoothed models for attack signal $f_a(t)$ in (9.5) with extrapolation coefficients $\alpha_0 = 0.5$, $\alpha_1 = 0.3$, $\alpha_2 = 0.1$, and $\alpha_3 = 0.1$ and the malicious attack signal $f_s(t)$ in (9.8) with extrapolation coefficients $\beta_0 = 0.6$, $\beta_1 = 0.2$, and $\beta_2 = \beta_3 = 0.1$, the nonlinear radar detection system of the incoming ballistic missile in (9.63) could be embedded with these two signal-smoothed models as the following augmented system:

$$d\bar{X} = [F(\bar{X}) + b_d(\bar{X})\bar{v}(t)]dt + g(\bar{X})dw(t) + j(\bar{X})dp(t)$$
$$Y(t) = c(\bar{X}) + b_n(\bar{X})\bar{v}(t) \tag{9.65}$$

where the augmented state $\bar{X} = [X^T(t), F_a^T(t), F_s^T(t)]^T \in \mathbb{R}^{14}$, and the augmented disturbance $\bar{v}(t) = [v(t)^T, n(t)^T, \xi_a(t)^T, \xi_s(t)^T]^T$.

Based on the global linearization technique with 48 local linearized missile systems, the incoming ballistic missile in (9.63) can be represented by (9.50) as the following interpolation of 48 local linearized missile detection systems:

$$d\bar{X}(t) = \sum_{i=1}^{48} \varpi_i(\bar{X}(t))[(\bar{A}_i\bar{X}(t) + \bar{B}_i\bar{v}(t))dt + \bar{G}_i\bar{X}(t)dw(t)$$
$$+ \bar{J}_i\bar{X}(t)dp(t)] \tag{9.66}$$
$$y(t) = \sum_{i=1}^{48} \varpi_i(\bar{X}(t))[\bar{C}_i\bar{X}(t) + \bar{D}_i\bar{v}(t)]$$

with the following interpolation functions:

$$\varpi_i(\bar{X}(t)) = \frac{\left(1/\left\|X_i^o - X(t)\right\|_2^2\right)}{\sum_{i=1}^{48}\left(1/\left\|X_i^o - X(t)\right\|_2^2\right)}, \text{for } i = 1,...,48$$

where $X_i^o \in S_i^n$ is the operation point of the ith local linear system.

The initial condition in the simulation is assumed to be $X(0) = [150000, 210000, 120000, -2500, -2500, -2500]^T$, and then the weighting matrices Q_2 and Q_∞ in (9.34) and (9.35) are designed as

$$Q_2 = diag\{Q_{2,x}, Q_{2,f_a}, Q_{2,f_s}\}$$
$$Q_{2,x} = I_6, Q_{2,f_a} = Q_{2,f_s} = 0.1I_4$$
$$Q_\infty = diag\{Q_{\infty,x}, Q_{\infty,f_a}, Q_{\infty,f_s}\}$$
$$Q_{\infty,x} = 2I_6, Q_{\infty,f_a} = Q_{\infty,f_s} = I_4$$

Based on the proposed reverse-order LMI-constrained MOEA, the LMI-constrained MOP in (9.58) can be solved. The design parameters of MOEA are chosen as follows: iteration number $N_i = 100$, search range $S = [10^8, 10^9] \times [0,6]$, crossover rate $R_c = 0.8$, mutation rate $R_m = 0.1$, and population number $N_p = 400$.

Once the iteration number N_i is achieved, we get (α^*, β^*), P^*, and $\bar{L}_i^* = (P_2^*)^{-1}W_i^*$ for $i = 1,...,48$. The Pareto front for the Pareto optimal solutions (α^*, β^*) of the MOP in (9.58) of the MO H_2/H_∞ SEF of the radar detection system of the incoming ballistic missile can be obtained in Figure 9.2 by the proposed design procedure. Among the Pareto optimal solutions on the Pareto front, the knee point solution is selected as

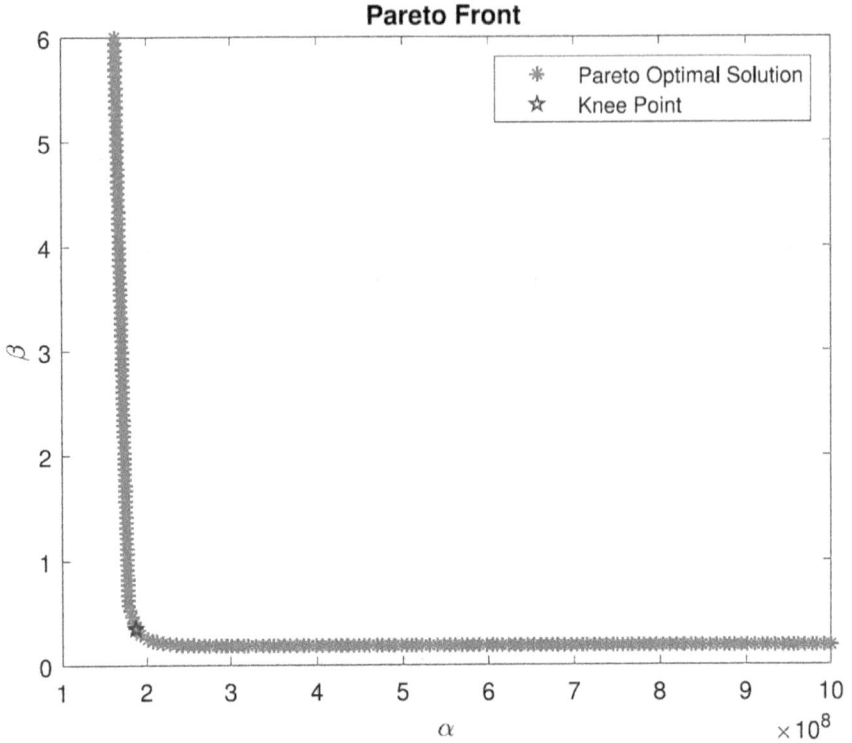

FIGURE 9.2 Pareto front obtained by the proposed LMI-constrained MOEA for the H_2/H_∞ SEF of the radar detection system.

the preferable solution in the Pareto optimal solutions set. In this case, the knee point solution of the MO H_2/H_∞ SEF $(\alpha^*, \beta^*) = (1.868 \times 10^8, 0.3522)$ is chosen. From the knee point solution, the Luenberger-type filter gain $\bar{L}_i^*$ in (9.51) could be constructed as $\bar{L}_i^* = (P_2^*)^{-1} W_i^*$ for $i = 1,...,48$.

In general, the incoming ballistic missile can transmit a malicious signal to interfere with the radar sensor and perform staggering side drift as equivalent malicious attack signals on the radar detection system via two-side maneuvering by the side jets. Therefore, the radar system could not estimate the trajectory of missile system precisely. In this simulation, the malicious attack signals $f_a(t)$ and $f_s(t)$ are shown as the aperiodic square signal and high frequency sine signal in Figure 9.3, respectively. To estimate the effect of unknown swaggering side acceleration and a malicious interference signal, the smoothed-model–based SEF design of the malicious attack signals $f_a(t)$ and $f_s(t)$ can be efficiently estimated in Figure 9.3. From Figure 9.3, due to the effect of initial condition, there exist some overshoots in estimating the malicious signals. After the effect of the initial condition is eliminated from $t = 8s$, the malicious signals on the system and sensor can be estimated by the proposed SEF and can be compensated by the Luenberger filter for missile trajectory estimation. For the proposed SEF design, the estimation of malicious attack signals $f_a(t)$

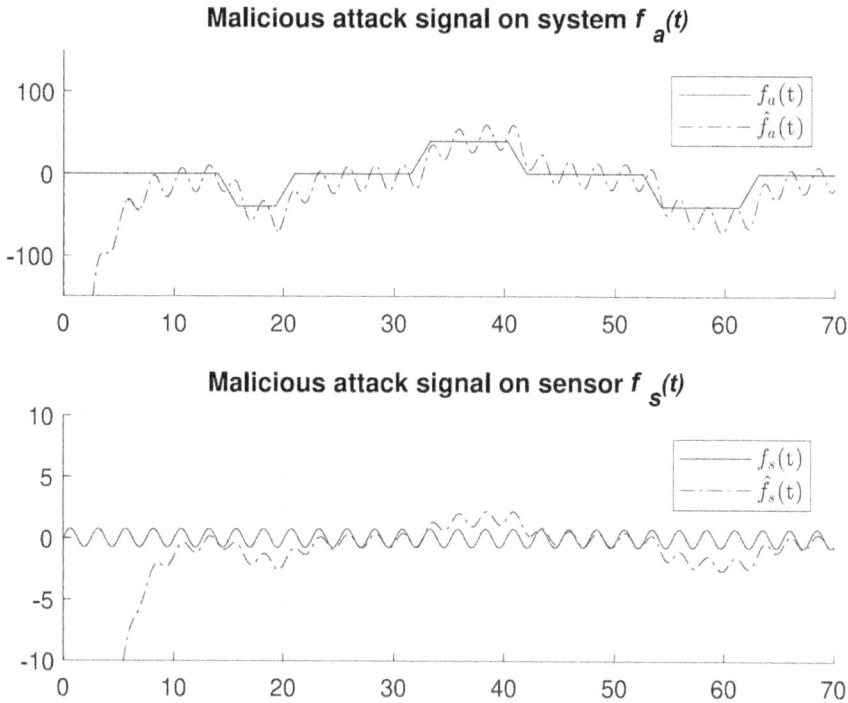

FIGURE 9.3 Malicious attack signals $f_a(t)$ and $f_s(t)$ and their estimation.

and $f_s(t)$ interact with each other. For example, when the malicious attack signal $f_a(t)$ enters the system at $t = 15s$, there is a small deviation on the estimated signal $\hat{f}_s(t)$. On the other hand, the estimated attack signal $\hat{f}_a(t)$ oscillates due to the high-frequency sine signal $f_s(t)$ from $Y(t)$ in (9.51). From Figure 9.3, the interacted effects on the estimation are efficiently reduced at the steady state. For example, once the malicious attack signal $f_a(t)$ vanishes, the high-frequency sine signal $f_s(t)$ is precisely estimated by the proposed SEF.

The states of the incoming ballistic missile system and the corresponding estimated states are shown in Figures 9.4–9.9. From the results of Figures 9.4–9.6, the incoming ballistic missile generates the malicious acceleration $f_a(t)$ and sensor interference $f_s(t)$ to dodge radar system detection. By using the estimated acceleration $\hat{f}_a(t)$ and malicious attack signals $\hat{f}_s(t)$, the three velocities of the missile system can be precisely estimated in Figures 9.4–9.6. Meanwhile, due to the good estimation of the three velocities, the proposed SEF can also precisely estimate the real position of the missile system in Figures 9.7–9.9. Moreover, the effect of external disturbance and sensor noise on the estimation is reduced with a minimum attenuation level $\beta^* = 0.3522$ by the proposed H_2/H_∞ SEF design.

To compare the effectiveness of the proposed SEF design with conventional approaches, the traditional mixed H_2/H_∞ robust filter design in [255] is carried out. In the traditional mixed H_2/H_∞ robust filter, the design is only concerned with

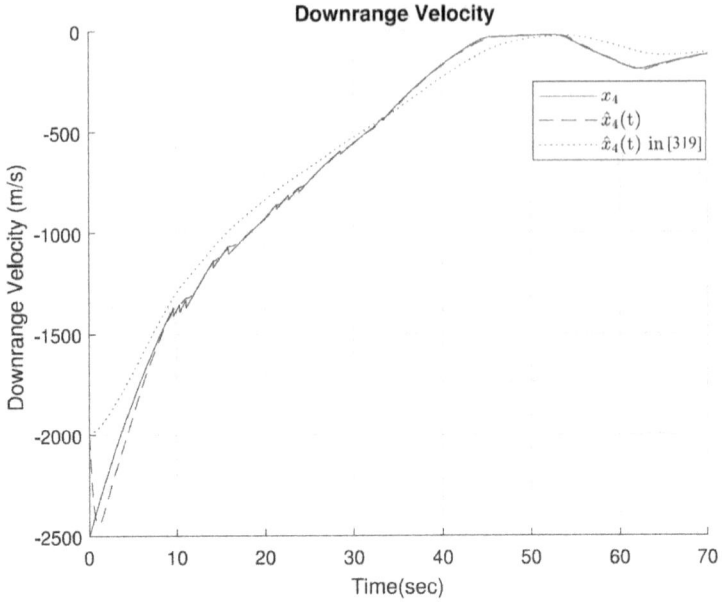

FIGURE 9.4 The downrange velocity trajectory and its estimation by the proposed MO H_2/H_∞ SEF and robust H_2/H_∞ filter in [255].

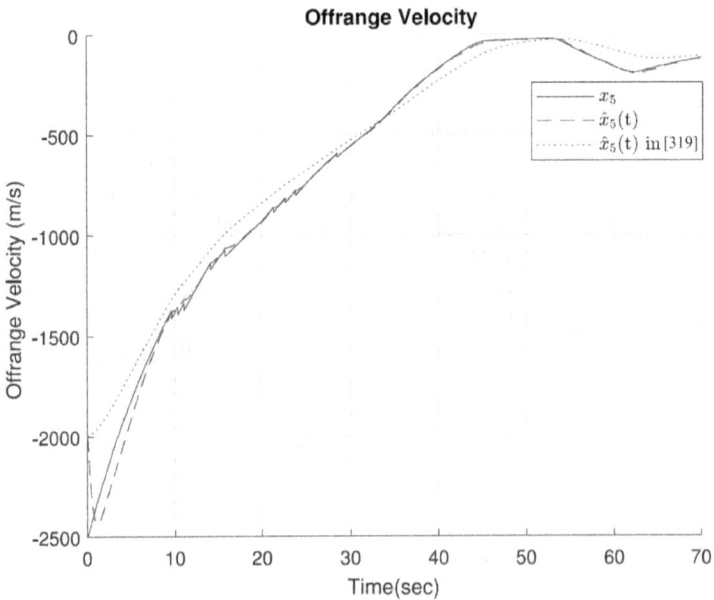

FIGURE 9.5 The offrange velocity trajectory and its estimation by the proposed MO H_2/H_∞ SEF and robust H_2/H_∞ filter in [255].

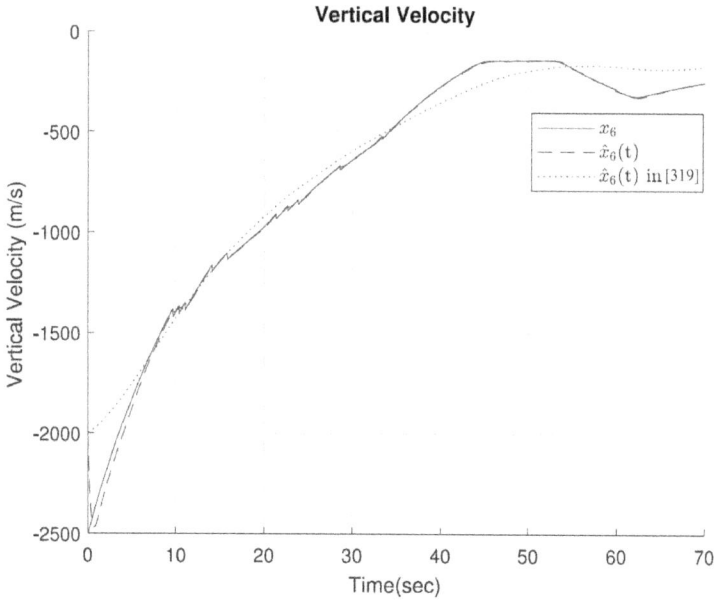

FIGURE 9.6 The vertical velocity trajectory and its estimation by the proposed MO H_2/H_∞ SEF and robust H_2/H_∞ filter in [255].

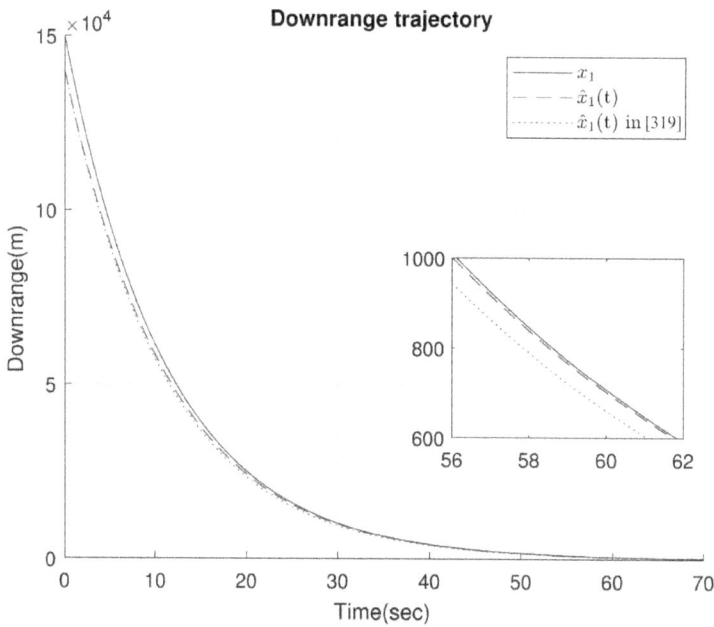

FIGURE 9.7 The downrange trajectory and its estimation by the proposed MO H_2/H_∞ SEF and robust H_2/H_∞ filter in [255].

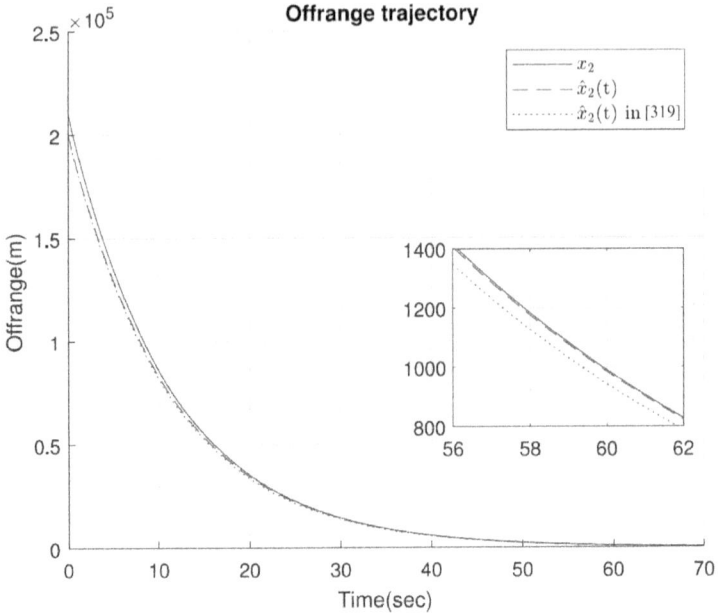

FIGURE 9.8 The offrange trajectory and its estimation by the proposed MO H_2/H_∞ SEF and robust H_2/H_∞ filter in [255].

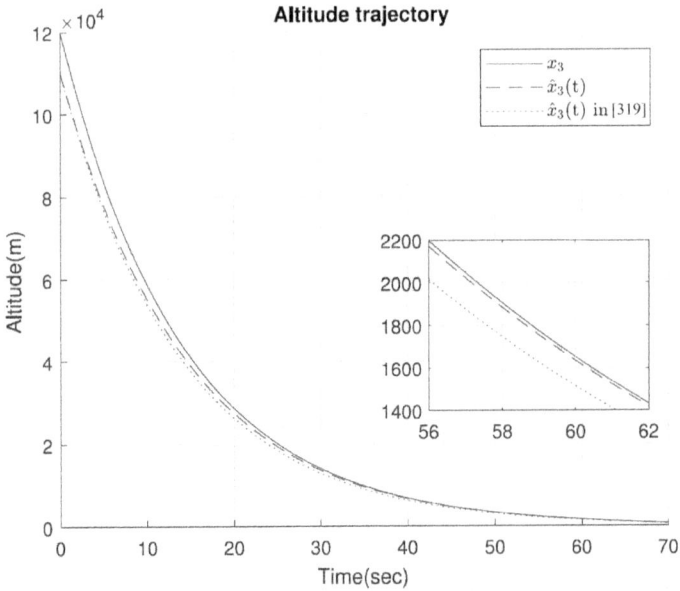

FIGURE 9.9 The altitude trajectory and its estimation by the proposed MO H_2/H_∞ SEF and robust H_2/H_∞ filter in [255].

attenuating the effect of the malicious attack, like the external disturbance and measurement noise on the optimal state estimation of missile. As a result, by the traditional mixed H_2/H_∞ robust filter, the state of missile will be more difficult to estimate when the malicious attack signals $f_a(t)$ and $f_s(t)$ occur. From the result of the traditional mixed H_2/H_∞ robust filter in Figures 9.4–9.6, the three estimated velocities have large deviations from the missile system when the malicious signals begin to attack. Moreover, from the results of the position estimation in Figures 9.7–9.9, the three position estimations using the mixed H_2/H_∞ robust filter have large gaps from the missile system. Especially when the last malicious attack signals occur at $t = 56s$, there is a 200-meter gap on the estimation of the altitude trajectory by the mixed H_2/H_∞ robust filter, and there is only a 20-meter gap by the proposed SEF. Thus, the mixed H_2/H_∞ robust filter is still very poor at missile trajectory detection from a weaponry resolution perspective.

9.5.2 SEF DESIGN FOR STOCHASTIC LINEAR MASS-SPRING SYSTEM

In this section, a fault estimation design proposed in [261] is carried out in comparison with our SEF design. Consider a mass-spring system with two masses and two springs as follows [261]:

$$
\begin{aligned}
\dot{x}_1(t) &= x_3(t) \\
\dot{x}_2(t) &= x_4(t) \\
\dot{x}_3(t) &= -\tfrac{k_1+k_2}{m_1} x_1(t) + \tfrac{k_2}{m_1} x_2(t) - \tfrac{c}{m_1} x_3(t) + u(t) \\
\dot{x}_4(t) &= \tfrac{k_2}{m_2} x_1(t) - \tfrac{k_2}{m_2} x_2(t) - \tfrac{c}{m_2} x_4(t)
\end{aligned}
\tag{9.67}
$$

where $x_1(t)$ is the position of mass 1; $x_2(t)$ is the position of mass 2; $x_3(t)$ is the velocity of mass 1; $x_4(t)$ is the velocity of mass 2; $u(t) = Kx(t)$ is the control input with state-feedback controller gain K; k_1 and k_2 are spring constants of springs 1 and 2, respectively; m_1 and m_2 are the masses of masses 1 and 2, respectively; and c is the viscous friction coefficient.

Under the concept of a networked control system, the mass-spring system in (9.67) is controlled by a remote control station. In this situation, a malicious attack signal will influence the control and filtering of the mechanical system through the attack on the system and the sensor of the wireless network. Furthermore, due to the intrinsic model uncertainties and unpredictable element deformation in physical plant, the networked-based mass-spring system in (9.67) should be modified as follows:

$$
\begin{aligned}
dx(t) &= [(A + BK)x(t) + B_d v(t) + B_a f_a(t)]dt + Gx(t)dw(t) \\
&\quad + Jx(t)dp(t) \\
y(t) &= Cx(t) + B_n n(t) + B_s f_s(t)
\end{aligned}
\tag{9.68}
$$

where $x(t) = [x_1(t) \quad x_2(t) \quad x_3(t) \quad x_4(t)]^T$ denotes the state vector, $y(t)$ is the measurement output, $v(t)$ denotes the finite-energy external disturbance, $f_a(t) \in \mathbb{R}^1$ is the malicious attack signal on the system, $f_s(t) \in \mathbb{R}^2$ is the malicious attack signal on the sensor, $n(t)$ denotes the finite-energy measurement noise, $w(t)$ is the 1-D Wiener process, and $p(t) \in \mathbb{R}^1$ is the Poisson counting process with jump intensity 0.2. For the stochastic mass-spring system in (9.68), $Gx(t)dw(t)$ is used to describe the effect of unmodeling parameters and intrinsic random continuous fluctuations, while $Jx(t)dp(t)$ denotes the effect of a sudden random event such as package dropout and element deformation.

Given the parameters in a mass-spring system as $k_1 = k_2 = 1$, $m_1 = 1$, $m_2 = 0.5$, and $c = 0.5$, the detailed system matrices in (9.68) can be written as follows:

$$A = \begin{bmatrix} 0 & 0 & 1 & 0 \\ 0 & 0 & 0 & 1 \\ -2 & 1 & -0.5 & 0 \\ 2 & -2 & 0 & -1 \end{bmatrix}, B = \begin{bmatrix} 0 \\ 0 \\ 1 \\ 0 \end{bmatrix}, C = \begin{bmatrix} 1 & 0 & 0 & 0 \\ 0 & 1 & 0 & 0 \end{bmatrix},$$

$$K = [-39.1, 1.3, -20.3, -12.3], G = diag\{0, 0, 0.2, 0.2\}, J = diag\{0, 0, 1, 1\},$$

$$B_a = diag\{0, 0, 1, 1\}, B_s = diag\{1, 1\}, B_d = [0, 0, 1, 1]^T, B_n = [1, 1]^T$$

To construct the third-order smoothed signal models in (9.5) and (9.8), the extrapolation parameters in (9.13) and (9.7) are selected as $\alpha_0 = 0.7$, $\alpha_1 = 0.2$, $\alpha_2 = \alpha_3 = 0.05$, $\beta_0 = 0.6$, $\beta_1 = 0.2$, and $\beta_2 = \beta_3 = 0.1$, and the time interval h is chosen as 10^{-3}.

In this simulation, the initial values $x(0)$, external disturbance $v(t)$, malicious attack signal on system $f_a(t)$, malicious attack signal on sensor $f_s(t)$, and measurement noise are given as:

$$x(0) = [2, 1, 3, 3]^T, v(t) = 0.2\sin(t), f_a(t) = 2\cos(t)e^{-0.1t}, n(t)$$

$$= 0.1N(0,1), f_s(t) = \begin{bmatrix} 1.5e^{-0.1t} \\ e^{-0.1t} \end{bmatrix}$$

where $N(0,1)$ denotes the normal distribution with mean 0 and variance 1.

Since the designer considers the filtering performance of position more important than the filtering performance of velocity, the weighting matrices for the stochastic H_2 filtering performance in (9.16) and the stochastic robust H_∞ filtering performance in (9.17) are given as:

$$Q_2 = diag\{Q_{2,x}, Q_{2,f_a}, Q_{2,f_s}\}$$
$$Q_{2,f_a} = I_4, Q_{2,f_s} = I_8 \qquad\qquad (9.69)$$
$$Q_{2,x} = diag\{1, 1, 0.5, 0.5\}$$

$$Q_\infty = diag\{Q_{\infty,x}, Q_{\infty,f_a}, Q_{\infty,f_s}\}$$
$$Q_{\infty,f_a} = I_4, Q_{\infty,f_s} = I_8 \tag{9.70}$$
$$Q_{\infty,x} = diag\{1,1,0.1,0.1\}$$

Before applying the proposed LMI-constrained MOEA, the detailed parameters in MOEA are given as: iteration number $N_i = 80$, search range $S = [0,50] \times [0,5]$, cross-over rate $R_c = 0.9$, mutation rate $R_m = 0.1$, and population number $N_p = 500$. Then, by solving the MOP in (9.30) with the aforementioned parameters in the LMI-constrained MOEA, we could get a set of Pareto optimal objective vector $(20.15, 3.21)$ selected with the corresponding Pareto optimal solution P_1^*, P_2^*, and filter gains L^*.

The simulation results are shown in Figures 9.10–9.13 with state variable estimations in Figs. 9.10–9.11 and malicious attack signal estimations in Figures 9.12–9.13. Moreover, the filter design of the malicious attack signal in [261] is carried out. From Figures 9.10 and 9.13, by the accurate estimation of the malicious attack signal on the sensor, the proposed robust SEF can achieve better filtering performance of position variables $x_1(t)$ and $x_2(t)$ in spite of the effect of the measurement noise and

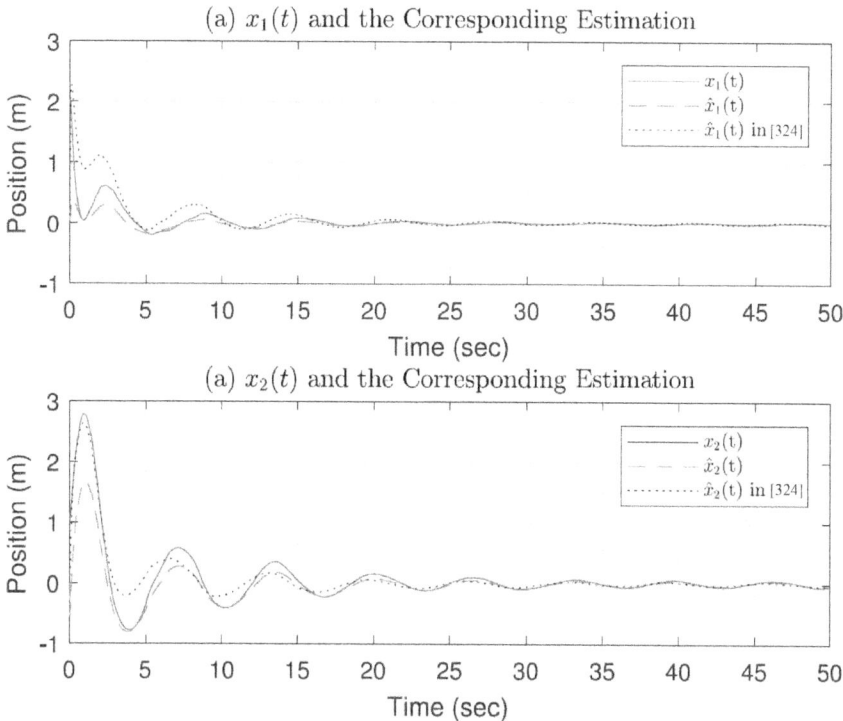

FIGURE 9.10 (a) The trajectory of $x_1(t)$ and its estimation. (b) The trajectory of $x_2(t)$ and its estimation.

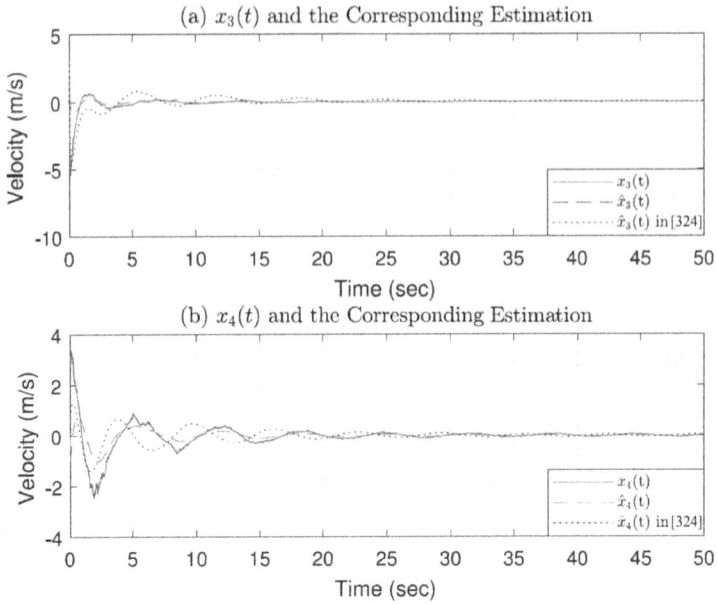

FIGURE 9.11 (a) The trajectory of $x_3(t)$ and its estimation. (b) The trajectory of $x_4(t)$ and its estimation.

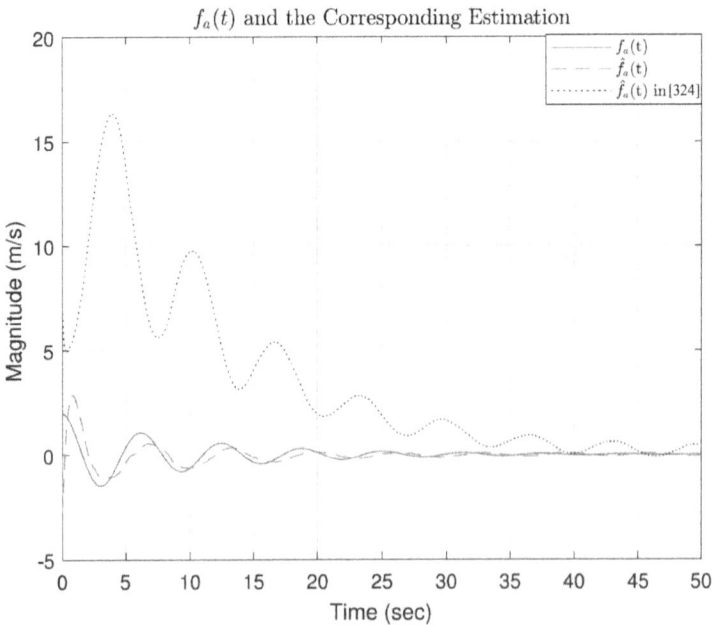

FIGURE 9.12 The trajectory of $f_a(t)$ and its estimation.

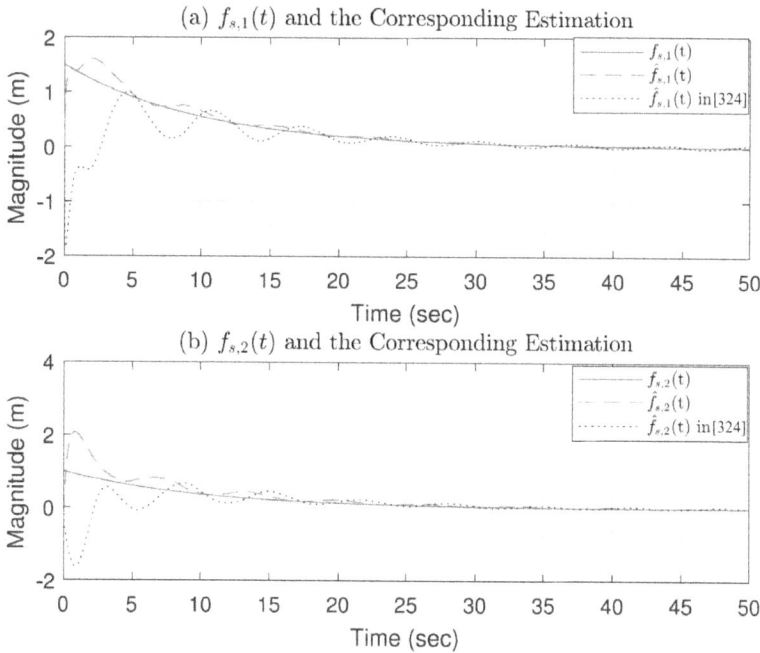

FIGURE 9.13   (a) The trajectory of $f_{s,1}(t)$ and its estimation. (b) The trajectory of $f_{s,2}(t)$ and its estimation.

malicious attack signal on the sensor. On the other hand, since the velocity variables $x_3(t)$ and $x_4(t)$ suffer from the intrinsic fluctuations and unknown malicious attack signal $f_a(t)$ on the actuator, the real velocities $x_3(t)$ and $x_4(t)$ fluctuate during the filtering process. By using the accurate estimation of malicious attack signal $f_a(t)$ in Figure 9.12, the proposed SEF can efficiently estimate the velocity variables $x_3(t)$ and $x_4(t)$ with compensation for the malicious attack signal $f_a(t)$. Besides, in the case of the design method in [261], it can be seen that the filter in [261] has a large transient response in estimating malicious fault signals. Hence, due to these large transient responses, the estimation of state variables by the filter design in [261] is severely affected by the malicious attack signals $f_a(t)$ and $f_s(t)$.

9.6   CONCLUSION

Security-enhanced filter design problems will become an important issue in signal processing systems under malicious attack signals in information transmission. For linear and nonlinear stochastic jump-diffusion systems suffering from system and sensor malicious attack signals and external disturbances, based on the proposed smoothed model of malicious attack signals, an SEF is proposed to efficiently esti-mate the system state and malicious attack signals on the signal system and sensor by the traditional Luenberger-type filter. The stochastic MO H_2/H_∞ SEF design is proposed to efficiently attenuate the effect of external disturbance to achieve security filtering performance. In the nonlinear system case, the stochastic MO H_2/H_∞ SEF design problem can be transformed into the HJI-constrained MOP. Nevertheless, the HJI-constrained MOP is not easy to solve yet. The global linearization method is employed to transform the HJI-constrained MOP to an LMI-constrained MOP. By using the proposed LMI-constrained MOEA, the LMI-constrained MOP for the MO H_2/H_∞ SEF design of a stochastic jump-diffusion system under malicious attack signals can be solved efficiently. Two simulation examples, a missile trajectory estimation problem by a ground radar system under malicious attack signals and estimation of a network-based mass spring system, are given to validate the effectiveness of the proposed method. In the future 5G wireless networked era, network-based security filter design to estimate trajectory/attack signals on receivers and sensors of wireless networks will be future research.

9.7 APPENDIX

9.7.1 Proof of Theorem 9.1

From [263], the augmented system $(\bar{A}, \bar{C})$ in (9.9) is observable if the following rank condition holds:

$$
rank \begin{bmatrix} sI_{n+(a+s)(k+1)} - \bar{A} \\ \bar{C} \end{bmatrix}
$$

$$
= rank \begin{bmatrix} sI_n - A & -B_a M_a & 0 \\ 0 & sI_{(k+1)a} - A_a & 0 \\ 0 & 0 & sI_{s(k+1)} - A_s \\ C & 0 & B_s M_s \end{bmatrix} \tag{9.71}
$$

$$
= n + (a+s)(k+1), \forall s \in C
$$

where C is the set of 1-dimensional complex numbers. To ensure the rank condition in (9.71) for $s \in C$, the proof can be split into four cases: (I) $s \notin eig(A) \cup eig(A_s) \cup eig(A_a)$, (II) $s \in eig(A_s)$ (III) $s \in eig(A_a)$ and (IV) $s \in eig(A)$.

In case (I), since $s \notin eig(A) \cup eig(A_s) \cup eig(A_a)$, we have

$$
rank \begin{bmatrix} sI_n - A & -B_a M_a & 0 \\ 0 & sI_{(k+1)a} - A_a & 0 \\ 0 & 0 & sI_{s(k+1)} - A_s \end{bmatrix}
$$

$$
= n + (a+s)(k+1), \forall s \notin eig(A) \cup eig(A_s) \cup eig(A_a) \tag{9.72}
$$

and it immediately shows that the rank condition in (9.71) holds for $s \notin eig(A) \cup eig(A_s) \cup eig(A_a)$.

In case (II), by the disjoint conditions $eig(A_s) \cap eig(A_a) = \varnothing$ and $eig(A) \cap eig(A_a) = \varnothing$ in (9.13), we have

$$
rank \begin{bmatrix} sI_n - A & -B_a M_a \\ 0 & sI_{(k+1)a} - A_a \end{bmatrix} = n + a(k+1), \forall s \in eig(A_s) \tag{9.73}
$$

which means that

$$
col \left\{ \begin{bmatrix} 0 \\ 0 \\ sI_{s(k+1)} - A_s \\ B_s M_s \end{bmatrix} \right\} \cap col \left\{ \begin{bmatrix} sI_n - A \\ 0 \\ 0 \\ C \end{bmatrix} \begin{bmatrix} -B_a M_a \\ sI_{(k+1)a} - A_a \\ 0 \\ 0 \end{bmatrix} \right\} = \varnothing
$$

for $s \in eig(A_s)$

and

$$
col \left\{ \begin{bmatrix} sI_n - A \\ 0 \\ 0 \\ C \end{bmatrix} \right\} \cap col \left\{ \begin{bmatrix} -B_a M_a \\ sI_{(k+1)a} - A_a \\ 0 \\ 0 \end{bmatrix} \right\} = \varnothing, \text{for } s \in eig(A_s) \tag{9.74}
$$

where $col\{A\}$ denotes the column space of matrix A.

Then, by (9.74), we immediately have

$$
rank \begin{bmatrix} sI_n - A & -B_a M_a & 0 \\ 0 & sI_{(k+1)a} - A_a & 0 \\ 0 & 0 & sI_{s(k+1)} - A_s \\ C & 0 & B_s M_s \end{bmatrix}
$$

$$
= n + a(k+1) + rank \begin{bmatrix} 0 \\ 0 \\ sI_{s(k+1)} - A_s \\ B_s M_s \end{bmatrix} \tag{9.75}
$$

$$
= n + (a+s)(k+1), \forall s \in eig(A_s)
$$

that is, the rank condition in (9.71) holds for $s \in eig(A_s)$.

In case (III), by the disjoint conditions $eig(A) \cap eig(A_a) = \varnothing$ and $eig(A_s) \cap eig(A_a) = \varnothing$ in (9.13), we have

$$rank \begin{bmatrix} sI_n - A & 0 \\ 0 & sI_{(k+1)s} - A_s \end{bmatrix} = n + s(k+1), \text{for } s \in eig(A_a) \qquad (9.76)$$

and the following conditions hold if the condition in (9.14) holds:

$$col \left\{ \begin{bmatrix} -B_a M_a \\ sI_{(k+1)a} - A_a \\ 0 \\ 0 \end{bmatrix} \right\} \cap col \left\{ \begin{bmatrix} sI_n - A \\ 0 \\ 0 \\ C \end{bmatrix} \begin{bmatrix} 0 \\ 0 \\ sI_{s(k+1)} - A_s \\ B_s M_s \end{bmatrix} \right\} = \varnothing,$$

$$\text{for } s \in eig(A_a) \qquad (9.77)$$

and

$$col \left\{ \begin{bmatrix} sI_n - A \\ 0 \\ 0 \\ C \end{bmatrix} \right\} \cap col \left\{ \begin{bmatrix} 0 \\ 0 \\ sI_{s(k+1)} - A_s \\ B_s M_s \end{bmatrix} \right\} = \varnothing, \text{ for } s \in eig(A_a)$$

As a result, with the help of condition in (9.77), the rank condition in (9.71) can be written as:

$$rank \begin{bmatrix} sI_n - A & -B_a M_a & 0 \\ 0 & sI_{(k+1)a} - A_a & 0 \\ 0 & 0 & sI_{s(k+1)} - A_s \\ C & 0 & B_s M_s \end{bmatrix}$$

$$= n + s(k+1) + rank \begin{bmatrix} -B_a M_a \\ sI_{(k+1)a} - A_a \\ 0 \\ 0 \end{bmatrix} \qquad (9.78)$$

$$= n + (a+s)(k+1), \forall s \in eig(A_a)$$

that is, the rank condition in (9.71) holds for $s \in eig(A_a)$.

In case (IV), by the disjoint conditions $eig(A) \cap eig(A_a) = \varnothing$ and $eig(A_s) \cap eig(A_a) = \varnothing$ in (9.13), we have

$$rank \begin{bmatrix} sI_{(k+1)a} - A_a & 0 \\ 0 & sI_{(k+1)s} - A_s \end{bmatrix} = (a+s)(k+1) \qquad (9.79)$$

for $s \in eig(A)$

which implies

$$
col\left\{\begin{bmatrix} sI_n - A \\ 0 \\ 0 \\ C \end{bmatrix}\right\} \cap col\left\{\begin{bmatrix} -B_a M_a \\ sI_{(k+1)a} - A_a \\ 0 \\ 0 \end{bmatrix}, \begin{bmatrix} 0 \\ 0 \\ sI_{s(k+1)} - A_s \\ B_s M_s \end{bmatrix}\right\} = \emptyset,
$$

for $s \in eig(A)$

and $\qquad\qquad\qquad\qquad\qquad\qquad\qquad\qquad\qquad\qquad\qquad$ (9.80)

$$
col\left\{\begin{bmatrix} -B_a M_a \\ sI_{(k+1)a} - A_a \\ 0 \\ 0 \end{bmatrix}\right\} \cap col\left\{\begin{bmatrix} 0 \\ 0 \\ sI_{s(k+1)} - A_s \\ B_s M_s \end{bmatrix}\right\} = \emptyset,
$$

for $s \in eig(A)$

Then, by (9.80), we have

$$
rank\begin{bmatrix} sI_n - A & -B_a M_a & 0 \\ 0 & sI_{(k+1)a} - A_a & 0 \\ 0 & 0 & sI_{s(k+1)} - A_s \\ C & 0 & B_s M_s \end{bmatrix}
$$

$$
= rank\begin{bmatrix} sI_n - A \\ 0 \\ 0 \\ C \end{bmatrix} + (a + s)(k + 1) \qquad (9.81)
$$

$$
= n + (a + s)(k + 1), \forall s \in eig(A)
$$

that is, the rank condition in (9.71) holds for $s \in eig(A)$. The proof is done.

9.7.2 PROOF OF THEOREM 9.2

This will be proven by the fact that the strict inequalities in (9.24) will disappear and only equalities hold when the Pareto optimal solution (α^*, β^*) of the MOP in (9.23) is achieved. We will prove the fact by contradiction. Suppose that a strict inequality still holds for any one inequality in (9.24) at the Pareto optimal solution. Given a solution L^* with corresponding objective value (α^*, β^*), without loss of generality, we assume $H_2(L^*) < \alpha^*$. Then there exists an α' such that $H_2(L^*) = \alpha' < \alpha^*$. This means (α', β^*) dominates the Pareto optimal solution (α^*, β^*), which is not dominated by other solutions. Based on the previous discussion, when the Pareto optimal solution of the MOP in (9.23) is reached, we can conclude that the inequalities in (9.24) become equalities. Thus, the MOP in (9.23) is definitely equivalent to the MOP in (9.22).

9.7.3   PROOF OF THEOREM 9.3

The proof of the theorem is divided into two parts: The first part is for $H_2(L) \leq \alpha$, and the second part is for $H_\infty(L) \leq \beta$.

(I) By utilizing Lemma 9.1, the Lyapunov function $V(\tilde{x}(t)) = \tilde{x}^T(t)P\tilde{x}(t)$ with a positive matrix P is selected for the augmented system in (9.19) with $\overline{v}(t) = 0$, and we get

$$
\begin{aligned}
H_2(L) = E[\int_0^{t_f} \tilde{x}^T(t)\tilde{Q}_2\tilde{x}(t)dt + \int_0^{t_f} dV(\tilde{x}(t)) + \tilde{x}^T(0)P\tilde{x}(0) \\
- \tilde{x}^T(t_f)P\tilde{x}(t_f)]
\end{aligned}
\tag{9.82}
$$

By substituting $V(\tilde{x}(t))$ in (9.25) into (9.82) with the fact $E[dw(t)] = 0$ and $E[dp(t)] = \lambda dt$, the H_2 optimization performance becomes:

$$
\begin{aligned}
H_2(L) = E[\int_0^{t_f} [\tilde{x}^T(t)\tilde{Q}_2\tilde{x}(t) + \tilde{x}^T(t)P\tilde{A}\tilde{x}(t) + \tilde{x}^T(t)\tilde{A}^T P\tilde{x}(t) \\
+ \tilde{x}^T(t)\tilde{G}^T P\tilde{G}\tilde{x}(t) + \lambda(\tilde{x}^T(t)\tilde{J}P\tilde{x}(t) + \tilde{x}^T(t)P\tilde{J}\tilde{x}(t) \\
+ \tilde{x}^T(t)\tilde{J}^T P\tilde{J}\tilde{x}(t))]dt + \tilde{x}^T(0)P\tilde{x}(0) - \tilde{x}^T(t_f)P\tilde{x}(t_f)] \\
\leq E[\int_0^{t_f} \tilde{x}^T(t)[\tilde{Q}_2 + P\tilde{A} + \tilde{A}^T P + \tilde{G}^T P\tilde{G} + \lambda(\tilde{J}P + P\tilde{J} \\
+ \tilde{J}^T P\tilde{J})]\tilde{x}(t)dt + \tilde{x}^T(0)P\tilde{x}(0)]
\end{aligned}
\tag{9.83}
$$

If (9.27) holds, then (9.83) becomes

$$
H_2(L) \leq E[\tilde{x}^T(0)P\tilde{x}(0)]
\tag{9.84}
$$

Besides, for the term $\tilde{x}^T(0)P\tilde{x}(0)$ on the right-hand side of the previous inequality, we have the following equality:

$$
E[\tilde{x}^T(0)P\tilde{x}(0)] = tr[PE[\tilde{x}^T(0)\tilde{x}(0)]] = tr[PE[R_0]]
\tag{9.85}
$$

where $R_0 = \tilde{x}(0)\tilde{x}^T(0)$.

Therefore, if the inequalities in (9.27) and (9.28) are satisfied, the inequality $H_2(L) \leq \alpha$ in (9.24) can be guaranteed.

(II) On the other hand, consider the numerator of the stochastic robust H_∞ filtering performance (9.21):

$$J'_\infty(L) = E\left[\int_0^{t_f} [\tilde{x}^T(t)\tilde{Q}_\infty \tilde{x}(t)dt - \tilde{x}^T(0)P\tilde{x}(0)\right] \tag{9.86}$$

By utilizing Lemma 9.1, we have:

$$J'_\infty(L) = E[\int_0^{t_f} [\tilde{x}^T(t)\tilde{Q}_\infty \tilde{x}(t)dt + \int_0^{t_f} dV(\tilde{x}(t)) + \tilde{x}^T(0)P\tilde{x}(0) \\ - \tilde{x}^T(t_f)P\tilde{x}(t_f) - \tilde{x}^T(0)P\tilde{x}(0)] \tag{9.87}$$

By substituting the Itô-Lévy formula in (9.25) into (9.87) with the fact $E[dw(t)] = 0$ and $E[dp(t)] = \lambda dt$, the previous equation becomes

$$J'_\infty(L) = E\left[\begin{matrix} \int_0^{t_f} [\tilde{x}^T(t)\tilde{Q}_\infty \tilde{x}(t) + \tilde{x}^T(t)P\tilde{A}\tilde{x}(t) + \tilde{x}^T(t)\tilde{A}^T P\tilde{x}(t) \\ + \tilde{x}^T(t)P\tilde{B}\bar{v}(t) + \bar{v}(t)^T \tilde{B}^T P\tilde{x}(t) + \tilde{x}^T(t)\tilde{G}^T P\tilde{G}\tilde{x}(t) \\ + \lambda(\tilde{x}^T(t)\tilde{J}P\tilde{x}(t) + \tilde{x}^T(t)P\tilde{J}\tilde{x}(t) \\ + \tilde{x}^T(t)\tilde{J}^T P\tilde{J}\tilde{x}(t))]dt - \tilde{x}^T(t_f)P\tilde{x}(t_f) \end{matrix}\right] \tag{9.88}$$

By using Lemma 9.2, we get the following result.

$$J'_\infty(L) = E\left[\begin{matrix} \int_0^{t_f} [\tilde{x}^T(t)\tilde{Q}_\infty \tilde{x}(t) + \tilde{x}^T(t)P\tilde{A}\tilde{x}(t) + \tilde{x}^T(t)\tilde{A}^T P\tilde{x}(t) \\ + \tilde{x}^T(t)P\tilde{B}\bar{v}(t) + \bar{v}(t)^T \tilde{B}^T P\tilde{x}(t) + \tilde{x}^T(t)\tilde{G}^T P\tilde{G}\tilde{x}(t) \\ + \lambda(\tilde{x}^T(t)\tilde{J}P\tilde{x}(t) + \tilde{x}^T(t)P\tilde{J}\tilde{x}(t) \\ + \tilde{x}^T(t)\tilde{J}^T P\tilde{J}\tilde{x}(t))]dt - \tilde{x}^T(t_f)P\tilde{x}(t_f) \end{matrix}\right] \tag{9.89}$$

for some $\beta > 0$.

Substituting the inequality (9.89) into (9.88), we have

$$J'_\infty(L) = E\left[\int_0^{t_f} [\tilde{x}^T(t)\tilde{Q}_\infty \tilde{x}(t) - x^T(0)P\tilde{x}(0)\right] \\ \leq E\left[\begin{matrix} \int_0^{t_f} \tilde{x}^T(t)\left[\begin{matrix} \tilde{Q}_\infty + P\tilde{A} + \tilde{A}^T P + \tilde{G}^T P\tilde{G} + \lambda(\tilde{J}P + P\tilde{J}) \\ + \tilde{J}^T P\tilde{J}) + \frac{1}{\beta} P\tilde{B}\tilde{B}^T P \end{matrix}\right]\tilde{x}(t)dt + \beta E\int_0^{t_f} \bar{v}^T(t)\bar{v}(t)dt \\ - \tilde{x}^T(t_f)P\tilde{x}(t_f) \end{matrix}\right] \tag{9.90}$$

If the following Riccati-like inequality is satisfied:

$$\tilde{Q}_\infty + P\tilde{A} + \tilde{A}^T P + \tilde{G}^T P\tilde{G} + \lambda(\tilde{J}P + P\tilde{J} + \tilde{J}^T P\tilde{J}) + \frac{1}{\beta} P\tilde{B}\tilde{B}^T P \leq 0 \tag{9.91}$$

which could be transformed into the equivalent BMI in (9.29); then (9.90) becomes:

$$E\left[\int_0^{t_f} \tilde{x}^T(t)\tilde{Q}_\infty\tilde{x}(t) - x^T(0)P\tilde{x}(0)\right] \le \beta E\left[\int_0^{t_f} \tilde{v}^T(t)\tilde{v}(t)dt\right] \tag{9.92}$$

which is equivalent to $H_\infty(L) \le \beta$ in (9.24). Therefore, if the BMI in (9.29) holds, then $H_\infty(L) \le \beta$ holds too.

In summary, based on the two parts of proof, the MOP in (9.23) and (9.24) could be transformed into the BMI-constrained MOP in (9.26)–(9.29).

9.7.4 PROOF OF THEOREM 9.5

The proof of the theorem is divided into two parts: The first part is for $H_2(l(\hat{\tilde{x}}(t))) \le \alpha$, and the second part is for $H_\infty(l(\hat{\tilde{x}}(t))) \le \beta$.

(I) To begin with, define a Lyapunov function $V(\tilde{x}(t))$ with $V(\tilde{x}(t)) \in C^2$, $V(\cdot) \ge 0$, and $V(0) = 0$. Thus, from (9.38), we have:

$$H_2(l(\hat{\tilde{x}}(t))) = E\left[\int_0^{t_f} \tilde{x}^T(t)\tilde{Q}_2\tilde{x}(t)dt + \int_0^{t_f} dV(\tilde{x}(t)) + V(\tilde{x}(0)) - V(\tilde{x}(t_f))\right] \tag{9.93}$$

By using $dV(\tilde{x}(t))$ in (9.43) with the case $\bar{v}(t) = 0$, (9.93) becomes

$$H_2(l(\hat{\tilde{x}}(t))) = E\left[\begin{matrix} \int_0^{t_f} [\tilde{x}^T(t)\tilde{Q}_2\tilde{x}(t) + \left(\frac{\partial V(\tilde{x}(t))}{\partial \tilde{x}(t)}\right)^T \tilde{f}(\tilde{x}(t)) \\ +\frac{1}{2}\tilde{g}(\tilde{x}(t))^T \frac{\partial^2 V(\tilde{x}(t))}{\partial \tilde{x}^2(t)} \tilde{g}(\tilde{x}(t)) + l(V(\tilde{x}(t) + \tilde{j}(\tilde{x}(t))) \\ -V(\tilde{x}(t)))]dt + V(\tilde{x}(0)) - V(\tilde{x}(t_f)) \end{matrix}\right] \tag{9.94}$$

by the fact $E[dw(t)] = 0$ and $E[dp(t)] = \lambda dt$.

By (9.45) and (9.46) with the fact $V(\tilde{x}(t_f)) \ge 0$, (9.94) becomes:

$$H_2(l(\hat{\tilde{x}}(t))) \le \alpha \tag{9.95}$$

Hence, by the HJIs in (9.45) and (9.46), the inequality $H_2(l(\hat{\tilde{x}}(t))) \le \alpha$ in (9.42) can be guaranteed.

(II) Further, the numerator of the stochastic robust H_∞ performance in (9.39) can be written as

$$J_\infty'(l(\hat{\tilde{x}}(t))) = E\left[\int_0^{t_f} \tilde{x}^T(t)\tilde{Q}_\infty\tilde{x}(t)dt - V(\tilde{x}(0))]\right]$$

$$= E\left[\begin{matrix} \int_0^{t_f} \tilde{x}^T(t)\tilde{Q}_\infty\tilde{x}(t)dt + \int_0^{t_f} dV(\tilde{x}(t)) + V(\tilde{x}(0)) \\ -V(\tilde{x}(t_f)) - V(\tilde{x}(0)) \end{matrix}\right] \tag{9.96}$$

By substituting the Itô-Lévy formula in (9.43) into (9.96) and with the fact $E[dw(t)] = 0$ and $E[dp(t)] = \lambda dt$, the previous equation becomes

$$J'_\infty(l(\hat{\tilde{x}}(t))) = E\left[\begin{array}{l} \int_0^{t_f} [\tilde{x}^T(t)\tilde{Q}_2\tilde{x}(t) + \left(\frac{\partial V(\tilde{x}(t))}{\partial \tilde{x}(t)}\right)^T \tilde{f}(\tilde{x}(t)) \\[2mm] \qquad + \tilde{b}(\tilde{x}(t))\overline{v}(t) + \frac{1}{2}\tilde{g}(\tilde{x}(t))^T \frac{\partial^2 V(\tilde{x}(t))}{\partial \tilde{x}^2(t)}\tilde{g}(\tilde{x}(t)) \\[2mm] \qquad + \lambda(V(\tilde{x}(t) + \tilde{j}(\tilde{x}(t))) - V(\tilde{x}(t)))]dt - V(\tilde{x}(t_f)) \end{array}\right] \tag{9.97}$$

By using Lemma 9.2, we get the following result

$$\left(\frac{\partial V(\tilde{x}(t))}{\partial \tilde{x}(t)}\right)^T \tilde{b}(\tilde{x}(t))\overline{v}(t)$$

$$= \frac{1}{2}\left(\frac{\partial V(\tilde{x}(t))}{\partial \tilde{x}(t)}\right)^T \tilde{b}(\tilde{x}(t))\overline{v}(t) + \frac{1}{2}\overline{v}^T(t)\tilde{b}^T(\tilde{x}(t))\frac{\partial V(\tilde{x}(t))}{\partial \tilde{x}(t)} \tag{9.98}$$

$$\leq \frac{1}{4\beta}\left(\frac{\partial V(\tilde{x}(t))}{\partial \tilde{x}(t)}\right)^T \tilde{b}(\tilde{x}(t))\tilde{b}^T(\tilde{x}(t))\frac{\partial V(\tilde{x}(t))}{\partial \tilde{x}(t)} + \beta\overline{v}^T(t)\overline{v}(t)$$

for some $\beta > 0$.

Substituting the inequality (9.98) into (9.97), we have

$$J'_\infty(l(\hat{\tilde{x}}(t)) \leq E\left[\begin{array}{l} \int_0^{t_f} [\tilde{x}^T(t)\tilde{Q}_\infty\tilde{x}(t) + \left(\frac{\partial V(\tilde{x}(t))}{\partial \tilde{x}(t)}\right)^T \tilde{f}(\tilde{x}(t)) \\[2mm] \qquad + \frac{1}{4\beta}\left(\frac{\partial V(\tilde{x}(t))}{\partial \tilde{x}(t)}\right)^T \tilde{b}(\tilde{x}(t))\tilde{b}^T(\tilde{x}(t))\frac{\partial V(\tilde{x}(t))}{\partial \tilde{x}(t)} + \beta\overline{v}^T(t)\overline{v}(t) \\[2mm] \qquad + \frac{1}{2}\tilde{g}(\tilde{x}(t))^T \frac{\partial^2 V(\tilde{x}(t))}{\partial \tilde{x}^2(t)}\tilde{g}(\tilde{x}(t)) + \lambda(V(\tilde{x}(t) + \tilde{j}(\tilde{x}(t))) \\[2mm] \qquad - V(\tilde{x}(t)))]dt - V(\tilde{x}(t_f)) \end{array}\right] \tag{9.99}$$

By the HJI in (9.47) and $V(\tilde{x}(t_f)) \geq 0$, we have

$$E\left[\int_0^{t_f} \tilde{x}^T(t)\tilde{Q}_\infty\tilde{x}(t) - V(\tilde{x}(0))\right] \leq \beta E\left[\int_0^{t_f} \overline{v}^T(t)\overline{v}(t)dt\right] \tag{9.100}$$

which guarantees $H_\infty(l(\hat{\tilde{x}}(t))) \leq \beta$ in (9.42).

Based on the proof of parts (I) and (II), if MOP in (9.44)–(9.47) holds, the MOP in (9.41) and (9.42) holds too.

9.7.5 PROOF OF THEOREM 9.6

We select the Lyapunov function $V(\tilde{x}(t)) = \tilde{x}^T(t)P\tilde{x}(t)$ for a positive matrix P. By the global linearization technique in (9.48)–(9.51) with the orthogonal interpolation function of $\alpha_i(\tilde{x})$ and $\alpha_j(\tilde{x})$ in Lemma 9.4, the nonlinear functions in (9.37) can be written as:

$$\tilde{f}(\tilde{x}(t)) = \sum_{i=1}^{N}\sum_{j=1}^{N} v_i(\overline{x}(t))\varpi_i(\hat{\overline{x}}(t))\left\{ \begin{bmatrix} \overline{A}_i & 0 \\ 0 & \overline{A}_j \end{bmatrix}\begin{bmatrix} \overline{x}(t) \\ \hat{\overline{x}}(t) \end{bmatrix} + \begin{bmatrix} 0 & 0 \\ L_j\overline{C}_i & -L_j\overline{C}_j \end{bmatrix}\begin{bmatrix} \overline{x}(t) \\ \hat{\overline{x}}(t) \end{bmatrix}\right\}$$

$$= \sum_{i=1}^{N}\sum_{j=1}^{N}\varpi_i(\overline{x}(t))\varpi_j(\hat{\overline{x}}(t))\begin{bmatrix} \overline{A}_i & 0 \\ L_j\overline{C}_i & \overline{A}_j - L_j\overline{C}_j \end{bmatrix}\begin{bmatrix} \overline{x}(t) \\ \hat{\overline{x}}(t) \end{bmatrix}$$

$$= \sum_{i=1}^{N}\sum_{j=1}^{N}\varpi_i(\overline{x}(t))\varpi_j(\hat{\overline{x}}(t))\tilde{A}_{ij}\tilde{x}(t)$$

$$\tilde{b}(\tilde{x}(t)) = \sum_{i=1}^{N}\sum_{j=1}^{N}\varpi_i(\overline{x}(t))\varpi_j(\hat{\overline{x}}(t))\begin{bmatrix} \overline{B}_i \\ L_j\overline{D}_i \end{bmatrix} = \sum_{i=1}^{N}\sum_{j=1}^{N}\varpi_i(\overline{x}(t))v_j(\hat{\overline{x}}(t))\tilde{B}_{ij}$$

$$\tilde{g}(\tilde{x}(t)) = \sum_{i=1}^{N}\varpi_i(\overline{x}(t))\begin{bmatrix} \overline{G}_i & 0 \\ 0 & 0 \end{bmatrix}\begin{bmatrix} \overline{x}(t) \\ \hat{\overline{x}}(t) \end{bmatrix} = \sum_{i=1}^{N}\varpi_i(\overline{x}(t))\tilde{G}_i\tilde{x}(t)$$

$$\tilde{j}(\tilde{x}(t)) = \sum_{i=1}^{N}\varpi_i(\overline{x}(t))\begin{bmatrix} \overline{J}_i & 0 \\ 0 & 0 \end{bmatrix}\begin{bmatrix} \overline{x}(t) \\ \hat{\overline{x}}(t) \end{bmatrix} = \sum_{i=1}^{N}\varpi_i(\overline{x}(t))\tilde{J}_i\tilde{x}(t)$$

According to the previous analysis with Lemma 9.4, the HJIIs in (9.45) and (9.47) are bounded above as follows:

$$\tilde{x}^T(t)\tilde{Q}_2\tilde{x}(t) + \left(\frac{\partial V(\tilde{x}(t))}{\partial\tilde{x}(t)}\right)^T(\tilde{f}(\tilde{x}(t))) + \tfrac{1}{2}\tilde{g}(\tilde{x}(t))^T\frac{\partial^2 V(\tilde{x}(t))}{\partial\tilde{x}^2(t)}\tilde{g}(\tilde{x}(t))$$

$$+l(V(\tilde{x}(t) + \tilde{j}(\tilde{x}(t))) - V(\tilde{x}(t)))$$

$$\le \sum_{i=1}^{N}\sum_{j=1}^{N}\varpi_i(\overline{x})\varpi_j(\hat{\overline{x}})\tilde{x}^T(t)\{\tilde{Q}_2 + P\tilde{A}_{ij} + \tilde{A}_{ij}^T P + \tilde{G}_i^T P\tilde{G}_i \qquad (9.101)$$

$$+l(\tilde{J}_i^T P + P\tilde{J}_i + \tilde{J}_i^T P\tilde{J}_i)\}\tilde{x}(t), \text{ for } i,j = 1,...,N$$

$$\tilde{x}^T(t)\tilde{Q}_{\infty 2}\tilde{x}(t) + \left(\frac{\partial V(\tilde{x}(t))}{\partial\tilde{x}(t)}\right)^T\tilde{f}(\tilde{x}(t)) + \tfrac{1}{2}\tilde{g}(\tilde{x}(t))^T\frac{\partial^2 V(\tilde{x}(t))}{\partial\tilde{x}^2(t)}\tilde{g}(\tilde{x}(t))$$

$$+\tfrac{1}{4\beta}\left(\frac{\partial V(\tilde{x}(t))}{\partial\tilde{x}(t)}\right)^T\tilde{b}(\tilde{x}(t))\tilde{b}^T(\tilde{x}(t))\frac{\partial V(\tilde{x}(t))}{\partial\tilde{x}(t)}$$

$$+l(V(\tilde{x}(t) + \tilde{j}(\tilde{x}(t))) - V(\tilde{x}(t))) \qquad (9.102)$$

$$\le \sum_{i=1}^{N}\sum_{j=1}^{N}\varpi_i(\overline{x})\varpi_j(\hat{\overline{x}})\tilde{x}^T(t)\{\tilde{Q}_\infty + P\tilde{A}_{ij} + \tilde{A}_{ij}^T P + \tfrac{1}{b}P\tilde{B}_{ij}\tilde{B}_{ij}^T P + \tilde{G}_i^T P\tilde{G}_i$$

$$+l(\tilde{J}_i^T P + P\tilde{J}_i + \tilde{J}_i^T P\tilde{J}_i)\}\tilde{x}(t), \text{ for } i,j = 1,...,N$$

By the Schur complement in [129], the inequalities in (9.102) are transformed into the BMIs in the following:

$$\begin{bmatrix} \begin{matrix}\tilde{Q}_\infty + P\tilde{A}_{ij} + \tilde{A}_{ij}^T P + \tilde{G}_i^T P\tilde{G}_i \\ +\lambda(\tilde{J}_i^T P + P\tilde{J}_i + \tilde{J}_i^T P\tilde{J}_i)\end{matrix} & P\tilde{B}_{ij} \\ \tilde{B}_{ij}^T P & -\beta I \end{bmatrix} < 0, \text{ for } i,j = 1,...,N \qquad (9.103)$$

Moreover, notice that $V(\tilde{x}(0))$ is reduced to $E\{\tilde{x}^T(0)P\tilde{x}(0)\}$. Thus, the constraint in (9.46) can be rewritten as:

$$P \le \frac{\alpha}{n} E[R_0^{-1}] \tag{9.104}$$

where $R_0 = \tilde{x}(0)\tilde{x}^T(0)$.

Therefore, if the equalities in (9.55)–(9.57) hold for $i, j = 1,...,N$, then the HJIs in (9.45)–(9.47) are guaranteed; that is, the HJI-constrained MOP in (9.44)–(9.47) can be transformed into the BMI-constrained MOP in (9.54)–(9.57).

10 Multi-Objective H_2/H_∞ Optimal Power Tracking Control for Interference-Limited Wireless Communication Systems

10.1 INTRODUCTION

For next-generation wireless systems, the ever-increasing demands by users sharing wireless services have made new standards (such as IEEE 802.16m [264] and 3GPP LTE-A [265]) to allow users to interfere with each other inevitable. Inter-user interference has hence become one of the bottlenecks limiting the performance of next-generation wireless systems. For example, in a reuse-1 cellular system where the same frequency band is used in all the cells and sectors, cell-edge users may experience strong interference from neighboring cells, resulting in poor quality of service (QoS). In a underlay cognitive radio (CR) network, a CR user is only allowed to coexist with the primary user (PU) if the interference level caused by the CR user can be managed under a tolerable threshold [266].

As a result, interference management has become an important topic of interest [267–276]. In particular, the issue of power control in interference-limited wireless systems has attracted enormous attention recently [273, 274, 276, 277]. Power control over frequency bands for an orthogonal frequency-division multiplexing (OFDM) system was presented in [278] to consider power optimization over orthogonal subcarriers. In [274], a cross-layer power allocation scheme for a code division multiple access (CDMA) system was considered. Both open-loop and closed-loop power control were then investigated.

In this chapter, we consider a closed-loop power control tracking problem for interference-limited wireless networks. In a closed-loop power control system, a reverse feedback link is available to the receiver to track a desirable target signal-to-interference-and-noise ratio (SINR) [275, 276, 279]. According to the measured SINR, the receiver can convey the information to the transmitter via the reverse feedback link to revise the transmission power in a closed-loop fashion. For instance, a robust power tracking control design was proposed to achieve a robust optimal target SINR tracking from the point of view of minimizing the worst-case interference upon tracking error [280]. In [281], a robust state feedback control via a desired pole placement and an l_1 optimal predictor was used to predict the tracking error to compensate for the effect of round-trip delay for power control. A two-step power control filter design was

      DOI: 10.1201/9781003362142-13

also proposed by utilizing H_2 optimal approximation theory to minimize the effect of uncertain noise and interference on the target SINR tracking [277].

Multiple-objective problems often arise in practical power control systems when tradeoffs between multiple objectives need to be considered in spite of most existing studies focusing on single-objective problems. Generally speaking, MOPs have two or more objectives, and the objectives may conflict with each other. As a result, there often does not exist a global solution that achieves the optimal values for all objective functions universally and hence will be dominant in all aspects. A legitimate solution to MOPs is therefore to seek a set of solutions, the so-called Pareto optimal solutions, such that each of these solutions is not dominated by any other solutions [211, 282].

There are two general approaches to MOPs. One is to use a classical algorithm such as the weighted sum method [283, 284]. The concept of the weighted sum method is to provide weighting parameters to each objective function and combine these objective functions to form a single objective function. But the resulting solution highly depends on the proper selection of the weighting parameters. Once given suitable weighting parameters, proper Pareto optimal solutions for the MOP can be found directly. However, accurately and precisely selecting these weights can be quite difficult in practice. On the other hand, a set of MOP Pareto optimal solutions can be obtained by multi-objective evolutionary algorithms, which consist of a niche Pareto genetic algorithm (NPGA) [285], nondominated sorting genetic algorithm II [2], and so on. MOEAs are designed based on two common goals, a fast and reliable convergence to the Pareto optimal set and a good distribution of solutions along the Pareto front. These algorithms are inspired by biological evolution, including the processes of reproduction, mutation, crossover, and selection. The advantages of these algorithms are that auxiliary information (differentiability or continuity) of the objective functions is not needed and that almost the entire Pareto optimal set can be obtained.

In this chapter, an evolution algorithm method based on NSGA-II is combined with the LMI convex optimization technique to deal with an LMI-constrained MOP for multi-objective H_2/H_∞ power tracking control for interference-limited wireless systems. In other words, the optimal H_2 tracking performance [277] and H_∞ robust performance are considered simultaneously as an MOP. For optimal H_2 tracking performance, the mean-square tracking error in the power control of the interference-limited wireless system should be minimized. For optimal H_∞ robust performance, the worst-case effect of channel delay and interference upon the SINR tracking error should also be minimized. Since H_2 and H_∞ tracking objectives have conflicting natures, they fit well into the MOP formulation. The resulting multi-objective H_2/H_∞ power tracking control is much more complex than conventional MOPs with algebraic functional constraints [2, 283–285], since for an interference-limited wireless network, the power control system is a stochastic dynamic system with interference and delay. To handle this complication, we propose an equivalent formulation by minimizing the upper bounds of two objective functions in the considered MOP. Furthermore, based on the stochastic state-space model of the power control system, the resulting constraints for the equivalent multi-objective H_2/H_∞ power tracking control problem are transformed into LMI constraints to simplify the design procedure; that is, we obtain an equivalent

LMI-constrained MOP. The LMI toolbox in MATLAB can be combined with an MOEA search based on NSGA-II to solve the LMI-constrained MOP, and a set of Pareto optimal solutions for controller gains is thus obtained. These Pareto optimal solutions are not dominated by any other solutions, and one particular solution can be selected based on the tradeoff determined by the system designer. With a similar procedure, the proposed method can be easily extended to other multi-objective designs for wireless communication systems.

This chapter is divided into six sections as follows: in Section 10.2, the system model and a closed-loop power control scheme are introduced for the interference-limited wireless system, and a state-space model is developed to describe the closed-loop power tracking control. The multi-objective H_2/H_∞ power control problem is formulated as an LMI-constrained MOP in Section 10.3. In Section 10.4, properties of the LMI-constrained MOP and the design procedure of the LMI-based MOEA are discussed. Simulation results and discussion are given in Section 10.5. Finally, conclusions are presented in Section 10.6.

Notation: We denote $\|\cdot\|$ as the vector Euclidean norm. The inequality $A \succeq 0$ or $A \preceq 0$ indicates that matrix A is positive semi-definite or negative semi-definite, respectively. The trace of matrix A is denoted as $Tr(A)$, and $E(\cdot)$ denotes the statistical mean.

10.2   SYSTEM MODEL FOR CLOSED-LOOP POWER TRACKING CONTROL OF WIRELESS COMMUNICATION SYSTEMS

10.2.1   INTERFERENCE-LIMITED WIRELESS CHANNEL MODEL

The architecture of an interference-limited wireless system is illustrated in Figure 10.1. We consider an N-user, frequency-flat, single-input single-output (SISO) interference channel where each transmitter and each receiver is equipped with a single antenna. Let $s_i(t)$ denote the signal sent from transmitter i to the intended receiver i. The received signal at receiver i can be expressed as

$$y_i(t) = h_{ii}(t)s_i(t) + \sum_{j \neq i}^{N} h_{ij}s_j(t) + n_i(t) \tag{10.1}$$

where $h_{ij} \in \mathbb{C}$ denotes the channel coefficient from transmitter j to receiver i, and n_i represents the additive white Gaussian noise at the receiver end, that is, $n_i(t) \sim CN(0, \sigma_n^2)$.

The received uplink SINR function at the ith receiver is then given by

$$SINR_i(t) = \frac{|h_{ii}(t)|^2 |s_i(t)|^2}{\sum_{j \neq i}^{N} |h_{ji}(t)|^2 |s_j(t)|^2 + \sigma_n^2} \tag{10.2}$$

To study the closed-loop power control for the SISO interference channel, we consider the log domain; that is, we use dB as the unit of reference. Hence, (10.2) can be rewritten as

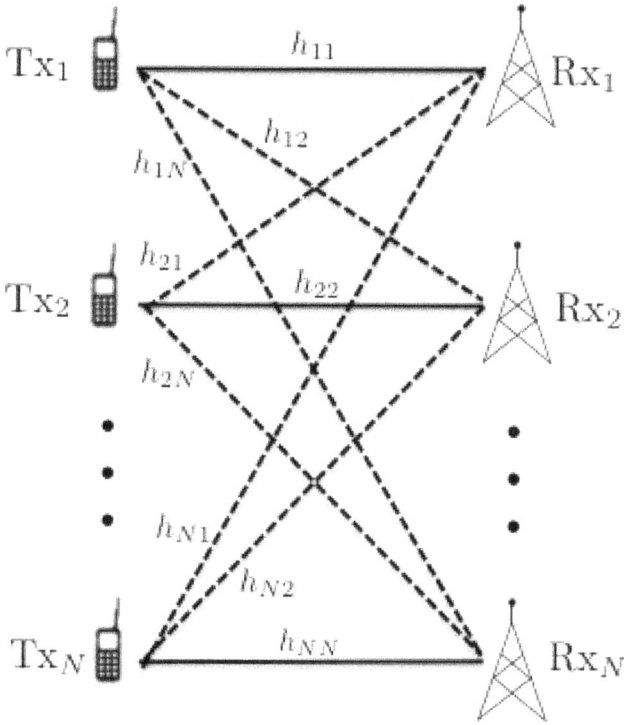

FIGURE 10.1 N-user SISO interference channel. Solid lines indicate user's signal link, and dotted lines represent interfering links.

$$\log_{10}(SINR(t)) = \log_{10}(|h_{ii}(t)|^2) + \log_{10}(|s_i(t)|^2)$$
$$- \log_{10}\left(\sum_{j \neq i}^{N}|h_{ji}(t)|^2 |s_j(t)|^2 + \sigma_n^2\right) \quad (10.3)$$

In the sequel, we focus on the closed-loop power control problem for the ith transceiver pair. For notational simplicity, we drop the user index i and define the following variables

$$y(t) \triangleq 10\log_{10}(SINR_i(t))$$
$$x(t) \triangleq 10\log_{10}(|s_i(t)|^2)$$
$$f(t) \triangleq 10\log_{10}(|h_{ii}(t)|^2) \quad (10.4)$$
$$w(t) \triangleq 10\log_{10}\left(\sum_{j \neq i}^{N}|h_{ji}(t)|^2 |s_j(t)|^2 + \sigma_n^2\right) + n_m(t)$$

where $y(t)$ denotes the measured SINR, and $x(t)$ represents the transmitted power. Also, $f(t)$ indicates the channel fading gain, and $w(t)$ denotes the multiple access interference (MAI) plus the SINR measurement error $n_m(t)$.

Remark 10.1: Although the channel model under consideration here only concerns a frequency-flat SISO interference channel, we can easily extend our power control scheme to a frequency-selective SISO interference channel by implementing orthogonal frequency division multiplexing both at the transmitter and receiver ends. In an OFDM-based wireless system, the frequency-selective channel is transformed into parallel narrow-band sub-channels (or subcarriers). Each sub-channel is now frequency flat, and the same formulation and analysis introduced in this chapter can be applied accordingly. Note that to study the optimal power control design for frequency-selective channels, OFDM systems have been considered and studied extensively [278, 286–290]. In [290], adaptive subcarrier assignment with power control was investigated. Power optimization over orthogonal subcarriers was also studied in [278].

10.2.2 Closed-Loop Power Control

The closed-loop power control scheme is depicted in Figure 10.2. Note that a round-trip delay of d is considered. The transmitted power $x(t)$ at the transmitter end can be described as

$$x(t) = x(t-1) + u(t-d) \tag{10.5}$$

where $u(t)$ denotes the power control command (PCC). Furthermore, the SINR measurement $y(t)$ at the receiver end can be represented by

$$y(t) = x(t) + f(t) - w(t) \tag{10.6}$$

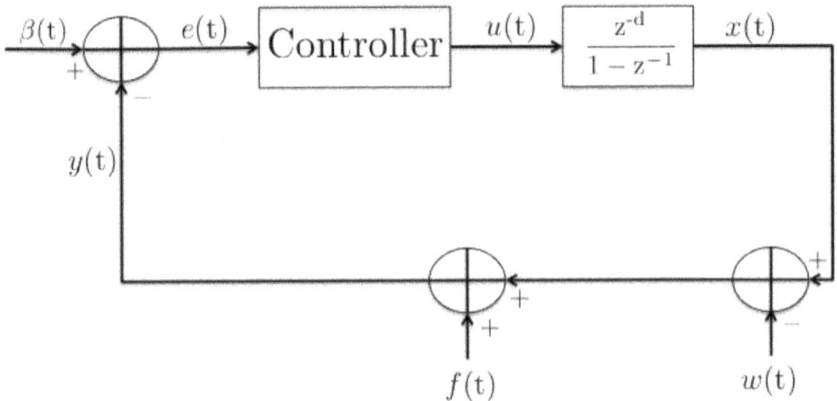

FIGURE 10.2   Closed-loop power control.

Note that (10.6) is a direct consequence of (10.3) and (10.4).

In order to maintain the received uplink SINR at a given target SINR $\beta(t)$, the SINR measurement $y(t)$ is compared with the target SINR to obtain the measured SINR tracking error $e(t)$ as follows:

$$e(t) = \beta(t) - y(t) \tag{10.7}$$

The SINR tracking error $e(t)$ is directly fed into the controller to generate the closed-loop power control update command $u(t)$, which is then quantized and sent back to the transmitter for power control.

10.2.3 STOCHASTIC STATE-SPACE MODEL

From (10.5), (10.6), and (10.7), the measured SINR tracking error $e(t)$ can be further derived as the following dynamic equation

$$e(t) = e(t-1) + v(t-1) - u(t-d) \tag{10.8}$$

where $v(t-1) = \beta(t) - \beta(t-1) - f(t) + f(t-1) + w(t) - w(t-1)$ indicates the changes of the target SINR and disturbances, including interference, fading, noise, and nonlinear effects.

For the purpose of designing the state feedback controller, the dynamic equation in (10.8) can be transformed into a stochastic state-space model by first defining a state vector

$$X(t) = [e(t) \quad u(t-d) \quad \cdots \quad u(t-2) \quad u(t-1)]^T \tag{10.9}$$

In (10.9), a sequence of past power command updates and measured SINR tracking errors is considered the state vector, and the effect of round-trip delay and channel disturbances can be compensated by the state-feedback control. Based on (10.8) and (10.9), a stochastic state-space model of the closed-loop power control scheme can be expressed as

$$X(t+1) = AX(t) + Bu(t) + Cv(t)$$
$$e(t) = DX(t) \tag{10.10}$$

where $A = \begin{bmatrix} 1 & 0 & -1 & \cdots & \cdots & 0 \\ 0 & 0 & 1 & \cdots & \cdots & 0 \\ 0 & 0 & 0 & 1 & \cdots & 0 \\ \vdots & \vdots & \vdots & \vdots & \ddots & \vdots \\ 0 & 0 & 0 & \cdots & 0 & 1 \\ 0 & 0 & 0 & \cdots & 0 & 0 \end{bmatrix} \in \mathbb{R}^{(d+1)\times(d+1)}$, $B = \begin{bmatrix} 0 \\ 0 \\ 0 \\ \vdots \\ 0 \\ 1 \end{bmatrix} \in \mathbb{R}^{(d+1)}$, $C = \begin{bmatrix} 1 \\ 0 \\ 0 \\ \vdots \\ 0 \\ 0 \end{bmatrix} \in \mathbb{R}^{(d+1)}$,

$D = \begin{bmatrix} 1 & 0 & \cdots & 0 \end{bmatrix} \in \mathbb{R}^{1\times(d+1)}$

The first state variable equation in (10.10) is a direct consequence of the error dynamic equation in (10.8). For the state space model of the power tracking control in the interference-limited wireless system presented in (10.10), we assume that there exists a state feedback controller with controller gain $K = [k_1, k_2, ..., k_{d+1}] \in \mathbb{R}^{1 \times (d+1)}$ at the receiver end. The power control command now becomes:

$$u(t) = KX(t) \qquad (10.11)$$

Since $X(t)$ is available at base station, if control gain K is specified based on the multi-objective H_2/H_∞ tracking performance, we can obtain the corresponding multi-objective H_2/H_∞ power command $u(t)$. By substituting (10.11) into (10.10), we obtain a state-space model of the closed-loop power control system with control gain K in the following form:

$$\begin{aligned} X(t+1) &= (A + BK)X(t) + Cv(t) \\ e(t) &= DX(t) \end{aligned} \qquad (10.12)$$

where $X(t) \in \mathbb{R}^{(d+1)}$ is a state vector; $v(t)$ is the disturbance in (10.10) with bounded power; and A, B, C, and D are constant matrices defined in (10.10) with appropriate dimensions. In the next section, based on the state-space model of the closed-loop power control system proposed here, we seek to properly formulate an MOP that considers both the H_2 and H_∞ tracking performance simultaneously.

10.3   PROBLEM FORMULATION

For a state feedback controller design (i.e., how to select the control gain K) in the closed-loop power control scheme in an interference-limited wireless system, we now consider both the optimal H_2 tracking performance and H_∞ robust tracking performance simultaneously. On one hand, the optimal H_∞ robust tracking performance can be expressed as the following optimization problem [291]

$$\min_{K} \max_{v(t)} \frac{E(e^T(t)R_1 e(t))}{E(v^T R_2 v(t))} \approx \min_{K} \max_{v(t)} \frac{\frac{1}{t_f}\sum_{t=0}^{t_f} e^T(t)R_1 e(t)}{\frac{1}{t_f}\sum_{t=0}^{t_f} v^T R_2 v(t)}$$

$$= \min_{K} \max_{v(t)} \frac{\|e(t)\|_2^2 R_1}{\|v(t)\|_2^2 R_2} \qquad (10.13)$$

(10.13) can be described as how to select the control gain K to minimize the worst-case power gain from the disturbance $v(t)$ to the SINR tracking error $e(t)$ for any possible uncertain external disturbance $v(t)$, that is, to minimize the maximum effect of $v(t)$ on the SINR tracking error or to achieve the optimal robust filtering of disturbance of $v(t)$. Note that t_f is the total transmission data length; R_1 and R_2 are two positive weighting factors. The minimax H_∞ scheme in (10.13) is quite suitable for filtering the unknown disturbance $v(t)$. On the other hand, the optimal H_2 tracking performance [200] can be described as the following optimization problem

$$\underset{K}{\text{minimize}}\, E[e^T(t+1)e(t+1)] \tag{10.14}$$

(10.14) represents the minimization of the mean square SINR tracking error with respect to the selection of control gain K, that is, to achieve optimal power tracking control. According to (10.13) and (10.14), the multi-objective H_2/H_∞ power tracking control problem for the closed-loop interference limited wireless system in (10.12) can be formulated as

$$\underset{K}{\text{minimize}}\{f_1(K), f_2(K)\} \tag{10.15}$$

where

$$f_1(K) = \underset{v(t)}{\text{maximize}}\, \frac{\|e(t)\|_2^2\, R_1}{\|v(t)\|_2^2\, R_2} \tag{10.16}$$

$$f_2(K) = E[e^T(t+1)e(t+1)] \tag{10.17}$$

However, because of the stochastic and dynamic nature of the state-space model involving the closed-loop power tracking control system in (10.12), the considered MOP for the multi-objective H_2/H_∞ power control problem in (10.15), (10.16), and (10.17) is highly complex and difficult to solve directly by conventional MOEA methods. Therefore, we propose an indirect method to solve the MOP in (10.15), (10.16), and (10.17) by the following upper bounds to each objective

$$f_1(K) = \underset{v(t)}{\text{maximize}}\, \frac{\|e(t)\|_2^2\, R_1}{\|v(t)\|_2^2\, R_2} \leq \gamma \tag{10.18}$$

$$f_2(K) = E[e^T(t+1)e(t+1)] \leq \alpha \tag{10.19}$$

for some positive scalar values γ and α, respectively. Based on (10.15), (10.18), and (10.19), the multi-objective H_2/H_∞ power control problem for the closed-loop power control system in (10.12) can be formulated in the following manner

$$\underset{K,\alpha,\gamma}{\text{minimize}}(\gamma,\alpha)$$
$$\text{subject to } f_1(K) \leq \gamma,\, f_2(K) \leq \alpha \tag{10.20}$$

And we have the following lemma:

Lemma 10.1: The optimization problem in (10.20) is equivalent to the MOP described in (10.15)–(10.17).

Proof: The proof of Lemma 10.1 is straightforward. One only needs to prove both inequality constraints in problem (10.20) become equality for Pareto optimal solutions. We will show this by contradiction.

Given a 3-tuple Pareto optimal solution $(K^*, \gamma^*, \alpha^*)$, we assume that either one of the inequalities in (10.20) remains a strict inequality at the optimal solution. Without loss of generality, suppose that $f_1(K^*) < \gamma^*$. As a result, there exists γ_1 such that $\gamma_1 < \gamma^*$ and that $f_1(K^*) = \gamma_1$. Now, for the same K^*, the solution (γ^*, α^*) is dominated by (γ_1, d^*), leading to a contradiction. This implies that both inequality constraints in problem (10.20) indeed become equality for Pareto optimal solutions. The optimization problem in (10.20) is hence equivalent to the MOP described in (10.15)–(10.17).

Now, the MOP for the multi-objective H_2/H_∞ power tracking control for the stochastic power control system in (10.12) can be equivalently described as the minimization of both γ and α under some inequality constraints. The solution to the reformulated MOP in (10.20) will be derived subsequently. Since the specification of the control gain K for solving the MOP in (10.20) is subject to the stochastic control system in (10.12), the MOP in (10.20) cannot be solved by conventional MOEAs, and additional efforts are needed. Before further analysis, we introduce the following lemma that will be useful for our solution of the MOP in (10.20).

Lemma 10.2 (Schur Complements) [291]:

Suppose that $P = P^T \succ 0$. The following inequality

$$D^T P^{-1} D - V \prec 0 \qquad (10.21)$$

is equivalent to the following LMI:

$$\begin{pmatrix} V & D^T \\ D & P \end{pmatrix} \prec 0 \qquad (10.22)$$

Proof: Please refer to [291].

Now, we proceed to state the following theorem.

Theorem 10.1: The multi-objective H_2/H_∞ power control problem in (10.20) for the stochastic dynamic power control system (10.12) can be formulated as the following MOP with three LMI constraints:

$$\underset{P_1 \succ 0, P_2 \succ 0, \gamma, \alpha, K}{\text{minimize}} \ (\gamma, \alpha) \qquad (10.23a)$$

subject to

$$\begin{pmatrix} P_1 - D^T R_1 D & 0 & (A+BK)^T P_1 \\ 0 & \gamma R_2 & C^T P_1 \\ P_1(A+BK) & P_1 C & P_1 \end{pmatrix} \succeq 0 \qquad (10.23b)$$

$$\begin{pmatrix} P_2 & P_2C & P_2(A+BK) \\ C^TP_2^T & I & 0 \\ (A+BK)^TP_2^T & 0 & P_2 \end{pmatrix} \succeq 0 \qquad (10.23c)$$

$$\frac{\alpha}{d+1}I_{(d+1)\times(d+1)} - P_2^{-1}D^TD \succeq 0 \qquad (10.23d)$$

where $P_1 \succ 0$ and $P_2 \succ 0$ are positive-definite matrices, and if the LMI-constrained MOP in (10.23a)–(10.23d) is solved, then the MOP problem in (10.20) is also solved.

Proof: Please refer to Appendix 10.7.1.

Based on Theorem 10.1, the multi-objective H_2/H_∞ power tracking control design problem can be transformed into the following equivalent LMI-constrained MOP

$$(\gamma^*, \alpha^*) = \underset{P_1 \succ 0, P_2 \succ 0, \gamma, \alpha, K}{\text{minimize}} (\gamma, \alpha)$$

$$\text{subject to LMIs (10.23b)} - (10.23d) \qquad (10.24)$$

Furthermore, by specifying $P_1 \succ 0$, $P_2 \succ 0$ to solve the LMI-constrained MOP in (10.24), we can obtain the solutions for the multi-objective H_2/H_∞ power control design problem. In the next section, we will introduce how to combine the LMI toolbox in MATLAB and an MOEA to efficiently solve the MOP in (10.24). In general, there does not exist a unique global solution for the LMI-constrained MOP in (10.24). However, we can obtain a set of Pareto optimal solutions (P_1^*, P_2^*, K^*) of the multi-objective H_2/H_∞ power control in an interference limited system from the viewpoint of MOP. Note that all derivations were based on the worst-case channel realization for H_∞ and channel covariance for H_2. We did not assume perfect channel estimation and rather only make use of statistical channel information.

Remark 10.2:

- From (10.24), the conventional H_∞ power control problem in (10.13) can be derived and forms the following SOP [280]

$$\gamma_0 = \underset{P_1 \succ 0, \gamma, K}{\text{minimize}} \gamma$$

$$\text{subject to LMIs (10.23b)} \qquad (10.25)$$

- The conventional H2 power control problem in (10.14) can be derived from (10.24) to form the following SOP

$$\alpha_0 = \underset{P_2 \succ 0, \alpha, K}{\text{minimize}} (\gamma, \alpha)$$

$$\text{subject to LMIs (10.23c)} - (10.23d) \qquad (10.26)$$

- In general, $\gamma^* \geq \gamma_0$ and $\alpha^* \geq \alpha_0$, where (γ^*, α^*) denotes the solutions of MOP in (10.24). Therefore, γ_0 and α_0 can be considered lower bounds of γ^* and α^*, respectively.
- Following a similar procedure, the result of Theorem 10.1 can be easily extended to other multi-objective designs of wireless communication systems with different LMI constraints.

10.4 PARETO OPTIMAL SOLUTIONS TO MULTI-OBJECTIVE POWER CONTROL DESIGN

In this section, we introduce some fundamental concepts of multi-objective optimization described in (10.24) for multi-objective H_2/H_∞ power tracking control design. Then an EA-based search method is proposed to efficiently solve the LMI-constrained MOP in (10.24). Finally, a step-by-step design procedure is presented for our proposed method. Before further discussion of the solutions for the LMI-constrained MOP in (10.24), some concepts and definitions of Pareto optimal solutions are given as follows.

10.4.1 CONCEPTS OF PARETO OPTIMAL SOLUTIONS

In contrast to the single-objective optimization problem, there may not exist a global solution that is optimal for all contradicting objectives; it is hence desirable to seek a set of good solutions that are no worse than other solutions. This set of solutions is called the Pareto optimal solutions from the MOP perspective [282]. These solutions are also termed noninferior, admissible, or efficient solutions.

For illustrative purposes, the proposed MOP is abstractly expressed as

$$(\gamma^*, \alpha^*) = \underset{(P_1 \succ 0, P_2 \succ 0, K) \in \Omega}{\text{minimize}} (\gamma, \alpha)$$

subject to LMIs (10.23b) – (10.23d)   (10.27)

where Ω represents the feasible set.

Some concepts regarding the Pareto optimal solution (P_1^*, P_2^*, K^*) with optimal objective values (γ^*, α^*) are introduced as follows [282].

Definition 10.1 (Pareto Dominance): For two feasible solutions (P_1^*, P_2^*, K^*) and (P_1, P_2, K) in Ω with two objective values (γ^*, α^*) and (γ, α) subject to LMIs (10.23b)–(10.23d), respectively. (P_1^*, P_2^*, K^*) is said to dominate (P_1, P_2, K) if $\gamma^* \leq \gamma$ and $\alpha^* \leq \alpha$ and at least one of the inequalities is a strict inequality.

Definition 10.2 (Pareto Optimality): A feasible solution $(P_1^*, P_2^*, K^*) \in \Omega$ is said to be Pareto optimal w.r.t. Ω if and only if there does not exist another feasible solution that dominates it.

Definition 10.3 (Pareto Optimal Solution Set): For a given MOP in (10.27), the Pareto optimal solution set ρ^* is defined as

$$\rho^* = \{(P_1^*, P_2^*, K^*) \in \Omega \big| (P_1^*, P_2^*, K^*) \text{ is Pareto optimal}\}$$

Definition 10.4 (Pareto Front): For a given MOP in (10.27) and Pareto optimal solution set ρ^*, the Pareto front f^* is defined as

$$f^* = \{(\gamma^*, \alpha^*) \big| (P_1^*, P_2^*, K^*) \in \rho^* \}$$

10.4.2 DESIGN PROCEDURE

Note that the concept of Pareto optimal solutions defined in Definitions 10.1, 10.2, 10.3, and 10.4 is established on a given feasible set and multiple objective function. In this chapter, an MOEA-based search is proposed to solve the Pareto optimal solution with the help of the LMI toolbox in MATLAB. First, we begin by choosing an initial Pareto population of (γ, α) in the feasible set (i.e., they should satisfy the LMIs in (10.23b)–(10.23d)); then the Pareto dominant solutions are found via MOEA-based search (i.e., to search for nondominant level of the initial population). Next, a new population is generated as the child population by crossover and mutation. The search for the nondominant level in the feasible set is then carried out. This procedure is conducted iteratively to find the final population that approaches the Pareto front. Once the set of the approximate Pareto front (γ^*, α^*) is obtained, we can solve the LMIs in (10.23b)–(10.23d) to obtain the Pareto optimal solutions (P_1^*, P_2^*, K^*), providing us the multi-objective H_2/H_∞ power control gains K^*.

Based on the reformulation and analyses in the previous section and the help of the LMI toolbox in MATLAB, a design procedure of the MOEA-based search method to solve the LMI-constrained MOP in (10.27) for the multi-objective H_2/H_∞ power control problem is proposed in Algorithm 10.1.

After the MOEA-based search converges in the proposed design procedure, the entire population (γ^*, α^*) of multi-objective H_2/H_∞ power control solutions will be close to the Pareto front f^* with corresponding Pareto optimal solutions (P_1^*, P_2^*, K^*) from the LMIs in (10.23b)–(10.23d) by the LMI toolbox in MATLAB.

10.5 SIMULATION RESULTS AND DISCUSSION

For the illustration of the proposed multi-objective H_2/H_∞ power tracking control method, we consider a typical interference limited wireless system, a direct-sequence code division multiple access (DS-CDMA) system, as a design example and provide numerical experiments to verify the effectiveness of the design procedure. We first introduce the simulation settings and demonstrate that the Pareto front (γ^*, α^*) is indeed closely approximated by the proposed MOEA-based search method through the combination of the LMI technique with the NSGA-II algorithm. Then we solve the Pareto optimal solutions (P_1^*, P_2^*, K^*), providing us the multi-objective H_2/H_∞ power tracking control gains K^*.

10.5.1 SIMULATION SETTINGS FOR MULTI-OBJECTIVE OPTIMIZATION

In this subsection, the simulation settings for the multi-objective evolution search algorithm in the design procedure are given as follows: we set the crossover probability $p_c = 0.9$, mutation probability $p_m = 0.1$, distribution index for cross-over $\eta_c = 20$, and mutation operator $\eta_m = 20$. In all our simulations, we use the proposed design procedure to search for a set of the Pareto front (γ^*, α^*), the corresponding Pareto optimal solutions (P_1^*, P_2^*, K^*), and hence the sate feedback controller gains K^*, which are then used as power control update command in (10.11).

The two weighting factors R_1 and R_2 in the LMI (10.23b) can be used as free design parameters to make a proper tradeoff between the system response time and the noise filtering capability. In our design example, the ratio R_1 / R_2 is chosen to be 0.0773, based on the optimal choice of parameters for the H_∞ power control in [280]. The optimal objectives (γ^*, α^*) and the corresponding power tracking control gains K^* are solved from (10.23a)–(10.23d), respectively.

From Figure 10.3, we observe that the Pareto optimal solutions obtained by Algorithm 10.1 are indeed distributed along the Pareto front. A particular Pareto optimal solution can then be selected based on the desirable tradeoff determined by the system designer. From Figure 10.3, we can also observe good solution diversity for the Pareto solutions generated by Algorithm 10.1. The effectiveness of the proposed MOEA-based search method is hence verified.

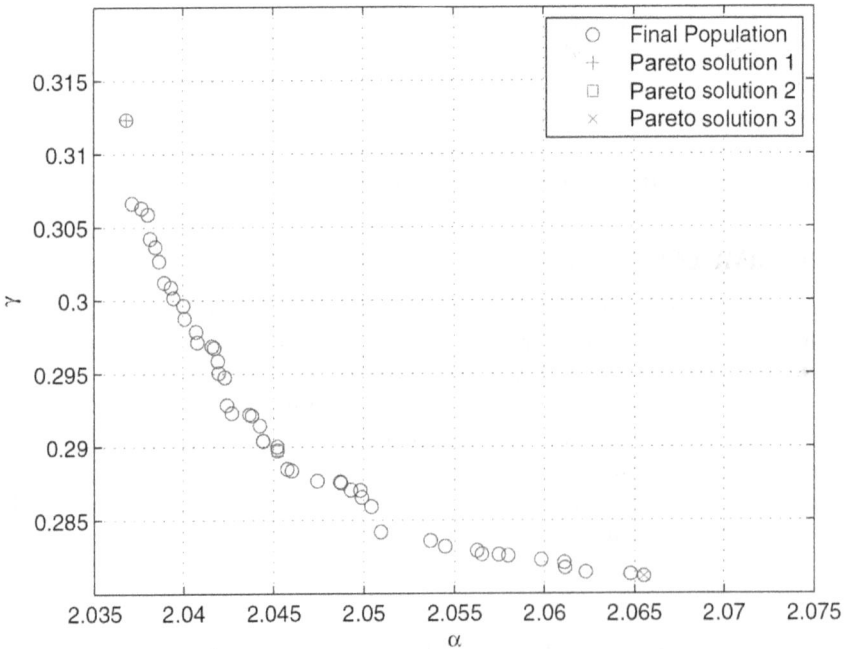

FIGURE 10.3 Pareto front for a given R_1 / R_2 and selection of three Pareto solutions for comparison.

Algorithm 10.1

Step 1:   Solve the SOP in (10.25) for γ_0 and the SOP in (10.26) for α_0 to find the lower bound of the H_2 and H_∞ objectives, respectively. Then the Pareto front (γ^*, α^*) is lower bounded as $\gamma^* \geq \gamma_0$ and $\alpha^* \geq \alpha_0$.

Step 2:   Set the size of population, the maximum number of generations, crossover, and mutation probability in the proposed MOEA search.

Step 3:   Initialize a population such that all pairs (γ, α) in the population are within $\gamma \geq \gamma_0$ and $\alpha \geq \alpha_0$. With the help of the LMI toolbox in MATLAB, determine whether there exist $P_1 \succ 0$ and $P_2 \succ 0$ that satisfy (10.23b)–(10.23d) for each pair (γ, α). If not, the corresponding pair is discarded and another pair (γ, α) is added to the initial population so that the corresponding LMIs have the solution $P_1 \succ 0$, $P_2 \succ 0$.

Step 4:   Assign a nondomination level to each pair (γ, α) within the population. Notice that pairs assigned the same nondomination level do not dominate each other.

Step 5:   Sort the initial population based on their nondomination levels. Each solution pair is assigned a fitness (rank) that is equal to its nondomination level. Once the nondominated sort is completed, a crowding distance is assigned.

Step 6:   Perform binary tournament selection by fitness and keep half of the population using a crowded-comparison operator based on the NSGA-II algorithm [2].

Step 7:   Reproduce the child population by the remaining half of the parent population through the crossover operator, called simulated binary crossover (SBX) [292], and the mutation operator, called polynomial mutation, to generate the child population based on the NSGA-II algorithm. For each solution pair (γ, α) in the child population, (P_1, P_2, K) is solved from LMIs (10.23b)–(10.23d) with the help of the LMI toolbox in MATLAB to guarantee a feasible child population for the MOP in (10.24).

Step 8:   Combine the better half of the parent population with the generated child population to form a new population (γ, α).

Step 9:   If the maximum number of generations is reached, halt the algorithm. If not, repeat steps 4–8 until the current population of (γ, α) approaches the Pareto front f^*.

10.5.2   PERFORMANCE OF THE MO H_2/H_∞ POWER CONTROL IN A DS-CDMA COMMUNICATION SYSTEM

Now, the performance of the multi-objective H_2/H_∞ power tracking control obtained from the proposed algorithm is evaluated by numerical simulation in a DS-CDMA communication system. In the sequel, numerical results are conducted using Monte Carlo simulations. The channels utilized are the uplink channels for the DS-CDMA system and are concisely described in the following.

The uplink channel is considered a time-varying power gain $f(t)$ consisting of a slow fading $f_{slow}(t)$ component (shadow fading) and a fast fading $f_{fast}(t)$ component (multipath). The slow fading power gain comprises a distance-dependent propagation loss and a slow power fluctuation (shadow fading) due to obstruction in the propagation link, and it can be expressed as [293]:

$$f_{slow}(t) = F_0 - 10n\log(D) + S_\sigma(t) \tag{10.28}$$

where F_0 is a constant, n is the path loss exponent, and D denotes the distance between the base station and the mobile station. S_σ is the shadow fading process that is typical of a log-normal Gaussian distribution with zero mean and standard deviation σ_s.

The fast fading power gain $f_{fast}(t)$ is the fast power fluctuation owing to channel fading and follows the typical Rayleigh distribution. Jakes' model is used to model the fast fading power gain. Apart from fading, the uplink channel is also corrupted by multiple access interference and SINR measurement errors.

We assumed 12 mobiles in the range of a base station. The operational frequency is approximately 2 GHz, and the bandwidth W for each channel is 3.6864 MHz in a CDMA 2000–3x system, where the simulated environment depends on the pseudo-noise chip rate [294]. The data rate R_d is set at 9.6 kHz, and thus the processing gain becomes 21 dB. The required $E_b / I_b = 7$ dB if the bit error rate (BER) is no more than 10^{-1}, where E_b and I_0 are the energy per information bit and the total interference and noise power spectral density, respectively. The SINR can be represented by $SINR(dB) = (E_b R_d) / (I_0 W)$, and we denote the target SINR by $SINR_{target}$ to ensure acceptable link quality, with a typical value of $SINR_{target} = -10$ dB. For simplicity, the measurement error $n_m(t)$ is ignored. According to the operational environment, the shadowing log-standard deviation σ is set to be 4.3 dB in urban areas, and a typical value of 8 dB is used for suburban areas [295, 296]. The power control update command is calculated by (10.11) during every update interval T_u using the adaptive quantization law to quantize and reconstruct the updated command. The adaptive quantization and reconstruction step size scalar $\delta(t)$ can be updated as follows [297]

$$\delta(t) = \sqrt{\left[\beta \delta^2 (t-1) + (1-\beta)\hat{\sigma}_e^2 \right]} \tag{10.29}$$

where β and $\hat{\sigma}_e$ is the forgetting factor and the sample standard deviation of power control error during the update window size, respectively. In the ensuing simulations, we assume $\beta = 0.95$, $T_u = 1.25$ ms and then define the standard deviation of SINR tracking error as the performance criterion

$$\hat{\sigma}_e \sqrt{\frac{1}{M} \sum_{t=1}^{M} (SINR_{target} - y(t))^2} \tag{10.30}$$

where $y(t)$ is given by (10.6), and M is the update window length and equals 12 in this simulation. For a given ratio $R_1 / R_2 = 0.0773$, we choose three corresponding Pareto optimal solutions as in Figure 10.3 and compare the corresponding power control performances. The optimal objectives (γ^*, α^*) and power control gains K^* of the three Pareto optimal solutions are given in Table 10.1, where Pareto solution 1 is close to the optimal H_∞ solution, Pareto solution 3 is close to the optimal H_2 solution, and Pareto solution 2 demonstrates a tradeoff in between.

Since the overall channel fading power control gain $f(t)$ depends on the mobile velocity, we consider 50~150 km/hr in mobile speeds, corresponding to a channel coherence time ranging from 0.36~1.08 μs. The round-trip delay is set to 2, the same as the sampling period T_s.

Table 10.1 The Pareto Solution (γ^*, α^*) and Control Gains K^*

	Pareto Solution 1	Pareto Solution 2	Pareto Solution 3
K^*	[0.944–0.986–0.6999]	[0.907–0.0527–0.6999]	[0.8567–0.0209–0.6999]
(γ^*, α^*)	(0.2812, 2.066)	(0.2897, 2.045)	(0.3123, 2.037)

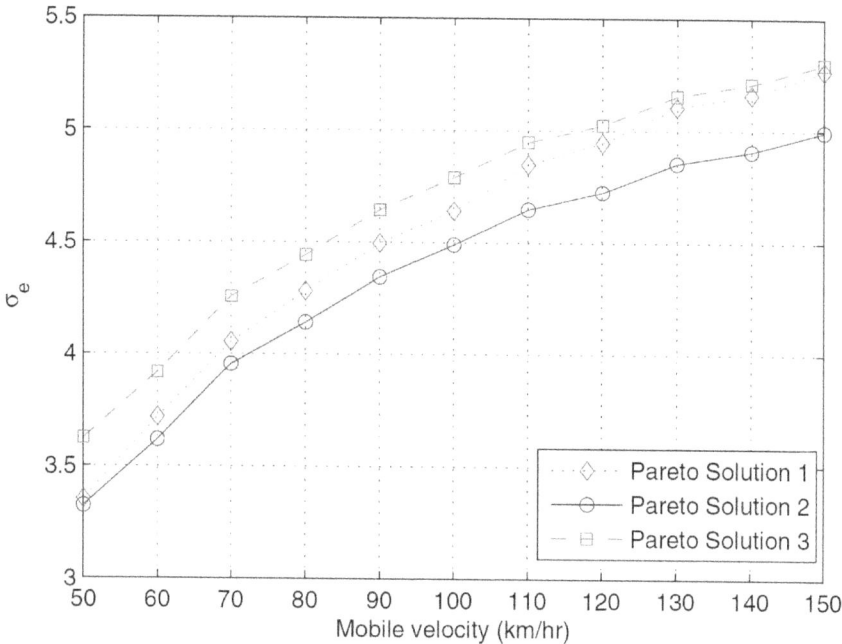

FIGURE 10.4 Comparison among three Pareto solutions based on the tracking performance $\hat{\sigma}_e$ versus v.

Figure 10.4 shows the standard deviations of the SINR tracking error $\hat{\sigma}_e$ versus the mobile velocity v for different Pareto optimal solutions. Interestingly, the optimal H_∞ solution (Pareto solution 1) achieves better performance than the optimal H_2 solution (Pareto solution 3) even when the comparison criterion $\hat{\sigma}_e$ in (10.30) is the standard deviation of the H_2 tracking error. This may result from the high sensitivity of the optimal H_2 control to modeling errors. Furthermore, by proper tradeoff between H_2 and H_∞ power control, we can obtain Pareto solution 3 that achieves even better performance than the other two Pareto solutions.

In Figure 10.5, we compare the performance of the proposed multi-objective H_2/H_∞ power tracking control (Pareto solution 2) with other power control schemes, such as the conventional power control with fixed step size (with both 1 and 2 power control bit (PCB)) [298], adaptive step size method (with both 1 and 2 PCB) [297], and Smith predictor scheme [299] in terms of the standard deviation of the power control

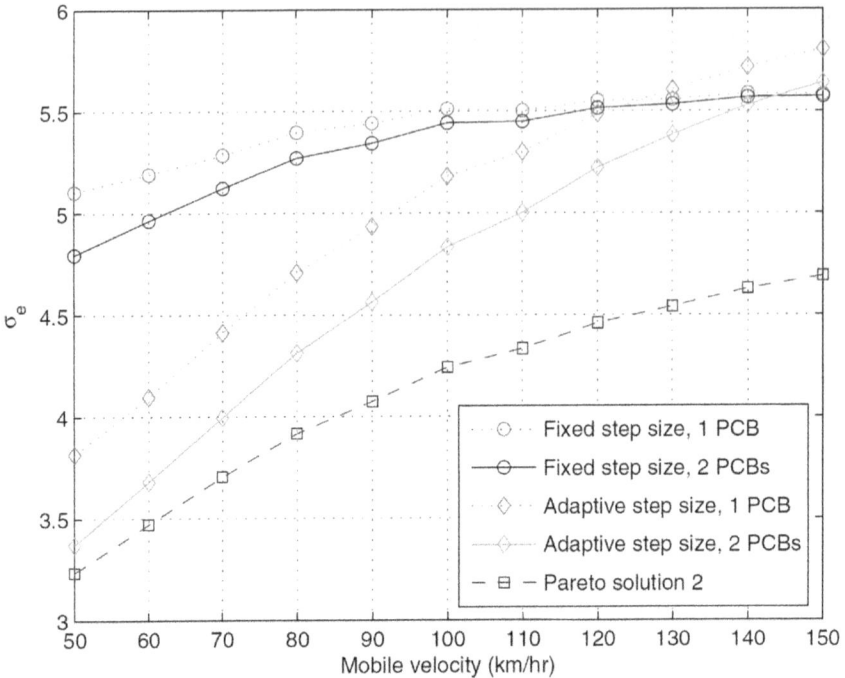

FIGURE 10.5 Comparison of the proposed MO scheme with conventional power control schemes based on the tracking performance $\hat{\sigma}_e$ versus v.

tracking error. Clearly, from Figure 10.5, we observe that the proposed scheme is more robust compared to other existing schemes. The robustness of the proposed scheme is due to the fact that the MO H_2/H_∞ power control is capable of compensating for the round trip delay and mitigating the effect of system uncertainties.

10.5.3 EFFECT ON OUTAGE PROBABILITY

Outages may occur for the communication link and lead to momentary deterioration of the system performance when the measured SINR goes below a minimum SINR. Here, the minimum SINR, denoted by $SINR_{min}$, is set to a typical value of -14 dB to ensure acceptable link quality. Consequently, the power control performance of communication system can also be measured by the outage probability denoted by P_o as follows

$$P_o = Prob\{SINR < SINR_{min}\} \qquad (10.31)$$

In general, the outage probability P_o depends on the uplink channel fading $f(t)$ and changes as a function of the mobile velocity and period of sampling. In this simulation, the operational environment we choose is based on the CDMA 2000–3x system, with the data rate R_d being set to 9.6 kHz and the sampling period T_s being set to the reciprocal of R_d. The outage probability P_o versus mobile velocity for different Pareto solutions from the proposed MO power tracking algorithm is given in Figure 10.6. It can be seen that Pareto solution 2 achieves a lower outage probability than other Pareto solutions; that is, a properly selected tradeoff between the optimal H_2 and the optimal H_∞ tracking controls can lead to a smaller outage probability. As can be observed from Figures 10.4 and 10.6, Pareto solution 2 actually outperforms the other two Pareto solutions both in tracking error and outage probability, demonstrating again the importance of a carefully selected tradeoff.

The proposed multi-objective H_2/H_∞ power tracking control method is also compared with other power control schemes, such as the conventional power control with fixed step size (with both 1 and 2 PCB) [298], adaptive step size method (with both 1 and 2 PCB) [297], and Smith predictor scheme [299] in terms of outage probability. From the Monte Carlo simulation results in Figure 10.7, it can be seen that the outage probability of the proposed multi-objective H_2/H_∞ method (Pareto solution 2) is indeed better than all other existing methods for mobile velocities higher than 80 (km/hr). For lower mobile velocities, Pareto solution 2 still outperforms all other schemes except the power control with adaptive step size and 2 PCB. This is reasonable since the proposed multi-objective H_2/H_∞ method is designed to achieve robust

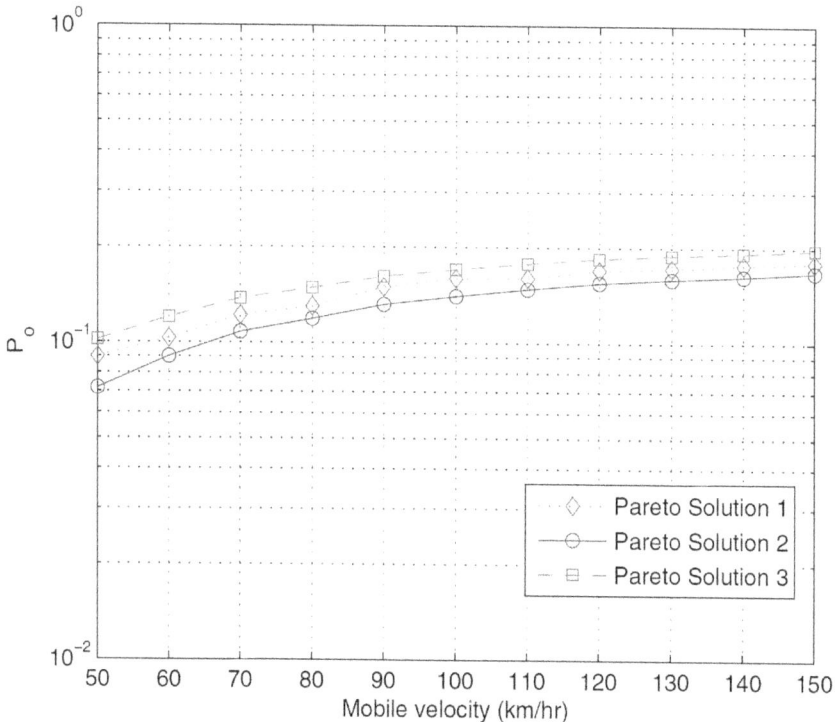

FIGURE 10.6 Comparison of outage probability P_o versus mobile velocity v among three Pareto optimal solutions.

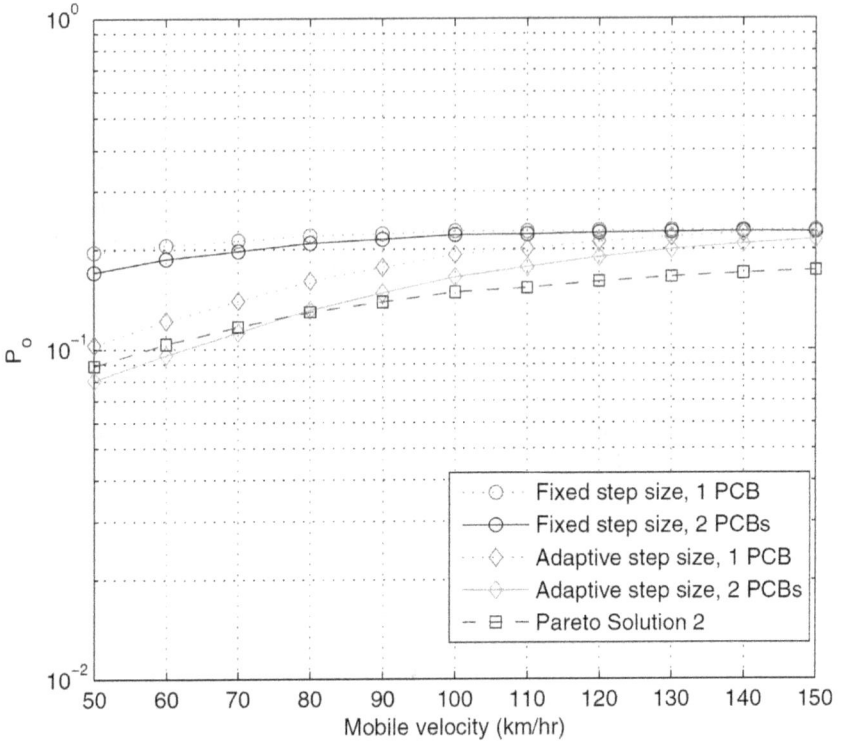

FIGURE 10.7 Comparison of outage probability P_o versus mobile velocity v of the proposed scheme with other conventional methods.

power tracking control, which should achieve better performance especially when system uncertainties and disturbances are high.

From Figure 10.6 and 10.7, we see that the outage probability increases as the velocity of the mobile unit increases. The outage probability is also affected by the margin between $SINR_{target}$ and $SINR_{min}$; that is,

$$SINR_{margin} = SINR_{target} - SINR_{min}$$

Figure 10.8 depicts the outage probability versus $SINR_{margin}$. We see that the higher the $SINR_{margin}$, the lower the outage probability. Achieving higher $SINR_{margin}$, however, requires higher transmission power when $SINR_{min}$ is fixed. A proper tradeoff between system performance (outage probability) and resource (transmission power) is hence essential. From Figure 10.8, we can also observe that our proposed optimal multi-objective H_2/H_∞ power control (Pareto solution 2) achieves the lowest outage probability.

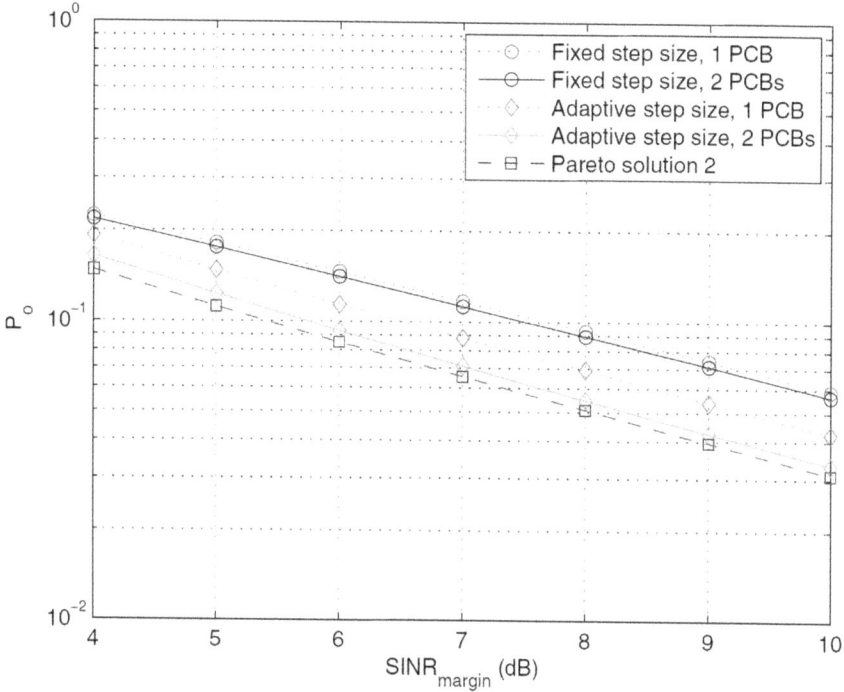

FIGURE 10.8 Outage probability P_o versus $SINR_{margin}$.

Figure 10.9 illustrates the time evolution of our proposed H_2/H_∞ power tracking control. Note that $SINR_{target}$ is set to -10 dB and that $SINR_{min}$ is set to -14 dB, as indicated in the figure. It is clear that our proposed H_2/H_∞ power tracking control indeed tracks $SINR_{target}$ well, and the received SINR is above $SINR_{min}$ for most of the time, demonstrating again the effectiveness and robustness of proposed Algorithm 10.1.

Remark 10.3: It is important to include complexity comparison of the proposed multi-objective H_2/H_∞ with other existing schemes. In fact, the major complexity of the proposed scheme comes from the acquisition of the optimal power control gain K^* that allows the transmitter to update its transmitted power accordingly. Once the optimal power gains are available, our SINR tracking procedure is similar to all other existing schemes. We would like to point out that all power control gains are obtained offline, that is, prior to the iterative procedure of the SINR tracking. Therefore, in terms of complexity, our proposed schemes do not sacrifice much compared with existing schemes. The slight increase in computational complexity of our proposed algorithm does not affect the time required by the algorithm to track the target SINR, which is the most crucial complexity incurred by all SINR tracking controls.

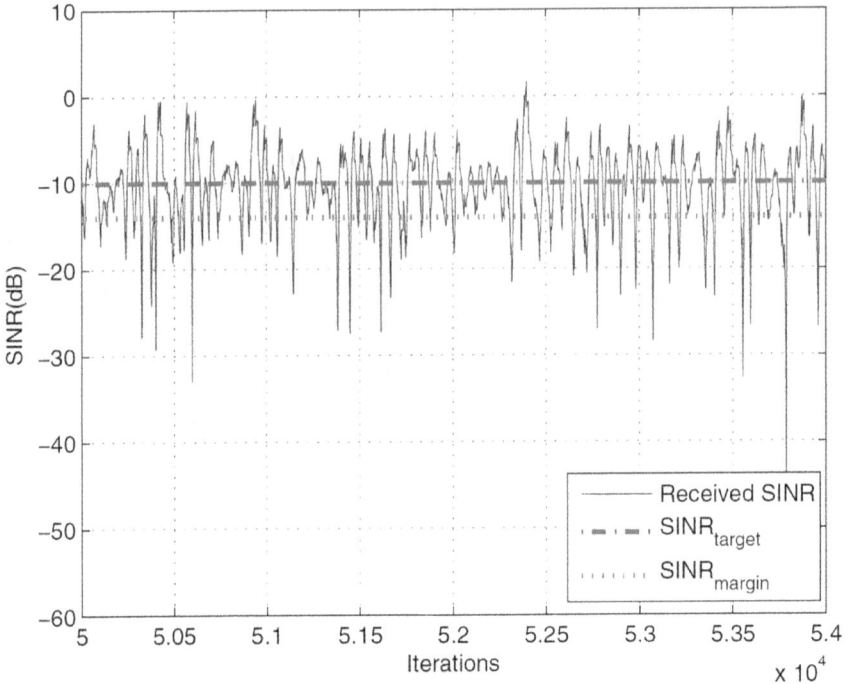

FIGURE 10.9　Time evolution of the proposed power tracking control.

10.6　CONCLUSION

This chapter introduced the multi-objective H_2/H_∞ closed-loop power control problem in an interference-limited wireless system. The rest of this chapter is described in the following: First, a stochastic closed-loop power control problem for an interference-limited wireless system with disturbances is represented by a stochastic state-space control model. Second, we formulated the optimal robust tracking control of the target SINR as an MOP with two conflicting objectives: (i) H_2 tracking performance and (ii) H_∞ tracking performance of the closed-loop power control scheme must be optimized simultaneously. Third, by the proposed equivalent formulation and the LMI technique, the multi-objective H_2/H_∞ power tacking control problem is transformed into an LMI-constrained MOP. Therefore, the multi-objective H_2/H_∞ power tracking problem can be efficiently solved by combining an MOEA search method (an NSGA-II-based algorithm) with the LMI toolbox in MATLAB to search for a set of the Pareto optimal solutions.

The simulation result demonstrates that the proposed design method can efficiently obtain a set of Pareto optimal solutions that approach the Pareto front of the concurrent optimization objectives. From the simulation example, the Pareto front for multi-objective H_2/H_∞ power tracking performance can be obtained in a single run, and the design time can be reduced. Based on the set of approximate Pareto optimal solutions thus obtained, proper tradeoff of the H_2/H_∞ power tracking performance can be determined based on the preference of system designers.

As a design example, we consider a typical interference-limited wireless system, a DS-CDMA cellular system. From the numerical experiments, we demonstrated that the proposed optimal H_2/H_∞ power control indeed outperforms all existing schemes both in terms of SINR tracking error and outage probability. We believe that the proposed multi-objective H_2/H_∞ power tracking control mechanism is suitable for future communication system design where multipurpose design may become a necessity.

10.7 APPENDIX

10.7.1 PROOF OF THEOREM 10.1

The proof procedure is divided into three steps: (a) The first step is to show that the H_∞ power tracking performance in (10.18) leads to the LMI in (10.23b). (b) The second step is to show that the H_2 power tracking performance in (10.19) leads to the LMIs in (10.23c) and (10.23d). (c) The third step is to show that the multi-objective H_2/H_∞ power tracking control problem in (10.20) is equivalent to solving the LMI-constrained MOP in (10.23a)–(10.23d).

(a) For the H_∞ power tracking performance, the inequality (10.18) is equivalent to the following inequality [300]

$$\sum_{t=0}^{t_f} e^T(t)R_1 e(t) \le \gamma \sum_{t=0}^{t_f} v^T(t)R_2 v(t), \forall v(t) \tag{10.32}$$

that is, if the inequality holds for every $v(t)$, then inequality (10.18) also holds.

The previous inequality is equivalent to

$$\sum_{t=0}^{t_f} e^T(t)R_1 e(t) = \sum_{t=0}^{t_f} X^T(t)D^T R_1 DX(t)$$

$$< \gamma \sum_{t=0}^{t_f} v^T(t)R_2 v(t), \forall v(t) \tag{10.33}$$

In fact, the previous H_∞ power tracking performance is under the assumption of initial condition $X(0) = 0$. If the effect of initial condition $X(0) \ne 0$ is considered in the H_∞ tracking performance, the H_∞ tracking inequality in (10.33) should be modified as [280, 281, 291]

$$\sum_{t=0}^{t_f} e^T(t)R_1 e(t) = \sum_{t=0}^{t_f} X^T(t)D^T R_1 DX(t)$$

$$\le X^T(0)PX(0) + \gamma \sum_{t=0}^{t_f} v^T(t)R_2 v(t), \forall v(t) \tag{10.34}$$

and for some $P_1 = P_1^T \succ 0$. Then we want to prove that if the LMI in (10.23b) holds, then the left-hand side of (10.33) or (10.34) will be less than the right-hand side of (10.33) or (10.34), respectively.

From the left-hand side of (10.33) or (10.34), we derive

$$
\begin{aligned}
&\sum_{t=0}^{t_f} e^T(t) R_1 e(t) \\
&= X^T(0) PX(0) - X^T(t_f + 1) PX(t_f + 1) \\
&\quad + \sum_{t=0}^{t_f} [X^T(t) D^T R_1 D X(t) + X^T(t+1) P_1 X(t+1) \\
&\quad - X^T(t) P_1 X(t)] \\
&\leq X^T(0) PX(0) + \gamma \sum_{t=0}^{t_f} v^T(t) R_2 v(t) \\
&\quad + \sum_{t=0}^{t_f} [X^T(t) D^T R_1 D X(t) + X^T(t+1) P_1 X(t+1) \\
&\quad - X^T(t) P_1 X(t) - \gamma v^T(t) R_2 v(t)]
\end{aligned}
\tag{10.35}
$$

where the first inequality is due to the equivalence between two sides and the second inequality comes from neglecting the term $X^T(t_f + 1) P_1 X(t_f + 1)$.

From (10.35), if the following inequality holds

$$
\begin{aligned}
&X^T(t) D^T R_1 D X(t) + X^T(t+1) P_1 X(t+1) \\
&- X^T(t) P_1 X(t) - \gamma v^T(t) R_2 v(t) \leq 0
\end{aligned}
\tag{10.36}
$$

then the inequality (10.34) holds for $X(0) \neq 0$ and the inequality (10.33) holds for $X(0) = 0$, $\forall v(t)$; that is, the upper bound γ of the H_∞ power tracking performance in (10.18) holds if the inequality in (10.36) holds. Applying (10.12) to (10.36) and rearranging it, we can obtain the formula in (10.37).

$$
\begin{aligned}
&\begin{pmatrix} X(t) \\ v(t) \end{pmatrix}^T \left\{ \begin{pmatrix} (A+BK)^T \\ C^T \end{pmatrix} P_1 \begin{pmatrix} (A+BK)^T \\ C^T \end{pmatrix}^T \right. \\
&\left. - \begin{pmatrix} P_1 - D^T R_1 D & 0 \\ 0 & \gamma R_2 \end{pmatrix} \right\} \begin{pmatrix} X(t) \\ v(t) \end{pmatrix} \leq 0
\end{aligned}
\tag{10.37}
$$

The matrix inequality in (10.37) holds if the following inequality holds

$$
\begin{pmatrix} (A+BK)^T \\ C^T \end{pmatrix} P_1 \begin{pmatrix} (A+BK)^T \\ C^T \end{pmatrix}^T - \begin{pmatrix} P_1 - D^T R_1 D & 0 \\ 0 & \gamma R_2 \end{pmatrix}
\tag{10.38}
$$

Hence, by using Lemma 10.2 again, we can obtain the following equivalent bilinear matrix inequality constraint:

$$\begin{pmatrix} P_1 - D^T R_1 D & 0 & (A+BK)^T P_1 \\ 0 & \gamma R_2 & C^T P_1 \\ P_1(A+BK) & P_1 C & P_1 \end{pmatrix} \succeq 0 \tag{10.39}$$

which is the LMI in (10.23b). Hence, if there exists a positive definite matrix $P_1 \succ 0$ and a matrix K such that (10.23b) is satisfied, then H_∞ tracking performance in (10.18) or (10.20) has an upper bound γ, that is, $f_1(K) \le \gamma$.

(b) For the H_2 tracking performance, the mean-square SINR tracking error is defined as

$$\begin{aligned} J &= E\{e^T(t+1)e(t+1)\} \\ &= E\{X^T(t+1)D^T DX(t+1)\} \\ &= E\{TrDX(t+1)X^T(t+1)D^T\} \\ &= Tr(DE\{X(t+1)X^T(t+1)D^T\}) \end{aligned} \tag{10.40}$$

From (10.12), the covariance matrix of $X(t+1)$ is given by

$$\begin{aligned} &E\{X(t+1)X^T(t+1)\} \\ &= E\{[(A+BK)X(t)+Cv(t)][(A+BK)X(t)+Cv(t)]^T\} \\ &= E\{(A+BK)X(t)X^T(t)(A+BK)^T + (A+BK)X(t)v^T(t)C^T \\ &\quad + Cv(t)X^T(t)(A+BK)^T + Cv(t)v^T(t)C^T\} \\ &= (A+BK)E\{X(t)X^T(t)\}(A+BK)^T + CC^T \end{aligned} \tag{10.41}$$

Equation (10.41) is obtained by assuming $X(t)$ and $v(t)$ to be mutually orthogonal and $E\{v(t)v^T(t)\} = I$. Denoting $R = E\{X(t)X^T(t)\}$ and substituting (10.41) into (10.40), we obtain

$$\begin{aligned} J &= Tr(D[(A+BK)R(A+BK)^T + CC^T]D^T) \\ &= Tr(D[(A+BK)R(A+BK)^T - R + CC^T]D^T) \\ &\quad + Tr(DRD^T) \end{aligned} \tag{10.42}$$

We then conclude that the mean-square SINR tracking error J is upper-bounded by

$$J \le Tr(DRD^T) \le \alpha \tag{10.43}$$

If the following inequality holds

$$(A+BK)R(A+BK)^T - R + CC^T \le 0 \tag{10.44}$$

Let $P_2 = R^{-1}$, and multiply P_2 from the left and right side of (10.44); then we get

$$P_2(A+BK)P_2^{-1}(A+BK)^T P_2 - P_2 + P_2 CC^T P_2^T \le 0 \qquad (10.45)$$

Applying Lemma 10.2 to (10.45), we obtain its equivalent BMI constraint as follows

$$\begin{pmatrix} P_2 & P_2 C & P_2(A+BK) \\ C^T P_2^T & I & 0 \\ (A+BK)^T P_2^T & 0 & P_2 \end{pmatrix} \succeq 0 \qquad (10.46)$$

which is equivalent to the LMI in (10.23c). This means that if the LMI constraint in (10.23c) holds, the H_2 tracking performance J has an upper bound α. From (10.43), and by the fact $Tr(\frac{\alpha}{d+1} I_{(d+1)\times(d+1)}) = \alpha$, we obtain

$$Tr(DP_2^{-1}D^T) = Tr(P_2^{-1}D^T D) \le Tr\left(\frac{\alpha}{d+1} I_{(d+1)\times(d+1)}\right) = \alpha \qquad (10.47)$$

From the inequality (10.47), a sufficient condition must be satisfied

$$P_2^{-1}D^T D < \frac{\alpha}{d+1} I_{(d+1)\times(d+1)} \qquad (10.48)$$

Hence, if there exists a positive-definite matrix $P_2 = P_2^T$ and K that satisfies (10.23c) and (10.23d), then the H_2 mean-square error has an upper bound α; that is, $f_2(K) \le \alpha$ holds in (10.20).

(c) From the derivation of (a) and (b), the MOP in (10.20) holds if the MOP in (10.23a)–(10.23b) holds. The proof of Theorem 10.1 is now completed.

11 Multi-Objective Power Minimization Design for Energy Efficiency in Multicell Multiuser MIMO Beamforming System

11.1 INTRODUCTION

Multiple-input multiple-output transmission has been considered a promising technique for future wireless cellular networks [301]. MIMO technologies exploit multiple antennas to provide spatial degrees of freedom (DoF), diversity, and multiplexing gains [302]. As a result, the MIMO technique can improve the transmission capacity and cell coverage and reduce the bit error rate without requiring extra time or frequency resources [303, 304]. Recently, massive MIMO transmission architecture with a large number of transmittal antennas has been widely studied for advanced fifth-generation technology [305]. The base stations (BSs) send out data streams to their multiple users in their own cells simultaneously. However, this concurrent communication will cause not only intracell interference but also intercell interference, leading to severe performance degradation.

For MIMO transmission, multicell multiuser MIMO beamforming design has been attracting a lot of attention in the research community recently. However, concurrent data transmission of multiple cells generates intercell and intracell interference. In multicell multiuser MIMO wireless systems, there are several previous studies investigating the quality of service or transmission rate maximization of the beamforming design problem [306–308]. An iteratively weighted sum transmission rate maximization algorithm through the minimum mean square error (MMSE) minimization scheme is proposed to maximize the total sum transmission rates [306]. A sum transmission rates maximization algorithm based on an iterative linear approximation and dirty paper coding is proposed in [307]. An algorithm of sum rate maximization with an outage probability constraint on users is proposed in [308]. There are also some cases considering different scenarios: MIMO broadcast channels and MIMO interference channels [309–312]. However, most of the previous algorithms focus on the sum rate maximization but ignore the QoS of each user, which ignores the user fairness of the wireless network. Also, the computational complexity is high in most

DOI: 10.1201/9781003362142-14

of these algorithms with lots of iterations. On the other hand, channel equalization could improve signal detection without extra bandwidth and power for training in multicell multiuser MIMO wireless systems. Several equalization designs have been proposed for signal detection, such as per-subchannel joint equalization for filter bank multicarrier systems [313], space time semi-blind equalizer design [314], and blind receivers with error propagation for robust detection [315]. For the next generation of wireless communication networks, an optimal equalization design could eliminate channel distortion and ambiguity to achieve optimal signal detection at receivers with less resource consumption.

From the 5G and green communication perspective, lower transmission power and green communication design are important upcoming trends. As the physical antenna components are inexpensive, the increase in number of radio frequency (RF) chains (i.e., mixers, D/A converters, etc.) will eventually increase power consumption and the associated complexity of multiple-antenna transmission. Therefore, current cellular networks result in inexpensive and power-hungry terminals. Recently, a large amount of work has been devoted to energy efficiency and green communication design. For the green energy issue in multicell multiuser MIMO transmission, an energy-efficient maximization beamforming design with the transmission rate constraints of each user is introduced in [316]. In [317], the beamforming design focuses on downlink energy efficiency maximization. Considering general MIMO interference broadcast channels, for example [318–320], a coordinated multicast beamforming power minimization algorithm is proposed in [318]. In [319], the authors propose a power minimization algorithm with gam theory. There are also some power minimization designs for other communication systems, such as MIMO-orthogonal frequency division multiple access wireless systems, nonorthogonal multiple access (NOMA) transmission, and full-duplex radio access networks [321–324]. Thus, we can see that power efficiency is an important issue in multicell multiuser MIMO wireless communication systems.

Generally, a multi-objective optimization problem involves several conflicting objectives. Since multi-objective optimization simultaneously achieves the requirements of multiple objectives, it has been applied to many research fields such as smart grids [325], communication [326–329], estimation [24], and power tracking control [28, 29]. We realize that the simultaneous power consumption minimization problem of multicell multiuser MIMO beamforming systems in different groups of cells is obviously a conflict multi-objective optimization problem and has no unique optimal solution. A multi-objective design could provide a set of Pareto optimal solutions to choose from, which can help the designer deal with an extra requirement, for example, the one with the best minimum mean square error equalization could be chosen as a preferable Pareto optimal solution from a set of Pareto optimal solutions of multi-objective beamforming design in the downlink powers of different groups of cells.

In this chapter, we consider a novel multi-objective power minimization design via downlink beamforming and equalization to achieve the minimum power consumption of different groups of cells, desired SINR constraints, and optimal MMSE equalization simultaneously. We provide the significance and difficulty of this power minimization MOP as follows: (1) power consumption minimization in different groups of cells of a multicell multiuser MIMO beamforming system is a novel and important design concept for energy-efficient communication in the future. With the advance of communication, different requirements are needed for different communication systems. For example, heterogeneous networks consist of macro and micro cells with wireless communication applications such as the Internet of Things and relay-based communication. Different applications of communication require various amounts of power consumption. With multi-objective design, the designer could choose a policy to save on power consumption of multicell multiuser MIMO beamforming systems in different groups of cells simultaneously. Thus, it helps with power scheduling of this energy-limited area in the era of energy-efficient communications. (2) When the number of different applications increases in the next generation of wireless systems, numerous different power consumption requirements will lead to a complicated optimization problem with a large number of conflicting objectives in different time periods. It is difficult to evaluate the weightings for each objective and select a specific Pareto optimal solution in this MOP. The aforementioned reasons motivate the multi-objective power minimization design in multicell multiuser MIMO systems. To the best of our knowledge, the intricate trade-off power minimization problem between different groups of cells has not been studied yet in the next generation of wireless communication systems.

To solve the aforementioned difficult multi-objective power minimization problem, a multi-objective power consumption design is proposed to first specify a set of optimal beamformings in the MOP to simultaneously minimize the power consumption of different groups of cells in the multicell multiuser MIMO network under the desired constraints of the SINR. In order to solve the multi-objective power consumption design problem globally, the original problem is transformed into an equivalent semidefinite programming (SDP)–constrained MOP. Then a convex multi-objective evolutionary algorithm is also proposed to solve the SDP-constrained MOP for the multi-objective power minimization problem of the multicell multiuser MIMO beamforming system. In general, the multi-objective power minimization design problem has no unique optimal solution. A set of Pareto optimal solutions of beamformings could be obtained for the multi-objective power minimization problem by the proposed SDP-constrained MOEA. From the set of Pareto optimal beamforming solutions, a solution with the least MMSE equalization is preferred; that is, the preferred beamforming equalizer could achieve the minimum power, the lowest MMSE equalization, and the SINR constraint simultaneously in the multicell multiuser MIMO beamforming system.

The main purposes of this chapter are summarized as follows:

(1) A novel multi-objective minimum power consumption beamforming design problem of different groups of cells with QoS constraints in a multicell multiuser MIMO network is introduced with imperfect channel state information (CSI).

(2) The original multi-objective power minimization problem is transformed into an equivalent convex SDP-constrained MOP, which can be efficiently solved by the proposed SDP-constrained MOEA with the help of the convex optimization toolbox in MATLAB.

(3) A lowest MMSE equalization is also achieved by a preferred Pareto optimal solution to improve the equalization performance for each user of the multicell multiuser MIMO beamforming system.

This chapter is divided into seven sections as follows: in Section 11.2, the system model of the multicell multiuser MIMO network is introduced. The problem formulation is also described in the same section. In Section 11.3, a multi-objective beamforming optimization for simultaneous minimum powers of different groups of cells of the multicell multiuser MIMO network is introduced and reformulated into a tractable SDP-constrained MOP in the same section. In Section 11.4, the SDP-constrained MOEA is introduced for solving the SDP-constrained MOP, and a design procedure is also proposed. The MMSE equalizer design procedure is also proposed. The MMSE equalizer design of the system is proposed in Section 11.5 as a strategy to select a preferred beamforming solution from the Pareto optimal solutions to achieve the lowest MMSE-equalization error (i.e., to achieve an optimal MMSE equalizer simultaneously). Simulation results are given in Section 11.6 to confirm the performance of the beamforming design for the multi-objective power minimization problem of the multicell multiuser MIMO communications system. Finally, some conclusions are summarized in Section 11.7.

11.2   SYSTEM MODEL

In this chapter, we consider a multicell multiuser MIMO interference broadcast network, as shown in Figure 11.1, that is, the so-called multicell multiuser MIMO communication system. The network is composed of Q separate cells, which operate on the same frequency band. At each cell, a multiple-antenna base station concurrently sends multiple independent data streams to multiple receiver antennas. We assume that the number of antennas in each BS and MS is N_t and N_r, respectively. The number of MS in each cell is K. Let $D = \min(N_r, N_t)$ be the number of data substreams for each MS. At a specific cell q, the transmitted data to the ith MS is $s_{q_i} \in \mathbb{C}^{D \times 1}$. Consequently, the received signal of user i in cell q is shown as

$$y_{q_i} = H_{qq_i} \sum_{j=1}^{K} V_{q_j} s_{q_j} + \underbrace{\sum_{r \neq q}^{Q} \sum_{j=1}^{K} H_{rq_i} V_{r_j} s_{r_j}}_{\text{intercell interference}} + \underbrace{n_{q_i}}_{\text{noise}} \qquad (11.1)$$

for $i = 1,.., K$, where $H_{rq_i} \in \mathbb{C}^{N_r \times N_t}$ is the channel matrix from cell r to user i to cell q. The matrix $V_{q_i} \in \mathbb{C}^{N_t \times D}$ is the beamforming matrix of user i to q. The vector $n_{q_i} \in \mathbb{C}^{N_r \times 1}$ is the additive Gaussian noise vector with the distribution $\mathbb{CN}(0, \delta^2 I)$. Without loss of generality, we assume that $E[s_{q_i} s_{q_i}^H] = I$, which is a basic assumption in multicell multiuser MIMO systems. In contrast to the studies focusing on the sum rate maximization problem, we pay attention to the quality of service of each user. Assume $\Xi = \{1, 2, 3, ..., Q\}$ denotes the set of all cells in the network. We separate the multicell network into N groups of cells $g_1, g_2, ..., g_N$ according to the required lower bound of QoS. In each group, different requirements are set, and we design the beamforming matrices to satisfy different requirements.

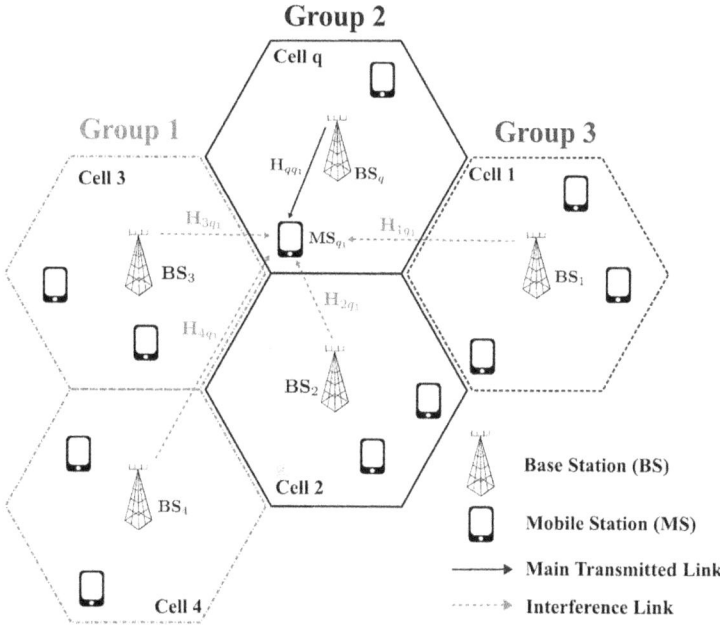

FIGURE 11.1   The scenario of the multicell multiuser MIMO system. In the example, multicell system are divided into three groups. Group 1 contains cells 3 and 4, group 2 contains cells q and 2, and group 3 contains cell 1.

In the real scenario, imperfect CSI should be taken into account, which is mainly due to the channel estimation error, the round trip delay, the measurement error, and so on. In this scenario, the channel uncertainty model is described in the following unbiased estimations:

$$H_{rq_i} = \hat{H}_{rq_i} + \Delta H_{rq_i} \tag{11.2}$$

$$H_{qq_i} = \hat{H}_{qq_i} + \Delta H_{qq_i} \tag{11.3}$$

with the assumption $E[Tr(\Delta H_{rq_i}, \Delta H_{rq_i}^H)] \le \varepsilon_{rq_i}^2$, $\forall (r,i)$ and $E[Tr(\Delta H_{qq_i}, \Delta H_{qq_i}^H)] \le \varepsilon_{qq_i}^2$, and $\forall (q,i)$, where $\varepsilon_{rq_i}^2$ and $\varepsilon_{qq_i}^2$ are the upper bounds of channel uncertainty. Based on the channel error model, the signal-to-interference plus noise ratio $SINR_{q_i}$ is expressed as follows [330]:

$$SINR_{q_i}$$

$$= \frac{E\left[\left\| \left(\hat{H}_{qq_i} + \Delta H_{qq_i} \right) V_{q_i} \right\|_F^2 \right]}{E\left[\sum_{j=1, j \ne i}^{K} \left\| \left(\hat{H}_{qq_i} + \Delta H_{qq_i} \right) V_{q_i} \right\|_F^2 + \sum_{r \ne q}^{Q} \sum_{j=1}^{K} \left\| \left(\hat{H}_{rq_i} + \Delta H_{rq_i} \right) V_{r_j} \right\|_F^2 + Tr(\delta^2 I) \right]} \tag{11.4}$$

Also, the total beamforming power consumption of cell q is denoted as

$$Power_q = \sum_{i=1}^{K} Tr(V_{q_i} V_{q_i}^H) \tag{11.5}$$

For some power shortage region, it is important to design a power-efficient beamform to achieve green communication. In the next section, we will introduce the problem to design the beamform by power minimization with a QoS constraint.

11.3 MULTI-OBJECTIVE POWER MINIMIZATION DESIGN FOR THE MULTICELL MULTIUSER MIMO BEAMFORMING SYSTEM

In this chapter, we consider a multi-objective power minimization design problem with different groups of cells to achieve green design for energy efficiency. In practice, a wireless communication system is composed of smaller distributed systems. This phenomenon is more general in advanced wireless communication (i.e., heterogeneous networks, macro-pico cellular networks, energy harvesting systems, etc.). Thus, it is reasonable to divide wireless communication systems into multiple groups with different requirements. In the proposed multi-objective design, the beamforming power consumption of different groups of cells is minimized simultaneously with the corresponding SINR constraints to achieve the following multi-objective power minimization design problem:

$$SINR_{q_i} \frac{Tr(\widehat{H}_{qq_i}^H \widehat{H}_{qq_i} + \delta_{qq_i}^2 I)Q_{q_i}}{\left[\begin{array}{c} \displaystyle\sum_{j=1, j\neq i}^{K} Tr(\widehat{H}_{qq_i}^H \widehat{H}_{qq_i} + \delta_{qq_i}^2 I)Q_{q_j} \\ + \displaystyle\sum_{r\neq q}^{Q}\sum_{j=1}^{K} Tr(\widehat{H}_{rq_i}^H \widehat{H}_{rq_i} + \delta_{rq_i}^2 I)Q_{r_j} + Tr(\delta^2 I) \end{array} \right]} \tag{11.6}$$

$$\min_{V_{q_i}, \forall(q,i)} \left(\sum_{(q,i)\in g_1} Tr\left(V_{q_i} V_{q_i}^H\right), ..., \sum_{(q,i)\in g_N} Tr\left(V_{q_i} V_{q_i}^H\right) \right) \tag{11.7a}$$

$$\text{subject to } SINR_{q_i} \geq \eta_{n,\min}, \forall(q,i) \in g_n, n = 1, ..., N \tag{11.7b}$$

where $\eta_{n,\min}$ denotes the minimum SINR requirement of cell $q \in g_n$ for $n = 1, ..., N$. To simplify the design procedure, we denote $Q_{q_i} = V_{q_i} V_{q_i}^H$ as the transmit covariance matrix intended for user i in cell q. In this study, all channels are assumed to be quasi-stationary flat fading. The channel uncertainty is assumed to be norm-bounded channel uncertainty. Also, since the channel estimation is unbiased, ΔH_{qq_i} is assumed to be a zero mean with covariance matrix $E[\Delta H_{qq_i}^H \Delta H_{qq_i}] = \delta_{qq_i}^2 I$. The radius of the channel uncertainty is modeled as $\varepsilon_{qq_i}^2 = \tau \delta_{qq_i}^2$, where the constant $\tau > 0$.

With the aforementioned assumption, we can present that $E[\Delta H_{qq_i}^H \widehat{H}_{qq_i}] = 0_{N_t}$, $E[\widehat{H}_{qq_i}^H \Delta H_{qq_i}] = 0_{N_t}$ from the independence between the estimated channel $\hat{H}_{qq_i}$ and ΔH_{qq_i}. Similarly, we can get $E[\Delta H_{qr_i}^H \Delta H_{qr_i}] = \delta_{qr_i}^2 I$, $E[\Delta H_{qr_i}^H \widehat{H}_{qr_i}] = 0_{N_t}$, $E[\widehat{H}_{qr_i}^H \Delta H_{qr_i}] = 0_{N_t}$, and the following equations:

$$E\left[\left\|(\widehat{H}_{qq_i} + \Delta H_{qq_i})V_{q_i}\right\|_F^2\right] = Tr((\widehat{H}_{qq_i}^H \widehat{H}_{qq_i} + \delta_{qq_i}^2 I)Q_{q_i}) \tag{11.8}$$

Thus, the multicell and multiuser interference with channel uncertainties can be expressed as follows:

$$E\left[\sum_{j=1, j \neq i}^{K} \left\|(\widehat{H}_{qq_i} + \Delta H_{qq_i})V_{q_j}\right\|_F^2\right]$$
$$= \sum_{j=1, j \neq i}^{K} Tr((\widehat{H}_{qq_i}^H \widehat{H}_{qq_i} + \delta_{qq_i}^2 I)Q_{q_j}) \tag{11.9}$$

$$E\left[\sum_{r \neq q}^{Q} \sum_{j=1}^{K} \left\|(\widehat{H}_{rq_i} + \Delta H_{rq_i})V_{r_j}\right\|_F^2\right]$$
$$= \sum_{r \neq q}^{Q} \sum_{j=1}^{K} Tr((\widehat{H}_{rq_i}^H \widehat{H}_{rq_i} + \delta_{qq_i}^2 I)Q_{r_j}) \tag{11.10}$$

According to (11.8)–(11.10), we have the multi-objective power minimization problem of the multicell multiuser MIMO beamforming system in (11.7a)–(11.7b) as follows:

$$\min_{Q_{q_i}, \forall (q,i)} \left(\sum_{(q,i) \in g_1} Tr(Q_{q_i}), ..., \sum_{(q,i) \in g_N} Tr(Q_{q_i})\right) \tag{11.11a}$$

$$\text{subject to inequality in (11.7b)} \tag{11.11b}$$

$$\forall (q,i) \in g_n, \forall n, Q_{q_i} \geq 0, \forall (q,i) \tag{11.11c}$$

Remark 11.1: For the original multi-objective power minimization problem of the multicell multiuser MIMO beamforming system in (11.7a)–(11.7b) and the MOP (11.11a)–(11.11c) after semi-definite relaxation (SDR), the quadratic term of the beamforming matrix $V_{q_i} \in \mathbb{C}^{N_t \times D}$ is relaxed to the covariance matrix $Q_{q_i} \in \mathbb{C}^{N_t \times N_t}$ according to the relation $Q_{q_i} = V_{q_i} V_{q_i}^H$. However, it remains to be understood whether (11.11a)–(11.11c) have an equivalent convex reformulation in general. If the obtained solution of Q_{q_i} has a rank higher than D, then it is difficult to recover V_{q_i} from Q_{q_i} without approximation. The simple recovery method is the use of singular value decomposition (SVD) to Q_{q_i}, which has a conservative solution with D higher singular values. At present, existing approximation methods can be used to recover the matrix with lower rank, such as the related research for the rank-constrained SDP [331–333].

In semidefinite programming, the constraint (11.11c) guarantees positive semi-definiteness due to the quadratic relation $Q_{q_i} = V_{q_i} V_{q_i}^H$. After some transformation, we can get the following multi-objective optimization problem of power minimization of the multicell multiuser MIMO system:

$$\min_{Q_{q_i}, \forall (q,i)} \left(\sum_{(q,i)\in g_1} Tr(Q_{q_i}), ..., \sum_{(q,i)\in g_N} Tr(Q_{q_i}) \right) \tag{11.12}$$

subject to $Tr(\widehat{H}_{qq_i}^H \widehat{H}_{qq_i} + \delta_{qq_i}^2 I)Q_{q_i}$

$$-\eta_{n,\min} \left(\sum_{j=1, j\neq i}^{K} Tr(\widehat{H}_{qq_i}^H \widehat{H}_{qq_i} + \delta_{qq_i}^2 I)Q_{q_j} \right.$$

$$\left. + \sum_{r\neq q}^{Q} \sum_{j=1}^{K} Tr(\widehat{H}_{rq_i}^H \widehat{H}_{rq_i} + \delta_{rq_i}^2 I)Q_{r_j} + Tr(\delta^2 I) \right) \geq 0 \tag{11.13}$$

$$\forall (q,i) \in g_n, \forall n, Q_{q_i} \geq 0, \forall (q,i)$$

In this chapter, the multi-objective power minimization design problem in (11.12)–(11.13) for the multicell multiuser MIMO beamforming system is to design the beamformings for different groups of cells to minimize their power consumption under the requirement of minimum SINR simultaneously. However, it is difficult to directly solve the multi-objective beamforming design problem in (11.12)–(11.13). Therefore, an indirect suboptimal method is proposed to solve the MOP in (11.12)–(11.13).

Assume α_n to be the upper bound of power consumption $\sum_{(q,i)\in g_n} Tr(Q_{q_i})$ of the nth group of cells. The MOP (11.12)–(11.13) of power minimization in the multicell multiuser MIMO system can be transformed into the following MOP:

$$\min_{Q_{q_i}, \forall (q,i)} \left(\alpha_1, ..., \alpha_n, ..., \alpha_N \right) \tag{11.14a}$$

subject to $\sum_{(q,i)\in g_n} Tr(Q_{q_i}) \leq \alpha_n, n = 1, ..., N$

$$Tr(\widehat{H}_{qq_i}^H \widehat{H}_{qq_i} + \delta_{qq_i}^2 I)Q_{q_i} - \eta_{n,\min}$$

$$\times \left(\sum_{j=1, j\neq i}^{K} Tr(\widehat{H}_{qq_i}^H \widehat{H}_{qq_i} + \delta_{qq_i}^2 I)Q_{q_j} \right.$$

$$+ \sum_{r\neq q}^{Q} \sum_{j=1}^{K} Tr(\widehat{H}_{rq_i}^H \widehat{H}_{rq_i} + \delta_{rq_i}^2 I)Q_{r_j} \tag{11.14b}$$

$$\left. + Tr(\delta^2 I) \right) \geq 0$$

$$\forall (q,i) \in g_n, n = 1, ..., N, Q_{q_i} \geq 0, \forall (q,i)$$

In general, the optimal solution of the MOP in (11.14a)–(11.14b) is not unique. Γ is the feasible solution set of the MOP in (11.14a)–(11.14b). We will show that the indirect MOP (11.14a)–(11.14b) is equivalent to the MOP (11.12)–(11.13) in the Pareto optimal sense.

Theorem 11.1: The indirect MOP in (11.14a)–(11.14b) is equivalent to the MOP in (11.12)–(11.13) when the Pareto optimal solution is achieved.

Proof: Please refer to the Appendix 11.8.1.

Next, we will show that the proposed MOP in (11.14a)–(11.14b) is a multi-objective convex optimization problem, which is also a semidefinite programming–constrained MOP. According to Jensen's inequality and the definition of the convex function in [34, 205, 334], we can derive the following results.

Remark 11.2: To show that the MOP in (11.14a)–(11.14b) is a convex MOP, the feasible set should become a convex set. Also, the objective and constraint functions must be convex functions [335]. First, the convex property of the feasible set of the MOP in (11.14a)–(11.14b) is linear functions by Jensen's inequality, which is well known as semidefinite programming constraints [335]. Furthermore, we know that the upper bound reformulation can reserve the convex property. Consequently, the MOP (11.14a)–(11.14b) is a convex MOP.

Remark 11.3: If the MOP to minimize power consumption under the power budget constraints of each cell is also considered for $n = 1,...,N$, the original MOP for the multicell multiuser MIMO beamforming system can be revised as the following MOP:

$$\min_{Q_{q_i}, \forall (q,i)} \left(\alpha_1, ..., \alpha_n, ..., \alpha_N \right)$$

$$\text{subject to } \sum_{(q,i) \in g_n} Tr(Q_{q_i}) \le \alpha_n, \sum_{i=1}^{K} Tr(Q_{q_i}) \le P_q, \forall q,$$

$$Tr(\widehat{H}_{qq_i}^H \widehat{H}_{qq_i} + \delta_{qq_i}^2 I) Q_{q_i} - \eta_{n,min}$$

$$\times \left(\sum_{j=1, j \ne i}^{K} Tr(\widehat{H}_{qq_i}^H \widehat{H}_{qq_i} + \delta_{qq_i}^2 I) Q_{q_j} \right. \tag{11.15}$$

$$+ \sum_{r \ne q}^{Q} \sum_{j=1}^{K} Tr(\widehat{H}_{rq_i}^H \widehat{H}_{rq_i} + \delta_{rq_i}^2 I) Q_{r_j}$$

$$\left. + Tr(\delta^2 I) \right) \ge 0$$

$$\forall (q,i) \in g_n, n = 1,...,N, Q_{q_i} \ge 0, \forall (q,i)$$

where P_q is the power budget constraint for BS q.

Remark 11.4: If the single-objective optimization problem to minimize the power consumption for a specific group of cells and the power budget constraints of each cell are also considered for $n = 1,...,N$, the original MOP for the multicell multiuser MIMO beamforming system can be revised as the following SOP:

$$\alpha_n^{*S} = \min_{Q_{q_i}, \forall (q,i)} \alpha_n^{S}$$

subject to $\sum_{(q,i)\in g_n} Tr(Q_{q_i}) \le \alpha_n^{S}, \sum_{(q,i)\in g_n} Tr(Q_{q_i}) \le P_{\max},$

$$\sum_{i=1}^{K} Tr(Q_{q_i}) \le P_q, \forall q,$$

$$\sum_{(q,i)\in g_n} Tr(Q_{q_i}) \le \alpha_n^{S}, \sum_{i=1}^{K} Tr(Q_{q_i}) \le P_q, \forall q,$$

$$Tr(\widehat{H}_{qq_i}^{H} \widehat{H}_{qq_i} + \delta_{qq_i}^{2} I)Q_{q_i} - \eta_{n,\min}$$ (11.16)

$$\times \left(\sum_{j=1,j\ne i}^{K} Tr(\widehat{H}_{qq_i}^{H} \widehat{H}_{qq_i} + \delta_{qq_i}^{2} I)Q_{q_j} \right.$$

$$+ \sum_{r\ne q}^{Q} \sum_{j=1}^{K} Tr(\widehat{H}_{rq_i}^{H} \widehat{H}_{rq_i} + \delta_{rq_i}^{2} I)Q_{r_j}$$

$$\left. + Tr(\delta^2 I) \right) \ge 0$$

$$\forall (q,i) \in g, Q_{q_i} \ge 0, \forall (q,i)$$

for $n = 1,...,N$, where P_q is the power budget constraint for BS P_q.

Remark 11.5: If only the power of one group of cells is to be minimized and the powers of other groups of cell are under their power budget constraints, the multi-objective optimization problem can be reduced to a single-objective optimization problem by fixing the other objectives with prescribed power budget constraints; that is, if we want to minimize the downlink power of group g_n, then the power minimization beamforming design problem for each group of cells becomes the SOP in (11.17) with $n = 1,...,N$. Normally, we have $\alpha_1^* \ge \alpha_1, \alpha_2^* \ge \alpha_2, ..., \alpha_N^* \ge \alpha_N$ of the Pareto optimal solutions $(\alpha_1^*,...,\alpha_N^*)$ for the MOP in (11.14a)–(11.14b). $(\underline{\alpha_1},...,\underline{\alpha_N})$ can be used as the lower bound of $(\alpha_1^*,...,\alpha_N^*)$ for the proposed SDP-constrained MOEA, respectively.

$$\underline{\alpha_n} = \min_{Q_{q_i}, \forall (q,i)} \alpha_n$$

subject to $\sum_{(q,i)\in g_n} Tr(Q_{q_i}) \le \alpha_n, \sum_{(q,i)\in g_m} Tr(Q_{q_i}) \le P_{\max}, \forall m \ne n,$

$$Tr(\widehat{H}_{qq_i}^{H} \widehat{H}_{qq_i} + \delta_{qq_i}^{2} I)Q_{q_i} - \eta_{n,\min}$$

$$\times \left(\sum_{j=1,j\ne i}^{K} Tr(\widehat{H}_{qq_i}^{H} \widehat{H}_{qq_i} + \delta_{qq_i}^{2} I)Q_{q_j} \right.$$ (11.17)

$$+ \sum_{r\ne q}^{Q} \sum_{j=1}^{K} Tr(\widehat{H}_{rq_i}^{H} \widehat{H}_{rq_i} + \delta_{rq_i}^{2} I)Q_{r_j}$$

$$\left. + Tr(\delta^2 I) \right) \ge 0$$

$$\forall (q,i) \in g, n = 1,...,N, Q_{q_i} \ge 0, \forall (q,i)$$

Once the MOP in (11.14a)–(11.14b) is solved, the corresponding precoding matrix $V_{q_i}^*$ can be recovered by using the Cholesky decomposition (Eigenvalue decomposition from $Q_{q_i}^*$) or the approximation method [331]. As a result, the precoding matrices can be computed as $V_{q_i}^* = L_{q_i}^*$ (note that $Q_{q_i}^* = L_{q_i}^* (L_{q_i}^*)^H$).

11.4 SDP-CONSTRAINED MOEA FOR MULTI-OBJECTIVE POWER MINIMIZATION BEAMFORMING DESIGN

The multi-objective evolutionary algorithm is a stochastic optimization algorithm. There are two advantages of the multi-objective evolutionary algorithm [205]. In contrast to conventional stochastic process, the multi-objective evolutionary algorithm enables finding a set of the Pareto optimal solutions in a single run instead of performing a series of conventionally separate runs by mimicking nature's evolutionary principles to drive its search towards optimal solutions. It reduces the considerable computational complexity. Besides, the multi-objective evolutionary algorithm can parallel search the Pareto optimal solutions globally and escape from the local optima. Second, the MOEA requires very little knowledge about the problem and is less susceptible to the shape or continuity of the Pareto front. As a result, the MOEA is easy to implement and apply to lots of multi-objective optimization problems. Thus, it is very suitable to solve the multi-objective beamforming design problem in (11.14a)–(11.14b). Moreover, according to Definition 11.5, the multi-objective optimization problem of the multicell multiuser MIMO system can guarantee global convergence by its convex constraints. In general, the conventional MOEA encodes the variables Q_{q_i} to obtain the Pareto optimal solutions of the MOP in (11.14a)–(11.14b). However, it is not easy to obtain Pareto optimal solutions by the conventional MOEA when the number of variables becomes larger. In this section, we will provide an SDP-constrained MOEA to search $(\alpha_1^*, ..., \alpha_N^*)$ instead to help us obtain the Pareto optimal solutions efficiently. Before further discussion, we have to introduce the following definitions for solving the MOP in (11.14a)–(11.14b).

Definition 11.1 (Pareto Dominance [205, 334]): Let $[Q_{11}^1, ..., Q_{QK}^1] \in \Gamma$ and $[Q_{11}^2, ..., Q_{QK}^2] \in \Gamma$ be the feasible solution of the MOP in (11.14a)–(11.14b) corresponding to the objective values $(\alpha_1^1, ..., \alpha_n^1, ..., \alpha_N^1)$ and $(\alpha_1^2, ..., \alpha_n^2, ..., \alpha_N^2)$, respectively. The solution $[Q_{11}^1, ..., Q_{QK}^1]$ is said to dominate $[Q_{11}^2, ..., Q_{QK}^2]$ if $\alpha_1^1 \le \alpha_1^2, ..., \alpha_n^1 \le \alpha_n^2, ..., \alpha_N^1 \le \alpha_N^2$ and at least one inequality is a strict inequality.

Definition 11.2 (Pareto Optimality [205, 334]): A solution $[Q_{11}^*, ..., Q_{QK}^*] \in \Gamma$ with the set of objective value $(\alpha_1^*, ..., \alpha_n^*, ..., \alpha_N^*)$ is said to be a Pareto optimal solution if and only if there does not exist another feasible solution that dominates it.

Definition 11.3 (Pareto Optimal Solution Set [205, 334]): For the given MOP (11.14a)–(11.14b) of power minimization in the multicell multiuser MIMO system, the Pareto optimal solution set ρ^* is defined as follows:

$$\rho^* = \left\{ \left[Q_{11}^*, ..., Q_{QK}^* \right] \in \Gamma \middle| \left[Q_{11}^*, ..., Q_{QK}^* \right] \text{ is Pareto optimal} \right\}$$

Definition 11.4 (Pareto Front [205, 334]): For the given MOP in (11.14a)–(11.14b) of power minimization in the multicell multiuser MIMO beamforming system, the Pareto front f^* is defined as

$$f^* = \{(\alpha_1^*,...,\alpha_n^*,...,\alpha_N^*)\big|[Q_{11}^*,...,Q_{QK}^*] \in \rho^*\}$$

Definition 11.5 (Multi-Objective Convex Optimization Problem [205, 334]): An MOP with the following form is convex if all the objective functions $f_i(x)$, $i = 1,...,M$ are convex and the feasible region is convex (or all inequality functions $g_j(x)$, $j = 1,...,J$ are convex and equality functions $h_k(x)$, $k = 1,...,K$ are linear).

$$\min_x(f_1(x),...,f_m(x),...,f_M(x))$$

subject to the following constraints:

$$g_j(x) \leq 0, j = 1,...,J$$
$$h_k(x) = 0, k = 1,...,K$$

It is clear that the proposed SDP-constrained MOP in (11.14a)–(11.14b) satisfies all of the properties in Definitions 11.1–11.5. The first step of the proposed MOEA is to specify the search region, which is a cuboid region including the ideal point as the lower bound and the power budget P_{max} as the upper bound. The definition of the ideal point is as follows.

Definition 11.6: The ideal point for the SDP-constrained MOP in (11.14a)–(11.14b) is defined as $(\alpha_1,...,\alpha_N)$, where α_1 is the single-objective problem solution in Remark 11.5; that is, α_1 is obtained by the single-objective problem in (11.17) with $n = 1$, and so on.

The proposed MOEA first initializes a number of objective vectors $(\alpha_1,...,\alpha_N)$ from the search region. If a feasible objective vector is obtained, it will be encoded into a corresponding feasible population C_i. Note that the search region of the traditional MOEA is a variable set (i.e., the proposed method is different from the traditional MOEA). In order to guarantee diversity and robustness to obtain the Pareto optimal solutions of the MOP in (11.14a)–(11.14b), the nondominated sorting method and crowding distance sorting are used before the evolution algorithm. As shown in Figure 11.2, after the EA generates child individuals by natural selection, crossover, and mutation, each child individual should be checked for whether it satisfies the SDP constraints of the MOP in (11.14a)–(11.14b). The feasible individuals will survive as candidate individuals. Otherwise, they will be deleted from the candidate sets. The current parent front will approach the actual Pareto front in the feasible region by the nondominated sorting method, crowding distance sorting, and the evolutionary algorithm, as shown in Figure 11.3. We can use MATLAB interior point solver CVX [336] to check the feasibility of the SDP constraints. As shown in Figure 11.2, the design procedure of the proposed SDP-constrained MOEA for solving the MOP in (11.14a)–(11.14b) of the multicell multiuser MIMO system is given as follows.

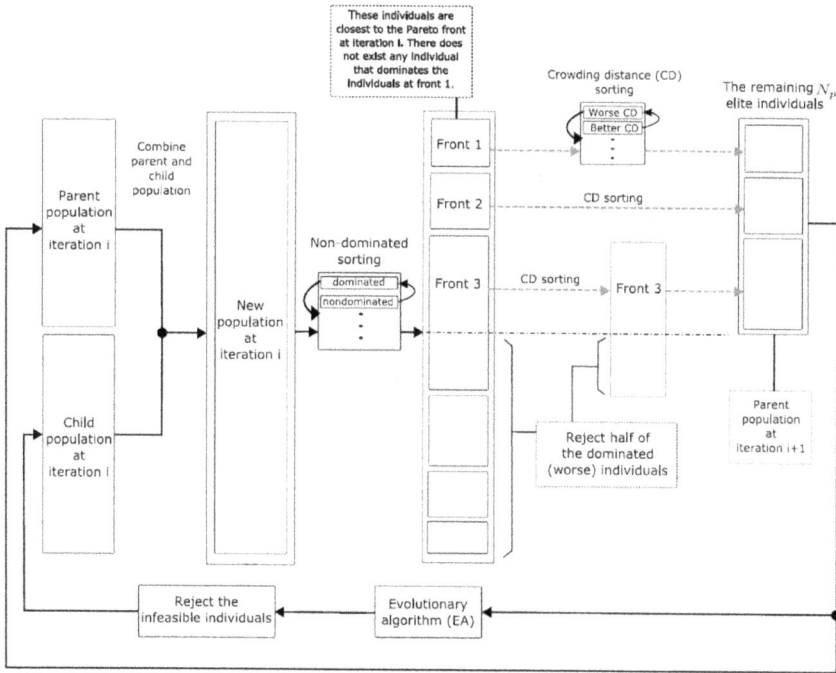

FIGURE 11.2 The procedure of the proposed SDP-constrained MOEA. The elite procedure will select half of the better individuals and reject the remaining dominated individuals in the original population.

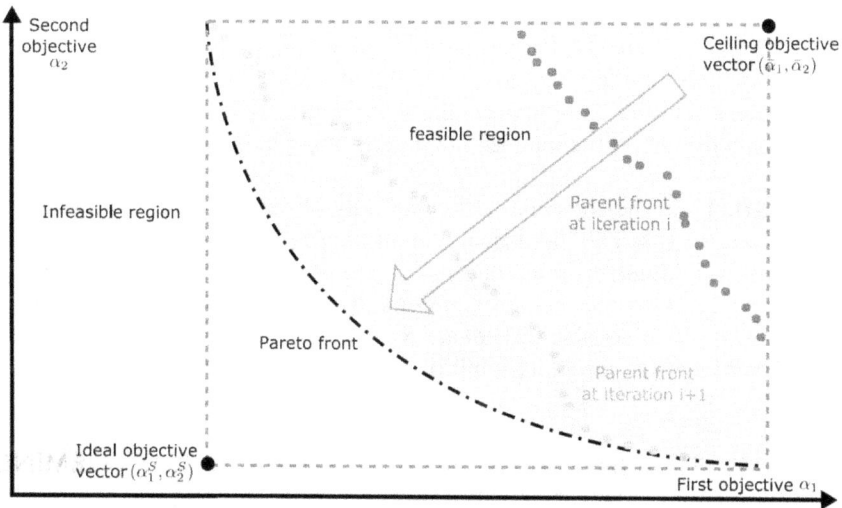

FIGURE 11.3 A schematic diagram of the proposed SDP-constrained MOEA for the simple example with two objectives. The current parent front will approach the actual Pareto front in the feasible region.

A. The Design Procedure of the Proposed SDP-Constrained MOEA for the MOP in (11.14a)–(11.14b)

(1) Solve the single objective optimization problem in (11.17) with $n = 1,...,n,...,N$ to specify the search region $(\underline{\alpha_1},...,\underline{\alpha_n},...,\underline{\alpha_N}) \times (\overline{\alpha_1},...,\overline{\alpha_n},...,\overline{\alpha_N})$, where $\overline{\alpha_n}$ is the power budget P_{max} for all $\overline{n}$.

(2) Set the generation number N_i, the population size N_p, the crossover probability P_c, and the mutation probability P_m.

(3) Initialize the population by randomly generating N_p feasible individuals as the first parent population in the search region; that is, satisfy the constraints in (11.14a)–(11.14b). Set the iteration index $i = 1$.

(4) Generate the child population by the parent population through the evolutionary algorithm.

(5) Combine the parent population with the generated child population to form a new population of $(\alpha_1,...,\alpha_n,...,\alpha_N)$ as the next iteration input.

(6) Assign a nondomination level to each individual $(\alpha_1,...,\alpha_n,...,\alpha_N)$ within the population. Notice that individuals assigned the same nondomination level do not dominate each other.

(7) Sort the initial population based on their nondomination levels. Each solution pair is assigned a fitness (rank) corresponding to its nondomination level. All the individuals in the same rank are assigned crowding distance (CD), respectively.

(8) Select half of the elite individuals according to nondomination level and CD by the NSGA-II algorithm [2], as shown in Figure 11.2. Set the iteration index $i = i + 1$.

(9) Repeat Steps 4 to 8 until the iteration index i reaching the generation number N_i is satisfied. We can get the Pareto front $f^* = F_{N_i}$, where F_{N_i} denotes the Pareto front obtained in the N_ith generation.

(10) Select the preferable Pareto optimal solution $(\alpha_1^*,...,\alpha_n^*,...,\alpha_N^*) \in f^*$ according to the designer's preference. Once the preferable Pareto optimal solution is selected, the solution of the beamforming $V_{q_i}^*, \forall(q,i)$ can be derived with the help of the semidefinite programming toolbox.

Remark 11.6: The computational complexity of the proposed SDP-constrained MOEA refers to [2, 337]. To acquire the complexity, we know that the number of objectives is N (different groups of cells) and the population size is N_p. For the MOP in (11.14a)–(11.14b), the complexity depends on the row size of inequalities and the number of decision variables. There are $N + QK + N_t QK$ rows and $N_t^2 QK$ decision variables. As a result, the total complexity is $O(NN_p^2(N + QK + N_t QK)(N_t^2 QK)^3)$.

11.5 MULTI-OBJECTIVE POWER MINIMIZATION BEAMFORMING DESIGN WITH THE BEST MMSE EQUALIZATION

For the SDP-constrained MOP in (11.14a)–(11.14b), we would like to find the preferable Pareto optimal solution in the Pareto front. In this chapter, we suggest finding the preferable optimal beamforming by evaluating the minimum mean square error equalization errors of the Pareto optimal solutions in the Pareto front and selecting a corresponding beamforming with the best MMSE equalization. After obtaining a set of optimal beamforming matrixes by MOP in (11.14a)–(11.14b), we evaluate their MMSE equalization errors for all Pareto optimal solutions of the MOP and select an optimal beamforming with the best MMSE equalization. According to the received signal of user i in cell q, the estimated signal $\hat{y}_{q_i}$ after equalization can be expressed as:

$$
\begin{aligned}
\hat{y}_{q_i} &= U_{q_i}^H y_{q_i} \\
&= U_{q_i}^H \left(H_{qq_i} \sum_{j=1}^{K} V_{q_j} s_{q_j} + \sum_{r\neq q}^{Q} \sum_{j=1}^{K} H_{rq_i} V_{r_j} s_{r_j} + n_{q_i} \right)
\end{aligned}
\tag{11.18}
$$

where $U_{q_i}^H \in \mathbb{C}^{D \times N_r}$ is the equalizer matrix for user i in cell q. According to the standard minimum MSE formulation, the objective of equalization is to minimize the sum mean square error (sum MSE). According to (11.1) to (11.18) with some rearrangement, we consider solving the MMSE equalization problem with a given Pareto optimal beamforming matrix $V_{q_i}^*$ as follows [306, 311, 337]:

$$
e^* = \min_{U_{q_i}, \forall (q,i)} E \left[\sum_{q=1}^{Q} \sum_{i=1}^{K} Tr(MSE_{q_i}) \right]
\tag{11.19}
$$

where

$$
\begin{aligned}
MSE_{q_i} &= \left(U_{q_i}^H \hat{H}_{qq_i} V_{q_i}^* - I \right)\left(U_{q_i}^H \hat{H}_{qq_i} V_{q_i}^* - I \right)^H \\
&+ U_{q_i}^H R_{q_i} U_{q_i}, \forall (q,i)
\end{aligned}
\tag{11.20}
$$

$$
\begin{aligned}
R_{q_i} &= \sum_{j=1, j\neq i}^{K} \hat{H}_{qq_i} V_{q_j} V_{q_j}^H \hat{H}_{qq_i}^H \\
&+ \sum_{r\neq q}^{Q} \sum_{j=1}^{K} \hat{H}_{rq_i} V_{r_j} V_{r_j}^H \hat{H}_{rq_i}^H + \delta^2 I, \forall (q,i)
\end{aligned}
\tag{11.21}
$$

and the equalizers U_{q_i} are employed to achieve the minimization of the sum MSE in (11.19). According to previous research [306, 311, 337], for the received signal in (11.1) and the given Pareto optimal solution of beamforming $V_{q_i}^*, \forall (q,i)$, the MMSE equalization problem in (11.19)–(11.21) could be solved by the following equations:

$$
U_{q_i}^* = (\hat{H}_{qq_i} V_{q_i}^* (V_{q_i}^*)^H \hat{H}_{qq_i}^H + R_{q_i})^{-1} \hat{H}_{qq_i} V_{q_i}^*, \forall (q,i)
\tag{11.21}
$$

$$
\begin{aligned}
e^* &= \sum_{q=1}^{Q} \sum_{i=1}^{K} Tr(((U_{q_i}^*)^H \hat{H}_{qq_i} V_{q_i}^* - I)((U_{q_i}^*)^H \hat{H}_{qq_i} V_{q_i}^* - I)^H \\
&+ (U_{q_i}^*)^H R_{q_i} U_{q_i}^*), \forall (q,i)
\end{aligned}
\tag{11.22}
$$

It is obvious that the optimal equalizers $U_{q_i}^*$ and the MMSE equalization error e^* depend on the beamforming solution $V_{q_i}^*$. Thus, we can choose a beamforming solution $V_{q_i}^*$ with a least sum-MSE e^* in (11.14a)–(11.14b) from the Pareto front as the preferable Pareto solution to achieve multi-objective power minimization and optimal MMSE equalization simultaneously.

11.6 SIMULATION EXAMPLE

This section presents the simulation results of the proposed multi-objective power minimization problem in a multicell multiuser MIMO beamforming system for energy-efficient design. We consider a multiuser MIMO interference broadcast channel network as indicated in (11.1) and simulate with the channel coefficients generated from the complex Gaussian distribution $\mathbb{CN}(0,1)$, which means all the channels are set as independent and identically distributed (i.i.d.) Rayleigh fading. The transmitted data symbols are assumed to be modulated using quadrature phase shift keying (QPSK) with unit-variance power. The system parameters are summarized in Table 11.1.

We show the simulation scenario as a simple diagram in Figure 11.4. The channel uncertainties are set the same as the distribution $\delta_{rq_i}^2 = \delta_e^2 = 10^{-8}$, $\forall r,q,i$ except for the case in Figure 11.9, in which the performance varies with different uncertainty levels. The additive Gaussian noise vector is generated by the distribution $\mathbb{CN}(0,\delta^2 I)$, where δ^2 changes with the signal-to-noise ratio (SNR). The SINR lower bound is $\eta_{n,\min} = P_{\max} = 15dB$ for each group of cells. By the proposed SDP-constrained MOEA, we can efficiently solve the MOP in (11.14a)–(11.14b) to obtain the Pareto optimal solutions in the Pareto front, as shown in Figure 11.5(a), in a single run. To compare the proposed Pareto solution with the least MMSE equalization and min-max fairness scheme, we consider the following Pareto optimal solutions within different regions of interest in the Pareto front. Pareto solution 1 is the knee point (minimum total transmittal power consumption) of the Pareto front, where the corresponding return of investments (RoI) is denoted as Region 1 in Figure 11.6. Pareto solution 2 tends to the minimum transmittal power consumption in the first group of cells, where the corresponding RoI is denoted as Region 2 in Figure 11.6. Pareto solution 3 tends to the minimum transmittal power consumption in the second

---

TABLE 11.1

The System Parameters in the Simulation

System bandwidth	**200 KHz**
Number of transmitting antennas	$N_t = 18$
Number of receiving antennas	$N_r = 2$
Number of groups	$N = 3$
Number of data streams	$D = 1$
Number of cells	$Q = 4$

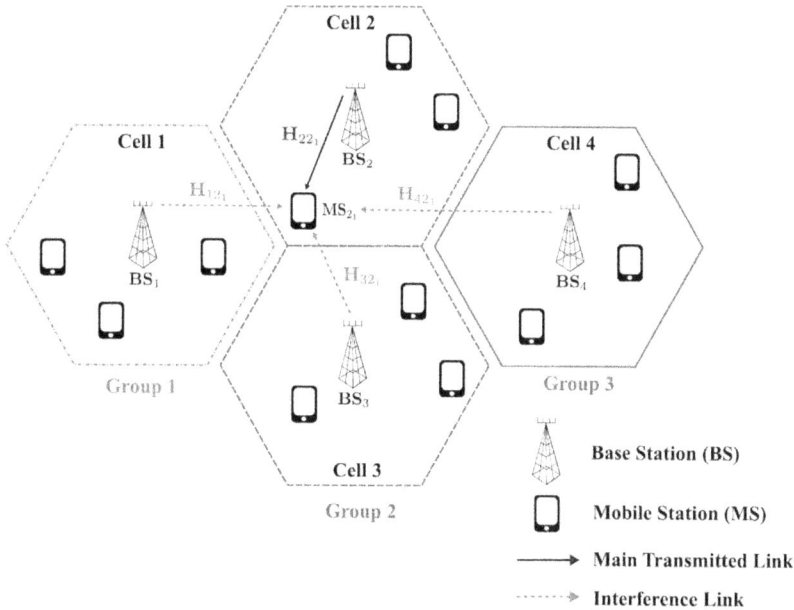

FIGURE 11.4 The scenario of the simulation example. Group 1 contains cell 1, group 2 contains cells 2 and 3, and group 3 contains cell 4. Note that the diagram only displays the main signal link of the first MS in cell 2 and the interference from other cells to the first MS in cell 2.

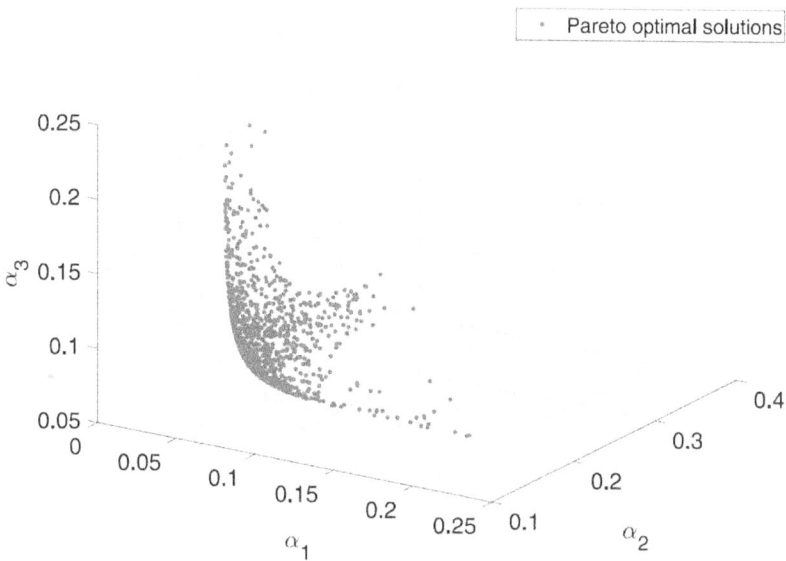

FIGURE 11.5 The Pareto fronts generated by the proposed SDP-constrained MOEA and the weighted sum method for single channel realization in SNR = 10dB. (a) The Pareto front generated by the proposed SDP-constrained MOEA with 50 generations and 1000 populations. (b) The Pareto front generated by the weighted sum method with 1000 solutions.

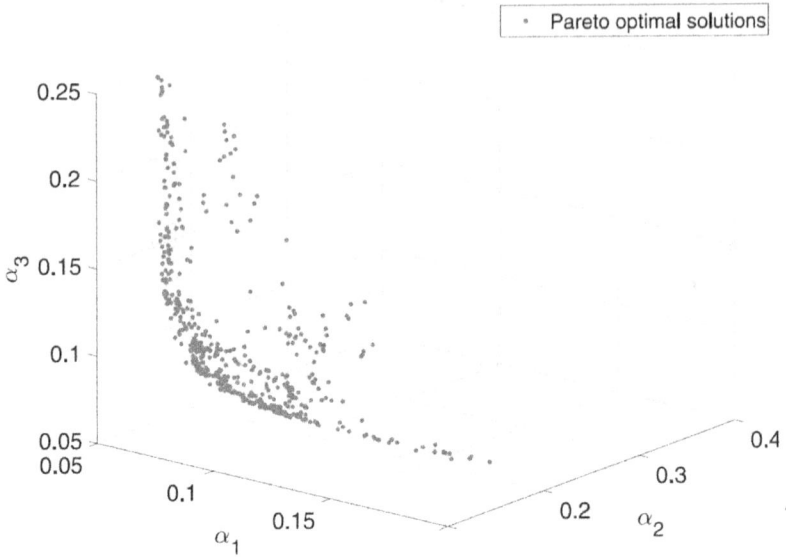

FIGURE 11.5   (Continued)

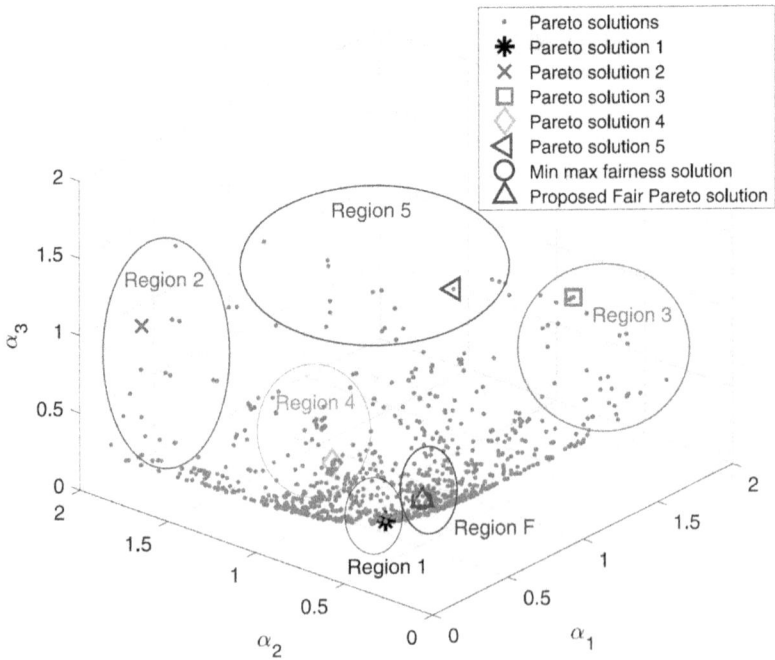

FIGURE 11.6   The Pareto front with 25 generations and 1000 populations in SNR = 10dB. Pareto solutions 1 to 5, the proposed fair Pareto solution, and the min-max fairness solution are also included in this figure with different return of investments (RoIs). The SINR constraints are set the same as $\eta_{n,\min} = 10dB$, $\forall n = 1,...,N$.

group of cells, where the corresponding RoI is denoted as Region 3 in Figure 11.6. Pareto solution 4 tends to the minimum transmittal power consumption in the third group of cells, where the corresponding RoI is denoted as Region 4 in Figure 11.6. Pareto solution 5 is the proposed preferable Pareto solution, which tends to the lowest MMSE equalization of the multicell multiuser MIMO beamforming system. The RoI of Pareto solution 5 is shown as Region 5 in Figure 11.6. Finally, the proposed fair Pareto optimal solution tends to the minimum power consumption gap (i.e., $\min \sqrt{(\alpha_1^* - \alpha_2^*)^2 + (\alpha_2^* - \alpha_3^*)^2 + (\alpha_3^* - \alpha_1^*)^2}$) across the three groups of cells, where the corresponding RoI is the same as the max-min fairness scheme. According to the general weighted-sum method [338], we compared the proposed SDP-constrained MOEA with the weighted sum optimization as follows:

$$\min_{Q_{q_i}, \forall (q,i)} W_1 \alpha_1 + W_2 \alpha_2 + \cdots + W_N \alpha_N \tag{11.24a}$$

$$\text{subject to the SDP constraints in } (11.12b) - (11.12c) \tag{11.24b}$$

where the positive weightings W_n, $\forall n = 1,...,N$ are randomly selected from the interval $[0,1]$ and satisfied with the weighting condition $\sum_{n=1}^{N} W_n = 1$. To generate the Pareto front, the solution number for the weighted-sum method is 1000, which is the same as the population number of the proposed SDP-constrained MOEA.

In Figures 11.5(a) and 11.5(b), we show the Pareto front obtained by the proposed SDP-constrained MOEA and the weighted sum method, respectively. It is obvious that the Pareto front obtained by the weighted sum method is more fragmented and unevenly distributed than the proposed method. The proposed SDP-constrained MOEA can generate the complete Pareto front with good dispersion and smooth curves, which is more suitable to find the possible preferable Pareto optimal solution for the designer. Also, the proposed SDP-constrained MOEA could solve the Pareto front in a single run. The weighted sum method of multi-objectives for the Pareto font needs to consider the optimization problems of all possible weightings of the objectives. It will become very complicated when the number of objectives becomes larger.

Although the min-max fairness scheme has its limitations for our multi-objective optimization design, we still can provide a comparison between the optimal solution of the min-max fairness scheme and the Pareto optimal solutions of the proposed MOEA. According to the min-max fairness scheme, we compared the proposed Pareto optimal solution with the min-max optimization as follows:

$$\min_{Q_{q_i}, \forall (q,i)} \max_n \sum_{(q,i) \in g_n} Tr(Q_{q_i}) \tag{11.25a}$$

$$\text{subject to the SDP constraints in } (11.13) \tag{11.25b}$$

where the corresponding RoI of (11.25a)–(11.25b) in the Pareto front is "Region F", as shown in Figure 11.6. In the following, we will compare the power consumption performance with the different RoIs between the aforementioned six Pareto solutions with the max-min fairness solution.

TABLE 11.2

The Corresponding Objective Values with Single-Channel Realization

Pareto Solutions	Objective Value α_1	Objective Value α_2	Objective Value α_3
Pareto solution 1	0.2925	0.5363	0.2538
Pareto solution 2	0.1012	1.7296	1.1227
Pareto solution 3	1.2851	0.2944	1.3065
Pareto solution 4	0.7717	1.2235	0.1136
Pareto solution 5	0.5308	0.3307	1.7147
Min-max fairness solution	0.3929	0.3929	0.3929
Proposed fair Pareto solution	0.3929	0.3931	0.3930

In Figure 11.6, we provide Pareto solutions 1 to 5, the proposed fair Pareto optimal solution, and the min-max fairness solution in the Pareto front with the different RoIs. The corresponding objective values are shown in Table 11.2. It is obvious that both the min-max fairness solution and the proposed fair Pareto solution can provide almost the same fair power consumption design across different groups of cells. For power consumption fairness design, the RoI of the MOP is Region F, as shown in Figure 11.6. However, if the RoI of the designer tends to the minimum power consumption of a specific group of cells (i.e., Region 2, 3, or 4 in Figure 11.6), then the min-max fairness solution and the proposed fair Pareto solution are not the best choices in this condition. On the contrary, Pareto solutions 2, 3, and 4 will be suitable candidates for Regions 2, 3, and 4.

For comparison with other design methods, our results are compared with (1) the distributed block diagonalization (i.e., distributed BD) [339], which is modified by considering the same SINR constraints in [319–321], and (2) the iteratively weighted minimum mean square error (i.e., iterative weighted MMSE) algorithm for the sum-utility maximization [306]. Both algorithms are designed for the same scenario (i.e., the MIMO interference broadcast channel), as mentioned. In the following, we consider $N_i = 50$ generations and $N_p = 20$ populations of MOEA with the average over 100 channel realizations to show the advantage and disadvantage of the proposed algorithm with respect to other algorithms.

11.6.1 Comparison of Power Consumption in Each Group

The power consumption minimization of the beamforming design under the required SINR is the key performance area of the proposed multi-objective beamforming design. In Figures 11.7 and 11.8, we can observe that the proposed multi-objective power consumption minimization method for each group is exactly effective to achieve their power minimization purposes. We transform the power into decibels to get a better view of the figures. The iterative weighted MMSE algorithm [306] could maximize the total sum transmission rates of all cells. Also, its power budget is fixed with respect to the SNR value. As a result, the power consumption of the weighted MMSE algorithm increases as the SNR increases. The distributed BD

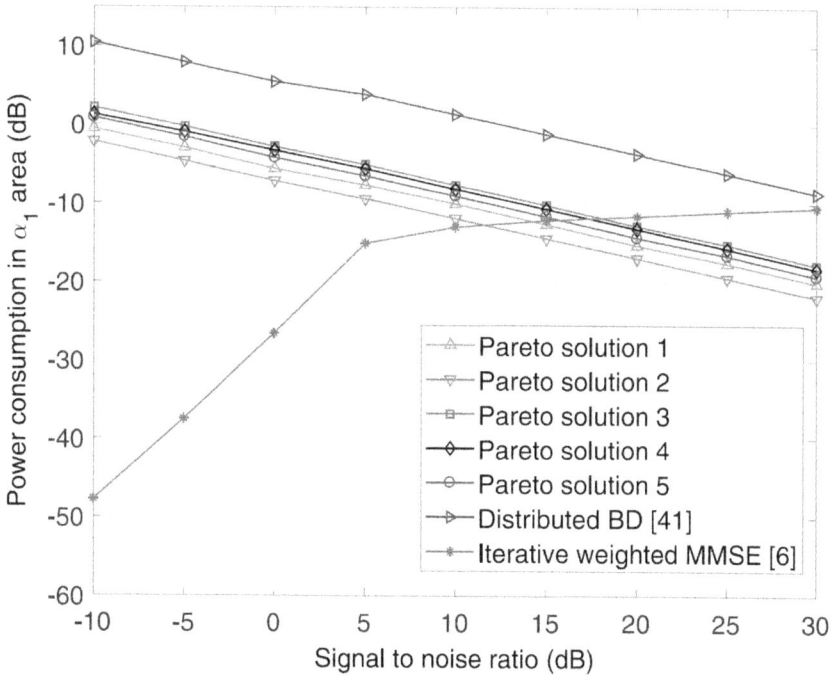

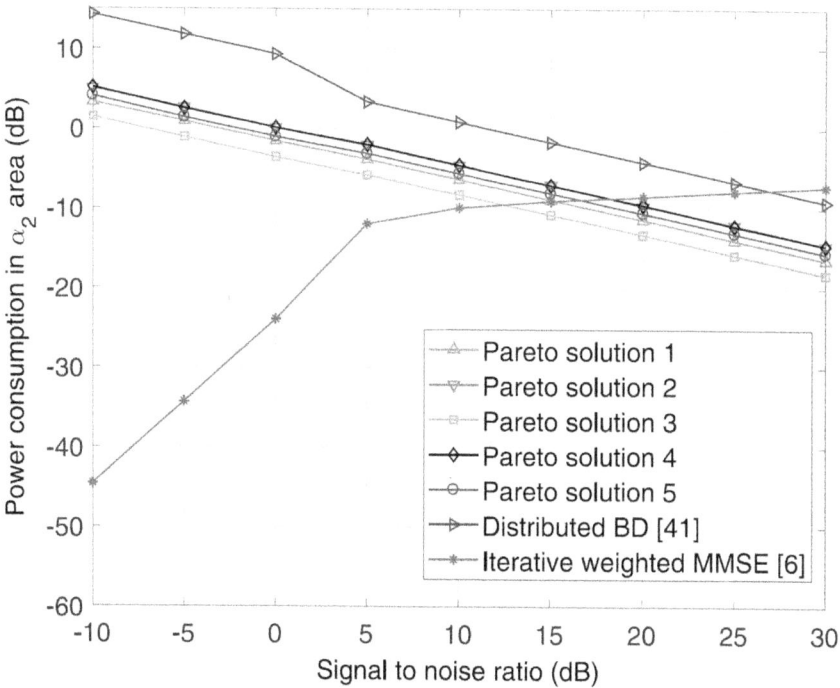

FIGURE 11.7 The power consumption of groups 1 and 2 versus different SNR levels. (a) The power consumption of group 1. (b) The power consumption of group 2.

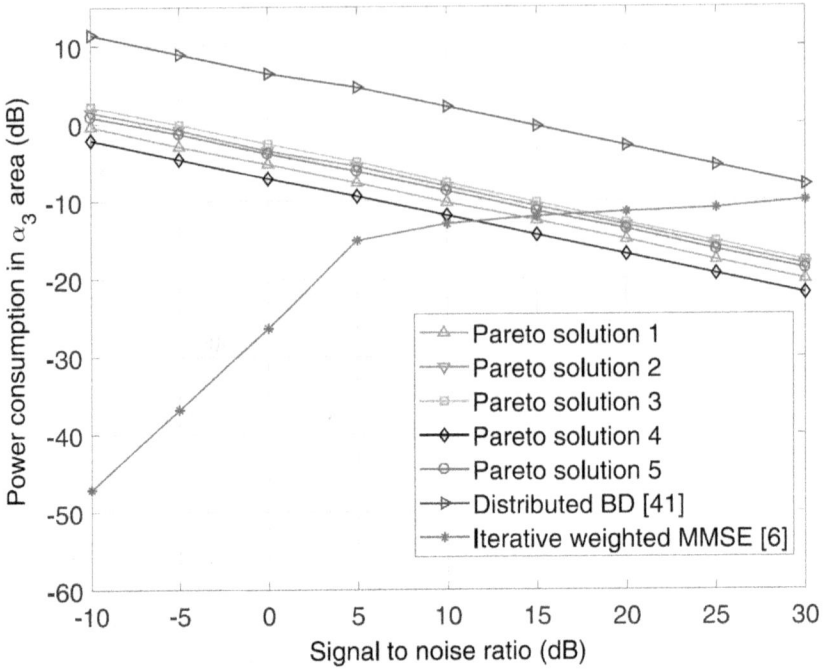

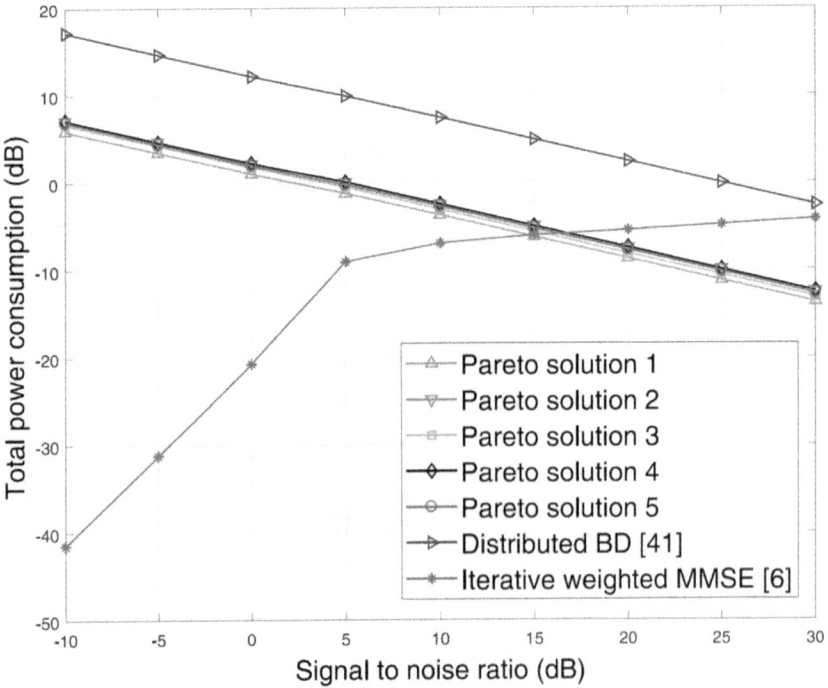

FIGURE 11.8   The power consumption of group 3 and total groups versus different SNR levels. (a) The power consumption of group 3. (b) The total power consumption of all groups.

[339] is modified with enough power to meet the SINR constraints. With different groups of cells, we can observe the difference between the Pareto solutions. In Figure 11.8(b), the total power consumption values of the Pareto solutions are close to each other. Since the knee point solution (Pareto solution 1) has the least total power consumption, it is reasonable that it consumes the least power compared to other Pareto solutions. We can observe that the more cells in the objective, the lower the power consumption of the solution.

11.6.2 TRANSMISSION CAPACITY

In MIMO wireless communication, it is important to have a better transmission rate to transmit data efficiently. We define the total transmission rate for the downlink data stream as follows:

$$\sum_{q=1}^{Q}\sum_{i=1}^{K}\log_2 \det(1+\gamma_{q_i}) \tag{11.26}$$

where γ_{q_i} is the estimated SINR matrix of the user i in the cell q. In Figure 11.9(a), we compare the total transmission rate with respect to different values of SNR. The throughput of the iterative weighted MMSE algorithm increases with the SNR. With

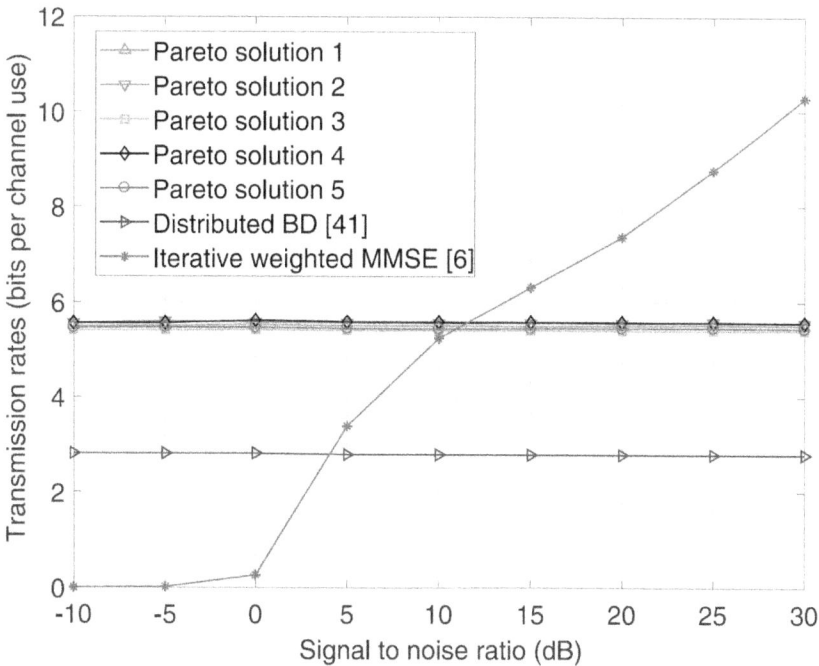

FIGURE 11.9   The total transmission rates of all groups versus different SNR levels and the total power consumption of all groups versus different uncertainty levels. (a) The power consumption of group 3. (b) The total power consumption of all groups.

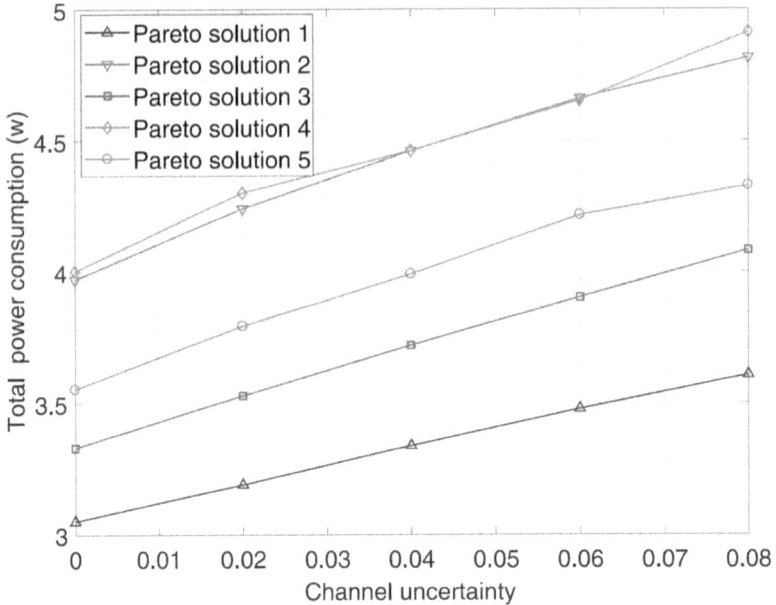

FIGURE 11.9 (Continued)

the SINR constraints on the proposed algorithm, the transmission rate of the proposed algorithm is kept at a certain level in all SNR conditions. On the contrary, the iterative weighted MMSE algorithm achieves a smaller transmission rate in a low SNR but a larger transmission rate in high SNR. In other words, the proposed algorithm is more reliable in all SNR conditions but lacks the drive to achieve the maximum transmission rate. Distributed BD is a simple and noniterative method. However, it suffers from performance degradation due to zero-forcing constraints. Thus, the distributed BD achieves the lowest sum transmission rate compared to the other schemes.

11.6.3 POWER CONSUMPTION UNDER DIFFERENT CHANNEL UNCERTAINTY LEVELS

This section discusses power consumption versus different levels of channel uncertainty in different Pareto solutions of the proposed multi-objective power minimization scheme as shown in Figure 11.9(b). We can observe that power minimization with less cells consumes more power; that is, Pareto solutions 2 and 4 consume more power than Pareto solution 3. Higher power consumption is required to achieve the desired SINR constraints with a larger channel uncertainty. This phenomena can also be observed in each Pareto optimal solution.

11.6.4 COMPARISON OF BIT ERROR RATES

To compare the equalization performance of the proposed MOP design with other methods, we consider the uncoded bit error rate [340] as the performance indicator in Figure 11.10 and Figure 11.11(a). Generally, the bit error rate is the error of

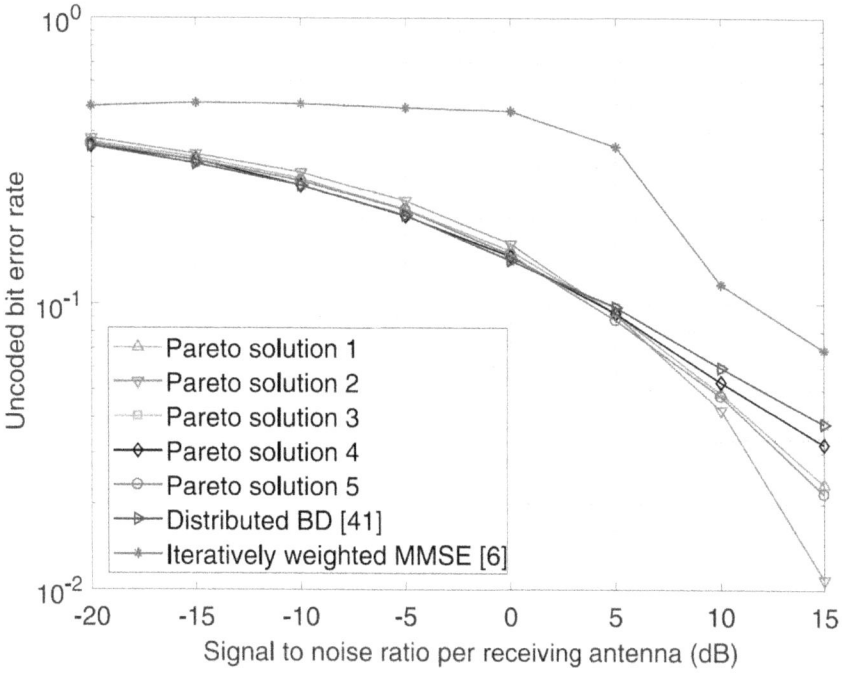

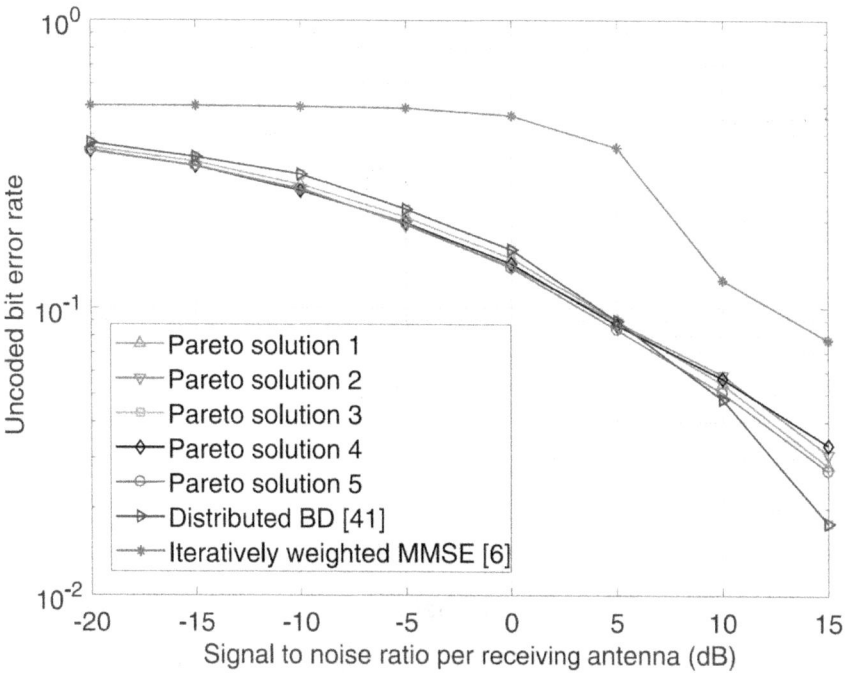

FIGURE 11.10   The average uncoded BEA in cells 1 and 2 versus different SNR levels. (a) The average uncoded BER in cell 1. (b) The average uncoded BER in cell 2.

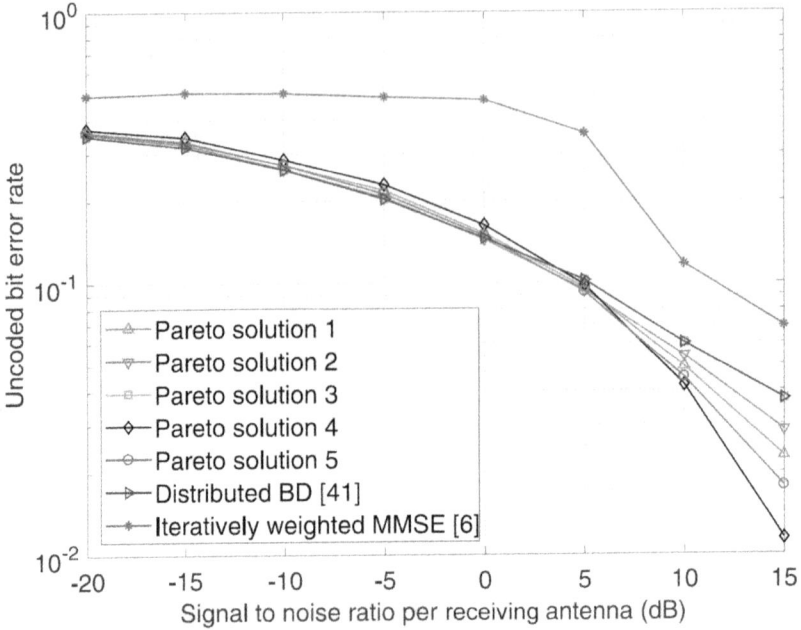

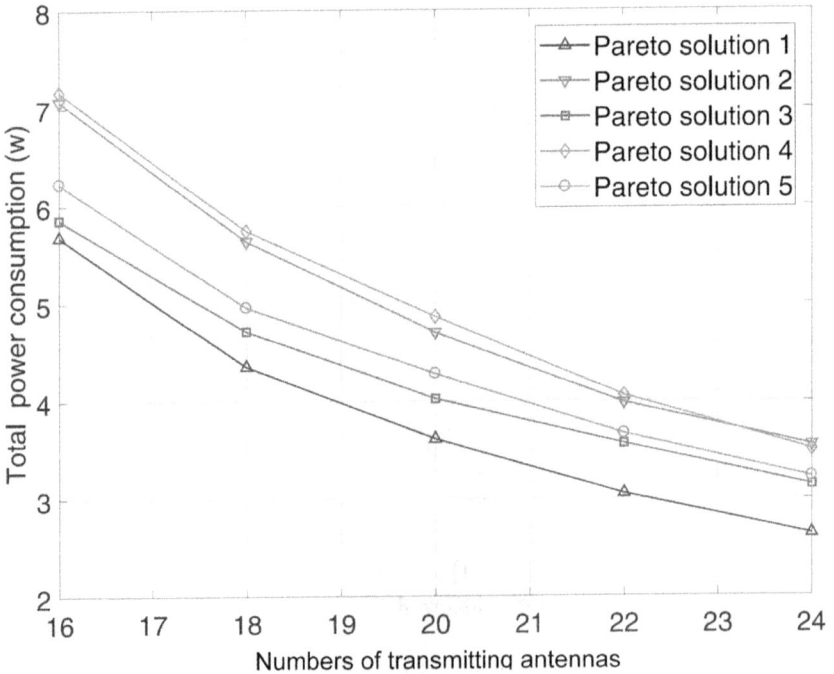

FIGURE 11.11 The average uncoded BER in cell 4 and the average power consumption dB of all groups in SNR = 0 dB. (a) The average uncoded BER in cell 4 versus different SNR levels. (b) The average power consumption in SNR = 0 dB of all groups versus different number of transmitting antennas.

the coded signal (digital signal). The uncoded bit error rate is the bit error rate for the signal before being coded (analog signal). We consider the maximum-likelihood detection scheme [340] to detect the received signals. For the users in each group, there is no difference with the BER at a low SNR. At a high SNR, Pareto solution 4 with the corresponding group achieves the best BER performance. Pareto solutions 5 (the proposed lowest MMSE equalization) and 1 (knee point) are the second. The Pareto solutions of the other groups get the worst BER. It is worth noting that the BER in group 2 is worse than that of the other groups. It seems that there is a trade-off between power consumption and BER. Generally, the iterative weighted MMSE algorithm achieves a worse bit error rate than the proposed Pareto solutions. It appears that the BER is influenced by the lack of SINR constraints. Affected by the zero forcing constraints, the distributed BD with the SINR constraint achieves a relatively high BER except in Figure 10.10(b).

11.6.5 EFFECT OF NUMBER OF TRANSMITTING ANTENNAS

In this section, we study the impact of the number of transmitting antennas on power consumption. Similar to Section 11.6.1, power minimization with more cells in a group results in less power consumption. Since the quantity of antennas has an impact on multiplex gains, we can observe the relation between the number of transmitting antennas and the total power consumption by the simulation result in Fig. 11.11(b). Although a large number of transmitting antennas decreases beamforming power consumption, it also increases the design complexity. In practical application, the trade-off between BER performance and complexity should be taken into account in the design procedure.

11.6.6 TRANSMISSION THROUGHPUTS

In Figure 11.12, we investigate the transmission throughput of the proposed algorithm at different SNR levels. According to related research [302, 341], the transmission throughput is defined as the average data bits (or packets) that are successfully received and decoded by the receiver per time slot. If we have the average transmission rate R and the bit error rate B (or packet error rate), then the general transmission throughput can be approximated as $(1 - B)R$. Based on this definition, both the low error rate and high transmission rate are important to achieve the maximum transmission throughput. According to Figures 11.9(a) and 11.11(a), we can observe that the iterative weighted MMSE algorithm can achieve a high system capacity but suffers from a high risk of losing information with large power consumption in the high SNR region. Also, the distributed BD has the worst performance due to the sum transmission rate degradation. On the contrary, the proposed algorithms keep the throughput at a fixed level with a lower BER, which can guarantee transmission reliability of the proposed algorithms. As a result, it is more recommendable to apply the proposed algorithms to receive information with a minimum power consumption in all SNR conditions.

As a conclusion, we analyze every aspect in the simulation example. Pareto solution 5 (the proposed lowest MMSE equalization) is more reliable than other

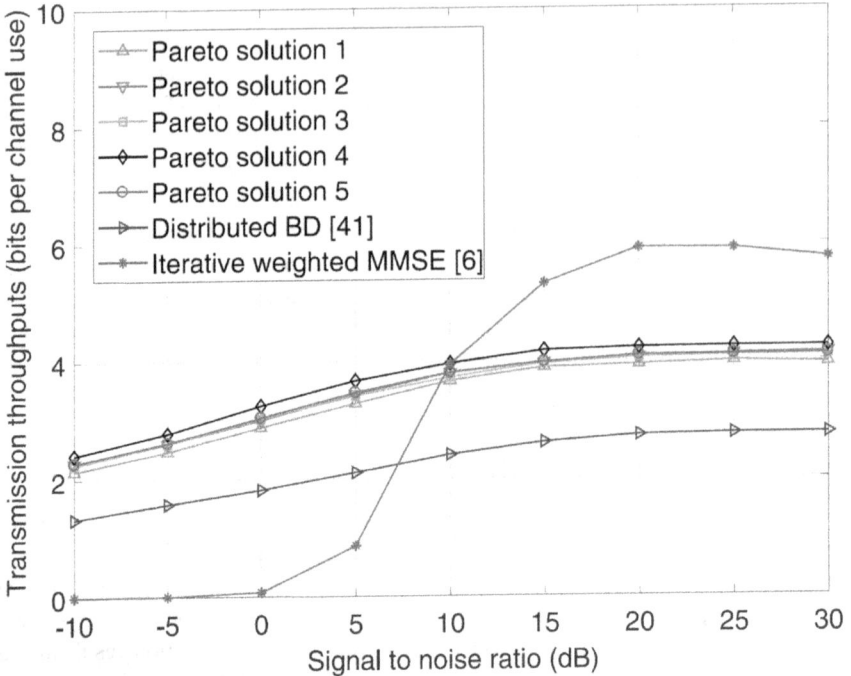

FIGURE 11.12 The total transmission throughputs of all groups versus different SNR levels.

algorithms. Although Pareto solution 5 consumes more power than Pareto solutions 3 and 1 with respect to transmitting antenna number and channel uncertainty in Figure 11.11(b) and Figure 11.9(b), Pareto solution 5 has a better transmission rate and low BER. Consequently, we should choose Pareto solution 5 as the preferred Pareto solution in the multi-objective power minimization design of the multicell multiuser MIMO beamforming system.

11.7 CONCLUSION

In this chapter, a multi-objective power minimization beamforming design is introduced for a multicell multiuser MIMO system. An indirect suboptimal MOP is proposed to solve the multi-objective power minimization problem by the proposed SDP-constrained MOEA. To search the Pareto optimal solutions in the Pareto front, the Pareto optimal solution with the best MMSE equalization is selected as a preferable solution among these Pareto optimal solutions. The simulation example shows that the beamforming of Pareto optimal solution with the best MMSE equalization of the proposed multi-objective power minimization design algorithm can achieve reliable performance with low power. Thus, the proposed SDP-constrained MOEA can get the Pareto optimal solution set and provide the designer with a preferred beamforming to achieve the minimum power consumption and signal detection error under a constrained SINR, which will be useful for energy-efficient communications in the future. For future research, we expect to expand the application of multi-objective optimization into another design aspect of wireless communication systems.

11.8 APPENDIX

11.8.1 PROOF OF THEOREM 11.1

To prove Theorem 11.1, we choose a straightforward way. The main idea is to prove that both inequality constraints in (11.14b) become equality for the Pareto optimal solution. We will prove this by contradiction. Given a Pareto optimal solution $(\alpha_1^*, ..., \alpha_n^*, ..., \alpha_N^*)$ of the MOP in (11.14a)–(11.14b), we assume that either of the inequalities in (11.14b) remains a strict inequality in the Pareto optimal solution. Without loss of generality, we assume that there exists a feasible solution β such that $\alpha_1^* > \beta$ and $\sum_{q_i \in g_1} Tr(Q_{q_i}) = \beta$. Then the solution $(\beta, ..., \alpha_n^*, ..., \alpha_N^*)$ dominates the original Pareto optimal solution $(\alpha_1^*, ..., \alpha_n^*, ..., \alpha_N^*)$, which leads to a contradiction. This implies that the inequality constraints in (11.14b) indeed become equalities for Pareto optimal solutions in the MOP in (11.14a)–(11.14b). Thus, the indirect MOP in (11.14a)–(11.14b) is indeed equivalent to the MOP in (11.12)–(11.13).

12 Multi-Objective Beamforming Power Control for Robust SINR Target Tracking and Power Efficiency in Multicell MU-MIMO Wireless Communication Systems

12.1 INTRODUCTION

For wireless communication systems in the next generation, the extreme demands of transmission throughput have triggered extensive investigation of coordinated communication and resource allocation to achieve performance enhancement in wireless cellular networks. Since multi-input multi-output transmission can provide an extra degree of freedom and spatial diversity, it has received significant attention from both academia and industry recently [342, 343]. On the other hand, it is well known that coordinated multi-point (CoMP) schemes, such as coordinated beamforming (CB) and joint processing (JP), could achieve the high quality of service requirements even if users suffer from multicell interference at the cell edge in the multicell multiuser MIMO (multicell MU-MIMO) system [344–346]. Among different coordinated communications, dirty paper coding (DPC) and block diagonalization (BD) (i.e., zero forcing beamforming (ZFBF)) are two kinds of conventional strategies to overcome intercell or intracell interference [347, 348]. However, the high complexity of DPC and the performance degradation due to noise amplification by ZFBF still remain a bottleneck in applications [349].

For most of the general optimization problems in coordinated beamforming algorithms for MIMO wireless systems, the implementation is to rely on the assumption of perfect channel state information from the ideal backhaul link. However, the applicability of these methods decreases with pilot contamination since the transmitter only obtains imperfect CSI with feedback delay, estimation error, quantization error, and so on [350, 351]. Therefore, it is necessary to consider robust design by taking

DOI: 10.1201/9781003362142-15

290

imperfect CSI and delay feedback into account [352]. For multicell or multiuser MIMO systems, a lot of research has concerned worst-case and probabilistic robustness [308, 349, 353–357]. Specifically, a study considered an outage probabilistic robust beamforming design for a multicell MIMO wireless system with a maximizing sum outage rate in [308]. Joint beamforming and broadcasting (JBB) methods have been investigated for massive MIMO systems in the case where one group of terminals cannot obtain CSI in the system [355]. In [356], with a single antenna receiver for each user, imperfect CSI is modelled within an upper bound with a Euclidean norm; then the authors use semidefinite relaxation and second-order cone (SOC) constraints to formulate a robust design in the MIMO relay system. A novel algorithm of precoding and detection methods for a nonorthogonal multiple access MIMO system is developed and the outage probability for performance enhancement is investigated in [357]. In [349], the delay effect in a feedback loop and mismatching CSI are modeled to achieve a robust signal-to-leakage-plus-noise ratio (SLNR) beamforming in the MU-MIMO system.

Moreover, to overcome interference, some research has paid attention to the design of power control problems for SINR target tracking in the past decade [330, 358–363]. For instance, energy-efficient robust transceiver design considering imperfect channels with the assumption of norm-bounded uncertainty for power control problems has been developed by minimizing interference leakage [330]. Generally, we can divide power control schemes for SINR target tracking into two categories: centralized [361] and decentralized [14, 359, 360, 362] autonomous power control schemes. Compared to the centralized scheme, decentralized power control only needs local measurements for each transmitter so that all users will satisfy the desired QoS requirement [330, 359]. In the proposed power control method, each transmitter only needs local CSI and measures the SINR to generate tracking control commands without the uncertainty of upper-bound information for SINR target tracking, which is more realistic than conventional methods for interference wireless systems.

Multi-Objective designs are more attractive than single-objective designs for the next generation of wireless communication networks to achieve multiple design purposes inherently [326, 327, 363–365]. Recently, two general algorithms have been developed to resolve multi-objective optimization problems. One is the classical algorithm using weighted methods [327, 364, 365], the another is the multi-objective heuristic algorithm [2, 325, 366–368]. The basic idea of the weighted method is to provide several weighting factors and combine the corresponding objective values into a single objective problem. For the multi-objective heuristic algorithm, some popular approaches can be divided into several categories, such as multi-objective evolution algorithms, multi-objective immune algorithms (MOIAs) [325], and multi-objective simulated annealing (MOSA), which consist of a nondominated sorting genetic algorithm [2], hybrid multi-objective immune algorithm [366], and archived multi-objective simulated annealing (AMOSA) [367]. MOEAs, MOIAs, and MOSAs provide us a set of Pareto optimal solutions based on common goals, which are reliable and fast converging to the Pareto optimal set with a uniform distribution of Pareto solutions along the Pareto front [369]. For the proposed MOPs to achieve SINR target tracking, a linear matrix inequality–constrained MOEA is proposed in this chapter to obtain the Pareto optimal solutions parallelly.

In this chapter, we consider a multi-objective robust beamforming power control scheme introduced to achieve optimal robust H_2/H_∞ SINR target tracking performance with the minimum power consumption for each user in the multicell MU-MIMO wireless system. The three objectives of the introduced design method with SINR target tracking are as follows: (1) achieving optimal H_∞ robust SINR target tracking performance to efficiently resist the worst-case effect of external disturbance (i.e., fluctuation of interference) without statistical information of uncertainty. (2) Achieving optimal H_2 SINR target tracking performance to minimize the mean square tracking error. (3) Achieving the minimum average power consumption performance for power efficiency to decrease the energy cost in power control command. For the robust optimal H_2/H_∞ SINR target tracking scheme [9], the worst-case effect of channel fading, channel uncertainty, SINR target levels, and round trip delay should be minimized. Most importantly, rather than the conventional robust design for imperfect CSI with norm-bounded uncertain information, the proposed robust H_2/H_∞ SINR target tracking scheme does not need to know the statistical property or the upper bound information of channel uncertainty. This method can transfer the proposed multi-objective robust H_2/H_∞ SINR target tracking problem into an offline LMI-constrained MOP to simplify the design procedure. Then an LMI-constrained MOEA is developed to obtain the Pareto optimal solutions for the MOP efficiently. Up to this, the offline MOP to achieve a robust optimal H_2/H_∞ SINR target tracking with the minimum average power consumption simultaneously for the improvement of QoS and power efficiency can be solved by the preferable Pareto optimal solution. Finally, the online SINR target tracking system can update the beamforming control command by the offline design Pareto optimal design. According to the foregoing beamforming control command, the system designer can acquire the corresponding beamforming matrices by solving the decentralized semi-definite programming process under the leakage interference constraints.

The main purposes of this chapter are as follows:

- Without the requirement of statistical property and upper bound information of channel uncertainty, the dynamic power control scheme with an offline design procedure is combined with an online beamforming optimization method in the multicell MU-MIMO wireless system to achieve robust SINR target tracking design.
- The Pareto front of the multi-objective H_2/H_∞ robust SINR target tracking design problem can be investigated with an optimal power efficiency, that is, a MOP to minimize (1) the worst-case effect of the channel uncertainty in the robust H_∞ sense, (2) the mean square error on SINR target tracking performance, and (3) minimum average power consumption simultaneously.
- The aforementioned MOP is difficult to solve in general due to the unknown channel uncertainty without the upper bound information. We transform the aforementioned MOP into an offline design LMI-constrained MOP, which can be solved efficiently with the help of an interior point method.
- An LMI-constrained MOEA is further developed to provide a parallel search scheme with convergence to the Pareto optimal solutions of the proposed multi-objective beamforming power control with good distribution along the Pareto front.

This chapter is divided into six sections as follows: in Section 12.2, we introduce the system model and the SINR target tracking model for the multicell MU-MIMO wireless system. In Section 12.3, the multi-objective optimization problem is turned into an LMI-constrained MOP. After we formulate the LMI-constrained MOP for the optimal robust H_2/H_∞ SINR target tracking problem with the optimal power efficiency, an LMI-constrained MOEA is developed to find the Pareto optimal solutions in Section 12.4. In Section 12.5, some simulation results are given for simulated scenarios of the proposed design procedure. Section 12.6 draws conclusions.

Notation: The scalar and the vector are indicated by italic lowercase letters and bold lowercase letters, respectively. The complex matrix A with dimensions $n \times m$ is denoted as a capital letter as $A \in \mathbb{C}^{n \times m}$. The real matrix A with dimensions $n \times m$ is presented as a capital letter as $A \in \mathbb{R}^{n \times m}$. The transposed operator of matrix A is presented as A^T. The symbol A^H stands for the complex conjugate operator of the matrix A. The matrix inequality $A \geq 0$ indicates that matrix A is positive semi-definite. The negative semi-definite matrix A is denoted as $A \leq 0$. The trace of matrix A is presented as $Tr(A)$. The notation $E\{A\}$ presents the statistical mean. The symbol I_N denotes the identity matrix with dimension N. $\|A\|_2$ and $\|A\|_F$ denote the spectral norm and the Frobenius norm of a matrix A, respectively. The complex Gaussian normal distribution with mean μ and covariance matrix Γ is presented as $CN(\mu, \Gamma)$.

12.2 SYSTEM MODEL FOR ROBUST BEAMFORMING POWER CONTROL DESIGN IN A WIRELESS COMMUNICATION SYSTEM

The proposed multi-objective robust beamforming power control design includes three objectives: (1) optimal H_∞ robust SINR target tracking performance, (2) optimal H_2 SINR target tracking performance, and (3) the minimum average power consumption in the multicell multiuser MIMO wireless system. After the beamforming power command is specified by the proposed SINR tracking system model, it will be transferred to a semidefinite programming block to find the beamforming matrix. In this section, the SINR tracking system model for the multi-objective SINR target tracking design is proposed to prepare for the beamforming control design in Section 12.3.

12.2.1 Multicell Multiuser MIMO Wireless System with Imperfect CSI

We consider the downlink multicell multiuser MIMO interference wireless system, as shown in Figure 12.1. All base stations serve multiple users at the same frequency [348] (i.e., each user will suffer from intracell and intercell interference). The multicell MU-MIMO wireless system includes Q cells. The qth cell is equipped with a multiple antenna base station, which transmits independent data streams to the K_q remote users in its coverage range. Each user has multiple receiving antennas. We denotes N_t and N_r as the transmitted antenna number of base station and the received antenna number of the user equipment, respectively. According to the related research [348], let $r_q(t) = \sum_{i=1}^{K_q} W_{q_i}(t) s_{q_i}(t) \in \mathbb{C}^{N_t \times 1}$ be the transmitted signal vector at time slot t from the qth base station, where $W_{q_i}(t) \in \mathbb{C}^{N_t \times L_{q_i}}$ is the beamforming matrix at time slot t with transmitted symbol L_{q_i} for the ith user at the qth base station (i.e., in the

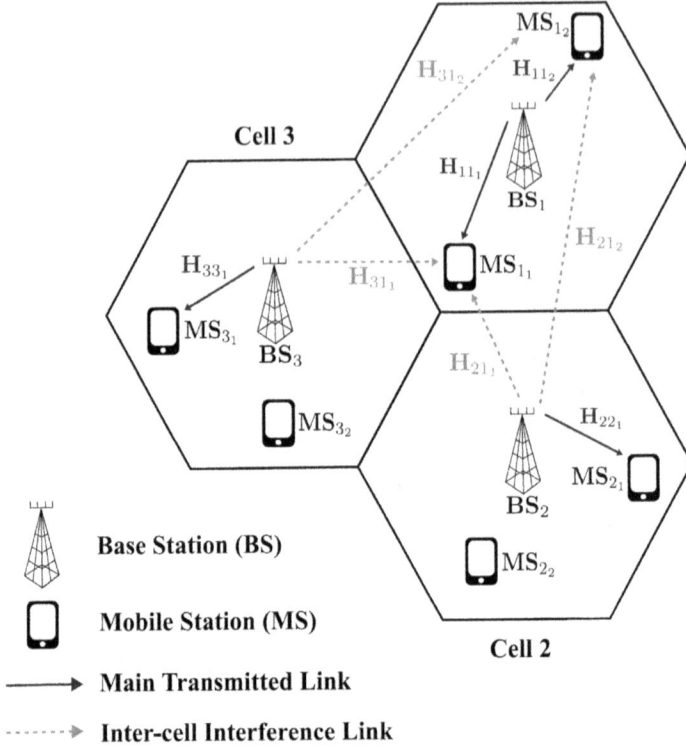

FIGURE 12.1 A three-hexagonal-cell example of the multicell MU-MIMO interference wireless communication system. The solid line represents the main transmitted link from the base station to the served users. The dotted line indicates the intracell interference link from other cells. Note that the figure only displays the interference from other cells for the users in the first cell.

qth cell). The symbol K_q is the number of users in the qth cell. $s_{q_i}(t) \in \mathbb{C}^{L_{q_i} \times 1}$ is the information symbol vector for the ith user at the qth cell with L_{q_i} transmitted symbols. The matrix $H_{kq_i} \in \mathbb{C}^{N_r \times N_t}$ denotes the channel response at the time slot t from the kth base station to the ith user in the qth cell. Thus, the received signal for the ith user at the qth cell can be expressed as follows [34, 348]:

$$
y_{q_i}(t) = \sum_{k=1}^{Q} H_{kq_i}(t) r_k(t) + n_{q_i}(t)
$$

$$
= \underbrace{H_{qq_i}(t) W_{q_i}(t) s_{q_i}(t)}_{\text{Desired signal}} + \underbrace{H_{qq_i}(t) \sum_{j \neq i}^{K_q} W_{q_j}(t) s_{q_j}(t)}_{\text{Intra-cell interference}} \tag{12.1}
$$

$$
+ \underbrace{\sum_{r \neq q}^{Q} H_{rq_i}(t) \sum_{j=1}^{K_r} W_{r_j}(t) s_{r_j}(t)}_{\text{Inter-cell interference}} + \underbrace{n_{q_i}(t)}_{\text{Noise}}
$$

where $s_{q_i}(t)$ is the signal of interest that we want to obtain under additive Gaussian noise and intracell and intercell interference. The Gaussian noise $n_{q_i}(t) \in \mathbb{C}^{N_r \times 1}$ is assumed to be a complex normal distribution with zero mean and covariance matrix

$\rho_{q_i}^2 I_{N_r}$, that is, $n_{q_i}(t) \sim CN(0, \rho_{q_i}^2 I_{N_r})$ with the positive scalar $\rho_{q_i}^2$. Without loss of generality, the required symbol vector $s_{q_i}(t)$ has $E\{s_{q_i}(t)s_{q_i}^H(t)\} = I$, $\forall i$, $\forall q$ [348]. According to (12.1) and the related study [370], we have the signal-to-interference-plus-noise ratio of the ith user in the qth cell with instantaneous perfect channel state information for the multicell MU-MIMO wireless system as follows:

$$SINR_{q_i}^{pcc}(t)$$

$$= \frac{\left\| H_{qq_i}(t)W_{q_i}(t) \right\|_F^2}{\sum_{r=1}^{Q}\sum_{j=1}^{K_r} \left\| H_{rq_i}(t)W_{r_j}(t) \right\|_F^2 - \left\| H_{qq_i}(t)W_{q_i}(t) \right\|_F^2 + \rho_{q_i}^2 N_r} \tag{12.2}$$

In fact, the base stations only obtain imperfect CSI in the multicell MU-MIMO wireless system due to feedback delay, estimation error, quantization error, and so on. Thus, channel imperfection exists in all the estimated channel matrices of the multicell MU-MIMO wireless system. The channel imperfection of all the channel matrices $H_{rq_i}(t)$, $\forall r,q,i$, can be expressed as:

$$H_{rq_i}(t) = \widehat{H}_{rq_i}(t) + \Delta H_{rq_i}(t), \forall r,q,i \tag{12.3}$$

where $\widehat{H}_{rq_i}(t)$ is the estimated channel matrix, and $\Delta H_{rq_i}(t)$ is the channel uncertain matrix. According to (12.3), we define the following SINR expression with instantaneous imperfect CSI of the ith user at the qth cell for the multicell MU-MIMO wireless system:

$$SINR_{q_i}(t) = \left\| \widehat{H}_{qq_i}(t)W_{q_i}(t) \right\|_F^2$$

$$\times \left(\sum_{r=1}^{Q}\sum_{j=1}^{K_r} \left\| (\widehat{H}_{rq_i}(t) + \Delta H_{rq_i}(t))W_{r_j}(t) \right\|_F^2 \right. \tag{12.4}$$

$$\left. - \left\| (\widehat{H}_{qq_i}(t) + \Delta H_{qq_i}(t))W_{q_i}(t) \right\|_F^2 + \rho_{q_i}^2 N_r \right)^{-1}$$

Some researchers may suggest that the main channel uncertain term $\left\| H_{qq_i}(t)W_{q_i}(t) \right\|_F^2$ could be included in the denominator of (12.4). In this chapter, we only consider the uncertainty other than the main channel uncertain term to have SINR expression with imperfect CSI in (12.4), which is similar to the existing robust beamforming design [333]. By using dB as the unit of reference, the SINR expression in (12.4) can be rewritten as:

$$10\log_{10}(SINR_{q_i}(t))$$

$$= 10\log_{10}\left(\left\| \widehat{H}_{qq_i}(t)W_{q_i}(t) \right\|_F^2 \right)$$

$$- 10\log_{10}\left(\sum_{r=1}^{Q}\sum_{j=1}^{K_r} \left\| (\widehat{H}_{rq_i}(t) + \Delta H_{rq_i}(t))W_{r_j}(t) \right\|_F^2 \right. \tag{12.5}$$

$$\left. - \left\| (\widehat{H}_{qq_i}(t) + \Delta H_{qq_i}(t))W_{q_i}(t) \right\|_F^2 + \rho_{q_i}^2 N_r \right)$$

For design simplicity, we have the following variables in decibel scale to formulate the SINR target tracking scheme [14, 362]:

$$y_{q_i}(t) \triangleq 10 \log_{10}(SINR_{q_i}(t)) \tag{12.6}$$

$$z_{q_i}(t) \triangleq 10 \log_{10}\left(\left\|\widehat{H}_{qq_i}(t)W_{q_i}(t)\right\|_F^2\right) \tag{12.7}$$

$$v_{q_i}(t) \triangleq 10 \log_{10}\left(\sum_{r=1}^{Q}\sum_{j=1}^{K_r}\left\|(\widehat{H}_{rq_i}(t) + \Delta H_{rq_i}(t))W_{r_j}(t)\right\|_F^2\right.$$
$$\left. - \left\|(\widehat{H}_{qq_i}(t) + \Delta H_{qq_i}(t))W_{q_i}(t)\right\|_F^2 + \rho_{q_i}^2 N_r\right) \tag{12.8}$$

where $z_{q_i}(t) \in \mathbb{R}$ is the beamforming control command to specify the power level of the main transmission link for the ith user in the qth cell; the variable $y_{q_i}(t) \in \mathbb{R}$ presents the measured SINR of the ith user in the qth cell; and the external disturbance $v_{q_i}(t) \in \mathbb{R}$ includes the power of channel uncertainty, interference, and the noise variance.

Remark 12.1: Note that after the beamforming control command $z_{q_i}(t)$ is generated by the SINR tracking system (i.e., the beamforming control gain k), as shown in Figure 12.2, we still need to find the beamforming matrix $W_{q_i}(t)$ to match the beamforming control command $z_{q_i}(t)$ in (12.7). The optimization method of semi-definite programming to find the beamforming matrix $W_{q_i}(t)$ will be discussed after the multi-objective optimization algorithm in Section 12.4.

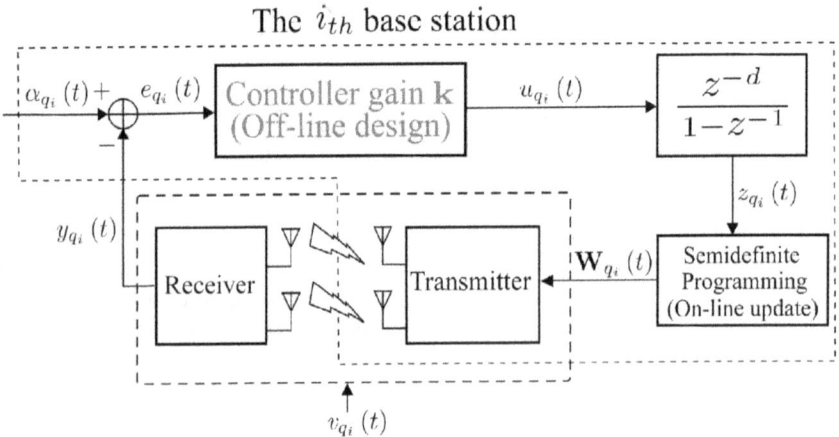

FIGURE 12.2 The simple closed-loop system for robust power control with the SINR target tracking scheme for the improvement of QoS. The beamforming control gain k and the beamforming matrix $W_{q_i}(t)$ are to be specified to achieve the multi-objective H_2/H_∞ SINR target tracking performance in (12.23) with the minimum average power consumption simultaneously.

By substituting (12.6)–(12.8) into (12.5), we can obtain

$$y_{q_i}(t) = z_{q_i}(t) - v_{q_i}(t) \tag{12.9}$$

As shown in Figure 12.2, the round-trip delay of the beamforming control command for $z_{q_i}(t)$ is modeled as [362]

$$z_{q_i}(t) = z_{q_i}(t-1) + u_{q_i}(t-d) \tag{12.10}$$

where $u_{q_i}(t-d) \in \mathbb{R}$ is the power control command with round-trip delay d, and $z_{q_i}(t)$ presents the combination of $z_{q_i}(t-1)$ and regulation effort $u_{q_i}(t-d)$ of the SINR tracking system. Note that the power control command $u_{q_i}(t)$ of each time slot is to be designed to achieve multiple objectives simultaneously for the beamforming control command $z_{q_i}(t)$ in Section 12.4, that is, the optimal H_∞ robust SINR target tracking performance, the optimal H_2 SINR target tracking performance, and the minimum average power consumption for each user in the multicell MIMO wireless systems.

12.2.2   SINR TARGET TRACKING SYSTEM MODEL

According to (12.9) and (12.10), the SINR measurement level $y_{q_i}(t)$ for the ith user in the qth cell can be represented as (12.9), and the updated beamforming control commands can be expressed as (12.10). Assume that we want to achieve the specific downlink SINR level $\alpha_{q_i}(t) \in \mathbb{R}$ to satisfy a required QoS for the measurement SINR, as shown in Figure 12.2, in the multicell MU-MIMO wireless system; the tracking error $e_{q_i}(t)$ for $y_{q_i}(t)$ to target SINR level $\alpha_{q_i}(t)$ can be expressed as:

$$e_{q_i}(t) = \alpha_{q_i}(t) - y_{q_i}(t) \tag{12.11}$$

Note that for the transmitter, the tracking error $e_{q_i}(t)$ should be regarded as the feedback signal and be used to calculate the power control command $u_{q_i}(t)$. From (12.9)–(12.11), we have the following SINR tracking control equation of the ith user in the qth cell:

$$\begin{aligned}
e_{q_i}(t) &= e_{q_i}(t-1) - e_{q_i}(t-1) + e_{q_i}(t) \\
&= e_{q_i}(t-1) + \alpha_{q_i}(t) - \alpha_{q_i}(t-1) + y_{q_i}(t-1) - y_{q_i}(t) \\
&= e_{q_i}(t-1) - u_{q_i}(t-d) + \zeta_{q_i}(t-1)
\end{aligned} \tag{12.12}$$

where $\zeta_{q_i}(t-1) = \alpha_{q_i}(t) - \alpha_{q_i}(t-1) + v_{q_i}(t) - v_{q_i}(t-1)$ denotes the changes of the target SINR level $\alpha_{q_i}(t)$ and the external disturbance $v_{q_i}(t)$. These fluctuations must be modeled as the disturbances in the SINR tracking error system in (12.12). We assume that all of the power control commands $u_{q_i}(t)$ in the past must be combined with the beamforming control command $z_{q_i}(t)$ and the error term $e_{q_i}(t)$ for the ith user in the qth cell as:

$$x_{q_i}(t) = [e_{q_i}(t), u_{q_i}(t-d), ..., u_{q_i}(t-1)]^T \tag{12.13}$$

Thus, the power control dynamic equation of the ith user in the qth cell for the SINR target tracking in the multicell MU-MIMO wireless system in (12.12) can be described as:

$$x_{q_i}(t+1) = Ax_{q_i}(t) + bu_{q_i}(t) + c\zeta_{q_i}(t)$$
$$e_{q_i}(t) = dx_{q_i}(t) \qquad (12.14)$$
$$u_{q_i}(t-1) = gx_{q_i}(t)$$

where

$$b = \begin{bmatrix} 0 \\ 0 \\ M \\ 1 \end{bmatrix} \in \mathbb{R}^{(d+1)\times 1}, \; c = \begin{bmatrix} 1 \\ 0 \\ M \\ 0 \end{bmatrix} \in \mathbb{R}^{(d+1)\times 1} \qquad (12.15)$$

$$d = \begin{bmatrix} 1 & 0 & \cdots & 0 \end{bmatrix} \in \mathbb{R}^{1\times(d+1)}$$
$$g = \begin{bmatrix} 0 & 0 & \cdots & 1 \end{bmatrix} \in \mathbb{R}^{1\times(d+1)} \qquad (12.16)$$

$$A = \begin{bmatrix} 1 & 0 & -1 & 0 & 0 & \cdots & 0 \\ 0 & 0 & 1 & 0 & 0 & \cdots & 0 \\ 0 & 0 & 0 & 1 & 0 & \cdots & 0 \\ \vdots & \vdots & \vdots & 0 & \ddots & & \vdots \\ \vdots & \vdots & \vdots & \vdots & & \ddots & 0 \\ \vdots & \vdots & \vdots & 0 & \cdots & 0 & 1 \\ 0 & 0 & 0 & 0 & \cdots & \cdots & 0 \end{bmatrix} \in \mathbb{R}^{(d+1)\times(d+1)} \qquad (12.17)$$

For the SINR tracking model of the beamforming power control problem in the multicell MU-MIMO wireless system, we assume that the feedback beamforming control gain $k \in \mathbb{R}^{1\times(d+1)}$ exists in each base station. Thus, the power control command $u_{q_i}(t)$ can be designed as the state feedback scheme by the offline design beamforming control gain k:

$$u_{q_i}(t) = kx_{q_i}(t) \qquad (12.18)$$

Since system matrices A, b, c, d, and g are all the same for all users, the designer can specify the same beamforming control gain k with offline design process to make the tracking error $e_{q_i}(t)$ as small as possible despite the disturbance $\zeta_{q_i}(t)$ in (12.14). By substituting (12.18) into (12.14), the SINR tracking system of the multicell MU-MIMO wireless system with control gain is presented as:

$$x_{q_i}(t+1) = (A+bk)x_{q_i}(t) + c\zeta_{q_i}(t)$$
$$e_{q_i}(t) = dx_{q_i}(t) \qquad (12.19)$$
$$u_{q_i}(t-1) = gx_{q_i}(t)$$

Up to this, the SINR tracking system model for robust beamforming power control design in the multicell MU-MIMO wireless system is completed. In the next section, the multi-objective optimal H_2/H_∞ SINR target tracking problem with the minimum power consumption is formulated to find the Pareto optimal solution of the multi-objective beamforming control gain k.

12.3   PROBLEM FORMULATION

The purpose of this section is to design the multi-objective beamforming control gain k to achieve optimal robust H_2/H_∞ SINR target tracking performance with the minimum average power consumption simultaneously of the SINR tracking system in (12.19) for the multicell MU-MIMO wireless system. First, the H_∞ robust SINR tracking system in (12.19) is defined to avoid the effect of the disturbance power on the SINR tracking performance as follows [9]:

$$J_\infty(u_{q_i}(t)) = \frac{E\{q_1 e_{q_i}^2(t) + r_1 u_{q_i}^2(t)\}}{E\{\zeta_{q_i}^2(t)\}} \tag{12.20}$$

where the weighted factors $q_i \in \mathbb{R}$ and $r_1 \in \mathbb{R}$ are positive constants and represent the trade-off between power control command $u_{q_i}(t)$ and tracking error $e_{q_i}(t)$. The meaning of (12.20) is that the effect of the disturbance $\zeta_{q_i}(t)$ on the tracking error $e_{q_i}(t)$ and the updated power command $u_{q_i}(t)$ is considered from the average energy of view. Since the disturbances of multipath fading are unpredictable in the MIMO wireless system, the H_∞ robust SINR target tracking design is considered from the worst-case effect of disturbance.

For the given SINR tracking system in (12.19), the H_2 optimal SINR target tracking performance index $J_2(u_{q_i}(t))$ [9] with external disturbance $\zeta_{q_i}(t) = 0$ for the SINR tracking system in (12.19) can be expressed as:

$$J_2(u_{q_i}(t)) = E\{q_2 e_{q_i}^2(t+1)\} \tag{12.21}$$

where the positive weighted factor $q_2 \in \mathbb{R}$. The expression of (12.21) is seen as the mean square SINR target tracking error $e_{q_i}(t+1)$. Moreover, the average power consumption performance index $J_p(u_{q_i}(t))$ for the given SINR tracking system in (12.19) is defined as the following to save the power cost in the multicell MU-MIMO wireless system:

$$J_p(u_{q_i}(t)) = E\{q_3 u_{q_i}^2(t)\} \tag{12.22}$$

where the positive weighted factor $q_3 \in \mathbb{R}$. The meaning of (12.22) is the updated power consumption from the average point of view by (12.10). According to (12.20)–(12.22), the design problem to achieve the multi-objective optimal robust H_2/H_∞ tracking performance with the minimum power consumption of the SINR tracking target system in (12.19) for the multicell MU-MIMO wireless beamforming system can be expressed as follows:

Problem 12.1: The multi-objective optimal H_2/H_∞ tracking problem with the minimum average power consumption of the given SINR tracking system in (12.19) is to design the admissible power control command $u_{q_i}(t)$ in (12.18) to achieve the following multi-objective tracking performance simultaneously:

$$\min_{u_{q_i}(t) \in \phi} \{J_\infty(u_{q_i}(t)), J_2(u_{q_i}(t)), J_p(u_{q_i}(t))\}$$

(12.23)

subject to (12.14)

where ϕ is the feasible set of all updated power commands $u_{q_i}(t)$ satisfying the state feedback scheme in (12.18) for the SINR target tracking system in (12.14).

Before further discussions of solving the MOP in (12.23), some fundamental concepts of MOPs need to be introduced.

Definition 12.1 ([368, 369]): (Pareto dominance) For two feasible solutions $u_{q_i}(t)$ and $u'_{qi}(t)$ corresponding to two objective values $(J_\infty(u_{q_i}(t)), J_2(u_{q_i}(t)),$ $J_p(u_{q_i}(t)))$ and $(J_\infty(u'_{qi}(t)), J_2(u'_{qi}(t)), J_p(u'_{qi}(t)))$, respectively, the solution $u_{q_i}(t)$ is said to dominate $u'_{qi}(t)$ if $J_\infty(u_{q_i}(t)) \le J_\infty(u'_{qi}(t)), J_2(u_{q_i}(t)) \le J_2(u'_{qi}(t)),$ and $J_p(u_{q_i}(t)) \le J_p(u'_{qi}(t))$, and at least one of the inequalities is a strict inequality.

Definition 12.2 ([368, 369]): (Pareto optimality) The feasible solution $u_{q_i}(t)$ is said to be Pareto optimality with respect to the feasible set ϕ if and only if there does not exist another feasible solution that dominates it.

Definition 12.3 ([368, 369]): (Pareto optimal set) For a given MOP in (12.23), the Pareto optimal solution set δ^*_{set} is defined as

$$\delta^*_{set} \triangleq \{u^*_{q_i}(t) \in \phi \,|\, u^*_{q_i}(t) \text{ is Pareto optimality}\}$$

Definition 12.4 ([368, 369]): (Pareto front) For a given MOP in (12.23) and the Pareto optimal solution set δ^*_{set}, the Pareto front F^* is defined as

$$F^* \triangleq \{(J_\infty(u^*_{q_i}(t)), J_2(u^*_{q_i}(t)), J_p(u^*_{q_i}(t))) \,|\, u^*_{q_i}(t)) \in \delta^*_{set}\}$$

In general, it is not easy to solve the multi-objective SINR target tracking problem in (12.23) by conventional methods directly. An indirect method is proposed to solve it in the following.

Theorem 12.1: Assume that β, γ, and δ are the upper bounds of the performance indices in (12.20)–(12.21), respectively. Then the multi-objective SINR target tracking problem in (12.23) is equivalent to the following suboptimal tracking problem in the Pareto optimal sense:

$$\min_{u_{q_i}(t) \in \phi} (\beta, \gamma, \delta)$$

(12.24)

subject to (12.14)

(12.25)

$$J_\infty(u_{q_i}(t)) \leq \beta \tag{12.26}$$

$$J_2(u_{q_i}(t)) \leq \gamma \tag{12.27}$$

$$J_p(u_{q_i}(t)) \leq \delta \tag{12.28}$$

Proof: To derive this result, we only need to prove that all of the inequality constraints in (12.26)–(12.28) become equalities when the MOP in (12.24) achieves Pareto optimal solutions. We will show the result by contradiction. Given a 4-tuple optimal solution vector $(u_{q_i}^*(t), \beta^*, \gamma^*, \delta^*)$ for the MOP in (12.24)–(12.28), we assume that the inequality in (12.26) remains the strict inequality at the Pareto optimal solution, that is, $J_\infty(u_{q_i}^*(t)) < \beta^*$. Therefore, there exists another lower bound $\hat{\beta}$ such that $\hat{\beta} < \beta^*$ and $J_\infty(u_{q_i}^*(t)) = \hat{\beta}$. It is obvious that the solution $(u_{q_i}^*(t), \hat{\beta}, \gamma^*, \delta^*)$ dominates the solution $(u_{q_i}^*(t), \beta^*, \gamma^*, \delta^*)$, leading to a contradiction. Similarly, the inequalities in (12.27)–(12.28) will become equalities when the Pareto optimal solutions are achieved. Thus, we can guarantee that the inequalities in (12.26)–(12.28) become equalities for the Pareto optimal solutions in (12.24). This implies that the suboptimal solution in (12.24)–(12.28) is equivalent to the Pareto optimal solution in (12.23).

Theorem 12.1 provides an indirect method to solve the multi-objective SINR target tracking problem in (12.23). According to (12.24)–(12.28), the inequality constraint on robust H_∞ tracking performance in (12.26) is equivalent to the following inequality:

$$\sum_{t=0}^{t_e} q_1 e_{q_i}^2(t) + r_1 u_{q_i}^2(t) \leq \beta \sum_{t=0}^{t_e} \zeta_{q_i}^2(t) \tag{12.29}$$

where t_e is the transmitted signal length. From (12.19), by substituting $e_{q_i}(t) = dx_{q_i}(t)$ and considering the effect of the initial condition, that is, $x_{q_i}(0) \neq 0$, (12.29) can be modified as:

$$\sum_{t=0}^{t_e} q_1 x_{q_i}^T(t) d^T dx_{q_i}(t) + r_1 u_{q_i}^2(t) \leq x_{q_i}^T(0) P x_{q_i}(0) + \beta \sum_{t=0}^{t_e} \zeta_{q_i}^2(t) \tag{12.30}$$

for the symmetric positive definite matrix $P > 0$. The following theorem will provide sufficient conditions for the control command $u_{q_i}(t)$ to solve the MOP in (12.23) or (12.24)–(12.28).

Theorem 12.2: Let us denote the Lyapunov function $V(x_{q_i}(t)) = x_{q_i}^T(t) P x_{q_i}(t)$ of the SINR tracking control system in (12.19) for $P = P^T > 0$. If we let $W = P^{-1}$ and $y = kW$, or $k = yW^{-1}$ subject to (12.14) and (12.18), then the inequality constraint in (12.26) could be transformed into (12.31), the inequality constraint in (12.27) could be transformed into (12.32) and (12.33), and the inequality constraint (12.28) could be transformed into the inequality constraints in (12.34) and (12.35).

$$\begin{bmatrix} W & Wd^T & y^T & 0 & WA^T + y^T b^T \\ dW & q_1^{-1} & 0 & 0 & 0 \\ y & 0 & r_1^{-1} & 0 & 0 \\ 0 & 0 & 0 & \beta & c^T \\ AW + by & 0 & 0 & c & W \end{bmatrix} \geq 0 \tag{12.31}$$

$$\begin{bmatrix} W & Wd^T & Wd^T + y^T b^T \\ dW & q_2^{-1} & 0 \\ AW + by & 0 & W \end{bmatrix} \geq 0 \tag{12.32}$$

$$\begin{bmatrix} (t_e + 1)\gamma & x_{q_i}^T(0) \\ x_{q_i}(0) & W \end{bmatrix} \geq 0 \tag{12.33}$$

$$\begin{bmatrix} W - q_3 cc^T & AW + by \\ WA^T + y^T b^T & q_3^{-1} W \end{bmatrix} \geq 0 \tag{12.34}$$

$$\frac{\delta}{d+1} I_{d+1} - Wg^T g \geq 0 \tag{12.35}$$

Proof: Please refer to Appendix 12.7.1.

According to Theorem 12.2, the inequality constraints in (12.26)–(12.28) can be replaced by the LMIs in (12.31)–(12.35). Thus, the multi-objective SINR target tracking problem in (12.23) can be transformed into the following LMI-constrained multi-objective optimization problem:

$$(\beta^*, \gamma^*, \delta^*) = \min_{\{W,y\} \in \eta} (\beta, \gamma, \delta) \tag{12.36}$$

subject to the LMIs in (12.31)–(12.35)

where η is the feasible set.

Remark 12.2: For the MOP in (12.36), the robust optimal H_∞ SINR target tracking problem to minimize (12.20) can be reduced as the following single-objective optimization problem:

$$\beta_L = \min_{\{W,y\}} \beta \tag{12.37}$$

subject to the LMI in (12.31)

The optimal H_2 SINR target tracking problem to minimize (12.21) can be reduced as the following SOP:

$$\gamma_L = \min_{\{W,y\}} \gamma \tag{12.38}$$

subject to the LMIs in (12.32)–(12.33)

The minimum average power consumption for the SINR target tracking problem to minimize (12.22) can be reduced as the following SOP:

$$\delta_L = \min_{\{W, y\}} \delta$$

subject to the LMIs in (12.34)-(12.35)

(12.39)

Normally, we have $\beta^* \geq \beta_L$, $\gamma^* \geq \gamma_L$, and $\delta^* \geq \delta_L$ of the Pareto optimal solutions $(\beta^*, \gamma^*, \delta^*)$ for the MOP in (12.36). β_L, γ_L, and δ_L can be used as the lower bounds of β^*, γ^*, and δ^*, respectively. Thus, we have the following definition:

Definition 12.5: (Ideal Objective Index) The ideal objective index $(\beta_L, \gamma_L, \delta_L)$ for the MOP in (12.36) is defined as the optimal objective indices of the LMI-constrained SOPs in (12.37)–(12.39), respectively.

Now, the feasible solutions of the LMI-constrained MOP in (12.36) are offline, obtained by the convex optimization toolbox in MATLAB for the given β, γ, and δ subject to the LMIs in (12.31)–(12.35) with W and y. We could design the multi-objective beamforming control gain $k = yW^{-1}$ offline to generate power control command $u_{q_i}(t) = kx_{q_i}(t)$ for each user and build the beamforming control command $z_{q_i}(t)$ from (12.10). Note that for the general MOP, the unique global solution does not exist for optimality. However, we can obtain a preferred Pareto optimal solution set of (12.36) in the sense of MOP. In the next section, we will introduce the LMI-constrained MOEA to obtain the Pareto optimal solutions for the MOP in (12.36).

12.4 PARETO OPTIMAL SOLUTIONS TO MULTI-OBJECTIVE BEAMFORMING CONTROL DESIGN

Up to this, the beamforming control gain k can achieve the multi-objective SINR target tracking performance by solving the LMI-constrained MOP in (12.36). As shown in Figure 12.2, the purpose of the SINR tracking system through the power control command $u_{q_i}(t) = kx_{q_i}(t)$ is to generate the beamforming control command $z_{q_i}(t)$ in (12.10), then transfer it into the online semi-definite programming part to obtain the corresponding beamforming matrix $W_{q_i}(t)$ from (12.7). In this section, we will introduce how to combine LMIs and MOEA to efficiently solve the MOP in (12.36).

12.4.1 LMI-CONSTRAINED MOEAs

A conventional MOEA is a practical scheme based on the evolution process and natural selection. For wireless communication networks, MOEAs have been seen as a possible candidate for solving the MOP of beamforming control design due to avoiding local optima with effective search of the Pareto optimal solutions. Different from the traditional algebra-constrained MOEA searches, the objective values $(\beta^*, \gamma^*, \delta^*)$ with the help of interior point methods, in this chapter, we choose NSGA-II as the structural basis of the algorithm [2]. The MOEA first encodes the individuals $(\beta_n, \gamma_n, \delta_n)$ for all $n = 1, ..., N_p$ as the initial parent population P in the feasible set, where the initial iteration number is set as $r = 1$. The symbol P_r is the population at the rth iteration. That is, the nth individual $(\beta_n, \gamma_n, \delta_n)$ must satisfy the LMIs in (12.31)–(12.35) with $\beta_U \geq \beta_n \geq \beta_L$, $\gamma_U \geq \gamma_n \geq \gamma_L$, and $\delta_U \geq \delta_n \geq \delta_L$ for all $n = 1,\dots,N_p$, where N_p is the population size. β_U, γ_U, and δ_U are the upper bounds of each objective value in (β, γ, δ), respectively. Note that infeasible individuals need to be excluded from the candidate set. Finally, the Pareto optimal solutions (W^*, y^*) with multi-objective beamforming control gain $k^* = y^*(W^*)^{-1}$ can also be obtained after achieving the set of approximate the Pareto front F^*. The algorithm of the LMI-constrained MOEA to solve the MOP in (12.36) is summarized in Algorithm 12.1. Once the preferable objective values $(\beta^*, \gamma^*, \delta^*)$ are selected, the corresponding Pareto optimal solution (W^*, y^*) can also be obtained with the help of the LMI toolbox in MATLAB.

Remark 12.3: The proposed LMI-constrained MOEA in Algorithm 12.1 indeed converges to the Pareto front (i.e., the Pareto optimal set) when the iteration number $N_m \to \infty$.

Proof: First, in each iteration of the proposed algorithm, the infeasible solutions of the LMI-constrained MOP in (12.36) will be deleted from the candidate set. Thus,

---

ALGORITHM 12.1
LMI-constrained MOEA.

Input: The LMIs in (12.31)–(12.35), maximum iteration number N_m, population size N_p, crossover probability rate M_c and mutation probability rate M_m.

Output: The Pareto front F^* and the preferable Pareto optimal solution $(\beta^*, \gamma^*, \delta^*)$.

1: Generate N_p feasible individuals randomly as the initial population P_r; that is, each individual $(\beta_n, \gamma_n, \delta_n)$ must satisfy the LMIs in (12.31)–(12.35) and $\beta_U \geq \beta_n \geq \beta_L$, $\gamma_U \geq \gamma_n \geq \gamma_L$, $\delta_U \geq \delta_n \geq \delta_L$, and $\forall n = 1,\dots,N_p$. Set iteration number $r = 1$.

2: **while** $r \leq N_m$ **do**

3: Perform the evolution algorithm to P_r with M_c and M_m to produce N_p feasible individuals $(\beta_n, \gamma_n, \delta_n)$, $\forall n = 1,\dots,N_p$ (i.e., by solving the LMIs in (12.31)–(12.35) for each individual) as the child population P_r^{EA}.

4: Combine P_r with P_r^{EA} as the new population P_r' with $2N_p$ feasible individuals, that is, $P_r' = \{P_r, P_r^{EA}\}$.

5: Assign the nondomination level and crowding distance based on $(\beta_n, \gamma_n, \delta_n)$, $\forall n = 1,\dots,2N_p$ in P_r'.

6: Make $2N_p$ feasible individuals in P_r' perform elitist strategies according to their fitness (rank) level, that is, the so-called crowded comparison operator, to select N_p elitist individuals.

7: Set iteration number $r = r+1$ and N_p elitist individuals as the population P_{r+1}.

8: **end while**

9: Set Pareto front $F^* = P_{N_m}$, and choose the preferable Pareto optimal solution in F^* as $(\beta^*, \gamma^*, \delta^*)$ according to the designer's preference.

---

the search region of the proposed algorithm is within the feasible set of the MOP. Denote $F(P_r)$ as the objective set of the rth iteration (i.e., the set consists of all the objective indices $(\beta_n, \gamma_n, \delta_n)$, $\forall n = 1, ..., N_p$ of the candidate solution in the rth iteration). The candidate population set P_r consists of N_p feasible solutions in the rth iteration. In each iteration, Algorithm 12.1 will combine N_p parent solutions in the previous iteration with N_p child solutions as a new population set with $2N_p$ solutions. Next, Algorithm 12.1 performs an elitist strategy to select half of the better N_p solutions in this new population set as the input of the next iteration [2]. "The better solution" means that the solution dominates another solution in the original population according to Definition 12.1. Thus, we have

$$F(P_r) \prec F(P_{r-1}), \forall r = 1, ..., N_m \tag{12.40}$$

where N_m is the iteration number. The operator $\prec$ in (12.40) means that at least one of the solutions in P_r dominates the solutions in P_{r-1}, that is, the nondominated sorting algorithm in the MOEA [2]. We can make sure that the objective set will be better than the previous iterations or keep the same in the proposed Algorithm 12.1 (i.e., $F(P_1) \prec F(P_0)$, $F(P_2) \prec F(P_1)$, $F(P_3) \prec F(P_2)$, and so on). According to the monotone sequence convergence theorem in real analysis, we can conclude that the proposed LMI-constrained MOEA in Algorithm 12.1 will indeed converge to the Pareto front (i.e., the Pareto optimal set) when $N_m \to \infty$.

Regarding the feasibility of the proposed multi-objective approach, we assume that there is a feasible solution (W, y) for the LMIs in (12.31)–(12.35) with the specific transmitted signal length $t_e = \underline{t}_e$.

Theorem 12.3: If the LMI-constrained MOPs in (12.36) with $t_e = \underline{t}_e$ are feasible, the proposed LMI-constrained MOEA presented in Algorithm 12.1 always produces feasible solutions for $t_e = \underline{t}_e$.

Proof: Let $(W, y, \beta, \gamma, \delta, \underline{t}_e)$ present a feasible solution of the LMI-constrained MOP in (12.36). Define

$$\Gamma(t_e) = (W, y, \beta, \gamma, \delta, t_e) \tag{12.41}$$

and hence $\Gamma(\underline{t}_e)$ indicates the feasible solution $(W, y, \beta, \gamma, \delta, \underline{t}_e)$. The proof of the theorem is to show that the solution $\Gamma(t_e)$ satisfies the constraints in (12.31)–(12.35) for all $t_e \geq \underline{t}_e$. The feasibility of $\Gamma(\underline{t}_e)$ implies that the constraints in (12.31)–(12.32) and (12.34)–(12.35) are satisfied by $\Gamma(t_e)$, as $\underline{t}_e$ is not contained in (12.31)–(12.32) and (12.34)–(12.35). According to (12.33), for all $t_e \geq \underline{t}_e$, we have the fact that

$$(t_e + 1)\gamma - x_{q_i}^T(0)W^{-1}x_{q_i}(0) \geq (\underline{t}_e + 1)\gamma - x_{q_i}^T(0)W^{-1}x_{q_i}(0) \geq 0 \tag{12.42}$$

The inequality in (12.42) implies that (12.33) is satisfied by $\Gamma(t_e)$, which completes the proof.

Theorem 12.3 provides us an efficient way to specify the transmitted signal length t_e. Once a feasible solution $\Gamma(\underline{t}_e)$ is chosen, the LMI-constrained MOP with $t_e \geq \underline{t}_e$ also guarantees feasibility.

Remark 12.4: After the offline design beamforming control gain k is generated by the proposed multi-objective optimization algorithm, the beamforming matrix $W_{q_i}(t)$ still needs to be specified to satisfy the equation in (12.7) after the beamforming control command $z_{q_i}(t)$ is generated by the equation in (12.10). Here we employ the following quadratic optimization method with a decentralized structure to solve $W_{q_i}(t)$ in (12.7) for the ith user in time slot t:

$$\min_{W_{q_i}(t)} \left\| W_{q_i}(t) \right\|_F^2$$

$$\text{subject to } \left\| \widehat{H}_{qq_i}(t) W_{q_i}(t) \right\|_F^2 = f_{q_i}(t), \tag{12.43}$$

$$\sum_{m \neq i}^{K_q} \left\| \widehat{H}_{qq_m}(t) W_{q_i}(t) \right\|_F^2 + \sum_{r \neq q}^{Q} \sum_{n=1}^{K_r} \left\| \widehat{H}_{qr_n}(t) W_{q_i}(t) \right\|_F^2 = E_{q_i}$$

where $f_{q_i}(t) = 10^{0.1 z_{q_i}(t)}$ is obtained from (12.7). Note that the power level $E_{q_i} \in \mathbb{R}^+$ in the second constraint of (12.43) is to weaken the leakage interference power for an acceptable link quality in the SINR target tracking. The meaning of the objective function in (12.43) is to minimize the beamforming power for the ith user in time slot t. This kind of quadratic optimization problem in (12.43) can be easily solved by semi-definite relaxation methods [36]. After some rearrangements with a new matrix variable $G_{q_i}(t) = W_{q_i}(t) W_{q_i}^H(t)$, we have the following decentralized SDP to solve the quadratic optimization problem in (12.43) [36]:

$$\min_{G_{q_i}(t)} Tr(G_{q_i}(t))$$

$$\text{subject to } Tr\left(G_{q_i}(t) \widehat{H}_{qq_i}^H(t) \widehat{H}_{qq_i}(t) \right) = f_{q_i}(t)$$

$$\sum_{m \neq i}^{K_q} Tr\left(G_{q_i}(t) \widehat{H}_{qq_m}^H(t) \widehat{H}_{qq_m}(t) \right) \tag{12.44}$$

$$+ \sum_{r \neq q}^{Q} \sum_{n=1}^{K_r} Tr\left(G_{q_i}(t) \widehat{H}_{qr_n}^H(t) \widehat{H}_{qr_n}(t) \right) = E_{q_i}$$

Note that the beamforming matrix $W_{q_i}(t)$ for the ith user in the qth cell can be obtained by using eigenvalue decomposition to the positive semidefinite matrix $G_{q_i}(t)$.

Remark 12.5: The proposed SINR target tracking scheme includes two design steps, the offline multi-objective controller gain design process and the online beamforming update process in the downlink transmission. We will analyze the computational complexity of these two steps point by point as follows:

(1) The LMI-constrained MOEA for the offline design controller gain K: The computational complexity of the proposed LMI-constrained MOEA (i.e., the Algorithm 12.1) is approximately estimated as $O((d+1)(d+2)N_p^2 N_m)$, including $O(\frac{(d+1)(d+2)}{2})$ for solving the LMIs and $O(2N_p^2 N_m)$ for the MOEA, where $d+1$ is the dimension of the system state $x_{q_i}(t) \in \mathbb{R}^{(d+1) \times 1}$ in (12.13).

(2) The online update semidefinite programming in (12.44): We assume that the transmitted symbol numbers L_{q_i} are set the same as L for all users ($L_{q_i} = L$, $\forall q,i$), and each cell has the same user number (i.e., $K_q = U$, $\forall q$). Thus, the computational complexity of the proposed SDP in (12.44) is estimated as $O(QU(N_t^2)\log(1/\varepsilon))$ including QU separated optimizations to obtain QU beamforming matrices, where ε is the desired accuracy of the solution [371].

Remark 12.6: According to Remark 12.5, online computational complexity is mainly from semi-definite programming. When the network dimension is large, the computational time of solving the SDP optimization problems may not be small enough within the coherence time to overcome fast small-scale fading. Although SDP optimization can be solved in polynomial time, the computational time will be high under a large number of variables. This is the inherent bottleneck of SDP optimization. But, even so, the proposed multi-objective beamforming power control algorithm is still a good solution to produce training data for a deep learning framework for a massive MIMO wireless system, such as the related research in [372]. According to [372], the outputs of the deep convolutional neural network (CNN) (which is named PowerNet) are the pilot and the data powers for the massive MIMO system. Thus, the proposed multi-objective beamforming power control algorithm can provide a suitable training data set for PowerNet. After PowerNet is completely trained, the computing unit of the proposed algorithm can be replaced by this deep CNN with less computational complexity.

12.5 SIMULATION RESULTS

To illustrate the proposed multi-objective SINR target tracking, a conventional multicell MU-MIMO interference wireless system is designed to guarantee the optimal H_∞ SINR target tracking and QoS performance with the minimum average power consumption simultaneously. The significant system parameters are summarized in Table 12.1. The path loss model is according to the non-light of sight (NLOS) urban micro-scenario in release document 9 from signal generation partnership project [373], which is defined as $36.7\log_{10}(D) + 22.7 + 26\log_{10}(f_c)$. The symbol D is the distance in meters, and f_c is the frequency in GHz. The multicell MU-MIMO wireless system includes two cells with $Q = 2$, $N_c = 2$, and $N_t = 10$. Each cell consists of four users, that is, $K_q = 4$, $\forall q$. The cell radius is 400m, and the reference distance is 50m. Each receiver is uniformly distributed between the reference distance and the cell radius in each cell. To evaluate the effect of noise power on the proposed SINR target tracking system, we define the normalized noise power level N_o as $N_o = \rho_{q_i}^2 / \left\| H_{qq_i}(t) \right\|_F^2$, $\forall q,i$. According to [354], the uncertain channel matrix $\Delta H_{rq_i}(t)$ can be generated by independent and identically distributed complex Gaussian distribution. Thus, we generate the uncertain channel matrix by $\Delta H_{rq_i}(t) = \Lambda^2 \left\| H_{rq_i}(t) \right\|_F \bar{H}_{rq_i}(t)$, where each element of the matrix $\bar{H}_{rq_i}(t)$ follows i.i.d. standard complex Gaussian distribution with unit variance. The scalar Λ^2 with $0 < \Lambda^2 < 1$ indicates the effect of channel uncertainty level from small to large. All of the actual channel matrices are generated

TABLE 12.1

System Parameters for Multicell MU-MIMO Wireless System

Path loss model	NLOS urban micro scenario, 3GPP [373]
Multipath fading distribution	i.i.d. Rayleigh fading
Carrier center frequency	2000 MHz
Number of transmitting antennas, N_t	10
Number of receiving antennas, N_r	2
Base station antenna height	10 meters
Mobile station antenna height	1.5 meters
Transmitting antenna gain	15 dBi
Cell radius	400 meters
Required interference power level, E_i, $\forall i$	−30 dBm
Fixed target SINR level, $\alpha_i(t)$, $\forall i,t$	10 dB

TABLE 12.2

Offline Design Control Parameters for the SINR Target Tracking System with Round Trip Delay $d = 2$

Pareto Optimal Solutions	Optimal Objective Values $(\beta^*, \gamma^*, \delta^*)$	Corresponding Control Gain k^*
Solution 1	$(0.0147, 0.0002, 163.2646)$	$[0.0977, -3.8536 \times 10^{-11}, -0.0967]$
Solution 2	$(0.0034, 0.0005, 449.2132)$	$[0.2783, -1.7069 \times 10^{-10}, -0.2749]$
Solution 3	$(0.0507, 0.0001, 444.9742)$	$[0.1060, -8.2897 \times 10^{-11}, -0.1063]$
Solution 4	$(0.0526, 0.0005, 25.8006)$	$[0.0548, -1.0194 \times 10^{-12}, -0.0515]$

by i.i.d. Rayleigh fading channels [374] (i.e., each element of the channel response follows complex Gaussian normal distribution $CN(0,1)$ with the 3GPP path loss model). The simulations are implemented by 1000 channel realizations to guarantee the reliability of the result. The weighting factors q_1, r_1, q_2, and q_3 in Theorem 12.2 can be used as the free design parameters, which represent the trade-off between the system response and the tracking error performance. In the simulation, we set $x_i(0) = [10, 0, 0]^T$, $q_1 = 0.00008$, $r_1 = 0.0003$, $q_2 = 0.00001$, $q_3 = 1$, and $t_e = 1000$, and the corresponding Pareto optimal control gain k^* can thus be obtained by solving the MOP in (12.36), as shown in Table 12.2. The interference power levels E_{q_i} for all $i = 1, ..., U$ are the same, with a value of −30dBm. We first introduce the simulation settings of the LMI-constrained MOEA and demonstrate that the Pareto front F^*, is indeed closely approached by the proposed LMI-constrained MOEA search. Then we can obtain the preferable objective value $(\beta^*, \gamma^*, \delta^*)$ to obtain the Pareto optimal solution (W^*, y^*) with $k^* = y^*(W^*)^{-1}$.

12.5.1 SIMULATION SETTINGS FOR THE MOEA

In this subsection, the simulation settings for the LMI-constrained MOEA in Section 12.4 are given as follows: we set the maximum iteration number $N_m = 100$, the population size $N_p = 3000$, the crossover probability $M_c = 0.9$, mutation probability $M_m = 0.3$, distribution index for simulating binary crossover $\eta_c = 10$, and the distribution index for mutation operator $\eta_m = 10$. In all of the simulations, we use the proposed LMI-constrained MOEA to search for a set of the Pareto front F^*, the preferable Pareto optimal solution (W^*, y^*), and hence the beamforming control gain k^*.

According to the proposed design procedure, the Pareto optimal solutions indeed approach the Pareto front efficiently, as shown in Figure 12.3. Moreover, the trade-off between each Pareto optimal solution becomes more pronounced when the Pareto front approaches the ideal objective index, as shown in Figure 12.3. The solution diversity also performs well, which guarantees the effectiveness of the proposed multi-objective optimization method. For performance comparison, we choose four Pareto optimal solutions as the preferable design schemes for the beamforming control design. We can say that solution 1 makes a compromise between the robust H_∞ solution, the optimal H_2 solution, and the average power consumption solution. Solution 2 tends to the robust H_∞ solution in (12.37), solution 3 tends to the optimal H_2 solution in (12.38), and solution 4 tends to the average power consumption solution in

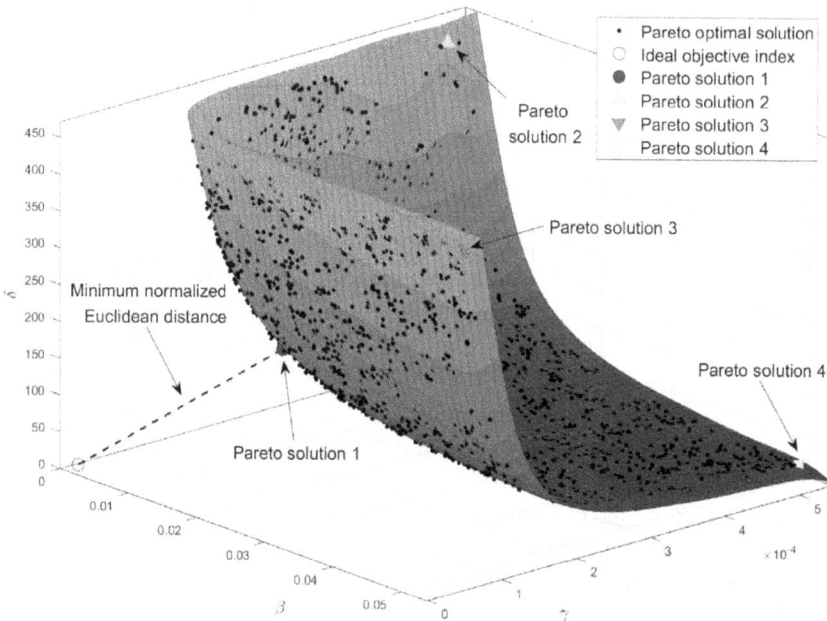

FIGURE 12.3   The Pareto front obtained from the proposed LMI-constrained MOEA with given weighted factors q_1, q_2, q_3, r_1, and round trip delay $d = 2$. The four Pareto optimal solutions are selected for comparison, as shown in the following figures. The curve fitting process by the MATLAB toolbox is also implemented to declare the trade-off region in the Pareto front.

(12.39). After the beamforming control gain k^* is obtained, we can use the MATLAB toolbox YALMIP with MOSEK to obtain the optimal beam-forming matrix $W_{q_i}^*(t)$ by solving the optimization problem in (12.44) for each user [375, 376].

12.5.2 PERFORMANCE STUDY

In this section, we shall compare the performance of different Pareto optimal solutions by the proposed multi-objective SINR target tracking scheme with other beamforming methods. Since the block diagonalization with water-filling method is a popular and conventional strategy for MIMO beamforming design, we choose BD with equal transmittal power for each user and the traditional water-filling block diagonalization (WFBD) [348] of the multicell case as two baseline schemes in the simulation with total transmittal power $P_t = 30$ dbm and $P_t = 40$ dbm. On the other hand, the practical QoS requirement in the next generation of wireless systems may require dynamic changes in different time periods. In view of this, we consider two cases to show the efficiency of the proposed methods, fixed target SINR and changed target SINR conditions. For the case of fixed target SINR, we denote the target SINR levels $\alpha_{q_i}(t)$ as the same for all users with a typical value $\alpha_{q_i}(t) = \alpha(t) = 10$ dB to ensure an acceptable QoS. For the case of changed target SINR, the target SINR level is set as a time-varying sinusoidal function, $\alpha_{q_i}(t) = 5 + 5\sin(0.05t), \forall q, \forall i$.

To compare SINR tracking ability with different methods, we define the standard deviation σ_e of tracking error with measured SINR $SINR(t)$ as the tracking error performance criterion:

$$\sigma_e = \sqrt{\frac{1}{T}\sum_{t=1}^{T}(\alpha(t) - SINR(t))^2} \qquad (12.45)$$

where T is the total transmitted length and is set as 1000 for all simulations. On the other hand, outages may arise for communication and result in instantaneous exacerbation of the QoS when the measured SINR goes below an outage threshold. Thus, we can also evaluate the quality of the communication link by the outage probability as follows:

$$P_{out} = \Pr\{SINR(t) \le SINR_{out}\} \qquad (12.46)$$

where $SINR_{out}$ is the outage threshold and is set as 0 dB in our performance evaluation.

Figure 12.4 shows the tracking error performance σ_e versus the uncertainty level Λ^2 of the four aforementioned Pareto optimal solutions with the fixed and changed target SINR conditions. For each case, the normalized noise power level N_o is set as −25 dB. Specifically, since the Pareto solution 2 (i.e., the robust H_∞ solution) prioritizes consideration of the worst-case effect on SINR target tracking, the tracking error performance (i.e., σ_e) is the top among the four Pareto solutions with the fixed target SINR condition. For Pareto solution 3 (i.e., the optimal H_2 solution), it is obvious that its tracking error performance will be better than that of Pareto solution 2 due to concern about the mean square SINR target tracking error performance.

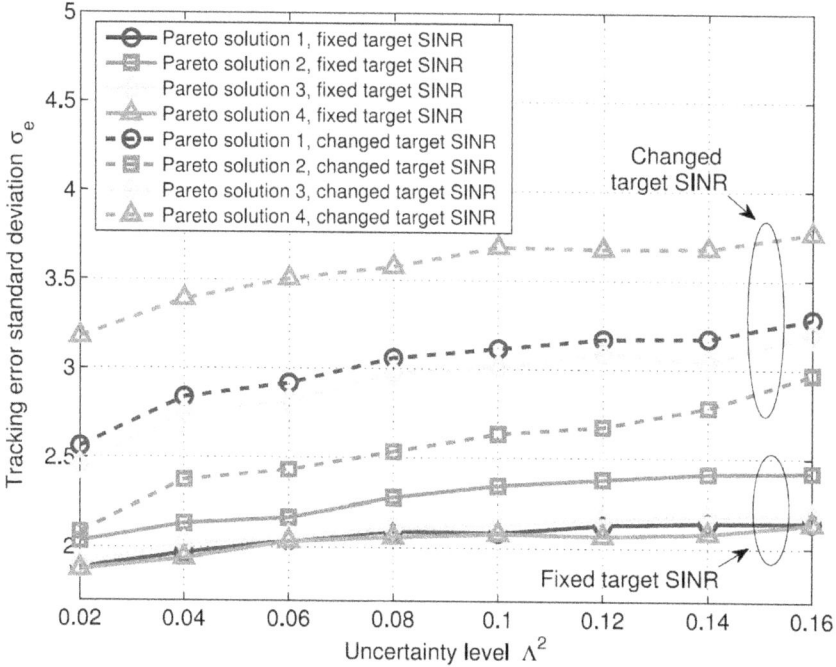

FIGURE 12.4 Comparison among four Pareto optimal solutions according to the tracking error performance σ_e versus different uncertainty levels Λ^2 for a fixed target SINR and changed target SINR. The normalized noise power level N_o is set as -25 dB.

Pareto solution 1 is a compromise between three objectives; thus its tracking error performance will naturally be lower than the aforementioned Pareto solutions in the fixed target SINR condition. Interestingly, the tracking error performance of Pareto solution 4 (i.e., the power efficient solution) is the greatest among the four Pareto solutions with a changed target SINR condition. The reason is that the SINR target tracking ability of Pareto solution 4 is difficult to catch up the change of dynamic target SINR level with the minimum power consumption. For the other three Pareto solutions with the changed target SINR condition, the tendency of the tracking error performance is similar to the cases with a fixed target SINR condition, which demonstrates the different design purposes among the four Pareto solutions on SINR target tracking.

For comparison between different normalized noise power levels and fast fading conditions on SINR target tracking, Figures 12.5 and 12.6 show the tracking error performance versus different normalized noise power levels N_o of the four Pareto optimal solutions with the fixed and changed target SINR conditions, respectively. Due to the change of noise power, it can be observed reasonably that the tracking error standard deviation rises dramatically when the normalized noise power level increases. Generally, the trend of four Pareto solutions with a fixed target SINR condition is similar to the cases in Figure 12.4; that is, the tracking error performance of Pareto solution 2, is the largest and Pareto solution 1 can be seen as the

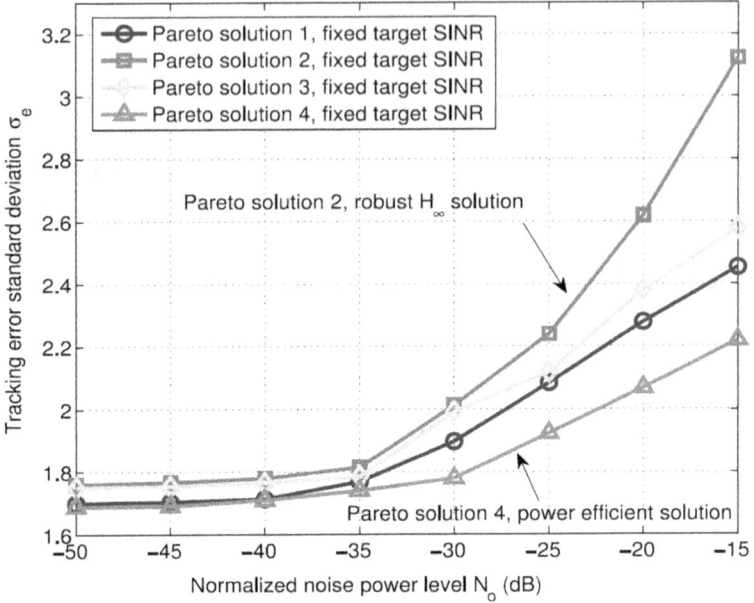

FIGURE 12.5　Comparison among four Pareto optimal solutions according to the tracking error performance σ_e with different normalized noise power levels N_o for a fixed target SINR condition. The uncertainty level $\Lambda^2 = 0.1$.

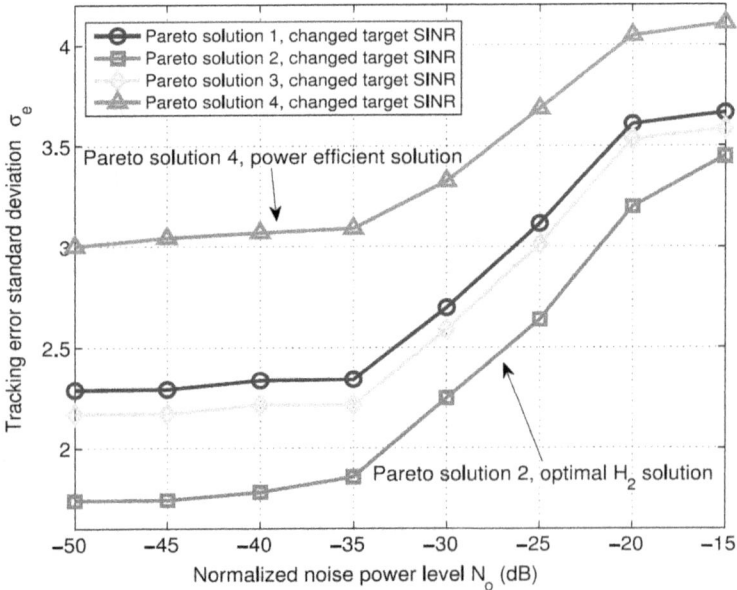

FIGURE 12.6　Comparison among four Pareto optimal solutions according to the tracking error performance σ_e with different normalized noise power levels N_o for the changed target SINR condition. The uncertainty level $\Lambda^2 = 0.1$.

moderate design criterion. For the changed target SINR condition, as shown in Figure 12.6, the tracking error performance of Pareto solution 4 also has the largest level between these Pareto solutions. This observation demonstrates that Pareto solution 3 can inhibit the external disturbance effect on SINR target tracking from an average energy point of view with the changed target SINR condition, but its suppressing ability for the worst-case effect on SINR target tracking is worse than that of Pareto solution 1, as shown in Figure 12.5.

For comparison, Figures 12.7 and 12.8 characterize the outage probability performance with a fixed target SINR condition of the proposed Pareto optimal solutions, equal power BD, and water-filling BD at different uncertainty levels and normalized noise power levels, respectively. Note that for all schemes, the outage threshold $SINR_{out}$ is set to 6 dB. As expected, the outage probability becomes large when the uncertainty level increases in each method. As can be seen from Figure 12.7, the proposed schemes all provide efficient performance gain when compared with equal power and water-filling BD. Since the baseline schemes do not consider imperfect CSI and SINR target tracking in their design procedures, their outage probability performs worse than the proposed methods. In Figure 12.8, for different normalized noise power levels, one can see that when the normalized noise power level increases, the outage probability of all the schemes increases naturally. Overall, the outage probability performance of all the proposed methods is better than the outage rate of the baseline schemes (i.e., equal power and water-filling BD).

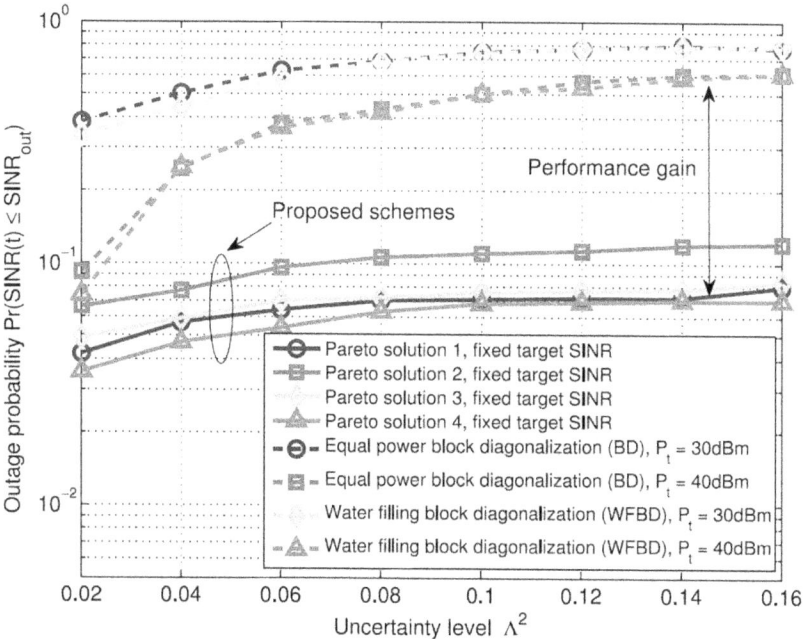

FIGURE 12.7 Comparison among four Pareto optimal solutions, equal power BD, and water-filling BD according to the outage probability with different uncertainty levels. The normalized noise power level N_o is set as −25 dB. The outage threshold $SINR_{out}$ is set as 6 dB.

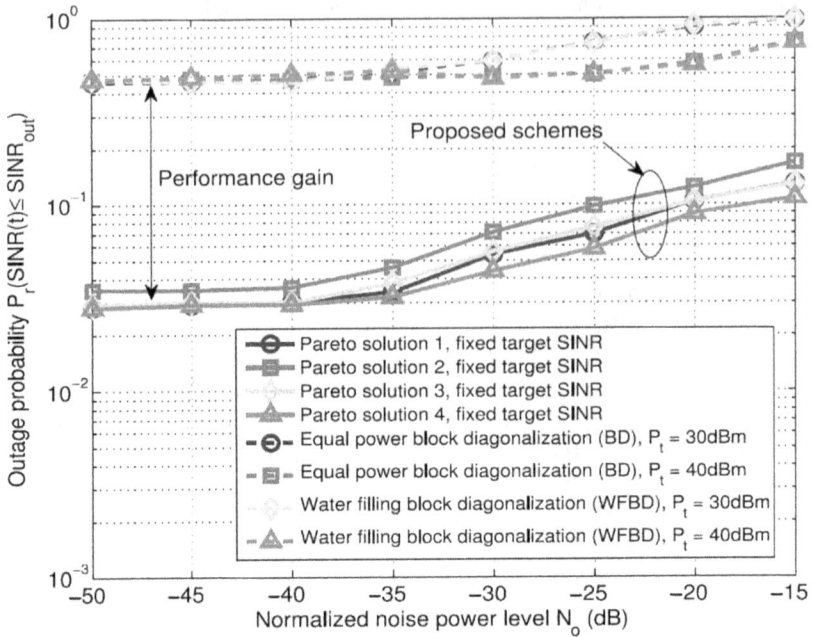

FIGURE 12.8 Comparison among four Pareto optimal solutions, equal power BD, and water-filling BD according to the outage probability versus different normalized noise power levels N_o. The uncertainty level $\Lambda^2 = 0.1$ and the outage threshold $SINR_{out}$ is set as 6 dB.

Figures 12.9 and 12.10 compare the total transmittal power consumptions obtained by the proposed Pareto optimal solutions and baseline schemes with a fixed target SINR condition at different uncertainty levels and different normalized noise power levels, respectively. Since the transmittal power constraints of the baseline schemes are set the same, the power consumption values of the baseline schemes are equal to 30 and 40dBm, as shown in Figures 12.9 and 12.10, respectively. As expected, the transmittal power of the proposed Pareto solutions is large when the uncertainty level increases due to the design concern of imperfect CSI as shown in Figure 12.9. Also, since Pareto optimal solution 4 emphasizes power-saving more than the others, the total transmittal power of Pareto optimal solution 4 is the smallest among the four Pareto solutions. We can observe that the proposed schemes can effectively save transmittal power compared with equal power and water-filling BD, as shown in Figure 12.9, which indicates the effectiveness of the proposed methods for power saving at different uncertainty levels. In Figure 12.10, one can see that the proposed Pareto solutions feature more power saving than the baseline schemes when the normalized noise power level is smaller than −20 dB. Since the proposed robust SINR target tracking schemes make meeting the fixed target SINR level a priority and consider imperfect CSI in design procedure, the transmittal power of the proposed Pareto optimal solutions might be greater than the baseline schemes at high normalized noise power levels with a large number of users in the multicell MU-MIMO wireless system. However, this deficiency

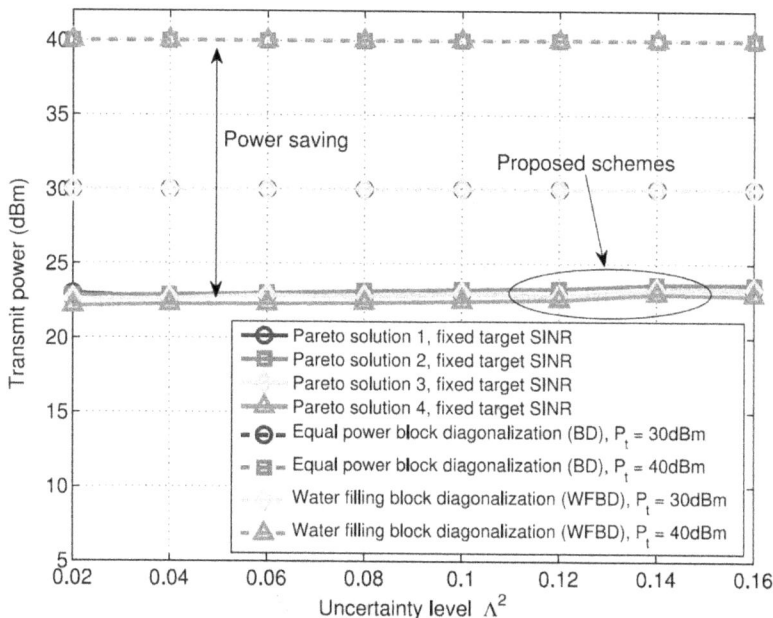

FIGURE 12.9 Comparison among four Pareto optimal solutions, equal power BD, and water-filling BD according to the transmittal power with different uncertainty levels. The normalized noise power level N_o is set as -25 dB.

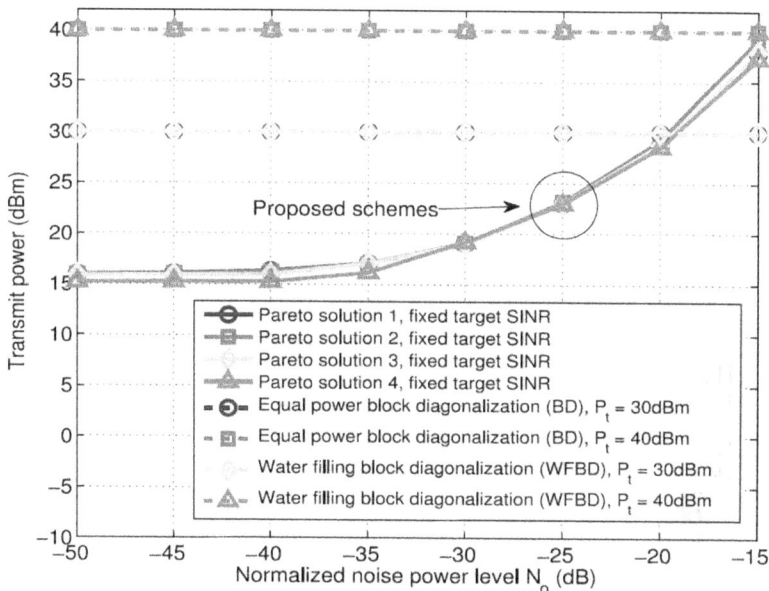

FIGURE 12.10 Comparison among four Pareto optimal solutions, equal power BD, and water-filling BD according to the transmittal power versus different normalized noise power levels. The uncertainty level is set as $\Lambda^2 = 0.1$.

can be mitigated by changing the target SINR level at different normalized noise power levels. Note that in this study, how to properly assign the target SINR level is not discussed, and this topic with deep learning methods can be future work.

For comparison between different numbers of users on SINR target tracking, Figure 12.11 shows the tracking error performance versus different numbers of users per cell of the four Pareto optimal solutions with the fixed and changed target SINR conditions, respectively. In the proposed SINR target tracking system model, the tracking error will be large mainly due to the variation $v_{q_i}(t) - v_{q_i}(t-1)$ of the interference power according to (12.12) instead of the magnification of the interference power. Thus, when the number of users is large, we can observe that the proposed SINR target tracking methods have a more robust ability to attenuate the effect of interference and channel estimation error from other cells, causing a significant performance improvement for SINR target tracking. The tendency of tracking error performance in Figure 12.11 is similar to the four Pareto optimal cases in Figure 12.4. In Figure 12.12, we show the effect of different numbers of users on the outage probability performance between the proposed four Pareto optimal solutions, equal power BD, and water-filling BD. Since the equal power BD and water-filling BD do not have the robust ability to attenuate the effect of channel imperfection, they will suffer from larger interference due to the increase in the number of intercell interference

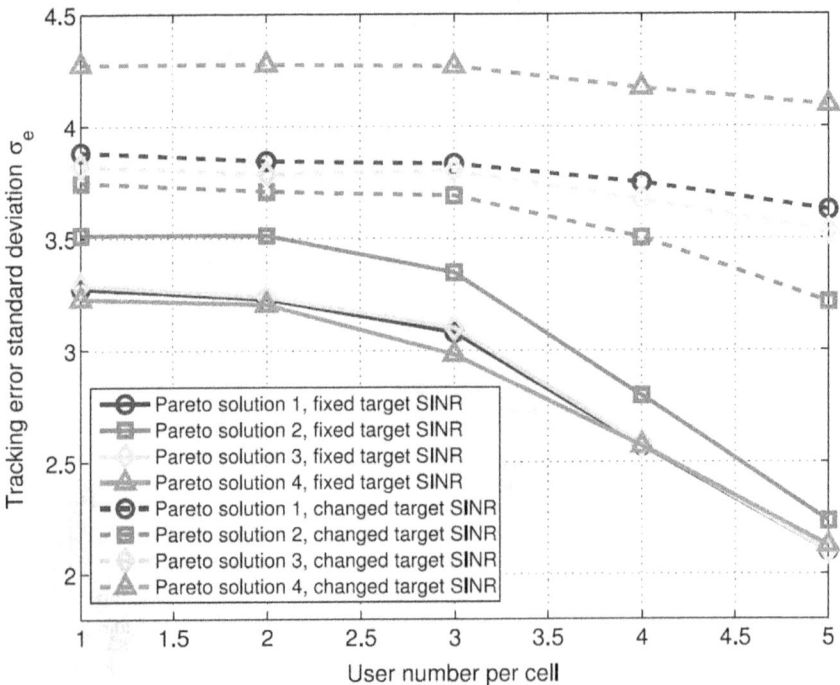

FIGURE 12.11 Comparison among four Pareto optimal solutions according to the tracking error performance σ_e versus different numbers of users per cell for fixed target SINR and changed target SINR conditions. In this figure, $N_t = 16$, $N_r = 2$, $L = 2$. The normalized noise power level N_o is set as -15 dB. The uncertainty level $\Lambda^2 = 0.1$.

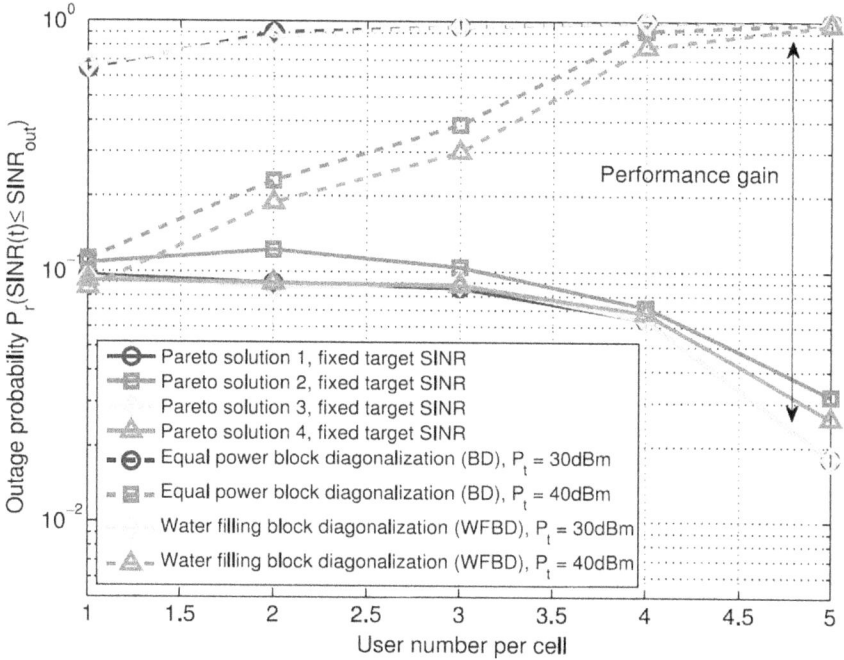

FIGURE 12.12 Comparison among four Pareto optimal solutions, equal power BD and water-filling BD according to the outage probability with different numbers of users per cell. In this figure, $N_t = 16$, $N_r = 2$, $L = 2$. The normalized noise power level N_o is set as -15 dB. The uncertainty level $\Lambda^2 = 0.1$, and the outage threshold $SINR_{out}$ is set as 7 dB.

channels. Conversely, the proposed tracking control methods can effectively maintain the target SINR level, even if it suffers from a large amount of channel uncertainty and intercell interference. Overall, the proposed SINR target tracking schemes still have an outstanding outage probability performance when the user number increases.

12.6 CONCLUSION

In this chapter, we introduced a multi-objective optimal H_2/H_∞ SINR target tracking design with a minimum average power consumption scheme for a multicell MU-MIMO wireless beamforming system with imperfect channel information. A design procedure with an SINR tracking system is considered to satisfy three objectives without the information of statistical characteristics and the upper bound information of channel uncertainty. First, the original MOP is equivalently transferred into another MOP by minimizing the corresponding upper bounds and is finally converted into an LMI-constrained MOP. Then a modified MOEA is proposed to efficiently solve the LMI-constrained MOP for control gain and the corresponding beamforming matrix. Simulation examples of SINR target tracking with imperfect channel coefficients are provided to illustrate the design aspect and to evaluate the SINR target tracking performance of the proposed multi-objective optimal robust H_2/H_∞ beamforming power control scheme with a minimum average power consumption in multicell MU-MIMO wireless beamforming systems. Further, how to properly assign the target SINR level in each case by deep learning methods should be investigated, and how to apply multi-objective power control schemes to different communication systems will become a potential research direction too.

12.7 APPENDIX

12.7.1 PROOF OF THEOREM 12.2

(1) For the robust H_∞ SINR target tracking performance in (12.26), according to (12.30), we have

$$
\sum_{t=0}^{t_e} q_1 x_{q_i}^T(t) d^T dx_{q_i}(t) + r_1 u_{q_i}^2(t)
$$

$$
\leq x_{q_i}^T(0) P x_{q_i}(0) + \sum_{t=0}^{t_e} [q_1 x_{q_i}^T(t) d^T dx_{q_i}(t) + r_1 u_{q_i}^2(t)
$$

$$
+ x_{q_i}^T(t+1) P x_{q_i}(t+1) - x_{q_i}^T(t) P x_{q_i}(t) - \beta \zeta_{q_i}^2(t)]
$$

$$
+ \beta \sum_{t=0}^{t_e} \zeta_{q_i}^2(t)
$$

(12.47)

The sufficient condition that the robust H_∞ SINR target tracking performance index $J_\infty(u_{q_i}(t))$ is bounded by β in (12.24) is to satisfy the following inequality.

$$
q_1 x_{q_i}^T(t) d^T dx_{q_i}(t) + r_1 u_{q_i}^2(t)
$$

$$
+ r_1 u_{q_i}^2(t) - x_{q_i}^T(t) P x_{q_i}(t) - \beta \zeta_{q_i}^2(t) \leq 0
$$

(12.48)

Thus, if the inequality (12.48) holds, then the robust H_∞ tracking performance index $J_\infty(u_{q_i}(t))$ is also bounded by β in (12.24). By substituting (12.18) and (12.19) into (12.48) and rearranging the equation, we have the equivalent matrix inequality as:

$$
\begin{bmatrix} x_{q_i}(t) \\ \zeta_{q_i}(t) \end{bmatrix}^T \left(\begin{bmatrix} (A+bk)^T \\ c^T \end{bmatrix} P \begin{bmatrix} (a+bk)^T \\ c^T \end{bmatrix}^T \right.
$$

$$
\left. - \begin{bmatrix} P - q_1 d^T d - r_1 k^T k & 0 \\ 0 & \beta \end{bmatrix} \right) \begin{bmatrix} x_{q_i}(t) \\ \zeta_{q_i}(t) \end{bmatrix} \leq 0
$$

(12.49)

It is obvious that equation (12.49) holds for all $x_{q_i}(t)$ and $\zeta_{q_i}(t)$ if and only if the following matrix inequality holds:

$$
\left(\begin{bmatrix} (A+bk)^T \\ c^T \end{bmatrix} P \begin{bmatrix} (a+bk)^T \\ c^T \end{bmatrix}^T - \begin{bmatrix} P - q_1 d^T d - r_1 k^T k & 0 \\ 0 & \beta \end{bmatrix} \right) \leq 0
$$

(12.50)

By multiplying the left and right sides of (12.50) by $diag(W,1)$ and substituting $y = kW$ into (12.50), the inequality in (12.50) is equivalent to the following equation:

$$-\begin{bmatrix} WA^T + y^T b^T \\ c^T \end{bmatrix} W^{-1} \begin{bmatrix} WA^T + y^T b^T \\ c^T \end{bmatrix}^T$$
$$+ \begin{bmatrix} W - q_1 W d^T dW - r_1 y^T y & 0 \\ 0 & \beta \end{bmatrix} \geq 0 \tag{12.51}$$

Using the Schur complement [90] to (12.51), we have the following equivalent linear matrix inequality:

$$\begin{bmatrix} W & Wd^T & y^T & 0 & WA^T + y^T b^T \\ dW & q_1^{-1} & 0 & 0 & 0 \\ y & 0 & r_1^{-1} & 0 & 0 \\ 0 & 0 & 0 & \beta & c^T \\ AW + by & 0 & 0 & c & W \end{bmatrix} \geq 0 \tag{12.52}$$

which is the LMI of (12.31) in Theorem 12.2.

(2) For the H_2 optimal SINR target tracking performance in (12.27), the performance index $J_2(u_{q_i}(t))$ can be indicated in the following for the SINR tracking model in (12.19) with disturbance $\zeta_{q_i}(t) = 0$ and the fact that $x_{q_i}^T(t_e + 1)Px_{q_i}(t_e + 1) \geq 0$:

$$J_2(u_{q_i}(t)) = E\{q_2 e_{q_i}^2(t)\}$$
$$\leq \frac{1}{t_e + 1} \sum_{t=0}^{t_e} [x_{q_i}^T(t)\{\Phi_{H_2}(k, P)\}x_{q_i}(t)] \tag{12.53}$$
$$+ \frac{1}{t_e + 1} x_{q_i}^T(0)Px_{q_i}(0)$$

where $\Phi_{H_2}(k, P) = q_2 d^T d + (A + bk)^T P(A + bk) - P$. Based on the equation in (12.53), if the following matrix inequalities hold

$$\frac{1}{t_e + 1} x_{q_i}^T(0)Px_{q_i}(0) \leq \gamma \tag{12.54}$$

$$\Phi_{H_2}(k, P)\} \leq 0 \tag{12.55}$$

then the H_2 optimal SINR target tracking performance index $J_2(u_{q_i}(t))$ is bounded by γ as follows:

$$J_2(u_{q_i}(t)) = E\{q_2 e_{q_i}^2(t)\}$$
$$\leq \frac{1}{t_e + 1} x_{q_i}^T(0)Px_{q_i}(0) \leq \gamma \tag{12.56}$$

Let the matrix variables $W = P^{-1}$ and $y = kW$. After multiplying both sides of the matrix inequality in (12.55) by W, we have

$$W - q_2 Wd^T dW - W(A+bk)^T W^{-1}(A+bk)W \geq 0 \qquad (12.57)$$

By applying the Schur complement [90] to (12.54) and (12.57) with matrix variables W and y, we have

$$\begin{bmatrix} W & Wd^T & Wd^T + y^T b^T \\ dW & q_2^{-1} & 0 \\ AW+by & 0 & W \end{bmatrix} \geq 0 \qquad (12.58)$$

$$\begin{bmatrix} (t_e+1)\gamma & x_{q_i}^T(0) \\ x_{q_i}(0) & W \end{bmatrix} \geq 0 \qquad (12.59)$$

which is the LMIs of (12.32)–(12.33) in Theorem 12.2.

(3) For the minimum average power consumption performance in (12.28), the performance index $J_p(u_{q_i}(t))$ in (12.22) can be expressed as the following equation according to the SINR tracking model in (12.19) and $u_{q_i}(t) = gx_{q_i}(t+1)$ in (12.14):

$$J_p(u_{q_i}(t)) = Tr(q_3 g E\{x_{q_i}(t+1)x_{q_i}^T(t+1)\}g^T) \qquad (12.60)$$

We assume that the state $x_{q_i}(t)$ and the disturbance $\zeta_{q_i}(t)$ are independent with $E[\zeta_{q_i}^2(t)] = 1$, that is, $E[\zeta_{q_i}(t)x_{q_i}(t)] = E[x_{q_i}(t)\zeta_{q_i}(t)] = 0$. According to the SINR tracking system in (12.19), the covariance matrix of $x_{q_i}(t+1)$ is given by

$$\begin{aligned} &E\{x_{q_i}(t+1)x_{q_i}^T(t+1)\} \\ &= (A+bk)E\{x_{q_i}(t)x_{q_i}^T(t)\}(A+bk)^T + cc^T \end{aligned} \qquad (12.61)$$

Let $E\{x_{q_i}(t)x_{q_i}^T(t)\} = R$. By substituting (12.61) into (12.60), we can obtain

$$J_p(u_{q_i}(t)) = Tr(g[\Omega_E(k,R)]g^T) + Tr(gRg^T) \qquad (12.62)$$

where $\Omega_E(k,R) = q_3(A+bk)R(A+bk)^T + q_3 cc^T - R$. If the following matrix inequalities hold:

$$\Omega_E(k,R) \leq 0 \qquad (12.63)$$

$$Tr(gRg^T) \leq \delta \qquad (12.64)$$

then the average power consumption performance index $J_p(u_{q_i}(t))$ is bounded by δ as follows:

$$J_2(u_{q_i}(t)) \leq Tr(gRg^T) \leq \delta \qquad (12.65)$$

For the matrix inequality in (12.63), let matrix variables $W = R$, $y = kW$ and rearrange the inequality; then we have

$$W - q_3 cc^T - q_3(AW + by)W^{-1}(AW + by)^T \geq 0 \qquad (12.66)$$

After applying the Schur complement [90] to (12.66), we have the following LMI constraint:

$$\begin{bmatrix} W - q_3 cc^T & AW + by \\ WA^T + y^T b^T & q_3^{-1} W \end{bmatrix} \geq 0 \qquad (12.67)$$

For (12.64), we have the following matrix inequality based on the fact $Tr(\frac{\delta}{d+1} I_{d+1}) = \delta$ with $W = R$:

$$Tr(gRg^T) = Tr(Wg^T g)$$

$$\leq Tr\left(\frac{1}{d+1} I_{d+1}\right) = \delta \qquad (12.68)$$

The sufficient condition of (12.68) is to satisfy the following LMI constraint:

$$\frac{\delta}{d+1} I_{d+1} - Wg^T g \geq 0 \qquad (12.69)$$

The LMIs in (12.67) and (12.69) are the LMIs of (12.34)–(12.35) in Theorem 12.2.

Part IV

Multi-Objective Optimization
Designs in Cyber-Social Systems

13 Multi-Objective Investment Policy for a Nonlinear Stochastic Financial System

13.1 INTRODUCTION

In the fields of finance, stocks, and social economics, because of the interaction between nonlinear factors, all kinds of economic problems have become more and more complicated today. Different types of mathematical models for financial systems have been widely studied. In recent years, researchers have focused on the system dynamic model to describe real economic and financial systems [132, 191, 377, 378]. However, in practical cases, the financial dynamic system is a nonlinear stochastic system and may suffer from continuous and discontinuous parametric fluctuations due to national and international situation changes, oil price changes, the surplus between investment and savings, variable interest rates, false economy policies, and so on. Thus, using a nonlinear stochastic dynamic model to describe a real financial system would be more appealing [129, 379–381]. Stochastic parametric fluctuations can be decomposed as continuous state-dependent fluctuations and discontinuous (jump) state-dependent fluctuations. Both of them will influence the stability of the stochastic financial dynamic system. Moreover, external investment disturbance due to unpredictable investment-environmental changes or a worldwide event, such as war, natural disaster, fatal epidemic disease, and so on, can also affect the fluctuations of financial systems. Thus, in this chapter, a nonlinear stochastic jump diffusion system with internal random fluctuation and external disturbance is employed to describe a nonlinear stochastic financial system.

Managers and investors always expect their investment policies to have not only higher return on investment but also lower risk. Because ROI and lower investment risks normally conflict with each other, investment policy can be regarded as a multi-objective optimization problem. Thus, multi-objective investment policies appear naturally and are also widely applied in financial systems [137]. For a financial system, the multi-objective H_2/H_∞ investment policy design for a nonlinear stochastic jump diffusion financial system can be seen as how to search a management policy to maximize ROI (H_2 management policy) and minimize investment risk (H_∞ management policy) simultaneously.

Robust regulation control is to design a state feedback controller to ensure that the trajectories of the controlled system gradually converge to desired trajectories

DOI: 10.1201/9781003362142-17

(target) despite intrinsic noise and extrinsic disturbance [32, 86, 382–384]. It is particularly appropriate to employ in financial systems. The most two important topics of robust optimal regulation control are how to minimize the regulation error to achieve the desired target despite intrinsic fluctuation and how to increase the system robustness to reject effects due to environmental disturbance. The optimal H_2 management policy is proposed to minimize the regulation error to achieve the desired reference target despite intrinsic fluctuation [385–387]. A robust H_∞ management policy is introduced to reduce the effects of external disturbances on the regulation performance. For a financial system, H_2 optimal regulation control can be regarded as how to design an investment policy to ensure that the financial system can gradually converge to the expected trajectories of managers or investors and minimize the investment cost, that is, a higher ROI policy. On the other hand, robust H_∞ regulation control for a financial system can be seen as how to design a manageable policy to minimize investment risk due to intrinsic continuous and discontinuous random fluctuations and external disturbance. However, both optimal H_2 and robust H_∞ management policy need to solve a difficult Hamilton–Jacobi inequality (a second-order nonlinear partial differential equation) to achieve a desired steady state (target) of the nonlinear stochastic financial system. To overcome this difficult problem, that is, solving HJI, the Takagi–Sugeno fuzzy model is employed to simplify the multi-objective H_2/H_∞ investment policy problem of nonlinear stochastic financial systems.

Recently, the T–S fuzzy model has been widely used to efficiently approximate nonlinear systems [171, 190, 382, 388], and a T–S fuzzy controller has also been successfully applied to stabilization control design for nonlinear systems. To avoid solving the annoying HJIs, the T–S fuzzy model is employed to approximate a nonlinear stochastic jump diffusion financial system by interpolating several local linear stochastic jump diffusion financial systems at different operation points. When a nonlinear stochastic jump diffusion financial system is represented by an interpolation of a set of linear stochastic jump diffusion financial systems, a T–S fuzzy model–based regulation method can also be developed to achieve the multi-objective H_2/H_∞ investment policy. Thus, the HJIs can be replaced by a set of linear matrix inequalities. These LMIs can be efficiently solved by a convex optimization method to complete the multi-objective H_2/H_∞ fuzzy investment policy design. Therefore, the multi-objective H_2/H_∞ investment policy problem of the nonlinear stochastic jump diffusion financial system becomes to how to solve an LMI-constrained MOP.

Although a number of multi-objective evolutionary algorithms have been discussed for their ability to efficiently search for the Pareto optimal solutions of MOP in a single run [2, 77, 206, 389, 390, 391], most of them focus on the multi-objective problem with algebraic functional systems or constraints. Few of these studies discuss the system dynamically constrained MOP for nonlinear stochastic financial systems, in which robust stability must be guaranteed beforehand. More effort is still needed to apply MOEAs to solve MOPs for the multi-objective H_2/H_∞ investment policy of nonlinear stochastic financial systems. Based on the T–S fuzzy model, the multi-objective H_2/H_∞ investment policy problem of the nonlinear stochastic jump diffusion financial system can be transformed into an LMI-constrained MOP. In general, there exists no unique solution for the LMI-constrained MOP. The multiple solutions of MOP are called Pareto optimal solutions. To efficiently solve the multi-objective H_2/H_∞ investment policy problem of a nonlinear financial system, an indirect method is introduced to help us approach the Pareto optimal solutions at the Pareto front of the LMI-constrained MOP by gradually decreasing the upper bound of H_2 and H_∞ performance indices; that is, a nondominating search scheme is employed to find the Pareto front for the Pareto optimal solutions of management policy without violating the stability of nonlinear stochastic system simultaneously.

The proposed LMI-constrained MOEA combines the LMI technique with MOEA to solve the LMI-constrained MOP for the multi-objective H_2/H_∞ investment policy of nonlinear stochastic jump diffusion financial systems. The LMI-constrained MOEA search algorithm using a number of genetic operations, such as selection, mutation, and crossover, is proposed with the help of the LMI toolbox in MATLAB to efficiently search for the Pareto front of Pareto optimal solutions and the corresponding fuzzy regulation gains $\{K_i\}$ from the LMI constraints to meet the robust stability and multi-objective H_2/H_∞ policy optimization of nonlinear stochastic jump diffusion financial systems simultaneously. As long as the Pareto front can be obtained, the multi-objective H_2/H_∞ investment policy problem for a nonlinear stochastic jump diffusion financial system can also be solved, and the manager (or investor) can select a mutual benefit policy according to their preference.

Notations: A^T: transpose of matrix A; $A \geq 0$ $(A > 0)$: symmetric positive semidefinite (symmetric positive definite) matrix A; $I_{n \times n}$: n-dimensional identity matrix; $\|x\|_2$: Euclidean norm for the given vector $x \in \mathbb{R}^n$; C^2: class of functions $V(x)$ twice continuously differential with respect to x; f_x: gradient column vector of n_x-dimensional twice continuously differentiable function $f(x)$(i.e., $\frac{\partial f(x)}{\partial x}$); f_{xx}: Hessian matrix with elements of the second partial derivatives of n_x-dimensional twice continuously differentiable function f_x, (i.e., $\frac{\partial^2 f(x)}{\partial x^2}$); $L^2_F(\mathbb{R}^+, \mathbb{R}^{n_y})$: space of nonanticipative stochastic processes $y(t) \in \mathbb{R}^l$ with respect to an increasing σ-algebra $F_t (t \geq 0)$ satisfying $\|y(t)\|_{L^2_F(\mathbb{R}^+, \mathbb{R}^{n_y})} \triangleq E\{\int_0^\infty y^T(t)y(t)dt\}^{\frac{1}{2}}$; $\|y(t)\|_{L^2_F(\mathbb{R}^+, \mathbb{R}^{n_y}, Q)} \triangleq E\{\int_0^\infty y^T(t)Qy(t)dt\}^{\frac{1}{2}}$; $B(\Theta)$: Borel algebra generated by Θ; E: expectation operator; $\bar{\tau}(M)$: maximum eigenvalue of real-value matrix M; ϕ: empty set.

13.2 FINANCIAL SYSTEM MODEL AND PROBLEM FORMULATION

In [132, 378], a financial dynamic model is given to illustrate the interaction between interest rate, x, investment demand, y, and price index z of a financial system as follows:

$$
\begin{aligned}
\dot{x}(t) &= z(t) + (y(t) - a)x(t) \\
\dot{y}(t) &= 1 - by(t) - (x(t))^2 \\
\dot{z}(t) &= -x(t) - cz(t)
\end{aligned}
\tag{13.1}
$$

where $a \geq 0$ is the savings amount, $b \geq 0$ is the per-investment cost, and $c \geq 0$ is the elasticity of commercial demands. In the nonlinear financial model (13.1), the interest rate $x(t)$ can be influenced by the surplus between investment and saving as well as adjustments of prices. The investment demand $y(t)$ is proportional to the rate of investment and inversely proportional to the cost of investment and the interest rate.

The price index $z(t)$ depends on the difference between supply and demand in the market, and it is also influenced by the inflation rate.

However, in practice, most financial systems are stochastic systems, and the dynamic model in (13.1) may suffer from parametric fluctuations due to the surplus between investment and savings, oil prices, rumors, or government policy changes. To mimic a real financial system, the dynamic model in (13.1) should be modified by continuous and discontinuous intrinsic random fluctuations and environmental disturbance. In the following, the Wiener process (i.e., Brownian motion) and the marked Poisson process are introduced to model the continuous and discontinuous random fluctuations in a real financial system (see Figure 13.1). More precisely speaking, a financial market, which refers to buying and selling behavior, can be

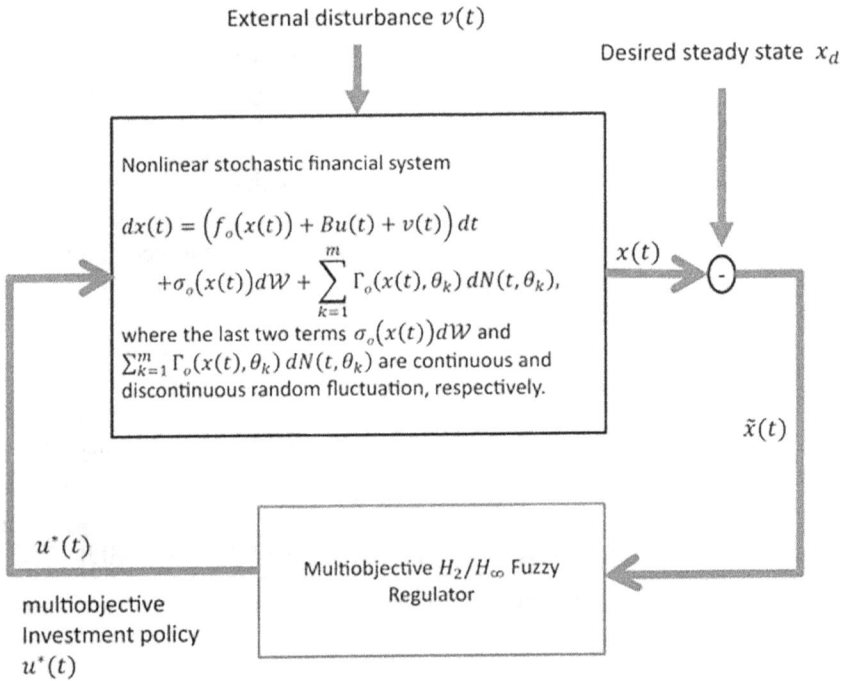

External disturbance $v(t)$

Desired steady state x_d

Nonlinear stochastic financial system

$$dx(t) = \Big(f_o(x(t)) + Bu(t) + v(t)\Big) dt$$
$$+\sigma_o(x(t))dW + \sum_{k=1}^{m} \Gamma_o(x(t), \theta_k) \, dN(t, \theta_k),$$

where the last two terms $\sigma_o(x(t))dW$ and $\sum_{k=1}^{m} \Gamma_o(x(t), \theta_k) \, dN(t, \theta_k)$ are continuous and discontinuous random fluctuation, respectively.

$x(t)$

$\tilde{x}(t)$

$u^*(t)$

multiobjective
Investment policy
$u^*(t)$

Multiobjective H_2/H_∞ Fuzzy Regulator

FIGURE 13.1 System diagram for the multi-objective investment policy problem of the nonlinear financial system. The nonlinear stochastic system in (13.3) is to be controlled to achieve the desired steady state x_d by the multi-objective H_2/H_∞ regulator. In order to achieve the desired steady state x_d, the original nonlinear stochastic financial system in (13.3) is shifted to x_d so that the regulation problem to x_d becomes a stabilization problem of the nonlinear stochastic system in (13.5). When the shifted stochastic financial system in (13.5) is completely approximated by the T-S fuzzy model, the multi-objective H_2/H_∞ investment policy problem in (13.18) can be regarded as the LMI-constrained MOP in (13.20) by employing the proposed indirect method in Lemma 13.2. Since the LMI-constrained MOP in (13.20) is difficult to solve by directly calculating, we proposed the LMI-constrained MOEA to efficiently solve the LMI-constrained MOP in (13.20). Thus, the multi-objective H_2/H_∞ investment policy problem for a nonlinear stochastic financial system can be efficiently solved by MATLAB.

described as a stochastic dynamic financial system. Thus, a stochastic dynamic system can be employed to describe the price changes of objects of transaction (like stocks, bonds, futures, and oil) or economic index changes (investment rate, price index, and investment demand) [392, 393]. One of the most famous stochastic financial systems is the stochastic stock price system [392].

Let (Ω, F, F_t, P) be a filtration probability space with $F = \{F_t : t \geq 0\}$. F is generated by the following two mutually independent stochastic processes: (1) one-dimensional standard Wiener process $W(t)$ and (2) marked Poisson processes $N(t; \theta_k)$ for $k = 1, 2, ..., m$, and the filtration F_t is also generated by σ-algebra of $W(s)$ and σ-algebra of marked Poisson processes $N(s; \theta_k)$, where $k = 1, 2, ..., m$ for $s < t$.

The nonlinear stochastic jump diffusion financial system with continuous and discontinuous random fluctuation in (13.1) can be described as follows:

$$dx(t) = (z(t) + (y(t) - a)x(t))dt + \sigma_1(x(t), y(t), z(t))dW$$
$$+ \sum_{k=1}^{m} \gamma_1(x(t), y(t), z(t), \theta_k)dN(t; \theta_k)$$
$$dy(t) = (1 - by(t) - (x(t))^2)dt + \sigma_2(x(t), y(t), z(t))dW$$
$$+ \sum_{k=1}^{m} \gamma_2(x(t), y(t), z(t), \theta_k)dN(t; \theta_k) \qquad (13.2)$$
$$dz(t) = (-x(T) - cz(t))dt + \sigma_3(x(t), y(t), z(t))dW$$
$$+ \sum_{k=1}^{m} \gamma_3(x(t), y(t), z(t), \theta_k)dN(t; \theta_k)$$

Thus, the stochastic nonlinear autonomous controlled system for the financial system in (13.2) can be written as follows (see system diagram in Figure 13.1):

$$dx(t) = (f_o(x(t)) + Bu(t) + v(t))dt + \sigma_o(x(t))dW$$
$$+ \sum_{k=1}^{m} \Gamma_o(x(t), \theta_k)dN(t; \theta_k) \qquad (13.3)$$

with

$$x(t) = [x(t), y(t), z(t)]^T, \; u(t) = [u_1(t), u_2(t), u_3(t)]^T, \; v(t) = [v_1(t), v_2(t), v_3(t)]^T,$$

$$f_1(x(t)) = (z(t) + (y(t) - a)x(t)), \; f_2(x(t)) = 1 - by(t) - (x(t))^2,$$

$$f_3(x(t)) = (-x(t) - cz(t)), \; f_o(x(t)) = [f_1(x(t)), f_2(x(t)), f_3(x(t))]^T,$$

$$\sigma_o(x(t)) = [\sigma_1(x(t)), \sigma_2(x(t)), \sigma_3(x(t))]^T,$$

$$\Gamma_o(x(t), \theta_k) = [\gamma_1(x(t), \theta_k), \gamma_2(x(t), \theta_k), \gamma_3(x(t), \theta_k)]^T$$

where $f_o : \mathbb{R}^3 \to \mathbb{R}^3$, $\sigma_o : \mathbb{R}^3 \to \mathbb{R}^3$, and $\Gamma_o : \mathbb{R}^3 \times \Theta \to \mathbb{R}^3$ are satisfied with Lipschitz continuity. B is a 3×3 real-value constant matrix. The càdlàg process $x(t) \in \mathbb{R}^3$ is the state vector, the initial state vector $x(0) = x_0$, the input vector $u(t) \in L_F^2(\mathbb{R}^+, \mathbb{R}^3)$ is the admissible regulation effort (i.e., investment policy) with respect to $\{F_t\}_{t \geq 0}$, and $v(t) \in L_F^2(\mathbb{R}^+, \mathbb{R}^3)$ is regarded as an unknown finite energy stochastic external disturbance and denotes the external disturbance caused by an international situation like war or natural disaster. Since 1-D standard Wiener process $W(t)$ is a continuous but nondifferentiable stochastic process, the term $\sigma_o(x(t))dW(t)$ can be regarded as a continuous state-dependent internal random fluctuation, that is, the financial system random fluctuation dependent on the current magnitude of investment rate, price index, and investment demand. The term $\Theta = \{\theta_1, \theta_2, ..., \theta_m\} \subset \mathbb{R}^1$ is the mark space of Poisson random processes and denotes the set of all possible financial emergencies, which will cause discontinuous changes of the system in (13.3); the mark θ_k denotes a financial emergency in the financial system such as financial tsunami, company merger, hot money inflow, or collapse of a bank; and the term $\Gamma_o(x(t), \theta_k)$ denotes a sudden violent change (magnitude jump) of investment rate, price index, or invest-ment demand of the stochastic financial system in (13.3) when a financial emergency happens. Thus, the summation term $\sum_{k=1}^m \Gamma_o(x(t), \theta_k)dN(t; \theta_k)$ denotes all the possi-ble nonlinear stochastic discontinuous changes of the system in (13.3) at the time instant t, whose jump amplitudes depend on the corresponding jump coefficient func-tion $\Gamma_o(x(t), \theta_k)$. It is necessary to note that in this chapter, we assume all emergen-cies are mutually exclusive events (i.e., when the jump occurs at time instant t, there is only one mark θ_k to be assigned). In this chapter, we use a 3-D nonlinear financial system (13.1) to illustrate the financial multi-objective design. However, the financial stochastic system in (13.2) or (13.3) can be extended to any n-dimensional nonlin-ear stochastic financial system. In this situation, $x(t) = (x_1(t), x_2(t), ..., x_n(t))^T \in \mathbb{R}^n$. In this case, we have $f_o : \mathbb{R}^n \to \mathbb{R}^n$, $\sigma_o : \mathbb{R}^n \to \mathbb{R}^n$, $\Gamma_o : \mathbb{R}^n \times \Theta \to \mathbb{R}^n$, and $B_o : \mathbb{R}^{n \times p}$ when $u(t) \in L_F^2(\mathbb{R}^+, \mathbb{R}^p)$.

Some important properties of the Poisson jump process are given as follows [379]:

(1) $E\{dN(t; \theta_k)\} = \lambda_k dt$ where the finite scalar number $\lambda_k > 0$ is the Poisson jump intensity for mark θ_k.

(2) $E\{[dN(t; \theta_k)]dt\} = 0$ for all k.

(3) $E\{[dN(t; \theta_k)]dW\} = 0$ for all k.

(4) $E\{[dN(t; \theta_{k_1})][dN(t; \theta_{k_2})]\} = 0$ for all $k_1 \neq k_2$.

(5) $E\{[dN(t; \theta_{k_1})][dN(t; \theta_{k_2})]\} = \lambda_{k_1} dt$ for all $k_1 = k_2$.

In fact, managers or investors always expect that the dynamic behaviors of the invested financial system satisfy their expectations. Thus, multi-objective optimal control theories are particularly suited for multi-objective investment policy prob-lems of financial systems.

From the system diagram in Figure 13.1, the regulation purpose of the sto-chastic financial system in (13.3) is to design regulation effort $u(t)$ so that (1) the intrinsic random continuous and discontinuous fluctuations $\sigma_o(x(t))dW(t)$ and $\sum_{k=1}^m \Gamma_o(x(t), \theta_k)dN(t; \theta_k)$ can be tolerated by the stochastic financial system; (2) the effect of external disturbance $v(t)$ on the regulation performance of the financial sys-tem should be as small as possible, that is, with less risk; and (3) the desired target x_d can finally be achieved with a low cost.

In order to regulate $x(t)$ to the desired steady state (target) x_d, for the convenience of control design, the origin of the nonlinear stochastic financial system in (13.3) should be shifted to x_d at first. In such a situation, if the shifted nonlinear stochastic financial system is robustly stabilized at the origin, then the robust regulation of $x(t)$ to the desired state(target) x_d will be equivalently achieved. This will simplify the regulation control design procedure of the stochastic nonlinear financial system. Let us denote

$$\tilde{x}(t) = x(t) - x_d \tag{13.4}$$

Then, we get the following shifted nonlinear stochastic financial system in (13.3):

$$d\tilde{x}(t) = (f(\tilde{x}(t)) + Bu(t) + v(t))dt + \sigma(\tilde{x}(t))dW$$
$$+ \sum_{k=1}^{m} \Gamma(\tilde{x}(t), \theta_k)dN(t; \theta_k) \tag{13.5}$$

where $f(\tilde{x}(t)) = f_o(\tilde{x}(t) + x_d)$, $\sigma(\tilde{x}(t)) = \sigma_o(\tilde{x}(t) + x_d)$, $\Gamma(\tilde{x}(t), \theta_k) = \Gamma_o(\tilde{x}(t) + x_d, \theta_k)$.

Thus, the origin $\tilde{x}(t) = 0$ of the nonlinear stochastic financial system in (13.5) is at the desired steady state (target) x_d of the original nonlinear stochastic financial system in (13.3); that is, the investment policy problem of regulating nonlinear stochastic financial system in (13.3) to the desired x_d is transformed into the stabilization problem of the shifted nonlinear stochastic financial system in (13.5).

In order to achieve the desired state x_d with less regulation effort $u(t)$ in spite of continuous and discontinuous random fluctuation, the H_2 control performance $J_2(u(t))$ index for the nonlinear stochastic jump diffusion system in (13.5) without consideration of $J_2(u(t))$ is defined as follows:

$$J_2(u(t)) = \left\| \tilde{x}(t) \right\|^2_{L^2_F(\mathbb{R}^+, \mathbb{R}^3, Q_1)} + \left\| u(t) \right\|^2_{L^2_F(\mathbb{R}^+, \mathbb{R}^3, R_1)} \tag{13.6}$$

where $Q_1 > 0$ and $R_1 > 0$ are weighting matrices to tradeoff between regulation error $\tilde{x}(t)$ and investment effort $u(t)$.

To avoid investment risk from the external disturbance, the H_∞ control performance index $J_\infty(u(t))$ of the nonlinear stochastic financial system in (13.5) is defined as follows:

$$J_\infty(u(t)) = \sup_{\substack{v(t) \in L^2_F(\mathbb{R}^+, \mathbb{R}^3) \\ v \neq 0, x_0 = 0}} \frac{\left\| \tilde{x}(t) \right\|^2_{L^2_F(\mathbb{R}^+, \mathbb{R}^3, Q_2)} + \left\| u(t) \right\|^2_{L^2_F(\mathbb{R}^+, \mathbb{R}^3, R_2)}}{\left\| v(t) \right\|^2_{L^2_F(\mathbb{R}^+, \mathbb{R}^3)}} \tag{13.7}$$

$$\text{if } \tilde{x}(0) = \tilde{x}_0 = 0$$

where $Q_2 > 0$, $R_2 > 0$, that is, the worst-case effect from the exogenous disturbance $v(t) \in L^2_F(\mathbb{R}^+, \mathbb{R}^3)$ to the regulation errors $\tilde{x}(t)$ and regulation effort $u(t)$ from the average energy point of view. The H_∞ performance index $J_\infty(u(t))$ in (13.7) can be considered the worst-case investment risk from external disturbance. Since the external disturbance is unpredictable, the worst-case effect of external disturbance is considered in the investment risk.

In this chapter, from the system diagram of the nonlinear stochastic financial system in Figure 13.1, we want to design a multi-objective H_2/H_∞ robust investment policy for system state $x(t)$ to achieve a desired state x_d from the perspective of minimum H_2 regulation error with minimum investment effort $u(t)$ in (13.6) and minimum H_∞ investment risk in (13.7) under intrinsic fluctuations and external disturbance. A more detailed design procedure of the multi-objective H_2/H_∞ robust investment policy $u(t)$ to simultaneously minimize $J_2(u(t))$ in (13.6) and $J_\infty(u(t))$ in (13.7) will be discussed in the following.

Remark 13.1: If the initial condition $\tilde{x}(0) = \tilde{x}_0 \neq 0$, then the H_∞ performance index $J_\infty(u(t))$ should be rewritten as

$$J_\infty(u(t)) = \sup_{\substack{v(t) \in L_F^2(\mathbb{R}_+, \mathbb{R}^3) \\ v \neq 0, x_0 = 0}} \frac{\left\| \tilde{x}(t) \right\|_{L_F^2(\mathbb{R}^+, \mathbb{R}^3, Q_2)}^2 + \left\| u(t) \right\|_{L_F^2(\mathbb{R}^+, \mathbb{R}^3, R_2)}^2}{\left\| v(t) \right\|_{L_F^2(\mathbb{R}^+, \mathbb{R}^3)}^2} \tag{13.8}$$

where $V(\cdot) \in C^2(\mathbb{R}^3)$ and $V(\cdot) \geq 0$; that is, the effect of initial condition $\tilde{x}_0$ should be deleted in order to obtain the real effect of $v(t)$ on the controlled output.

Lemma 13.1 ([379]): Let $V : \mathbb{R}^3 \to \mathbb{R}$, $V(\cdot) \in C^2(\mathbb{R}^3)$, and $V(\cdot) \geq 0$. For the nonlinear stochastic jump diffusion system in (13.5), the Itô–Lévy formula of $V(\tilde{x}(t))$ is given as follows:

$$\begin{aligned} dV(\tilde{x}(t)) = &[V_{\tilde{x}}^T f(x(t)) + V_{\tilde{x}}^T Bu(t) + V_{\tilde{x}}^T v(t) \\ &+ \frac{1}{2}\sigma^T(\tilde{x}(t))V_{\tilde{x}\tilde{x}}\sigma(\tilde{x}(t))]dt + V_{\tilde{x}}^T \sigma(\tilde{x}(t))dW(t) \\ &+ \sum_{k=1}^{m}\{V(\tilde{x}(t) + \Gamma(\tilde{x}(t), \theta_k)) - V(\tilde{x}(t))\}dN(t; \theta_k) \end{aligned} \tag{13.9}$$

13.3 MULTI-OBJECTIVE H_2/H_∞ INVESTMENT POLICY DESIGN FOR NONLINEAR STOCHASTIC FINANCIAL JUMP SYSTEMS VIA FUZZY INTERPOLATION METHOD

In general, construction of the multi-objective H_2/H_∞ regulation control for a nonlinear stochastic system needs to solve an HJI-constrained MOP, which is difficult to solve analytically or numerically. To overcome this difficult problem, the T–S fuzzy interpolation method is employed here to approximate the nonlinear stochastic jump diffusion financial system in (13.5). The T–S fuzzy dynamic model is proposed to approximate the nonlinear stochastic system in (13.5) by interpolating several local linearized stochastic jump diffusion financial systems around some operation points [171, 190, 382, 388, 394–396]. The T–S fuzzified model is described by a group of if–then rules and is used to simplify the multi-objective H_2/H_∞ investment policy design problem of the nonlinear stochastic jump diffusion financial system. The ith rule of the T–S fuzzy model for the nonlinear stochastic jump diffusion financial system in (13.5) is described by

$$
\begin{aligned}
&\text{System Rule :} \\
&\text{If } z_1 \text{ is } G_{i1} \text{ and ... and } z_g \text{ is } G_{ig} \\
&\text{then} \\
&d\tilde{x}(t) = [A_i\tilde{x}(t) + Bu(t) + v(t)]dt + C_i\tilde{x}(t)dW(t) \\
&\qquad + \sum_{k=1}^{m} E_i(\theta_k)\tilde{x}(t)dN(t;\theta_k), \text{for } i = 1,2,...,l
\end{aligned}
\tag{13.10}
$$

where l is the number of fuzzy rules, G_{ij} is the fuzzy set, the matrices $A_i, B, C_i \in \mathbb{R}^{3\times3}$ are constant, $E_i(\theta_k)$ is a constant matrix for $k = 1,2,...,m$, and $z_1 \cdots z_g$ are premise variables.

The overall fuzzy system in (13.10) can be inferred as follows [190]:

$$
\begin{aligned}
d\tilde{x}(t) = \sum_{i=1}^{l} h_i(z)\{[A_i\tilde{x}(t) + Bu(t) + v(t)]dt \\
+ C_i\tilde{x}(t)dW(t) + \sum_{k=1}^{m} E_i(\theta_k)\tilde{x}(t)dN(t;\theta_k)\}
\end{aligned}
\tag{13.11}
$$

where $z = [z_1^T, z_2^T, ..., z_g^T]^T$

$$
\mu_i(t) = \prod_{j=1}^{g} G_{ij}(z_j) \geq 0, h_i(z) = \frac{\mu_i(z)}{\sum_{i=1}^{l}\mu_i(z)} \geq 0
\tag{13.12}
$$

and $G_{ij}(z_j)$ is the membership grade of z_j in G_{ij}. From the definitions given previously, we have

$$
\sum_{i=1}^{l} h_i(z) = 1
\tag{13.13}
$$

The physical meaning of the fuzzy model in (13.11) is that the following l locally linearized financial systems

$$
\begin{aligned}
d\tilde{x}(t) = [A_i\tilde{x}(t) + Bu(t) + v(t)]dt + C_i\tilde{x}(t)dW(t) \\
+ \sum_{k=1}^{m} E_i(\theta_k)\tilde{x}(t)dN(t;\theta_k), \text{for } 1 \leq i \leq l
\end{aligned}
\tag{13.14}
$$

at different operation points (different fuzzy set G_{ij}) are interpolated smoothly via the fuzzy certainty function $h_i(z)$ to approximate the original nonlinear stochastic jump diffusion financial system in (13.5).

Similarly, the multi-objective H_2/H_∞ investment policy $u(t) = K(\tilde{x}(t))$ for the nonlinear stochastic jump diffusion financial system could be approximated by the following fuzzy investment policy:

$$
\begin{array}{l}
\text{Investment Policy } i: \\
\text{For } i = 1,2,...,l \\
\text{if } z_1 \text{ is } G_{i1} \text{ and...and } z_g \text{ is } G_{ig} \\
\text{then } u(t) = K_i \tilde{x}(t), \text{for } i = 1,2,...,l
\end{array}
\tag{13.15}
$$

The overall multi-objective H_2/H_∞ investment policy $u(t)$ for the nonlinear stochastic jump diffusion financial system can be represented by

$$
u(t) = \sum_{i=1}^{l} h_i(z) K_i \tilde{x}(t) \tag{13.16}
$$

where $h_i(z)$ is designed as (13.11), and K_i is the regulation gain for the ith fuzzy linearized system for $i = 1,2,...,l$.

The fuzzy parameters A_i, C_i, and $E_i(\theta_k)$ of l local linear financial systems in (13.11) can be easily identified by the system identification toolbox in MATLAB. If an adequate number of fuzzy rules is used in the T–S fuzzy system in (13.11), then the fuzzy approximation error could be considered one kind of parametric fluctuation, and the proposed robust H_2/H_∞ multi-objective investment policy can efficiently override these parametric fluctuations due to fuzzy approximation. Therefore, the multi-objective H_2/H_∞ investment policy problem can be regarded as how to design fuzzy investment policy $u(t)$ in (13.16) so that the T–S stochastic financial system in (13.11) can achieve the multi-objective H_2/H_∞ investment policy in (13.6) and (13.7).

Higher-benefit investment behavior always comes with higher risk and capital cost. The optimal H_2 regulation in (13.6) expects the energy of regulation error $\tilde{x}(t)$ and admissible investment policy $u(t)$ to be as low as possible, that is, use minimum capital to earn a desired benefit. However, for robust H_∞ regulation design, a better (lower) H_∞ performance index is always coupled with a higher capital cost to reject intrinsic continuous and discontinuous fluctuation as well as external disturbance to achieve the desired target x_d with less risk. It is clear that the optimal H_2 and robust H_∞ regulation design are mutually conflicting. Thus, designing a controller (investment policy) to optimize the H_2 performance index $J_2(u(t))$ and the H_∞ performance index $J_\infty(u(t))$ is indeed an MOP of financial regulation systems. The definition of the multi-objective H_2/H_∞ investment policy for the nonlinear stochastic jump diffusion financial system is given in the following.

Definition 13.1: The multi-objective H_2/H_∞ investment policy of a given nonlinear stochastic Poisson jump diffusion financial system in (13.11) is to design an admissible investment policy $u(t)$ in (16), which can make the H_2 and H_∞ performance indices minimum in the Pareto optimal sense simultaneously, that is

$$
\min_{u(t) \in U} \left(J_2(u(t)), J_\infty(u(t)) \right) \text{s.t. (13.11)} \tag{13.17}
$$

where U is the set of all the admissible investment policies for the given nonlinear stochastic jump diffusion financial system; the objective functions $J_2(u(t))$ and $J_\infty(u(t))$ are defined in (13.6) and (13.7), respectively; the vector of the objective functions $(J_2(u(t)), J_\infty(u(t)))$ is the objective vector of $u(t)$.

Lemma 13.2 ([15]): Suppose α and β are the upper bounds of the H_2 and H_∞ performance indices, respectively, that is, $J_2(u(t)) \le \alpha$ and $J_\infty(u(t)) \le \beta$. The MOP in (13.17) is equivalent to the MOP given in the following:

$$\min_{u(t) \in U} (\alpha, \beta) \text{ s.t. } J_2(u(t)) \le \alpha \text{ and } J_\infty(u(t)) \le \beta \tag{13.18}$$

Proof: The proof of this lemma is straightforward. One only needs to prove that both inequalities contained in the multi-objective problem in (13.18) become equal for Pareto optimal solutions. We will show this by contradiction. Given a 3-tuple Pareto optimal solution $(u^*(t), \alpha^*, \beta^*)$ of the MOP in (13.18), we assume that either of the inequalities in (13.18) remains a strict inequality at the Pareto optimal solution. Without loss of generality, suppose that $J_2(u(t)) < \alpha^*$. As the result, there exists α_1 such that $\alpha_1 < \alpha^*$ and $J_2(u(t)) = \alpha_1$. Now, for the same $u(t)$, the solution (α, β^*) dominates the Pareto optimal solution (α^*, β^*), leading to a contradiction. This implies that both inequality constraints in MOP (13.18) indeed become equality for Pareto optimal solutions. The optimization problem in (13.18) is, hence, equivalent to the MOP described in (13.17).

In this study, Lemma 2 provides an indirect method to solve the multi-objective H_2/H_∞ investment policy problem of the nonlinear stochastic jump diffusion financial system.

Lemma 13.3 ([90]): For any two real matrices A, B with appropriate dimensions, we have

$$A^T B + B^T A \le \gamma^2 A^T A + \frac{1}{\gamma^2} B^T B$$

where γ is any nonzero real number.

Lemma 13.4 ([190]): For any matrix M_i with appropriate dimensions and scheduling functions $h_i(z)$ with $0 \le h_i(z) \le 1$, for $i \in \mathbb{N}^+$, $1 \le i \le m$, $P > 0$, and $\sum_{i=1}^{m} h_i(z) = 1$, we have

$$\left(\sum_{j=1}^{l} h_j(z) M_j \right)^T P \left(\sum_{i=1}^{l} h_i(z) M_i \right) \le \sum_{i=1}^{l} h_i(z) M_i^T P M_i \left(\sum_{j=1}^{l} h_j(z) M_j \right)$$

Then the following theorems will provide sufficient conditions for the fuzzy investment policy $u(t)$ in (13.16) to solve the multi-objective H_2/H_∞ investment policy problem for the nonlinear stochastic jump financial system in (13.5). In the following, according to two kinds of Poisson noise in stochastic financial systems, marked Poisson process $N(t; \theta_k)$ and marked compensation Poisson processes $\hat{N}(t; \theta_k)$, the multi-objective H_2/H_∞ investment policy problem will be solved separately.

13.3.1 MULTI-OBJECTIVE H_2/H_∞ INVESTMENT POLICY PROBLEM FOR THE NONLINEAR STOCHASTIC JUMP DIFFUSION FINANCIAL SYSTEM DRIVEN BY THE MARKED POISSON PROCESS $N(t;\theta_k)$

Theorem 13.1: If the following LMI-constrained MOP can be solved:

$$\min_{\{P,K_1,K_2,\dots,K_l\}} (\alpha,\beta) \tag{13.19}$$

s.t. the following LMIs, for all $i,j = 1,2,\dots,l$

$$P \le \alpha[Tr(R_{\tilde{x}_0})]^{-1} I \tag{13.20}$$

$$\begin{bmatrix} \Psi_{ij}^2 & W & Y_j^T & WC_i^T & WE_i^T(\theta_1) & \cdots & WE_i^T(\theta_m) \\ * & -Q_1^{-1} & 0 & 0 & 0 & 0 & 0 \\ * & * & -R_1^{-1} & 0 & 0 & 0 & 0 \\ * & * & * & -W & 0 & 0 & \vdots \\ * & * & * & * & -\lambda_1^{-1}W & 0 & 0 \\ * & * & * & * & * & \ddots & 0 \\ * & * & * & * & * & * & -\lambda_m^{-1}W \end{bmatrix} \le 0 \tag{13.21}$$

$$\begin{bmatrix} \Psi_{ij}^\infty & W & Y_j^T & WC_i^T & WE_i^T(\theta_1) & \cdots & WE_i^T(\theta_m) \\ * & -Q_2^{-1} & 0 & 0 & 0 & 0 & 0 \\ * & * & -R_2^{-1} & 0 & 0 & 0 & 0 \\ * & * & * & -W & 0 & 0 & \vdots \\ * & * & * & * & -\lambda_1^{-1}W & 0 & 0 \\ * & * & * & * & * & \ddots & 0 \\ * & * & * & * & * & * & -\lambda_m^{-1}W \end{bmatrix} \le 0 \tag{13.22}$$

where $W = P^{-1}$, $Y_j = K_j W$, $\Psi_{ij}^2 = A_i W + WA_i^T + BY_j + Y_j^T B^T + \sum_{k=1}^m \lambda_k[WE_i^T(\theta_k)+, E_i(\theta_k)W]$ and $\Psi_{ij}^\infty = A_i W + WA_i^T + BY_j + Y_j^T B^T + \frac{1}{\beta}I + \sum_{k=1}^m \lambda_k[WE_i^T(\theta_k) + E_i(\theta_k)W]$, then the multi-objective H_2/H_∞ investment policy problem for the fuzzy stochastic jump financial systems in (13.5) can be solved.

Proof: Let $V(x(t)) = \tilde{x}^T(t)P\tilde{x}(t)$ be the Lyapunov function for the nonlinear stochastic jump financial system in (13.5), where $P = P^T > 0$ is a positive-definite matrix.

We derive the sufficient condition for $J_2(u(t)) \leq \alpha$ of the MOP in (13.18) first. Add and subtract the term $dV(\tilde{x}(t))$ to the intergrand of $E\{\int_0^\infty \tilde{x}(t)^T Q_1 \tilde{x}(t) + u^T(t)R_2 u(t))dt\}$. By the fact of $\lim_{t\to\infty} V(\tilde{x}(t)) > 0$ and Lemma 13.1, we have

$$J_2(u(t)) = E\{\int_0^\infty \tilde{x}(t)^T Q_1 \tilde{x}(t) + u^T(t)R_1 u(t))dt\}$$

$$\leq E\{V(\tilde{x}_0)\} + E\{\int_0^\infty [\tilde{x}(t)^T Q_1 \tilde{x}(t) + u^T(t)R_1 u(t))dt + dV(\tilde{x})\}$$

$$= E\{V(\tilde{x}_0)\} + E\{\int_0^\infty (\tilde{x}(t)^T Q_1 \tilde{x}(t) + u^T(t)R_1 u(t)) + 2\tilde{x}(t)^T Pf(\tilde{x}(t))$$

$$+2\tilde{x}(t)PBu(t) + \sigma^T(\tilde{x}(t))P\sigma(\tilde{x}(t)) + \sum_{k=1}^m \lambda_k \{V(\tilde{x}(t) + \Gamma(\tilde{x}(t),\theta_k)) - V(\tilde{x}(t))\})dt\}$$

If both the following Hamilton–Jacobi–Bellman inequality

$$\tilde{x}(t)^T Q_1 \tilde{x}x(t) + u^T(t)R_1 u(t) + 2\tilde{x}(t)^T Pf(\tilde{x}(t))$$
$$+2\tilde{x}(t)^T PBu(t) + \sigma^T(\tilde{x}(t))P\sigma(\tilde{x}(t)) \qquad (13.23)$$
$$+\sum_{k=1}^m \lambda_k \{V(\tilde{x}(t)) + \Gamma(\tilde{x}(t),\theta_k) - V(\tilde{x}(t))\} \leq 0$$

and the initial mean inequality

$$E\{V(\tilde{x}_0)\} \leq \alpha \qquad (13.24)$$

hold, we have

$$J_2(u(t)) = E\{\int_0^\infty \tilde{x}(t)^T Q_1 \tilde{x}(t) + u^T(t)R_1 u(t))dt\} \leq E\{V(\tilde{x}_0)\} \leq \alpha$$

By applying Lemma 13.4, we obtain the following inequalities:

$$2\tilde{x}^T(t)Pf(x(t)) = \sum_{i=1}^l h_i(z)\tilde{x}^T(t)[A_i^T P + PA_i]\tilde{x}(t) \qquad (13.25)$$

$$2\tilde{x}^T(t)PBu(t) = \sum_{j=1}^l h_j(z)\tilde{x}^T(t)[PBK_j + (PBK_j)^T]\tilde{x}(t) \qquad (13.26)$$

$$\sigma^T(\tilde{x}(t))V_{\tilde{x}\tilde{x}}(x(t))\sigma(\tilde{x}(t)) \leq \sum_{i=1}^l h_i(z)\tilde{x}^T(t)C_i^T PC_i\tilde{x}(t) \qquad (13.27)$$

and

$$V(\tilde{x}(t)) + \Gamma(\tilde{x}(t), \theta_k) - V(\tilde{x}(t))$$
$$= [\tilde{x}(t) + \Gamma(\tilde{x}(t), \theta_k)]^T P[\tilde{x}(t) + \Gamma(\tilde{x}(t), \theta_k)] - \tilde{x}^T(t)P\tilde{x}(t)$$
$$\leq \sum_{i=1}^{l} h_i(z)\tilde{x}^T(t)\{[E_i^T(\theta_k)PE_i(\theta_k)] + [E_i^T(\theta_k)P]$$
$$+ [PE_i(\theta_k)]\}\tilde{x}(t) \tag{13.28}$$

It is clear that

$$E\{V(\tilde{x}_0)\} \leq \overline{\tau}(P)E\{Tr(\tilde{x}_0^T\tilde{x}_0)\} = \overline{\tau}(P)Tr(R_{\tilde{x}_0}) \leq \alpha \tag{13.29}$$

and $P \leq \alpha[Tr(R_{\tilde{x}_0})]^{-1}I$ which is (13.20).
Thus, the HJBI in (13.23) can be replaced by

$$\tilde{x}(t)^T Q_1 \tilde{x}(t) + u^T(t)R_1 u(t) + 2\tilde{x}(t)^T Pf(\tilde{x}(t))$$
$$+ 2\tilde{x}(t)^T PBu(t) + \sigma(\tilde{x}(t))^T P\sigma(\tilde{x}(t))$$
$$+ \sum_{k=1}^{m} \lambda_k \{V(\tilde{x}(t)) + \Gamma(\tilde{x}(t), \theta_k) - V(\tilde{x}(t))\}$$
$$\leq \sum_{i=1}^{l} \sum_{j=1}^{l} h_i(z)h_j(z)\tilde{x}^T(t)(Q_1 + K_j^T R_1 K_j + A_i^T P + PA_i \tag{13.30}$$
$$+ PBK_j + (PBK_j)^T + C_i^T PC_i + \sum_{k=1}^{m} \lambda_k[E_i^T(\theta_k)PE_i(\theta_k)$$
$$+ E_i^T(\theta_k)P + PE_i(\theta_k)])\tilde{x}(t)$$

Now, we derive the sufficient condition for $J_\infty \leq \beta$ of the MOP in (13.18). By using Lemmas 13.3 and 13.4, we get

$$E\{\int_0^\infty \tilde{x}(t)^T Q_2 \tilde{x}(t) + u^T(t)R_2 u(t))dt\}$$
$$\leq E\{\int_0^\infty [\tilde{x}(t)^T Q_2 \tilde{x}(t) + u^T(t)R_2 u(t))dt + dV(\tilde{x})\} + E\{V(\tilde{x}_0)\}$$
$$= E\{\int_0^\infty [\overline{x}(t)^T \tilde{Q}_2 \overline{x}(t) + u^T(t)R_2 u(t)) + 2\tilde{x}(t)^T Pf(\tilde{x}(t))$$
$$+ 2\tilde{x}(t)PBu(t) + 2\tilde{x}(t)Pv(t) + \sigma^T(\tilde{x}(t))P\sigma(\tilde{x}(t))$$
$$+ \sum_{k=1}^{m} \lambda_k \{V(\tilde{x}(t)) + \Gamma(\tilde{x}(t), \theta_k)) - V(\tilde{x}(t))\})dt\} + E\{V(\tilde{x}_0)\}$$
$$\leq E\{V(\tilde{x}_0)\} + E\{\int_0^\infty [\tilde{x}(t)^T Q_2 \tilde{x}(t) + u^T(t)R_2 u(t)) + 2\tilde{x}(t)^T Pf(\tilde{x}(t))$$
$$+ 2\tilde{x}(t)PBu(t) + \frac{1}{\beta}\tilde{x}^T(t)PP\tilde{x}(t) + \sigma^T(\tilde{x}(t))P\sigma(\tilde{x}(t)) + \beta v^T(t)v(t)]dt$$
$$+ \sum_{k=1}^{m} \lambda_k \{V(\tilde{x}(t)) + \Gamma(\tilde{x}(t), \theta_k)) - V(\tilde{x}(t))\})dt\}$$

If the following Hamilton–Jacobi–Isaacs inequality is satisfied:

$$\tilde{x}^T(t)Q_2\tilde{x}(t) + u^T(t)R_2u(t) + 2\tilde{x}(t)^T Pf(\tilde{x}(t))$$

$$+2\tilde{x}(t)^T PBu(t) + \frac{1}{\beta}\tilde{x}^T(t)PP\tilde{x}(t) + \sigma^T(\tilde{x}(t))P\sigma(\tilde{x}(t)) \tag{13.31}$$

$$+\sum_{k=1}^{m}\lambda_k\{V(\tilde{x}(t)) + \Gamma(\tilde{x}(t),\theta_k) - V(\tilde{x}(t))\} \leq 0$$

then $J_\infty(u(t)) \leq \beta$ for all possible $v(t) \in L_F^2(\mathbb{R}^+;\mathbb{R}^3)$.

By introducing the T-S fuzzy model, we obtain the following inequality:

$$\tilde{x}(t)^T Q_2\tilde{x}(t) + u^T(t)R_2u(t) + 2\tilde{x}(t)^T Pf(\tilde{x}(t))$$

$$+2\tilde{x}(t)^T PBu(t) + \frac{1}{\beta}\tilde{x}^T(t)PP\tilde{x}(t) + \sigma^T(\tilde{x}(t))P\sigma(\tilde{x}(t))$$

$$+\sum_{k=1}^{m}\lambda_k\{V(\tilde{x}(t)) + \Gamma(\tilde{x}(t),\theta_k) - V(\tilde{x}(t))\}$$

$$\leq \sum_{i,j=1}^{l} h_i(z)h_j(z)\tilde{x}^T(t)(Q_2 + K_j^T R_1 K_j + A_i^T P + PA_i \tag{13.32}$$

$$+ PBK_j + (PBK_j)^T + C_i^T PC_i + \frac{1}{\beta}PP$$

$$+\sum_{k=1}^{m}\lambda_k[E_i^T(\theta_k)PE_i(\theta_k) + E_i^T(\theta_k)P + PE_i(\theta_k)])\tilde{x}(t)$$

From the inequalities in (13.30) and (13.32), if the following two algebraic Riccati-like inequalities are satisfied:

$$Q_1 + K_j^T R_1 K_j + A_i^T P + PA_i + PBK_j + (PBK_j)^T + C_i^T PC_i$$

$$+\sum_{k=1}^{m}\lambda_k[E_i^T(\theta_k)PE_i(\theta_k) + E_i^T(\theta_k)P + PE_i(\theta_k)] \leq 0 \tag{13.33}$$

$$Q_2 + K_j^T R_1 K_j + A_i^T P + PA_i + PBK_j + (PBK_j)^T + C_i^T PC_i$$

$$+\frac{1}{\beta}PP + \sum_{k=1}^{m}\lambda_k[E_i^T(\theta_k)PE_i(\theta_k) + E_i^T(\theta_k)P + PE_i(\theta_k)] \leq 0 \tag{13.34}$$

then we have $J_2(u(t)) \leq \alpha$ and $J_\infty(u(t)) \leq \beta$, respectively.

Let $W = P^{-1}$ and $Y_j = K_j W$; then the inequalities in (13.33) and (13.34) are equivalent to the following two inequalities, respectively:

$$WQ_1W + Y_j^T R_1 Y_j + WA_i^T + A_i W + BY_j + (BY_j)^T$$

$$+WC_i^T W^{-1}C_i W + \sum_{k=1}^{m}\lambda_k[WE_i^T(\theta_k)W^{-1}E_i(\theta_k)W \tag{13.35}$$

$$+WE_i^T(\theta_k) + E_i(\theta_k)W] \leq 0$$

$$WQ_2W + Y_j^T R_2 Y_j + WA_i^T + A_i W + BY_j + (BY_j)^T$$

$$+ WC_i^T W^{-1} C_i W + \frac{1}{\beta} I + \sum_{k=1}^{m} \lambda_k [WE_i^T (\theta_k) W^{-1} E_i (\theta_k) W \qquad (13.36)$$

$$+ WE_i^T (\theta_k) + E_i (\theta_k) W] \le 0$$

By Schur complement [90], the quadratic inequalities are equivalent to the LMIs in (13.21) and (13.22).

Definition 13.2 ([397]): The nonlinear stochastic jump diffusion financial system in (13.5) or (13.11) is said to be exponentially mean square stable if, for some positive constants $A > 0$ and $m > 0$, the following inequality holds:

$$E\{\|\tilde{x}(t)\|_2^2\} \le A \exp(-mt)$$

Theorem 13.2: For the nonlinear stochastic jump diffusion financial system in (13.5) or (13.11), if the external noise $v(t) = 0$, and $u(t)$ is a feasible solution of the MOP in (13.19), then $u(t)$ stabilizes the nonlinear stochastic Poisson jump diffusion financial system in (13.5) or (13.11) exponentially in the mean square sense, that is, $x(t) \to x_d(t)$ exponentially in the mean square sense.

Proof: Since the given Lyapunov function $V(\tilde{x}(t)) = \tilde{x}^T (t) P \tilde{x}(t)$ is satisfied by the following two inequalities:

$$m_1 \|\tilde{x}(t)\|_2^2 \le V(\tilde{x}(t)) \le m_2 \|\tilde{x}(t)\|_2^2 \qquad (13.37)$$

where $m_1 > 0$ and $m_2 > 0$, by the Itô–Lévy formula of $V(\tilde{x}(t))$ in (13.9) and the fact that $u(t)$ is a feasible solution of the LMI-constrained MOP in (13.19), we obtain

$$dE\{V(\tilde{x}(t))\} = E\{dV(\tilde{x}(t))\}$$

$$= E\{(2\tilde{x}^T (t) Pf(\tilde{x}(t)) + 2\tilde{x}^T (t) PBu(t) + \sigma^T (\tilde{x}(t)) P\sigma(\tilde{x}(t))$$

$$+ \sum_{k=1}^{m} \lambda_k [V(\tilde{x}(t) + \Gamma(\tilde{x}(t), \theta_k)) - V(\tilde{x}(t))])dt\} \qquad (13.38)$$

$$\le E\{-\tilde{x}^T (t) Q_1 \tilde{x}(t)\}dt \le E\{-m_3 \|\tilde{x}(t)\|_2^2\}dt$$

$$\le \frac{-m_3}{m_2} E\{V(\tilde{x}(t))\}dt < 0$$

where m_3 is the smallest eigenvalue value of positive-definite matrix $Q_1 > 0$.

By using the inequalities in (13.38), we get

$$\frac{d}{dt} E\{V(\tilde{x}(t))\} \le \frac{-m_3}{m_2} E\{V(\tilde{x}(t))\} \qquad (13.39)$$

which implies the following two inequalities:

$$E\left\{\|\tilde{x}(t)\|^2\right\} \leq E\left\{\frac{V(\tilde{x}(t))}{m_1}\right\} \leq E\left\{\frac{V(\tilde{x}_0)}{m_1}\right\}\exp\left(\frac{-m_3 t}{m_2}\right) \qquad (13.40)$$

It is obvious that

$$\lim_{t\to\infty} E\{\|\tilde{x}(t)\|_2^2\} = 0$$

and we obtain $\lim_{t\to\infty} \tilde{x}(t) = 0$ exponentially in the mean square sense.

13.3.2 Multi-Objective H_2/H_∞ Investment Policy Problem for the Nonlinear Stochastic Jump Diffusion Financial System Driven by Marked Compensation Poisson Processes $\hat{N}(t;\theta_k)$

Moreover, in some situations, nonlinear stochastic jump diffusion financial systems are driven by compensation Poisson processes [379]:

$$\hat{N}(t,\theta_k) \triangleq N(t;\theta_k) - \lambda_k t, \text{ for } k = 1,2,...,m \qquad (13.41)$$

Thus, the stochastic nonlinear autonomous controlled system for the financial system in (13.5) can be replaced as follows:

$$d\tilde{x}(t) = (f(\tilde{x}(t)) + Bu(t) + v(t))dt$$
$$+ \sigma(\tilde{x}(t))dW + \sum_{k=1}^{m}\Gamma(\tilde{x}(t),\theta_k)d\hat{N}(t;\theta_k) \qquad (13.42)$$

The overall fuzzy system in (13.10) should be modified as

$$d\tilde{x}(t) = \sum_{i=1}^{l} h_i(z)\{[A_i\tilde{x}(t) + Bu(t) + v(t)]dt$$
$$+ C_i\tilde{x}(t)dW(t) + \sum_{k=1}^{m}E_i(\theta_k)\tilde{x}(t)d\hat{N}(t;\theta_k) \qquad (13.43)$$

The Itô–Lévy formula for the nonlinear stochastic jump diffusion financial system driven by compensation Poisson process $d\hat{N}(t,\theta_k)$ is given as

Lemma 13.5 ([129]): Let $V : \mathbb{R}^3 \to \mathbb{R}$, $V(\cdot) \in C^2(\mathbb{R}^3)$, and $V(\cdot) \geq 0$. For the non-linear stochastic jump diffusion system in (13.42), the Itô–Lévy formula of $V(\tilde{x}(t))$ is given as follows:

$$dV(\tilde{x}(t)) = [V_{\tilde{x}}^T f(x(t)) + V_{\tilde{x}}^T Bu(t) + V_{\tilde{x}}^T v(t)$$

$$+ \frac{1}{2}\sigma(\tilde{x}(t))^T V_{\tilde{x}\tilde{x}}\sigma(\tilde{x}(t))$$

$$+ V_{\tilde{x}}^T \sigma(\tilde{x}(t))dW(t) \tag{13.44}$$

$$+ \sum_{k=1}^{m}\{V(\tilde{x}(t) + \Gamma(\tilde{x}(t),\theta_k)) - V(\tilde{x}(t))\}d\widehat{N}(t;\theta_k)$$

Thus, the sufficient condition for the fuzzy investment policy $u(t)$ in (13.16) to solve the multi-objective H_2/H_∞ investment policy problem for the nonlinear stochastic jump diffusion financial system in (13.42) is given by the following theorem.

Theorem 13.3: If the following LMI-constrained MOP can be solved

$$\min_{\{P,K_1,K_2,...,K_l\}} (\alpha,\beta) \tag{13.45}$$

s.t. the following LMIs, for all $i, j = 1, 2, ..., l$

$$P \le \alpha[Tr(R_{\tilde{x}_0})]^{-1} I \tag{13.46}$$

$$\begin{bmatrix} \Psi_{ij}^2 & W & Y_j^T & WC_i^T & WE_i^T(\theta_1) & \cdots & WE_i^T(\theta_m) \\ * & -Q_1^{-1} & 0 & 0 & 0 & 0 & 0 \\ * & * & -R_1^{-1} & 0 & 0 & 0 & 0 \\ * & * & * & -W & 0 & 0 & \vdots \\ * & * & * & * & -\lambda_1^{-1}W & 0 & 0 \\ * & * & * & * & * & \ddots & 0 \\ * & * & * & * & * & * & -\lambda_m^{-1}W \end{bmatrix} \le 0 \tag{13.47}$$

$$\begin{bmatrix} \Psi_{ij}^\infty & W & Y_j^T & WC_i^T & WE_i^T(\theta_1) & \cdots & WE_i^T(\theta_m) \\ * & -Q_2^{-1} & 0 & 0 & 0 & 0 & 0 \\ * & * & -R_2^{-1} & 0 & 0 & 0 & 0 \\ * & * & * & -W & 0 & 0 & \vdots \\ * & * & * & * & -\lambda_1^{-1}W & 0 & 0 \\ * & * & * & * & * & \ddots & 0 \\ * & * & * & * & * & * & -\lambda_m^{-1}W \end{bmatrix} \le 0 \tag{13.48}$$

where $W = P^{-1}$, $Y_j = K_j W$, $\hat{\Psi}_{ij}^2 = A_i W + WA_i^T + BY_j + Y_j^T B^T$, and $\hat{\Psi}_{ij}^\infty = A_i W + WA_i^T + BY_j + Y_j^T B^T + \frac{1}{\beta}I$, then the multi-objective H_2/H_∞ investment policy problem for the fuzzy stochastic jump diffusion financial system in (13.42) can be solved.

Proof: This proof is similar to Theorem 13.1. By applying Lemma 13.5, the HJIs in (13.23) and (13.31) will be replaced as

$$\tilde{x}^T(t)Q_1\tilde{x}(t) + u^T(t)R_1u(t) + 2\tilde{x}(t)^T Pf(\tilde{x}(t))$$
$$+2\tilde{x}(t)^T PBu(t) + \sigma(\tilde{x}(t))^T P\sigma(\tilde{x}(t)) \qquad (13.49)$$
$$+\sum_{k=1}^{m}\lambda_k\{V(\tilde{x}(t)) + \Gamma(\tilde{x}(t),\theta_k) - V(\tilde{x}(t))\} \leq 0$$

and

$$\tilde{x}^T(t)Q_2\tilde{x}(t) + u^T(t)R_2u(t) + 2\tilde{x}(t)^T Pf(\tilde{x}(t))$$
$$+2\tilde{x}^T(t)PBu(t) + \frac{1}{\beta}\tilde{x}^T(t)PP\tilde{x}(t) + \sigma^T(\tilde{x}(t))P\sigma(\tilde{x}(t)) \qquad (13.50)$$
$$+\sum_{k=1}^{m}\lambda_k\{V(\tilde{x}(t)) + \Gamma(\tilde{x}(t),\theta_k) - V(\tilde{x}(t))\} \leq 0$$

respectively.

Based on analyses in (13.49) and (13.50), the inequalities in (13.35) and (13.36) are replaced by

$$WQ_1W + Y_j^T R_1 Y_j + WA_i^T + A_i W + BY_j + (BY_j)^T$$
$$+WC_i^T W^{-1}C_i W + \sum_{k=1}^{m}\lambda_k[WE_i^T(\theta_k)W^{-1}E_i(\theta_k)W] \leq 0 \qquad (13.51)$$

and

$$WQ_1W + Y_j^T R_1 Y_j + WA_i^T + A_i W + BY_j + (BY_j)^T$$
$$+WC_i^T W^{-1}C_i W + \frac{1}{\beta}I + \sum_{k=1}^{m}\lambda_k[WE_i^T(\theta_k)W^{-1}E_i(\theta_k)W] \leq 0 \qquad (13.52)$$

respectively.

Remark 13.2: In this chapter, we use a 3-D nonlinear stochastic financial system to illustrate the proposed multi-objective H_2/H_∞ investment theories. However, the proposed theories can be extended to an n-dimensional nonlinear stochastic financial system; that is, the system state $x(t)$ of the nonlinear stochastic financial system in (13.5) can be of arbitrary n dimensions. In this situation (n-dimensional nonlinear stochastic financial system), the nonlinear functions are defined as $f : \mathbb{R}^n \to \mathbb{R}^n$, $\sigma : \mathbb{R}^n \to \mathbb{R}^n$, and $\Gamma : \mathbb{R}^n \times \Theta \to \mathbb{R}^n$, which are nonlinear Borel measurable continuous functions. Moreover, these constant matrices A_i, C_i, and $E_i(\theta_k)$ for the T–S fuzzy model in (13.11) and (13.43) should be $\mathbb{R}^{n \times n}$ matrices and $B \in \mathbb{R}^{n \times p}$, where $u(t) \in \mathbb{R}^{n \times p}$. Once the n-dimensional nonlinear stochastic financial system can be approximated by the T–S fuzzy model, the proposed LMI-constrained MOEA can efficiently solve the multi-objective H_2/H_∞ investment policy problem, too.

Remark 13.3: Theorem 13.3 is similar to Theorem 13.1, except marked Poisson process $N(t;\theta_k)$ is replaced by marked compensation Poisson processes $\hat{N}(t;\theta_k)$ and some modifications in $\hat{\Psi}_{ij}^2$ and $\hat{\Psi}_{ij}^\infty$; that is, the term $\sum_{k=1}^m \lambda_k [WE_i^T(\theta_k) + E_i(\theta_k)W]$ is eliminated from $\hat{\Psi}_{ij}^2$ and Ψ_{ij}^∞ in Theorem 13.3.

Remark 13.4: The existence and uniqueness conditions for the stochastic differential equation in (13.42) are, respectively, given as follows [129]:

The existence condition is

for all $\tilde{x}(t) \in \mathbb{R}^3$

$$\left\| f(\tilde{x}(t)) \right\|^2 + \left\| \sigma(\tilde{x}(t)) \right\|^2 + \left\| u(\tilde{x}(t)) \right\|^2 + \sum_{k=1}^m \lambda_i \left\| \Gamma(\tilde{x}(t),\theta_k) \right\|^2 \le K_1 (1 + \left\| \tilde{x}(t) \right\|^2)$$

where $K_1 < \infty$.

The uniqueness condition is for all $\tilde{x}, \tilde{y} \in \mathbb{R}^3$

$$\left\| f(\tilde{x}(t)) - f(\tilde{y}(t)) \right\|^2 + \left\| \sigma(\tilde{x}(t)) - \sigma(\tilde{y}(t)) \right\|^2 + \left\| u(\tilde{x}(t)) - u(\tilde{y}(t)) \right\|^2$$
$$+ \sum_{k=1}^m \lambda_i \left\| \Gamma(\tilde{x}(t),\theta_k) - \Gamma(\tilde{y}(t),\theta_k) \right\|^2 \le K_2 \left\| \tilde{x}(t) - \tilde{y}(t) \right\|^2$$

where $K_2 < \infty$.

13.4 MULTI-OBJECTIVE H_2/H_∞ INVESTMENT POLICY OF NONLINEAR STOCHASTIC FINANCIAL SYSTEM DESIGN VIA LMI-CONSTRAINED MOEA

The MOEA is a stochastic search method simulating natural selection in natural evolution. It is also particularly suitable for solving the multi-objective investment optimization problem of nonlinear stochastic financial systems due to its population-based nature, permitting a set of Pareto optimal solutions to be obtained in a single run. For a multi-objective H_2/H_∞ investment policy problem of a nonlinear stochastic jump diffusion financial system, the MOPs in Theorems 13.1 and 13.3 with LMI constraints to guarantee the robust stability of the nonlinear stochastic financial system in (13.5) or (13.11) under continuous and discontinuous parametric fluctuations and external disturbance. It is not easy to solve the MOP in Theorem 13.1 or 13.3 directly. In this section, an LMI-constrained MOEA search algorithm is developed to help us solve the MOP in Theorem 13.1 or 13.3 iteratively. Before further discussion, some important definitions of Pareto optimality of an LMI-constrained MOP for nonlinear stochastic financial systems are given as follows:

Definition 13.3 ([77, 205]): Consider the LMI-constrained MOP in (13.19). A feasible objective vector (α^1, β^1) is said to dominate another feasible objective vector (α^2, β^2) if and only if $\alpha^1 \le \alpha^2$ and $\beta^1 \le \beta^2$ for at least one inequality as a strict inequality.

Definition 13.4 ([77, 205]): Let $(P^1, K_1^1,..., K_l^1)$ and $(P^2, K_1^2,..., K_l^2)$ be the feasible solution corresponding to the objective value (α^1, β^1) and (α^2, β^2) subject to the LMIs in (13.20)–(13.22) for all $i, j = 1,...,l$, respectively. $(P^1, K_1^1,..., K_l^1)$ is said to dominate $(P^2, K_1^2,..., K_l^2)$ if $\alpha^1 \leq \alpha^2$ and $\beta^1 \leq \beta^2$ for at least one inequality as a strict inequality.

Definition 13.5 ([77, 205]): A solution $(P^*, K_1^*,..., K_l^*)$ with objective value (α^*, β^*) is said to be a Pareto optimal solution of (13.19) if there does not exist another feasible solution $(P, K_1,..., K_l)$ with objective value (α, β), such that (α, β) dominates (α^*, β^*).

Definition 13.6 ([77, 205]): For the LMI-constrained MOP of the nonlinear stochastic financial system in (13.19), the Pareto front P_F is defined as

$$P_F = \left\{ (\alpha^*, \beta^*) \middle| (P^*, K_1^*,..., K_l^*) \text{ is a Pareto optimal solution and } (\alpha^*, \beta^*) \text{ is} \right.$$
generated by $(P^*, K_1^*,..., K_l^*)$ subject to the LMIs in (13.20), (13.21), and (13.22) for $i, j = 1,...,l\}$

The LMI-constrained MOP for the nonlinear stochastic financial system concerns an evolution algorithm. For the proposed LMI-constrained MOEA, the EA operates on a number of encoding feasible objective vectors called a population so that, for a feasible objective vector (α^k, β^k), it should be encoded into a chromosome C_k. A feasible chromosome C_k is defined as a coded feasible objective vector. The chromosome of the EA for the LMI-constrained MOEA employs the real-value representation to avoid long binary strings and a large search space in the EA for the multi-objective H_2/H_∞ investment policy of the nonlinear stochastic jump diffusion financial system. In general, EA takes a population to be the algorithm input and returns the chromosomes with fit performance to be the next population. The LMI constraints in (13.20)–(13.22) to guarantee the robust stability of the financial system and the two upper bounds α and β are just like the environments in natural evolution. Only the adaptive chromosomes survive. To guarantee that all of these chromosomes can be decoded as feasible objective vectors for the MOP in (13.19), each chromosome C_k should be examined for the existence of a feasible solution $(P^k, K_1^k,..., K_l^k)$ in LMIs in (13.20)–(13.22) with the LMI toolbox in MATLAB after the mating operation. If some chromosomes are not feasible, these chromosomes need to be removed from the candidate chromosomes. It is worth mentioning that the LMI constraints in (13.20)–(13.22) impose more restrictions on searching the feasible chromosomes than the conventional MOEA approach. Besides, using the LMI toolbox in MATLAB can help efficiently examine whether these chromosomes satisfy the LMIs in (13.20)–(13.22), which accelerates the selection speed of the initial populations via the initialization scheme.

Design procedure of multi-objective H_2/H_∞ investment policy of nonlinear stochastic financial systems:

Step 1: Select the search range $S = (\alpha_0, \beta_0) \times (\bar{\alpha}, \bar{\beta})$ for the feasible objective vector (α, β) and set the iteration number $\bar{N}$, the population number N_p, the crossover ration N_c, and the mutation ratio N_m in the LMI-constrained MOEA.

Step 2: Select N_p feasible chromosomes from the feasible chromosome set randomly to be the initial population P_1.

Step 3: Set iteration index $N_i = 1$.

Step 4: Operate the EA with the crossover ratio N_c and mutation ratio N_m and generate $2N_p$ feasible chromosomes by examining whether their corresponding objective vectors are feasible objective vectors for the LMIs in (13.20)–(13.22).

Step 5: Set the iteration index $N_i = N_i + 1$ and select N_p chromosomes from the $2N_p$ feasible chromosomes in Step 4 through the nondominated sorting method as the population P_{N_i}.

Step 6: Repeat Steps 4 and 5 until iteration number $\bar{N}$ is reached. If iteration number $\bar{N}$ is satisfied, then we set $P_{N_i} = P_F$.

Step 7: Select a preferable feasible objective individual $(\alpha^\dagger, \beta^\dagger) \in P_F$ according to the designer's preference. Once the preferable feasible objective individual is selected, the corresponding Pareto optimal solution

$$\xi^\dagger = \{W^\dagger, K_1^\dagger, K_2^\dagger, ..., K_l^\dagger\}$$

is obtained. By using $\xi^\dagger$, the proposed multi-objective H_2/H_∞ fuzzy investment policy $u(t) = \sum_{i=1}^{l} h_i(z)(K_i^\dagger \tilde{x}(t))$ with $K_i^\dagger = Y_i^\dagger W^{*-1}$ in (13.16) can be constructed and the multi-objective H_2/H_∞ investment policy problem in (13.5) or (13.11) can be solved with $J_2 = \alpha^\dagger$ and $J_\infty = \beta^\dagger$ simultaneously.

Note that the system diagram of the multi-objective H_2/H_∞ investment policy problem is given in Figure 13.1, and the system diagram of the proposed LMI-constrained MOEA is given in Figure 13.2, used in the design procedure.

Remark 13.5: The computational complexity of the proposed LMI-constrained MOEA is approximately $O(n(n+1)lN_p^2\bar{N})$, including $O(\frac{n(n+1)l}{2})$ for solving the LMIs and $O(2N_p^2\bar{N})$ for the MOEA, where n is the dimension of system state $x(t)$, $\bar{N}$ is the iteration number of the MOEA, l is the number of local linear models of the nonlinear stochastic financial system in (13.11), and N_p is the population number of the LMI-constrained MOEA.

13.5 SIMULATION RESULTS

To illustrate the design procedure and to confirm the performance of the proposed optimal investment policy for the nonlinear stochastic jump diffusion financial system, we introduce a nonlinear stochastic jump diffusion financial system to mimic an emerging market in (13.2). In general, an emerging market always suffers from continuous and discontinuous intrinsic random fluctuations due to national and international situation changes, oil price changes, surplus between investments and savings, variable interest rates, and so on. External investment disturbance is caused by unpredictable investment changes or worldwide events such as war, nature disaster, fatal epidemic, and so on.

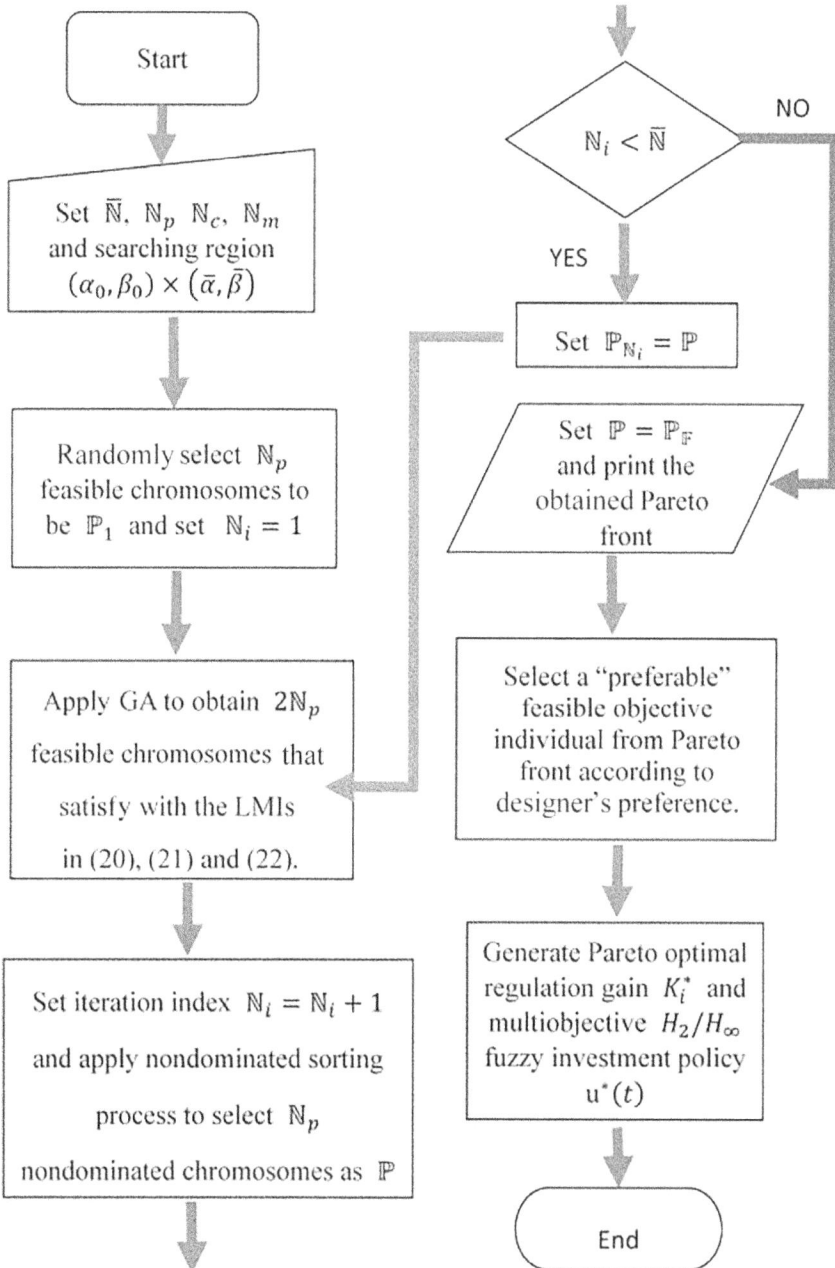

FIGURE 13.2   Flowchart of the proposed LMI-constrained MOEA.

The related parameters and intrinsic continuous and discontinuous fluctuation are, respectively, given as follows:

$$a = 1.5, \ b = 0.2, \ c = 0.25, \ \Theta = \{\theta_1, \theta_2, ..., \theta_6\}$$

$$\begin{cases} \lambda_i = 0.3, \text{for } i = 1, 2, 3, 4 \\ \lambda_i = 0.2, \text{for } i = 5, 6 \end{cases}$$

$$\sigma_1(x(t), y(t), z(t)) = 0.03 \times [z(t) + (y(t) - a)x(t)]$$

$$\sigma_2(x(t), y(t), z(t)) = 0.01 \times [1 - by(t) - (x(t))^2]$$

$$\sigma_3(x(t), y(t), z(t)) = 0.02 \times [-x(t) - cz(t)]$$

$$B = I_{3\times3}$$

$$\gamma_1(x(t), y(t), z(t), \theta_i) = \begin{cases} 0.3x(t), \text{if } \theta_i = \theta_1 \\ -0.3x(t), \text{if } \theta_i = \theta_2 \\ 0, \text{else} \end{cases}$$

$$\gamma_2(x(t), y(t), z(t), \theta_i) = \begin{cases} 0.05y(t), \text{if } \theta_i = \theta_3 \\ -0.05y(t), \text{if } \theta_i = \theta_4 \\ 0, \text{else} \end{cases}$$

$$\gamma_3(x(t), y(t), z(t), \theta_i) = \begin{cases} -0.1z(t), \text{if } \theta_i = \theta_5 \\ 0.1z(t), \text{if } \theta_i = \theta_6 \\ 0, \text{else} \end{cases}$$

Suppose the initial states are given as

$$\tilde{x}_0 = (0.37, -3.06, 0.71)$$

The external investment disturbance is assumed to be

$$v(t) = [0.01\sin(2t), -0.02\sin(2t), -0.01\sin(2t)]$$

Figure 13.3 is used to describe the dynamic behavior of the nonlinear stochastic jump diffusion financial system without an investment policy $u(t)$ in (13.2) with

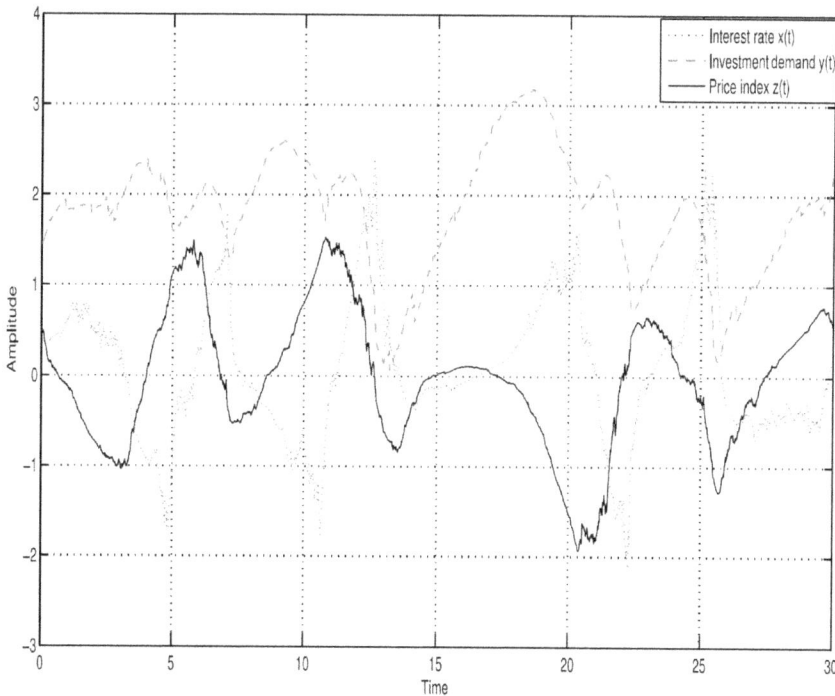

FIGURE 13.3 Trajectories of the interest rate $x(t)$, the investment demand $y(t)$, and the price index $z(t)$ for the nonlinear stochastic jump diffusion system in (13.2), that is, the system dynamic behaviors for nonlinear stochastic jump diffusion financial system in (13.3) without introducing investment policy $u(t)$.

numerous fluctuations in a real situation. It is seen that the three states $x(t)$, $y(t)$, and $z(t)$ of the nonlinear stochastic financial system fluctuate with random jumps. Therefore, the multi-objective H_2/H_∞ investment policy of the MOP in (13.17) is employed by the government to design an investment policy to regulate the stochastic financial system to achieve the following desired steady state:

$$x_d = (0.1, 4.5, -0.2) \qquad (13.53)$$

that is, the government of the emerging market expects its financial system to be regulated to achieve a desired financial steady state with interest rate 0.1, investment demand 4.5, and price index −0.2 to coordinate inflation rate, supply, and demand in the emerging market to stimulate the recovery of emerging market under continuous and discontinuous random fluctuation.

In this example, we assume the vector of premise variables $z = [x_1(t), x_2(t)]^T = [z_1(t), z_2(t)]^T$ in (13.11) and (13.16) is available in the regulation design. Since there are eight fuzzy sets associated with premise variable $z_1(t)$ and eight fuzzy sets associated with premise variable $z_2(t)$, there are 64 fuzzy if–then rules in total in the T-S fuzzy financial system. The operation points of the T-S fuzzy model and the linearized local linear stochastic models are shown in the Appendix. Suppose the weighting matrices in the multi-objective H_2/H_∞ investment policy in (13.17) are used to regulate the stochastic financial system with the following weighting matrices:

$$Q_1 = I_{3\times3}, R_1 = I_{3\times3}, Q_2 = I_{3\times3}, R_2 = 0.5I_{3\times3}$$

Based on the proposed design procedure of the multi-objective H_2/H_∞ optimal investment policy, the MOEA is employed to solve the MOP in (19). In the MOEA, the search region Γ is set as $\Gamma = [0 \quad 70] \times [0.4 \quad 1.2]$, the maximum number of individuals $N_p = 80$, iteration number $N_i = 100$, crossover rate $N_c = 0.8$, and mutation ratio $N_m = 0.2$.

Once the iteration number $\bar{N} = 100$ is achieved, the Pareto front P_F for the Pareto optimal solutions of the MOP for the investment policy of the nonlinear stochastic jump diffusion financial system of the emerging market in (5) or (11) can be obtained as shown in Figure 13.4.

In order to illustrate how to select the preferable solution of the investment policy, we choose three Pareto optimal solutions from the Pareto front for comparison, whose Pareto objective vectors are given in Table 13.1. Moreover, in Figures 13.5–13.7, the simulation results of the nonlinear stochastic jump diffusion financial system for the emerging market are given to illustrate the performance of the multi-objective H_2/H_∞ optimal fuzzy investment policy of the three chosen Pareto

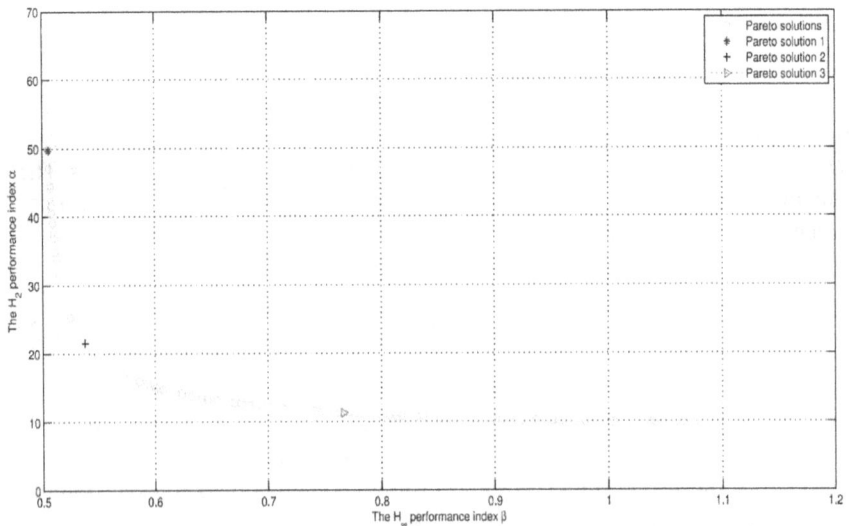

FIGURE 13.4 The Pareto front for the Pareto optimal solutions of the MOP in (13.19) can be obtained by the proposed LMI-constrained MOEA. Three marked Pareto optimal solutions are the preferable multi-objective H_2/H_∞ investment policy with their simulations in Figures 13.5 and 13.7.

TABLE 13.1

Pareto Objective Vectors of the Three Chosen Individuals in the Mean Sense

Pareto solution 1	Pareto solution 2	Pareto solution 3
(49.7177, 0.5055)	(21.5353, 0.5380)	(11.3162, 0.7667)

TABLE 13.2

Positive-Definite Matrix of the Three Chosen Pareto Optimal Solutions in Figure 13.2

Pareto solution 1

$$P_{solution1} = \begin{bmatrix} 68.54 & 1.56 & 10.49 \\ 1.56 & 2.28 & 1.99 \\ 10.49 & 1.99 & 24.13 \end{bmatrix}$$

Pareto solution 2

$$P_{solution2} = \begin{bmatrix} 31.44 & 0.18 & -0.77 \\ 0.18 & 1.65 & 0.71 \\ -0.77 & 0.71 & 10.35 \end{bmatrix}$$

Pareto solution 3

$$P_{solution3} = \begin{bmatrix} 10.57 & -0.21 & -1.94 \\ -0.21 & 1.03 & 0.41 \\ -1.94 & 0.41 & 4.79 \end{bmatrix}$$

optimal solutions for the nonlinear stochastic jump diffusion financial system in (13.5) or (13.11). The corresponding P matrices in (13.20) of the three chosen Pareto solutions are also given in Table 13.2. In Figures 13.5–13.7, Pareto optimal solution 1 has the best J_∞ performance index of the three chosen Pareto optimal solutions so that the trajectory of Pareto optimal solution 1 has the minimum perturbation but has the maximum J_2 performance index. However, Pareto optimal solution 3 has the minimum J_2 performance index, but its trajectory has the maximum perturbation of the three because it has the maximum J_2 performance index. It is easy to observe that Pareto solution 2 is the preferable regulation solution of the three chosen Pareto solutions because Pareto solution 2 is a compromise solution for J_2 and J_∞ in the multi-objective investment policy problem of the nonlinear stochastic jump diffusion financial system. However, these three Pareto optimal solutions have satisfactory regulation results to achieve the desired steady states of the financial system in spite of intrinsic continuous and discontinuous random fluctuation and external disturbance, respectively.

13.6 CONCLUSION

This chapter has investigated the multi-objective H_2/H_∞ investment policy for nonlinear stochastic jump diffusion financial systems via the T-S fuzzy model interpolation

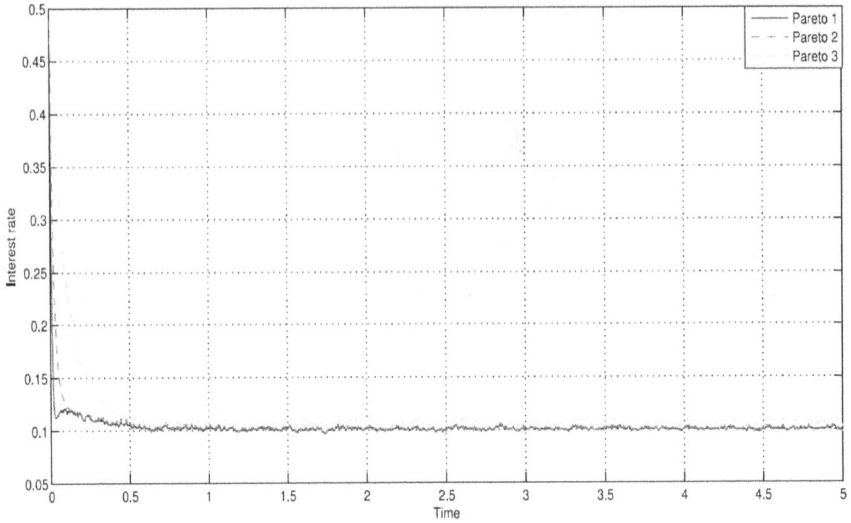

FIGURE 13.5   Interest rate trajectories $x(t)$ of the three chosen Pareto solutions.

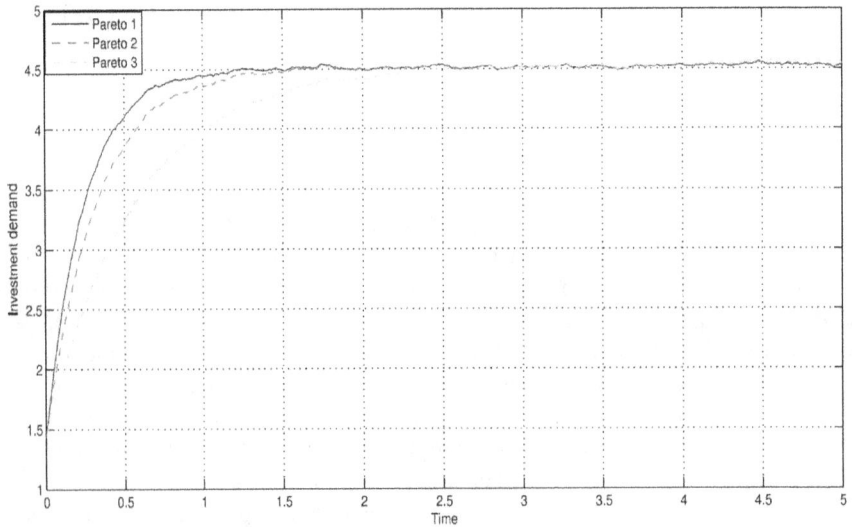

FIGURE 13.6   Investment demand trajectories $y(t)$ of the three chosen Pareto solutions.

method. Unlike most multi-objective problems only focusing on algebraic systems via the indirect method in Lemma 13.2, the proposed multi-objective H_2/H_∞ investment policy can solve the optimal robustness and regulation problems of a nonlinear stochastic jump diffusion financial system to achieve a desired target with less risk and less regulation of investment effort simultaneously. To avoid solving a nonlinear

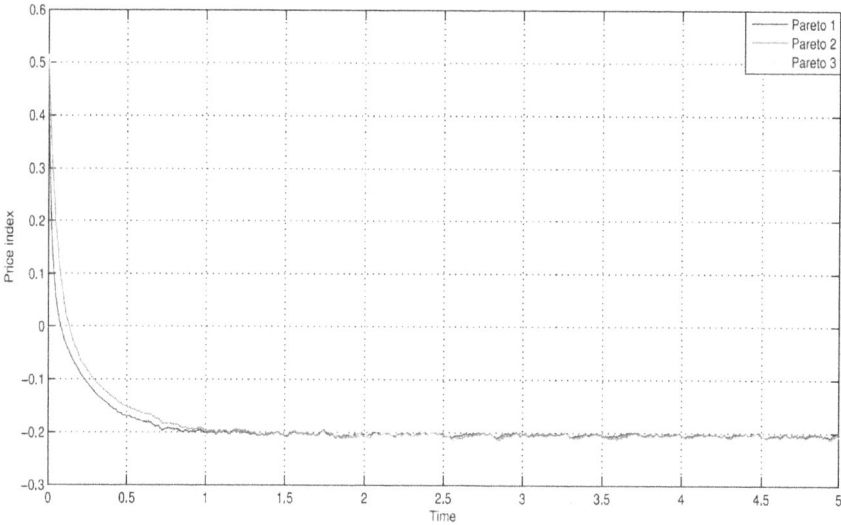

FIGURE 13.7 Price index trajectories $z(t)$ of the three chosen Pareto solutions.

system dynamic-constrained MOP, the T–S fuzzy model interpolation method is employed such that the nonlinear system dynamic constraints can be replaced by two sets of LMI constraints. If there exist feasible solutions for the LMI constraints in (13.20)–(13.22), the LMI-constrained MOP can be solved more easily by the proposed LMI-constrained MOEA. Thus, the multi-objective H_2/H_∞ investment policy for nonlinear stochastic jump diffusion financial systems can be solved efficiently with the help of MATLAB via the LMI toolbox. When the Pareto front is obtained for the Pareto optimal solutions of the MOP in (13.19), the manager can select a fuzzy investment policy from the set of Pareto optimal fuzzy investment policies according to their preference and finish the multi-objective H_2/H_∞ investment policy design of the nonlinear stochastic financial system. Finally, an example is given to confirm the satisfactory performance of the proposed multi-objective H_2/H_∞ investment policy design for nonlinear stochastic jump diffusion financial systems through computer simulation.

13.7 APPENDIX

The eight operation points of x_1 are given at

$$x_1^1 = -1.50, \ x_1^1 = -1.50, \ x_1^3 = -0.72, \ x_1^4 = -0.33$$

$$x_1^5 = 0.07, \ x_1^6 = 0.46, \ x_1^7 = 0.85, \ x_1^8 = 1.24$$

and the eight operation points of x_2 are given at

$$x_1^1 = -3.30, \ x_1^1 = -3.10, \ x_1^3 = -2.90, \ x_1^4 = -2.70$$

$$x_1^5 = -2.50, \ x_1^6 = -2.30, \ x_1^7 = -2.10, \ x_1^8 = -1.90$$

The qth rule of the T–S fuzzy model for the nonlinear stochastic jump diffusion financial system in (13.5) is described as

$$\text{System Rule } q = 8(j-1)+i, \text{for } i,j = 1,...,8$$
$$\text{if } z_1 \text{ is } x_1^i \text{ and } z_2 \text{ is } x_2^j \text{ then}$$
$$d\tilde{x}(t) = (f(\tilde{x}(t)) + Bu(t) + v(t))dt + \sigma(\tilde{x}(t))dW$$
$$+ \sum_{k=1}^{m} \Gamma(\tilde{x}(t), \theta_k) dN(t; \theta_k)$$

(13.54)

where

$$A_q = \begin{bmatrix} M_1^q & x_d & 1 \\ M_2^q & M_3^q & 0 \\ M_4^q & 0 & -c \end{bmatrix}, \ C_q = \begin{bmatrix} 0.3M_1^q & 0.3x_d & 0.3 \\ 0.1M_2^q & 0.1M_3^q & 0 \\ 0.2M_4^q & 0 & -0.2c \end{bmatrix}, \ M_1^q = -a + x_2^j + y_d + \frac{\Delta_1}{x_1^i},$$

$$M_2^q = -x_1^i - 2x_d, \ M_3^q = -b + \frac{\Delta_2}{x_2^i}, \ M_4^q = -1 - \frac{\Delta_3}{x_1^i}, \ \Delta_3 = x_d + cz_d, \ \Delta_1 = 1 - by_d - (x_d)^2$$

$$\Delta_3 = x_d + cz_d$$

$$E_q(\theta_1) = diag([0.21,0,0]), \ E_q(\theta_2) = diag([-0.3,0,0])$$

$$E_q(\theta_3) = diag([0,0.05,0]), \ E_q(\theta_4) = diag([0,-0.05,0])$$

$$E_q(\theta_5) = diag([0,0,-0.1]), \ E_q(\theta_6) = diag([0,0,0.15])$$

14 Multi-Objective Optimal H_2/H_∞ Dynamic Pricing Management Policy of a Mean Field Stochastic Smart Grid Network

14.1 INTRODUCTION

In recent years, power demand is increasing along with the deployment of technology and industry, and there have arisen a lot of challenges for traditional power systems, such as the integration of renewable energy sources (RESs), energy efficiency, and the reliability of power systems. In this situation, issues of smart grids have attracted much attention. Smart grid networks have changed the conventional operation of power systems from power generation, transmission, and distribution in the last several years by integrating advanced control technology, sensing technology, and the topology of communication. The main characteristics of a smart grid are consumer friendliness, self-healing, support of multiple types of generation, storage options, efficient electricity market-based operation, and high power quality [398].

In a smart grid network, energy management methods play an important role and can help solve many problems of energy management with various methods, such as power flow management [399, 400] and demand side management [325, 401–404]. Smart pricing, which is an energy management method, has been intensively investigated [325, 401–404], because it can effectively control the balance between power supply and demand based on the electricity market mechanism. It is worth pointing out that although RESs can provide extra clean energy in a grid network, the generation of energy from RESs is usually unpredictable and intermittent given weather conditions. This will influence the reliability of the power system or the quality of the power supply, and it could even lead to blackouts in the case of an event in which the power supply cannot meet the demand. Therefore, integration of RESs is also an important issue for smart pricing in smart grid networks. A practical pricing scheme [401] is proposed to encourage different consumers to participate in smart grid demand response with a nonlinear programming problem that aims at minimizing the overall operation cost. A robust dynamic pricing scheme [402] is proposed to minimize the imbalance between power supply and demand in a smart grid network under a worst-case disturbance due to the uncertainty of RESs. An advanced metering infrastructure (AMI) is the deployment of interconnected smart meters

DOI: 10.1201/9781003362142-18

that enable two-way communication to exchange information among the agents in a power system, and the interactions of agents in the grid are not like the traditional ones [405]. With a smart meter in a smart grid, the users of microgrids access their real-time information such as the price policy of the grid. Moreover, smart meters have a certain amount of computation capability. Users can actively participate in the operation of the power market. Thus, the consideration of these interactions in the smart grid is important. There have been several studies to discuss these interactions in smart grid networks [406, 407].

In general, a manager of a grid network expects to design a pricing policy to balance the stored energy of the microgrid at a desired level for operation in emergencies and also that the smart grid network can have robust energy management against external disturbances from RESs. For the energy management systems in smart grid networks, the H_2 pricing policy management scheme can be regarded as how to design a pricing policy to ensure that the stored energy of the microgrids could be driven to a desired reference level with a minimum mean square price perturbation. In general, H_2 pricing policy ignores the effect of external disturbance on the desired reference-level tracking of the management policy of the smart grid network. On the other hand, a robust H_∞ pricing policy in a smart grid network could be seen as a management scheme to design a dynamic pricing policy to minimize the worst-case effect of intermittent and unpredictable power input from RESs on the tracking performance of the desired energy storage reference level. The H_∞ pricing management policy can effectively attenuate the worst-case effect of external disturbance from RESs on the reference tracking performance of energy management, but it usually leads to a higher price perturbation, which is not wanted by the market operator. Because there is a partial conflict between the objectives of the H_2 and H_∞ pricing management policies, the pricing management policy to simultaneously achieve multi-objective H_2/H_∞ pricing management policy can be regarded as an MOP.

To start up or shut down the current load, the users of a smart grid may pay extra when they adjust their power demands according to the price signal. The average of the power demand can provide valuable information for the price forecast of the users in the smart grid network. Thus, the collective (average) demand behaviors will also influence the users' power demand behaviors in these situations. Mean field theory has attracted a considerable attention and has studied collective behaviors resulting from mutual interaction in various research areas [408–411]. In general, in the mean field game, there exist the coupled Hamilton–Jacobi–Bellman (HJB) equation and Fokker–Planck–Kolmogorov (FPK) equation to represent the optimal response of players and the dynamic evolution of the population distribution, respectively [410]. Mean field theory has also been applied in the field of smart grids [412–415]. In [414], agents in the smart grid can submit their graphs of supply and demand to bid according to the price signal and the average effect calculated by statistical information.

In previous work on mean field smart grid network systems [412–415], they do not consider the integration of RESs in the smart grid. The integration of RESs is an important issue for a smart grid network. On the other hand, in the real world, the power market system may also suffer from continuous or discontinuous parametric fluctuations due to changes in natural or economic situations [416, 417]. In this situation, the smart grid network system should be modeled as a stochastic jump diffusion system with the consideration of RESs as an uncertain external disturbance. We propose a dynamic stochastic mean field smart grid network system to describe the interacting and collective behavior of microgrids equipped with energy storage devices and RESs. Moreover, when each microgrid changes its own demand behavior, it interacts with the others through the dynamic price policy to influence the overall demand. Thus, by using the mean field theory, the total influence of microgrid actions can be considered collective behaviors (mean field term) of microgrids in the smart grid network.

In this chapter, a multi-objective H_2/H_∞ dynamic pricing scheme is proposed to design a pricing strategy to manage the energy in a mean field stochastic smart grid network, with an optimal H_2 tracking strategy to achieve the desired stored energy working levels of microgrids with parsimonious management effort and the optimal H_∞ tracking strategy to attenuate the worst-case effect of external disturbance due to unpredictable and intermittent RESs. Since the initial producer and consumer's marginal benefit is constant and cannot be easily attenuated by the conventional H_∞ state feedback tracking control method, an integration-based state feedback control scheme is proposed to eliminate constant disturbances due to the constant initial producer marginal cost and initial consumer marginal benefit to achieve asymptotic tracking of the desired stored energy working level of the microgrids. However, it is still difficult to directly solve the multi-objective H_2/H_∞ dynamic pricing management problem for the desired reference tracking control of microgrids in the mean field stochastic smart grid network system. An indirect method is proposed to solve the design problem of the multi-objective H_2/H_∞ pricing management policy for the mean field stochastic smart grid network by simultaneously minimizing the upper bounds of the H_2 and H_∞ dynamic pricing management performance to achieve Pareto multi-objective optimality. By the indirect method, the multi-objective H_2/H_∞ dynamic pricing management problem can be transformed into a linear matrix inequality–constrained MOP from which we can find a set of control gains for integration-based state feedback control to minimize the upper bound of the H_2 and H_∞ pricing management policy performance subject to a set of LMI constraints in the Pareto multi-objective optimal sense. These LMIs can be efficiently solved by the LMI toolbox in MATLAB.

Furthermore, we propose an LMI-constrained multi-objective evolutionary algorithm to efficiently solve the LMI-constrained MOP for the multi-objective H_2/H_∞ dynamic pricing management scheme for the mean field stochastic smart grid network system. The LMI-constrained MOEA search via a number of genetic operations includes selection, mutation, and crossover to efficiently search the Pareto front of the Pareto optimal solutions and its corresponding control strategy to achieve the H_2 and H_∞ dynamic pricing management policy with the help of the LMI toolbox in MATLAB for the mean field stochastic smart grid network system simultaneously.

The objectives of this chapter are summarized as follows:

- A dynamic stochastic mean field smart grid network system, in which each microgrid is influenced by the collective demand behaviors and unpredictable and intermittent RESs, is introduced with continuous and discontinuous random fluctuation to mimic a realistic networked power system.

- Since the initial producer and consumer marginal benefit are constant and cannot be easily attenuated by the conventional state feedback tracking control method, an integration-based state feedback control strategy is proposed to eliminate constant disturbances due to constant initial producer marginal cost and initial consumer marginal benefit in the mean field stochastic smart grid network system.
- To achieve the desired energy storage level of each microgrid, a multi-objective H2/H∞ dynamic pricing management policy is proposed based on the state and integrative state feedback control for the smart grid network to achieve the optimal desired reference storage tracking with a parsimonious price policy and optimal attenuation of external disturbance mainly due to unpredictable and intermittent RESs and constant initial producer marginal cost and initial consumer marginal benefit.
- An indirect method is introduced to transform the multi-objective H_2/H_∞ dynamic pricing management problem of the stochastic mean field smart grid network system into an LMI-constrained MOP, and an LMI-constrained MOEA is also proposed to efficiently solve the LMI-constrained MOP with the help of the MATLAB toolbox.

This chapter is divided into six sections as follows: in Section 14.2, the mean field stochastic smart grid network system is introduced, and the MO H_2/H_∞ dynamic pricing management problem is formulated. In Section 14.3, using the proposed indirect method, the MO H_2/H_∞ state feedback controller design problem is transformed into an equivalent LMI-constrained MOP. In Section 14.4, an LMI-constrained MOEA is developed to efficiently solve the LMI-constrained MOP. A design example with the simulation results is given in Section 14.5. The chapter is concluded in Section 14.6.

Notation: A^T: transpose of matrix A; $A \geq 0(A > 0)$: symmetric positive semi-definite (symmetric positive definite) matrix A; C^2: class of function $V(x)$ that is twice continuously differentiable with respect to x; V_x: gradient column vector of function $V(x)$; V_{xx}: Hessian matrix with elements of the second partial derivatives of function $V(x)$; E: expectation operator.

14.2 SYSTEM DESCRIPTION AND PROBLEM FORMULATION

14.2.1 MODEL OF MEAN FIELD STOCHASTIC SMART GRID NETWORK SYSTEM

We consider the following smart grid network model consisting of a power grid and N microgrids [5]:

$$\tau_n^d dp_n^d(t) = (b_n^d + c_n^d p_n^d(t) - u(t))dt$$

$$\tau_g dp_n^g(t) = (\tfrac{1}{N}u(t) - (\tfrac{b_g}{N} + c_g p_n^g(t)) - \tau_k s_n(t))dt \qquad (14.1)$$

$$ds_n(t) = (p_n^g(t) - p_n^d(t) + r_n(t))dt, \text{ for } n = 1, 2, ..., N$$

where $p_n^d(t)$ denotes the power demand of the nth microgrid; $p_n^g(t)$ denotes the distributed power from the power grid to the nth microgrid; $u(t)$ denotes the dynamic pricing policy; $b_n^d > 0$ is the initial marginal benefit of the nth microgrid; $c_n^d > 0$ is the demand elasticity of the nth microgrid; $\frac{b_g}{N} > 0$ is the initial marginal cost of the distributed power; $c_g > 0$ is the demand elasticity of the distributed power; τ_g and τ_n^d are time constants; $s_n(t)$ denotes the stored energy state of the nth microgrid; and $\tau_k s_n(t)$ with $\tau_k > 0$ is considered the additional cost for the excess power supply $p_n^g(t)$.

In the model in (14.1), the nth microgrid would adjust one's own power demand $p_n^d(t)$ according to the pricing policy $u(t)$. When the marginal benefit of the nth microgrid $b_n^d + c_n^d p_n^d(t)$ exceeds the pricing policy $u(t)$, the user of the nth microgrid would increase their power demand rate, and vice versa. The power grid adjusts the distributed power of the nth microgrid $p_n^g(t)$ in the way that increases/decreases the distributed power rate of the nth microgrid, when the pricing policy $\frac{1}{N}u(t)$ exceeds/less than the sum of the distributed power marginal cost function $\frac{b_g}{N} + c_g p_n^g(t)$ and the additional cost $\tau_k s_n(t)$ of the excess power supply of the nth microgrid to ensure the stability of the smart grid network system. For storage energy $s_n(t)$ in the nth microgrid, each microgrid is equipped with a stored energy device and local energy harvesting device so that each microgrid can receive power from distributed power $p_n^g(t)$ and the RES $r_n(t)$ from the local energy harvesting device. Since the solar panels or wind turbine are influenced by weather disturbance, the RES $r_n(t)$ is always unpredictable and intermittent, so it can be regarded as an external disturbance for each microgrid in the smart grid system [403].

In the real world, a power market system always suffers from random parametric fluctuations due to the behaviors of consumers and sudden political or economic events in the market. The power demand behavior of each microgrid will impact the pricing policy, and users of the microgrid will adjust their power demands according to the pricing policy. However, the users of a smart grid may pay high costs for the startup or shutdown of the current load when they adjust their power demand according to the pricing policy. We assume that the average of the power demands of all microgrids can provide valuable information on the trend of collective power demand and the price forecast for the users of the smart grid. Thus, the user of each microgrid will also consider the mean behavior of power demands of microgrids to adjust their own power demand. The dynamic of the power demand of the nth microgrid can be reformulated as follows

$$
\begin{aligned}
\tau_n^d dp_n^d(t) = {} & (b_n^d + c_n^d p_n^d(t) - u(t) + \sum_{i=1}^{N} \eta_{n,i} E p_i^d(t)) dt \\
& + (\sigma_n p_n^d(t) + \sigma_n^\mu E p_n^d(t)) dW_n(t) \\
& + (\rho_n p_n^d(t) + \rho_n^\mu E p_n^d(t)) dN_n(t), \text{for } n = 1,2,\dots,N
\end{aligned}
\tag{14.2}
$$

where $\eta_{n,1}, \eta_{n,2}, \dots, \eta_{n,N}$ are aggregation parameters for the nth microgrid that can be determined by the users of microgrids in consideration of the contribution of mean (collective) power demand, the term $\sum_{i=1}^{N} \eta_{n,i} E p_i^d(t)$ is the effect of average demand on the behavior of the nth microgrid, $W_n(t) \in \mathbb{R}^1$ denotes the standard one-dimensional Wiener process, and $N_n(t) \in \mathbb{R}^1$ denotes the Poisson random processes with jump intensity $\lambda_n > 0$.

To mimic a real power market system, the power market dynamic model should be modified by continuous and discontinuous intrinsic random fluctuations [416, 417]. In (14.2), the term $(\sigma_n p_n^d(t) + \sigma_n^\mu Ep_n^d(t))dW_n(t)$ can be deemed a continuous state-dependent intrinsic random fluctuation to describe the users of the microgrid changing their demand behavior because of random factors like rumors in the power market and weather changes; the Poisson random processes $(\rho_n p_n^d(t) + \rho_n^\mu Ep_n^d(t))dN_n(t)$ are discontinuous sudden random fluctuations such as policy changes or instantaneous loads when the users of the smart grid use high-power appliances.

In order to maintain normal operation of the microgrid in an emergency, the energy storage system of the microgrid must be maintained at a working level; that is, the market operator needs to have the energy storage state track a desired working level. Suppose a desired energy storage state of the nth microgrid can be generated by the following reference model

$$ds_n^d(t) = (a_n s_n^d(t) + b_n(t))dt, n = 1, 2, ..., N \tag{14.3}$$

where $s_n^d(t)$ is the desired energy state to be tracked by $s_n(t)$, a_n is a specified constant to model the transient behavior of $s_n^d(t)$, and $b_n(t)$ is a bounded reference input. If we specify $a_n = -1$ at the steady state $s_n^d(t) = b_n(t)$, $b_n(t)$ denotes the desired steady working state of energy storage in the nth microgrid.

The error state dynamic of the energy storage of microgrid n can be expressed as

$$de_n(t) = (p_n^g(t) - p_n^d(t) - a_n s_n^d(t) + r_n(t) - b_n(t))dt \tag{14.4}$$
$$\text{for } i = 1, 2, ..., N$$

where the error state $e_n(t) = s_n(t) - s_n^d(t)$.

The distributed power $p_n^g(t)$ from the power grid to each microgrid can be reformulated as follows

$$\tau_g dp_n^g(t) = (\tfrac{1}{N}u(t) - (\tfrac{b_g}{N} + c_g p_n^g(t)) - \tau_k e_n(t))dt \tag{14.5}$$
$$\text{for } i = 1, 2, ..., N$$

Thus, the dynamic system of the overall mean field stochastic smart grid network in (14.2), (14.3), (14.4), and (14.5) can be expressed in the following:

$$dx(t) = (Ax(t) + A^\mu Ex(t) + b + Bu(t) + v(t))dt$$
$$+ \sum_{r=1}^{n}[C_r x(t) + C_r^\mu Ex(t)]dW_r(t)$$
$$+ \sum_{q=1}^{n}[D_q x(t) + C_q^\mu Ex(t)]dN_q(t) \tag{14.6}$$
$$e(t) = Mx(t)$$

with

$$x = [g^T(t) \quad d^T(t) \quad e^T(t) \quad s_d(t)^T]^T, \, g(t) = [g_1(t) \quad g_2(t) \quad \cdots \quad g_N(t)]^T,$$

$$d(t) = [p_1^d(t) \quad p_2^d(t) \quad \cdots \quad p_N^d(t)]^T, \, e(t) = [e_1(t) \quad e_2(t) \quad \cdots \quad e_N(t)]^T,$$

$$s_d(t) = [s_1^d(t) \quad s_2^d(t) \quad \cdots \quad s_N^d(t)]^T,$$

$$A = \begin{bmatrix} -\frac{c_g}{\tau_g} I_N & 0_{N \times N} & -\frac{\tau_k}{\tau_g} I_N & 0_{N \times N} \\ 0_{N \times N} & diag(\frac{c_1^d}{\tau_1^d}, ..., \frac{c_N^d}{\tau_N^d}) & 0_{N \times N} & 0_{N \times N} \\ I_N & -I_N & 0_{N \times N} & -diag(a_1, ..., a_N) \\ 0_{N \times N} & 0_{N \times N} & 0_{N \times N} & diag(a_1, ..., a_N) \end{bmatrix},$$

$$M = [0_{N \times N} \quad 0_{N \times N} \quad I_N \quad 0_{N \times N}], \, C_r = diag(0, ..., 0, \underbrace{\sigma_r}, 0, ..., 0),$$
$$\underbrace{}_{N+r-1} \quad \underbrace{}_{3N-r}$$

$$C_r^\mu = diag(0, ..., 0, \underbrace{\sigma_r^\mu}, 0, ..., 0), \text{for } r = 1, ..., N, \, D_q = diag(0, ..., 0, \underbrace{\sigma_q}, 0, ..., 0),$$
$$\underbrace{}_{N+r-1} \quad \underbrace{}_{3N-r} \qquad \underbrace{}_{N+q-1} \quad \underbrace{}_{3N-q}$$

$$D_q^\mu = diag(0, ..., 0, \underbrace{\sigma_q}, 0, ..., 0), \text{for } q = 1, ..., N, \, A^\mu = diag(0_{N \times N}, L, 0_{N \times N}, 0_{N \times N}),$$
$$\underbrace{}_{N+q-1} \quad \underbrace{}_{3N-q}$$

$$L = \begin{bmatrix} \frac{\eta_{1,1}}{\tau_1^d} & \frac{\eta_{1,2}}{\tau_1^d} & & \frac{\eta_{1,N}}{\tau_1^d} \\ \frac{\eta_{2,1}}{\tau_2^d} & \frac{\eta_{2,2}}{\tau_2^d} & & \frac{\eta_{2,N}}{\tau_2^d} \\ & & & \\ \frac{\eta_{1,N}}{\tau_N^d} & \frac{\eta_{2,N}}{\tau_N^d} & & \frac{\eta_{N,N}}{\tau_N^d} \end{bmatrix}, B = \begin{bmatrix} (\frac{1}{N\tau_g})_{N \times 1} \\ -\frac{1}{\tau_1^d} \\ -\frac{1}{\tau_2^d} \\ \vdots \\ -\frac{1}{\tau_N^d} \\ 0_{N \times 1} \\ 0_{N \times 1} \end{bmatrix}, b = \begin{bmatrix} (-\frac{b_g}{N\tau_g})_{N \times 1} \\ \frac{b_1^d}{\tau_1^d} \\ \frac{b_2^d}{\tau_2^d} \\ \vdots \\ \frac{b_N^d}{\tau_N^d} \\ 0_{N \times 1} \\ 0_{N \times 1} \end{bmatrix}, v(t) = \begin{bmatrix} 0_{N \times 1} \\ 0_{N \times 1} \\ r_1(t) - b_1(t) \\ \vdots \\ r_N(t) - b_N(t) \\ b_1(t) \\ \vdots \\ b_N(t) \end{bmatrix}.$$

where $x(t)$ denotes the system state vector of the mean field stochastic smart grid network, $Ex(t)$ is the mean of the system state, $u(t)$ denotes the admissible control input for the price policy, and $v(t) \in L_F^2(\mathbb{R}^+; \mathbb{R}^{n_v})$ denotes the unknown finite energy external disturbance from the RESs. Let $W_n(t)$, $N_n(t)$ on (Ω, F, F_t, P) be a filtration probability space with $F = \{F_t : t \geq 0\}$. F is generated by the following two stochastic processes: (i) the Wiener process $W_n(t)$ and (ii) the Poisson processes $N_n(t)$, and the filtration F_t is also generated by σ-algebra of $N_n(t)$. It is assumed that the Wiener process $W_n(t)$ and the Poisson processes $N_n(t)$ are independent for $n = 1, 2, ..., N$.

Remark 14.1: Some important properties of the Wiener process and Poisson jump process are given as follows:

(1) $E\{W_i(t)\} = E\{dW_i(t)\} = 0$, for $i = 1, 2, ..., N$.
(2) $E\{dN_i(t)dt\} = 0$, for $i = 1, 2, ..., N$.
(3) $E\{dN_i(t)\} = \lambda_i dt$, where $\lambda_i > 0$ is the Poisson jump intensity for $i = 1, 2, ..., N$.

Since the term $Ex(t)$ appears in (14.6), robust tracking control design problems for mean field stochastic smart grid systems are more difficult than the classical tracking control design problem of stochastic smart grid systems. To overcome this difficult problem, a decouple method is proposed to solve the robust tracking control problem of the mean field stochastic smart grid system by separating the smart grid network system into two orthogonal subsystems, a mean subsystem and a variation subsystem.

By applying the expectation operator to (14.6), we have the mean subsystem as follows

$$dEx(t) = [(\hat{A} + \sum_{q=1}^{N} \lambda_q \hat{D}_q)Ex(t) + BEx(t) + Ev(t) + b]dt \tag{14.7}$$

where $\hat{A} = (A + A^{\mu})$, $\hat{D}_q = (D_q + D_q^{\mu})$, $Eu(t) \triangleq E\{u(t)\}$, and $Ev(t) \triangleq E\{v(t)\}$.

Subtracting the system in (14.6) from the mean subsystem in (14.7), the variation subsystem $\tilde{x}(t) = x(t) - Ex(t)$ can be obtained as follows

$$d\tilde{x}(t) = [(A\tilde{x}(t) - \sum_{q=1}^{N} \lambda_q \hat{D}_q Ex(t) + B\tilde{u}(t) \mp v(t)]dt$$

$$+ \sum_{r=1}^{N} [C_r \tilde{x}(t) + C_r^{\mu} Ex(t)]dW_r(t) \tag{14.8}$$

$$+ \sum_{q=1}^{N} [D_q \tilde{x}(t) + C_q^{\mu} Ex(t)]dN_q(t)$$

$$d\tilde{e}(t) = M\tilde{x}(t) \tag{14.9}$$

where $\hat{C}_r = C_r + C_r^{\mu}$, $\tilde{u}(t) = u(t) - Eu(t)$ and $\tilde{v}(t) = v(t) - Ev(t)$.

The effect of constant vector b due to the initial marginal cost and benefit on the power grid and microgrids is still in (14.7), and it cannot be easily attenuated asymptotically by conventional state feedback control alone. Thus, an additional integration state feedback control of (14.7) is necessary to achieve asymptotic tracking. Let $z(t)$ denote the integration of $Ee(t)$

$$dz(t) = Ee(t)dt = MEx(t)dt \tag{14.10}$$

For convenience of design, the following augmented mean subsystem is introduced as follows

$$d\xi(t) = [A_{\xi}\xi(t) + B_{\xi}Eu(t) + S(Ev(t) + b)]dt \tag{14.11}$$

where

$$\xi(t) = \left[Ex(t)^T \quad z^T(t) \right]^T, \quad A_\xi = \begin{bmatrix} \hat{A} + \sum_{q=1}^n \lambda_q \hat{D}_q & 0 \\ M & 0 \end{bmatrix}, \quad B_\xi = \begin{bmatrix} B \\ 0 \end{bmatrix}, \quad S = \begin{bmatrix} I \\ 0 \end{bmatrix}.$$

Denote $\bar{X}(t) = [\tilde{x}(t)^T \quad \xi(t)^T]^T$, $\bar{v}(t) = [\tilde{v}(t)^T \quad (Ev(t)+b)^T]^T$, and $\bar{u}(t) = [\tilde{u}(t)$ $Eu(t)^T]^T$, and the augmented mean field stochastic smart grid network system can be obtained by (14.7) and (14.11) as follows:

$$d\bar{X}(t) = (F\bar{X}(t) + G\bar{u}(t) + L\bar{v}(t))dt + \sum_{r=1}^N H_r \bar{X}(t)dW_r(t)$$

$$+ \sum_{q=1}^N T_q \bar{X}(t)dN_q(t)$$

(14.12)

where $F = \begin{bmatrix} A & -\sum_{q=1}^n \lambda_q \hat{D}_{\xi,q} \\ 0 & A_\xi \end{bmatrix}$, $G = \begin{bmatrix} B & 0 \\ 0 & B_\xi \end{bmatrix}$, $H_r = \begin{bmatrix} C_r & \hat{C}_{\xi,r} \\ 0 & 0 \end{bmatrix}$, $T_q = \begin{bmatrix} D_q & \hat{D}_{\xi,q} \\ 0 & 0 \end{bmatrix}$,

$L = \begin{bmatrix} I \\ S \end{bmatrix}$, $\hat{D}_{\xi,q} = \begin{bmatrix} \hat{D}_q & 0 \end{bmatrix}$, and $\hat{C}_{\xi,r} = \begin{bmatrix} \hat{C}_r & 0 \end{bmatrix}$.

The dynamic pricing policy $u(t)$ for the mean field stochastic smart grid system can be represented by

$$u(t) = \tilde{u}(t) + Eu(t)$$
$$\tilde{u}(t) = K_1 \tilde{x}(t)$$
$$Eu(t) = K_\xi \xi(t) = K_2 Ex(t) + K_3 z(t)$$

(14.13)

where K_1 is state feedback control gain for the variation subsystem of $\tilde{x}(t)$, and $K_\xi = \begin{bmatrix} K_2 & K_3 \end{bmatrix}$ is state feedback control gain for the augmented mean subsystem of $\xi(t)$. From (14.10), it should be noted that $K_3 z(t) = K_3 \int_0^t Ee(s)ds$ denotes the integration mean tracking error state feedback control to eliminate the steady state tracking error due to the constant b in (14.7).

14.2.2 PROBLEM FORMULATION

In general, market managers always want to design a pricing policy so that the stored energy state $s_n(t)$ can track a desired energy state $s_n^d(t)$; that is, to be on the safe side, the stored energy state $s_n(t)$ can track a desired energy state $s_n^d(t)$ in (14.3) with a parsimonious price policy $u(t)$ to get a better profit in the mean field smart grid network. Thus, for the mean field stochastic jump diffusion system (MFSJD) of the smart grid network in (14.6), in order to achieve the desired state $s_d(t)$ with a parsimonious price policy $u(t)$ in spite of continuous and discontinuous random fluctuation, the H_2 optimal tracking performance $J_2(u(t))$ index for the stochastic mean field jump diffusion system in (14.6) without consideration of $v(t)$ can be proposed as follows:

$$
\begin{aligned}
J_2(u(t)) &= E\{\int_0^{t_f} [\tilde{e}^T(t)Q_1\tilde{e}(t) + Ee(t)^T Q_1^\mu Ee(t) + z^T(t)Q_1^z z(t) \\
&\quad + \tilde{u}^T(t)R_1\tilde{u}(t) + Eu^T(t)R_1^\mu Eu(t)]dt\} \\
&= E\{\int_0^{t_f} [\tilde{x}(t)^T M^T Q_1 M\tilde{x}(t) + Ex(t)^T M^T Q_1^\mu MEx(t) \\
&\quad + z^T(t)Q_1^z z(t) + \tilde{u}^T(t)R_1\tilde{u}(t) + Eu^T(t)R_1^\mu Eu(t)]dt\} \\
&= E\{\int_0^{t_f} [\overline{X}^T(t)\overline{M}^T \overline{Q}_1 \overline{M}\overline{X}(t) + \overline{u}^T(t)\overline{R}_1\overline{u}(t)]dt\}
\end{aligned}
\tag{14.14}
$$

where $\overline{M} = \begin{bmatrix} M & 0 \\ 0 & M_\xi \end{bmatrix}$, $M_\xi = \begin{bmatrix} M & 0 \\ 0 & I \end{bmatrix}$, $\overline{Q}_1 = diag(Q_1, Q_1^\xi)$, $Q_1^\xi = diag(Q_1^\mu, Q_1^z)$, and $\overline{R}_1 = diag(R_1, R_1^\mu)$ are the corresponding weighting matrices on the tracking error desired state and the price, respectively, which are specified as a tradeoff between $\overline{X}(t)$ and $u(t)$. t_f is the terminal time.

Even if the H_2 optimal quadratic tracking design could achieve the optimal reference tracking of desired stored energy state, it could not treat the external disturbance due to unpredictable and intermittent RESs and the constant vector b. To attenuate the effect of external disturbances from RESs and b in the smart grid system, the H_∞ tracking control performance index $J_\infty(u(t))$ of the mean field stochastic system in (14.6) can be defined as follows:

$$
J_\infty(u(t))
$$

$$
= \sup_{\substack{\tilde{v}(t),Ev(t) \\ \in L_F^2([0,t_f];\mathbb{R}^{n_\omega})}} \frac{E\{\int_0^{t_f} [\tilde{e}^T(t)Q_2\tilde{e}(t) + Ee(t)^T Q_2^\mu Ee(t) \\ + z^T(t)Q_2^z z(t) + \tilde{u}^T(t)R_2\tilde{u}(t) \\ + Eu^T(t)R_1^\mu Eu(t)]dt\}}{E\{\int_0^{t_f} [\tilde{v}^T(t)\tilde{v}(t) + (Ev(t)+b)^T (Ev(t)+b)]dt\} \\ + E\{\overline{X}^T(0)P\overline{X}(0)\}}
\tag{14.15}
$$

$$
= \sup_{\substack{\tilde{v}(t),Ev(t) \\ \in L_F^2([0,t_f];\mathbb{R}^{n_\omega})}} \frac{E\{\int_0^{t_f} [\overline{X}^T(t)\overline{M}^T \overline{Q}_1 \overline{M}\overline{X}(t) + \overline{u}^T(t)\overline{R}_1\overline{u}(t)]dt\}}{E\{\int_0^{t_f} [\tilde{v}^T(t)\tilde{v}(t)]dt\} + E\{\overline{X}^T(0)P\overline{X}(0)\}}
$$

where $\overline{Q}_2 = diag(Q_2, Q_2^\xi)$, $Q_2^\xi = diag(Q_2^\mu, Q_2^z)$, and $\overline{R}_2 = diag(R_2, R_2^\mu)$ are the weighting matrices on the desired reference tracking error and the electric price policy, respectively. The term $E\{\overline{X}^T(0)P\overline{X}(0)\}$ denotes the effect of initial condition with the positive definite matrix $P = diag(P_1, P_2)$. The physical meaning of H_∞ tracking control performance $J_\infty(u(t))$ means the worst-case effect external disturbance of $\overline{v}(t)$ on the tracking error and the price policy.

The multi-objective H_2/H_∞ dynamic pricing management problem can be regarded as how to design the pricing policy $\bar{v}(t)$ in (14.13) so that the mean field stochastic smart grid system in (14.6) can achieve the multi-objective H_2/H_∞ dynamic pricing optimization performance in (14.14) and (14.15) simultaneously. The optimal H_2 pricing policy in (14.14) expects the imbalanced energy in each microgrid and admissible pricing policy $u(t)$ as lower as possible. However, for robust H_∞ pricing management design, the H_∞ pricing policy is always coupled with a higher price $u(t)$ to reject external disturbance from RESs and constant b to achieve the desired energy storage $s_d(t)$ of the working level for each microgrid. It is clear that the optimal H_2 management pricing policy and robust H_∞ management pricing policy are partly conflicting. Thus, how to design a pricing management policy to optimize the H_2 tracking performance $J_2(u(t))$ and robust H_∞ performance $J_\infty(\lambda(t))$ simultaneously is an MOP of the mean field stochastic smart grid system. The multi-objective H_2/H_∞ dynamic pricing management problem for the mean field stochastic smart grid system is formulated as follows

> **Definition 14.1:** The multi-objective H_2/H_∞ dynamic pricing management for the augmented mean field stochastic smart grid system in (14.12) is to design an admissible pricing policy $u(t)$ in (14.13) to simultaneously minimize the H_2 and H_∞ management performance indices in the Pareto optimal sense, that is,
>
> $$\min_{u(t) \in U} (J_2(u(t)), J_\infty(u(t)))$$
>
> subject to (14.12)
>
> (14.16)

where U is the set of all the admissible pricing policies of $u(t)$ for the given mean field stochastic smart grid systems, and the objective functions $J_2(u(t))$ and $J_\infty(u(t))$ are defined in (14.14) and (14.15), respectively. The vector of the objective functions $(J_2(u(t)), J_\infty(u(t)))$ is called the objective vector of the MOP.

14.3 MULTI-OBJECTIVE H_2/H_∞ DYNAMIC PRICING POLICY DESIGN FOR MEAN FIELD STOCHASTIC SMART GRID SYSTEMS

In this section, we want to design a multi-objective H_2/H_∞ optimal dynamic pricing policy $u(t)$ for the mean field stochastic smart grid network system from the optimal H_2 working level tracking perspective to achieve the minimum quadratic tracking error under intrinsic random fluctuation and from the optimal robust H_∞ working level tracking perspective to minimize the worst-case effect of external disturbance from RESs and the constant vector b simultaneously.

However, it is still difficult to solve the multi-objective H_2/H_∞ dynamic pricing policy design problem in (14.16) directly. An indirect method is proposed to minimize the corresponding upper bounds simultaneously.

Theorem 14.1: Suppose α, β are the upper bounds of the H_2 and H_∞ tracking performance, respectively, that is, $J_2(u(t)) \leq \alpha$ and $J_\infty(u(t)) \leq \beta$. The MOP in (14.16) is equivalent to the following MOP:

$$\min_{u(t)\in U}(\alpha,\beta)$$

(14.17)

subject to $J_2(u(t))\leq\alpha$ and $J_\infty(u(t))\leq\beta$

Before the proof of the theorem, the following definitions for the MOP are given:

Definition 14.2: ([15]) Consider the MOP in (14.17). A feasible objective vector (α_1,β_1) is said to dominate another feasible objective vector (α_2,β_2) if and only if $\alpha_1\leq\alpha_2$ and $\beta_1\leq\beta_2$ for at least one inequality that is a strict inequality.

Definition 14.3: ([15]) The solution $u^*(t)$ of the MOP in (14.17) is called a Pareto optimal solution with the corresponding objective vector (α^*,β^*) if (α^*,β^*) is not dominated by another feasible objective vector (α,β).

Proof: The proof of this theorem is straightforward. One only needs to prove that both inequality constraints on the MOP in (14.17) become equality for the Pareto optimal solution. We will show this by contradiction. Given a Pareto optimal solution $u^*(t)$ of the MOP in (14.17) with objective vector (α^*,β^*), we assume that any one of the inequalities in (14.17) remains a strict inequality in the optimal solution. Without loss of generality, suppose $J_2(u(t))=\alpha'$, such that $\alpha'<\alpha^*$. As a result, (α',β^*) dominates the Pareto optimal solution (α^*,β^*), leading to a contradiction of the previous definitions. This implies that both inequality constraints in (14.17) indeed become equalities for the Pareto optimal solution. The MOP in (14.17) is equivalent to the MOP in (14.16).

Lemma 14.1: ([90]) For any real matrices A and B with appropriate dimensions, we have the following inequality for any nonzero real positive number γ

$$A^T B+B^T A\leq\gamma A^T A+\frac{1}{\gamma}B^T B$$

Lemma 14.2: ([129]) Consider a Lyapunov function $V(\bar{X}(t))=\bar{X}^T(t)P\bar{X}(t)$ for the energy of the augmented mean field stochastic smart grid network system in (14.12), and $P=P^T>0$. Then, the Itô-Lévy formula of $V(\bar{X}(t))$ is given as follows:

$$E\{dV(\bar{X}(t))\}=E\{(\bar{X}^T(t)P(\bar{X}(t)+G\bar{u}(t)+L\bar{v}(t))+(F\bar{X}(t)+G\bar{u}(t)+L\bar{v}(t))^T P\bar{X}(t)$$
$$+\sum_{r=1}^{N}\bar{X}^T(t)H_r^T PH_r\bar{X}(t)+\sum_{q=1}^{N}\lambda_q[(\bar{X}(t)+T_q\bar{X}(t))^T P(\bar{X}(t)+T_q\bar{X}(t))$$
$$-\bar{X}^T(t)P\bar{X}(t)])dt\}$$

Then the following theorem will provide sufficient conditions for the dynamic pricing policy $u(t)$ to solve the multi-objective H_2/H_∞ dynamic pricing management problem in (14.17) for the mean field stochastic smart grid system in (14.12).

Theorem 14.2: The MOP for the multi-objective H_2/H_∞ dynamic pricing policy design in (14.17) could be transformed into the following BMI-constrained MOP:

$$\min_{\{P_1>0,P_2>0\}} (\alpha,\beta)$$

$$\text{subject to } (14.19)-(14.22)$$

(14.18)

$$E\{\bar{X}^T(0)P\bar{X}(0)\} \le \alpha$$

$$M^T Q_1 M + K_1^T R_1 K_1 + P_1 A + A^T P_1 + P_1 B K_1 + K_1^T B^T P_1$$

$$+ \sum_{r=1}^{N} C_r^T P_1 C_r + \sum_{q=1}^{N} \lambda_q (D_q^T P_1 D_q + P_1 D_q + D_q^T P_1) < 0$$

(14.19)

$$M_\xi^T Q_1^\xi M_\xi + K_\xi^T R^\mu K_\xi + P_2 A_\xi + A_\xi^T P_2 + P_2 B_\xi K_\xi + K_\xi^T B_\xi^T P_2$$

$$+ \sum_{r=1}^{N} \hat{C}_{\xi,r}^T P_1 \hat{C}_{\xi,r} + \sum_{q=1}^{N} \lambda_q D_{\xi,q}^T P_1 D_{\xi,q} < 0$$

(14.20)

$$M^T Q_2 M + K_1^T R_2 K_1 + P_1 A + A^T P_1 + P_1 B K_1 + K_1^T B^T P_1$$

$$+ \frac{1}{\beta} P_1 P_1 + \sum_{r=1}^{N} C_r^T P_1 C_r + \sum_{q=1}^{N} \lambda_q (D_q^T P_1 D_q + P_1 D_q + D_q^T P_1) < 0$$

(14.21)

$$M_\xi^T Q_2^\xi M_\xi + K_\xi^T R_2^\mu K_\xi + P_2 A_\xi + A_\xi^T P_2 + P_2 B_\xi K_\xi + K_\xi^T B_\xi^T P_2$$

$$+ \frac{1}{\beta} P_2 S S^T P_2 + \sum_{r=1}^{N} \hat{C}_{\xi,r}^T P_1 \hat{C}_{\xi,r} + \sum_{q=1}^{N} \lambda_q D_{\xi,q}^T P_1 D_{\xi,q} < 0$$

(14.22)

Proof: Please refer to Appendix 14.7.1.

It is still difficult to solve the BMI-constrained MOP in (14.18)–(14.22) directly. Thus, we need to transform the BMI-constrained MOP into an LMI-constrained MOP by the following theorem.

Theorem 14.3: The MOP for the multi-objective H_2/H_∞ dynamic pricing policy design in (14.18) could be transformed into the following LMI-constrained MOP:

$$\min_{\{W_1>0,W_2>0,Y_1,Y_2\}} (\alpha,\beta)$$

$$\text{subject to LMIs in } (14.24)-(14.28)$$

(14.23)

$$diag(W_1,W_2) \ge \alpha^{-1} tr(E\{\bar{X}^T(0)\bar{X}(0)\})I$$

(14.24)

$$
\begin{bmatrix}
\Lambda_1 & W_1 M^T & Y_1^T & W_1 C_1^T & \cdots & W_1 C_N^T & W_1 D_1^T & \cdots & W_1 D_N^T \\
* & -Q_1^{-1} & 0 & 0 & \cdots & 0 & 0 & \cdots & 0 \\
* & * & -R_1^{-1} & 0 & \cdots & \cdots & \cdots & \cdots & \vdots \\
* & * & * & -W_1 & \ddots & \ddots & \ddots & \ddots & \vdots \\
* & * & * & * & \ddots & 0 & \ddots & \ddots & \vdots \\
* & * & * & * & * & -W_1 & 0 & \ddots & \vdots \\
* & * & * & * & * & * & -\lambda_1^{-1} W_1 & \ddots & \vdots \\
* & * & * & * & * & * & * & \ddots & 0 \\
* & * & * & * & * & * & * & * & -\lambda_N^{-1} W_1
\end{bmatrix} < 0 \quad (14.25)
$$

$$
\begin{bmatrix}
\Lambda_2 & W_2 M_\xi^T & Y_2^T & W_2 \hat{C}_{\xi,1}^T & \cdots & W_2 \hat{C}_{\xi,N}^T & W_2 \hat{D}_{\xi,1}^T & \cdots & W_2 \hat{D}_{\xi,N}^T \\
* & -(Q_1^\xi)^{-1} & 0 & 0 & \cdots & 0 & 0 & \cdots & 0 \\
* & * & -(R_1^\mu)^{-1} & 0 & \cdots & \cdots & \cdots & \cdots & \vdots \\
* & * & * & -W_1 & \ddots & \ddots & \ddots & \ddots & \vdots \\
* & * & * & * & \ddots & 0 & \ddots & \ddots & \vdots \\
* & * & * & * & * & -W_1 & 0 & \ddots & \vdots \\
* & * & * & * & * & * & -\lambda_1^{-1} W_1 & \ddots & \vdots \\
* & * & * & * & * & * & * & \ddots & 0 \\
* & * & * & * & * & * & * & * & -\lambda_N W_1
\end{bmatrix} < 0 \quad (14.26)
$$

$$
\begin{bmatrix}
\Xi_1 & W_1 M^T & Y_1^T & W_1 C_1^T & \cdots & W_1 C_N^T & W_1 D_1^T & \cdots & W_1 D_N^T \\
* & -Q_2^{-1} & 0 & 0 & \cdots & 0 & 0 & \cdots & 0 \\
* & * & -R_2^{-1} & 0 & \cdots & \cdots & \cdots & \cdots & \vdots \\
* & * & * & -W_1^{-1} & \ddots & \ddots & \ddots & \ddots & \vdots \\
* & * & * & * & \ddots & 0 & \ddots & \ddots & \vdots \\
* & * & * & * & * & -W_1^{-1} & 0 & \ddots & \vdots \\
* & * & * & * & * & * & -\lambda_1^{-1} W_1^{-1} & \ddots & \vdots \\
* & * & * & * & * & * & * & \ddots & 0 \\
* & * & * & * & * & * & * & * & -\lambda_N^{-1} W_1^{-1}
\end{bmatrix} < 0 \quad (14.27)
$$

$$\begin{bmatrix} \Xi_2 & W_2 M_\xi^T & Y_2^T & W_2 \hat{C}_{\xi,1}^T & \cdots & W_2 \hat{C}_{\xi,N}^T & W_2 \hat{D}_{\xi,1}^T & \cdots & W_2 \hat{D}_{\xi,N}^T & S^T \\ * & -(Q_2^\xi)^{-1} & 0 & 0 & \cdots & 0 & 0 & \cdots & 0 & 0 \\ * & * & -(R_2^\mu)^{-1} & 0 & \cdots & & & \cdots & & 0 \\ * & * & * & -W_1 & 0 & \cdots & & & \cdots & \vdots \\ * & * & * & * & \ddots & \ddots & \ddots & \ddots & \ddots & \vdots \\ * & * & * & * & * & -W_1 & \ddots & \ddots & \ddots & \vdots \\ * & * & * & * & * & * & -\lambda_1^{-1} W_1 & \ddots & \ddots & \vdots \\ * & * & * & * & * & * & * & \ddots & \ddots & 0 \\ * & * & * & * & * & * & * & * & -\lambda_N^{-1} W_1 & 0 \\ * & * & * & * & * & * & * & * & * & -\beta \end{bmatrix} < 0 \quad (14.28)$$

where $Y_1 = K_1 W_1$, $Y_2 = K_2 W_2$, $\Lambda_1 = AW_1 + W_1 A^T + BY_1 + Y_1^T B^T + \sum_{q=1}^{N} \lambda_q (D_q W_1 + W_1 D_q^T)$,

$\Lambda_2 = A_\xi W_2 + W_2 A_\xi^T + B_\xi Y_2 + Y_2^T B_\xi^T$, $\quad \Xi_1 = AW_1 + W_1 A^T + BY_1 + Y_1^T B^T + \frac{1}{\beta} I + \sum_{q=1}^{N} \lambda_q$

$(D_q W_1 + W_1 D_q^T)$, and $\Xi_2 = A_\xi W_2 + W_2 A_\xi^T + B_\xi Y_2 + Y_2^T B_\xi^T$.

Proof: Please refer to Appendix 14.7.2.

14.4 THE REVERSE-ORDER LMI-CONSTRAINED MOEA FOR MULTI-OBJECTIVE H_2/H_∞ DYNAMIC PRICING POLICY OF MEAN FIELD STOCHASTIC SMART GRID SYSTEMS

For the multi-objective H_2/H_∞ dynamic pricing problem of the mean field stochastic smart grid network system, it is not easy to solve the MOP in (14.23)–(14.28) by the conventional MOEA directly. In this section, a reverse-order LMI-constrained MOEA search is developed to help us solve the MOP in (14.23)–(14.28), that is, to update (α, β) to search for $\{W_1 > 0, W_2 > 0, Y_1, Y_2\}$ using the LMI toolbox in MAT-LAB instead of updating the complex $\{W_1 > 0, W_2 > 0, Y_1, Y_2\}$ to search for (α, β) in the conventional MOEA. Some important definitions of Pareto optimality are given as follows:

Definition 14.4: ([15]) Let (Y_1, W_1, Y_2, W_2) and (Y_1', W_1', Y_2', W_2') be the feasible solutions of MOP in (14.23) corresponding to the objective value (α, β) and (α', β') subject to the LMIs in (24)–(28), respectively. If $\alpha \le \alpha'$ and $\beta \le \beta'$ with at least one inequality being a strict inequality, the solution (Y_1, W_1, Y_2, W_2) is said to dominate the solution (Y_1', W_1', Y_2', W_2').

Definition 14.5: ([15]) A solution $(Y_1^*, W_1^*, Y_2^*, W_2^*)$ with the objective value (α^*, β^*) is said to be a Pareto optimal solution of (14.23) if there does not exist any feasible solution (Y_1, W_1, Y_2, W_2) with the objective value (α, β) such that (α, β) dominates (α^*, β^*).

Definition 14.6: ([15]) For the LMI-constrained MOP in (14.23)–(14.28) of the mean field stochastic smart grid network system in (14.12), the Pareto front P_F is defined as:

$$P_F = \left\{ (\alpha^*, \beta^*) \middle| \begin{array}{l} (Y_1^*, W_1^*, Y_2^*, W_2^*) \text{ is the Pareto optimal solution of MOP} \\ \text{in (14.23) with the corresponding objective vector } (\alpha^*, \beta^*) \end{array} \right\}$$

For the proposed LMI-constrained MOEA, the EA operates on a number of feasible encoding objective vectors called a population so that a feasible objective vector (α^k, β^k) can be encoded into a chromosome C_k. The chromosome for the reverse-order LMI-constrained MOEA uses the real-value representation to avoid long binary strings and a large search space. In general, the chromosomes perform the EA with fitness to generate the next population. To guarantee that all of these chromosomes can be decoded as feasible objective vectors for the MOP in (14.23), each chromosome C_k should be examined for the existence of a feasible solution $(W_1 > 0, W_2 > 0, Y_1, Y_2)$ in (14.24)–(14.28) with the LMI toolbox in MATLAB after the mating operators. If the chromosomes are not feasible so that they cannot satisfy the LMIs in (14.24)–(14.28), the chromosomes will be deleted from the candidate chromosomes.

The design procedure of the reverse-order LMI-constrained MOEA for the MO H_2/H_∞ dynamic pricing policy of the mean field stochastic smart grid system is proposed as follows:

Step 1: Select the search range $S = [\alpha_{min}, \alpha_{max}] \times [\beta_{min}, \beta_{max}]$ for the feasible objective vector (α, β) and set the iteration number N_{max}, the population N_p, the crossover rate N_c, and the mutation rate N_m in the reverse-order LMI-constrained MOEA.

Step 2: Randomly select N_p feasible chromosomes from search region S as the initial population P_1.

Step 3: Set the iteration index $N_i = 1$.

Step 4: Execute the EA operation with the crossover rate N_c and the mutation rate N_m to P_{N_i} and generate N_p feasible chromosomes by examining whether their corresponding objective vectors are feasible objective vectors for the LMIs in (14.24)–(14.28).

Step 5: Set the iteration index $N_i = N_i + 1$ and select N_p chromosomes from the $2N_p$ feasible chromosomes in Step 4 through nondominated sorting method as the population P_{N_i}.

Step 6: Repeat Steps 4 and 5 until the iteration number N_{max} is reached. If the iteration number N_{max} is satisfied, then we set $P_{N_i} = P_F$.

Step 7: Select a preferable feasible objective individual $(\alpha^*, \beta^*) \in P_F$ according to the designer's preference. Once the preferable feasible objective individual is selected, the corresponding Pareto optimal solution is obtained. The proposed multi-objective H2/H∞ dynamic pricing policy $\bar{u}(t) = \bar{K}^* \bar{X}(t)$ with $\bar{K}^* = Y^* W^{*-1}$ in (14.12).

14.5 SIMULATION RESULTS

To illustrate the design procedure and to confirm the performance of the proposed MO H_2/H_∞ dynamic pricing policy for the stochastic mean field smart grid network system, we introduce a stochastic mean field smart grid network with continuous and discontinuous random fluctuation to mimic the real-world power market in (14.6). We consider the smart network consisting of a power grid and three microgrids in industrial, residential, and commercial areas, respectively. In order to mimic the real situation, we assume microgrids $n = 1,2,3$ are industrial, residential, and commercial areas, respectively, to construct a mean field smart grid network. In general, the users of microgrids in industrial and commercial areas will consider a higher marginal benefit function than the users of the microgrid in residential areas. Therefore, the related smart grid network parameters of the stochastic mean field smart grid network are given in Table 14.1.

To mimic a real power market system, the power market dynamic model should be modified by continuous and discontinuous intrinsic random fluctuations [416, 417]. The Wiener process is used to describe the users of the microgrid randomly changing their demand behavior due to random factors like rumors in the power market and weather changes, and the Poisson counting process is used to describe sudden random fluctuations like economic events and policy changes, with the jump intensity of three microgrids as: $\lambda_1 = 0.05$, $\lambda_2 = 0.03$, $\lambda_3 = 0.05$.

Due to continuous and discontinuous fluctuations, the power demand in the power market system will be disturbed, and power demand information cannot be directly used for pricing policy design. In the proposed mean field stochastic smart grid network system, collective behavior is introduced to mimic the average (mean) behavior of each microgrid. As a result, by using the information of states and mean states, the designed MO H_2/H_∞ pricing policy can not only achieve great reference tracking performance but also have a certain attenuation of the stochastic effect. The matrix for the effect of mean demand on each microgrid in (14.6) is given as follows:

$$
L = \begin{bmatrix} -0.1 & 0 & 0 \\ -0.03 & -0.05 & -0.02 \\ -0.03 & -0.02 & -0.05 \end{bmatrix}
$$

TABLE 14.1
Network Parameters of Power Grid $b_g = 2, c_g = 0.4, \tau_g = 0.1, \tau_k = 0.1$

Microgrid 1 (Industrial)	Microgrid 2 (Residential)	Microgrid 3 (Commercial)
$b_1^d = 11$	$b_2^d = 9$	$b_3^d = 10$
$c_1^d = 0.55$	$c_2^d = 0.5$	$c_3^d = 0.55$
$\tau_1^d = 0.25$	$\tau_2^d = 0.25$	$\tau_3^d = 0.25$

In the real situation, each microgrid may need its own desired working level to maintain operation in a stage of emergency. For example, the microgrids in the residential area do not need a higher working level between 8 a.m. and 6 p.m. because residential users of the microgrid are working in other areas. However, commercial and industrial areas need to maintain a higher working level between 8 a.m. and 6 p.m. because they are at the peak of electricity consumption. The users of microgrids in industrial, residential, and business areas have their own desired working levels, which are generated by the reference model in (14.3) with $a_1 = a_2 = a_3 = -1$ and

$$
b_1(t) = \begin{cases} -3+\frac{5}{3}t, t \in [6,9) \\ 12, t \in [9,18) \\ 42-\frac{5}{3}t, t \in [18,21) \\ 7, o.w. \end{cases}
$$

$$
b_2(t) = \begin{cases} 17-\frac{5}{3}t, t \in [6,9) \\ 2, t \in [9,18) \\ -28+\frac{5}{3}t, t \in [18,21) \\ 7, o.w. \end{cases}
$$

$$
b_3(t) = \begin{cases} -5+\frac{5}{3}t, t \in [6,9) \\ 10, t \in [9,18) \\ 40-\frac{5}{3}t, t \in [18,21) \\ 5, o.w. \end{cases}
$$

The MO H_2/H_∞ pricing policy of the MOP in (14.6) is employed by the manager of the smart grid to design a pricing policy to make the stored energy $s_n(t)$ of each microgrid track the desired work level $s_n^d(t)$ in (14.3) such that the microgrids can maintain operation in an emergency. The weighting matrices for the H_2 performance in (14.14) and H_∞ performance in (14.15) are given as follows:

$$
\begin{aligned}
Q_1 &= Q_1^\mu = Q_1^z = diag(1,0,1,1), R_1 = R_1^\mu = 0.1 \\
Q_2 &= Q_2^\mu = Q_2^z = diag(1,0,1,1), R_2 = R_2^\mu = 0.05
\end{aligned}
\tag{14.29}
$$

According to the weighting matrices in (14.29), the weighting parameters in the first and third states are greater than the weighting parameter in the second state. This reveals the pricing policy design strategy will focus more on the tracking error of the microgrids in the industrial and commercial areas.

The initial values of the smart grid network system are given as

$$
\begin{aligned}
P_1^g(0) &= P_2^g(0) = P_3^g(0) = 10 \\
P_d^1(0) &= P_d^2(0) = P_d^3(0) = 14 \\
s_1 &= s_2 = s_3 = 0
\end{aligned}
$$

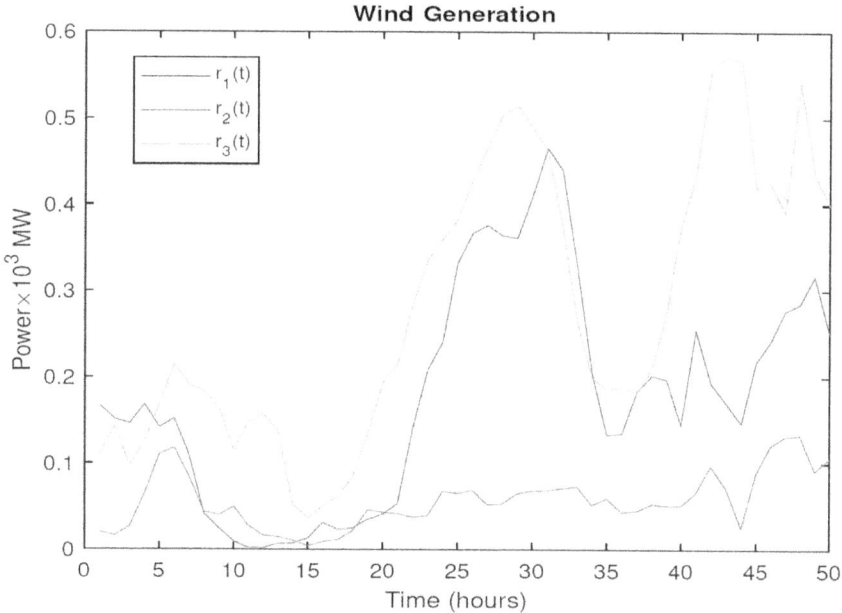

FIGURE 14.1 The wind generation $r_n(t)$ in mean field stochastic smart grid network in 50 hours.

Wind generation is considered for the generation of RESs in each microgrid. To mimic the real-world wind generation of the RESs, the data of the wind generation in three different regions are employed from PJM Interconnection [418], which is a regional transmission organization that coordinates the movement of wholesale electricity. The generation of wind power of $r_n(t), n = 1, 2, 3$ MW in 50 hours is shown in Figure 14.1.

Based on the proposed design procedure of MO H_2/H_∞ dynamic pricing policy, the reverse-order LMI-constrained MOEA is employed to solve the MOP in Theorem 14.3. In the reverse-order LMI-constrained MOEA, the search region $S = [900, 1000] \times [1000, 1200]$, the iteration number $N_I = 50$, the population number $N_p = 500$, crossover rate $N_c = 0.9$, and mutation rate $N_m = 0.1$ are given. When the iteration number is achieved, the Pareto front P_F for the MOP of the pricing policy for the mean field smart grid network in (14.6) can be obtained, as shown in Figure 14.2.

In this simulation, the knee solution in Pareto front is chosen as the dynamic pricing policy of the mean field smart grid network system since the knee point has balanced performance between H_2 and H_∞, and the dynamic pricing policy in (14.13) can be obtained. The corresponding controller gain K_1^*, K_2^*, K_3^* can be given as

$$K_1^* = [0.65 \quad 0.3 \quad -0.8 \quad 9.8 \quad 4.91 \quad -13.97 \quad -4.6 \quad -2.31$$
$$6.5 \quad -1.5 \times 10^{-9} \quad -5.3 \times 10^{-10} \quad -9.7 \times 10^{-10}]$$

$$K_2^* = [2.92 \quad 0.7 \quad 0.49 \quad -1.8 \quad 4.8 \quad 5.69 \quad -10.9 \quad -4.00$$
$$-3.4 \quad 1.57 \times 10^{-9} \quad -5.3 \times 10^{-10} \quad -9.7 \times 10^{-10}]$$

$$K_3^* = [-17.17 \quad -4.10 \quad -2.72]$$

The dynamic pricing policy plays an important role in managing imbalanced energy in the mean field smart grid network. In general, it is expected that the pricing policy has an appropriate effect on the power demand and supply of each microgrid to meet their desired stored energy levels. In general, in the conventional power market, the power price is a positive value depending on the buying and selling behavior of the power market. However, in the smart grid system, due to the higher cost of shutdown and ramp-up of a power plant, the power grid will send surplus power that cannot be stored to the microgrid with a negative price to encourage consumers to consume the microgrid surplus power [419–421]. Due to the uncertain initial conditions in each microgrid, the effect of negative prices appears in Figure 14.3 at the beginning; that is, the pricing policy oscillates at the beginning. Then the proposed pricing policy achieves the desired reference working level due to the proposed integrator-based state feedback control but with price fluctuation due to the uncertainty of the RESs and random fluctuation. It is worth pointing out that the pricing policy for the mean field smart grid network system and price vibration are used to against external disturbance from RESs and continuous and discontinuous random fluctuations.

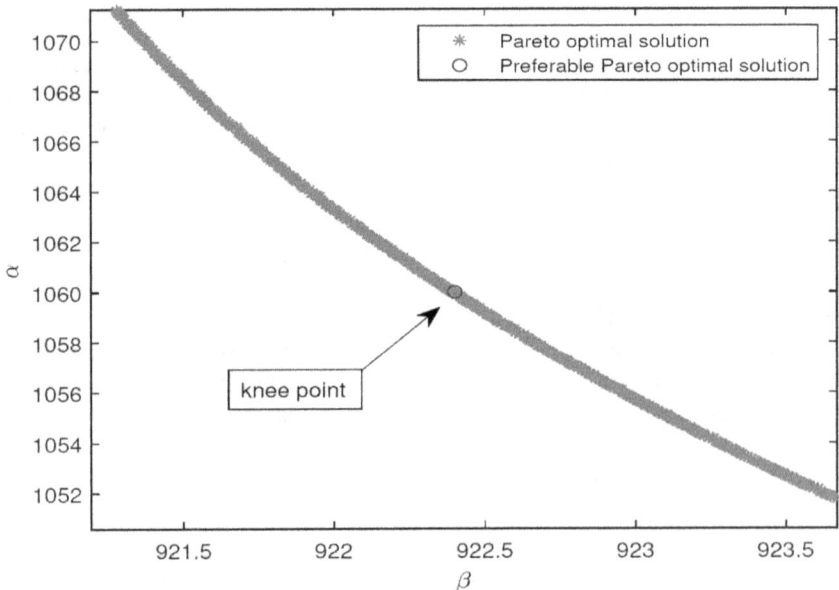

FIGURE 14.2 After the iteration number of the dynamic H_2/H_∞ pricing policy is achieved, the corresponding Pareto front of the Pareto optimal solution set is plotted. Each point in the Pareto front is a Pareto optimal solution. The marked point is the knee point.

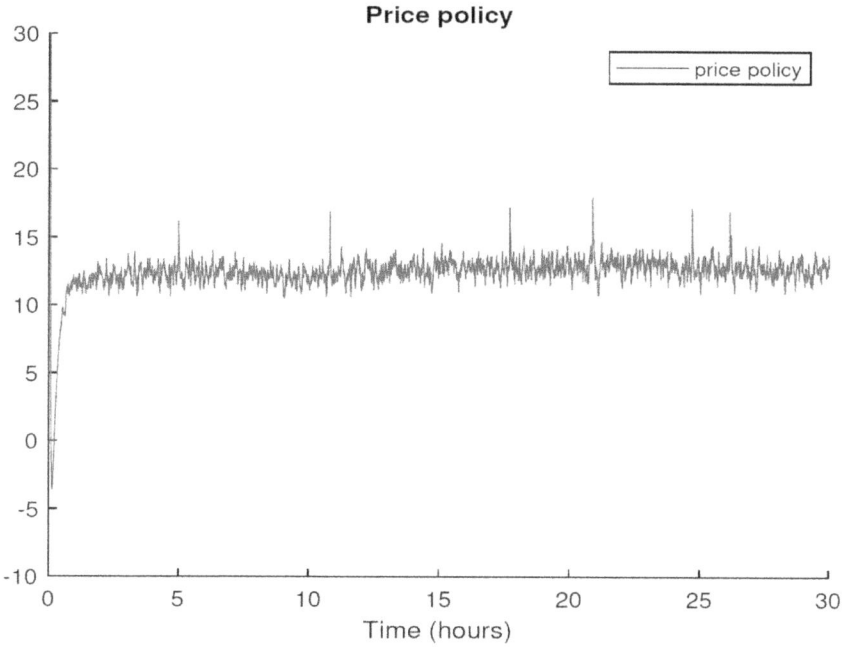

FIGURE 14.3 The proposed MO H_2/H_∞ dynamic pricing policy for the mean field stochastic smart grid network.

Figures 14.4–14.6 show the tracking performance of proposed pricing policy for the mean field smart grid network. Due to the effect of negative price by the designed MO H_2/H_∞ pricing policy, the three microgrids receive high power at first. For the microgrids in industrial and commercial areas, the designed MO H_2/H_∞ pricing policy can achieve great tracking performance to achieve the desired reference stored energy; that is, the stored energy states $s_1(t)$ and $s_3(t)$ can achieve their desired reference stored energy states $s_1^d(t)$ and $s_3^d(t)$, respectively. Thus, the proposed MO H_2/H_∞ pricing policy can effectively adjust the power demand and supply of the microgrids in the industrial and commercial areas to achieve the desired stored energy level. For the microgrid in the residential area, there exists a large transient tracking response during the tracking process. This phenomenon is caused by the weighting matrices in (14.29) since the designed pricing policy puts much emphasis on the microgrids in the industrial and commercial areas.

On other hand, if we pay more attention to the desired work tracking performance of the microgrid in the residential area than to those of the other microgrids, the weighting matrices in (14.29) can be redesigned as follows:

$$Q_1 = Q_1^\mu = Q_1^z = diag(0.1,1,0,1), R_1 = R_1^\mu = 0.1$$
$$Q_2 = Q_2^\mu = Q_2^z = diag(0.1,1,0,1), R_2 = R_2^\mu = 0.05$$

(14.30)

Based on the weighting matrices in (14.30), the pricing policy design strategy will focus more on the tracking error of the microgrids in the residential area.

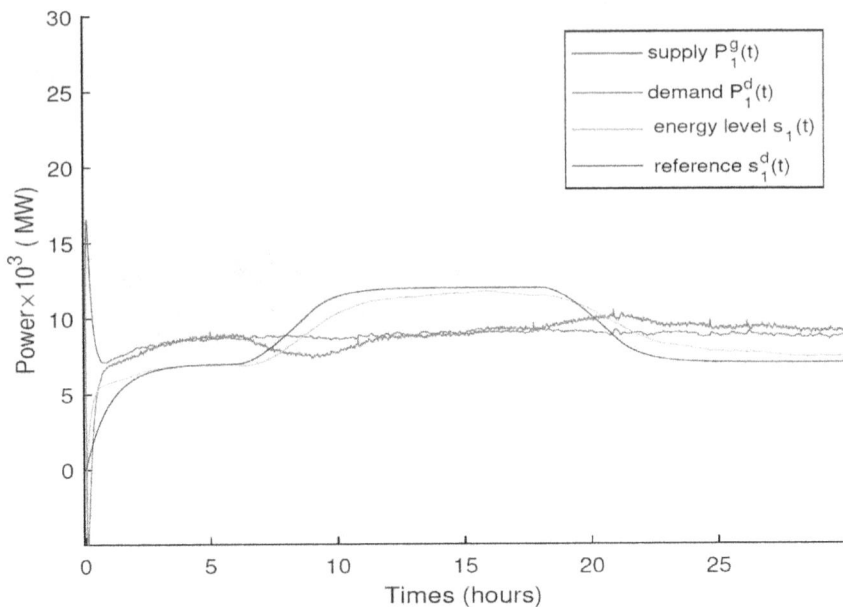

FIGURE 14.4 The distributed power, power demand, stored energy level, and desired working level for the industrial area of microgrid 1.

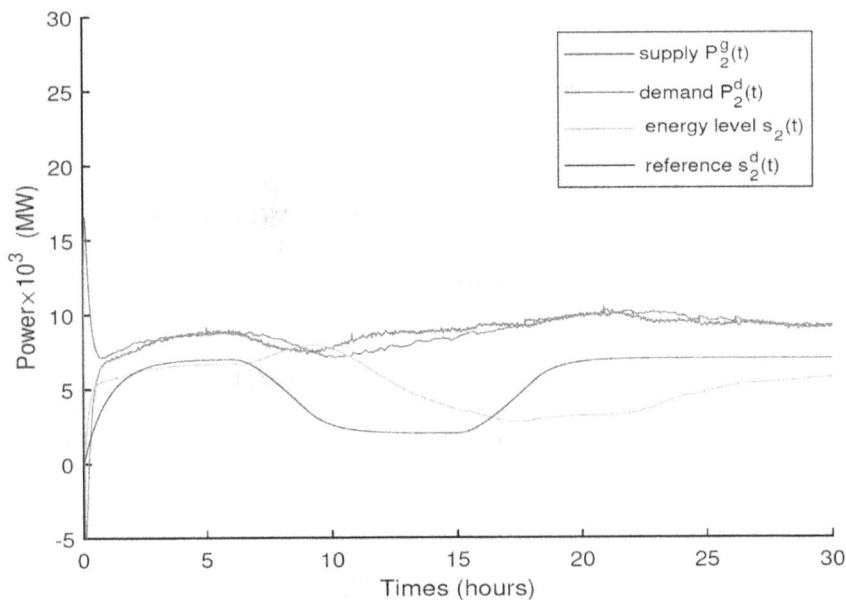

FIGURE 14.5 The distributed power, power demand, stored energy level, and desired working level for the residential area of microgrid 2.

Figures 14.7–14.9 demonstrate the simulation results of the weighting matrices in (14.30) for the three microgrids. Due to the weighting matrices in (14.30),

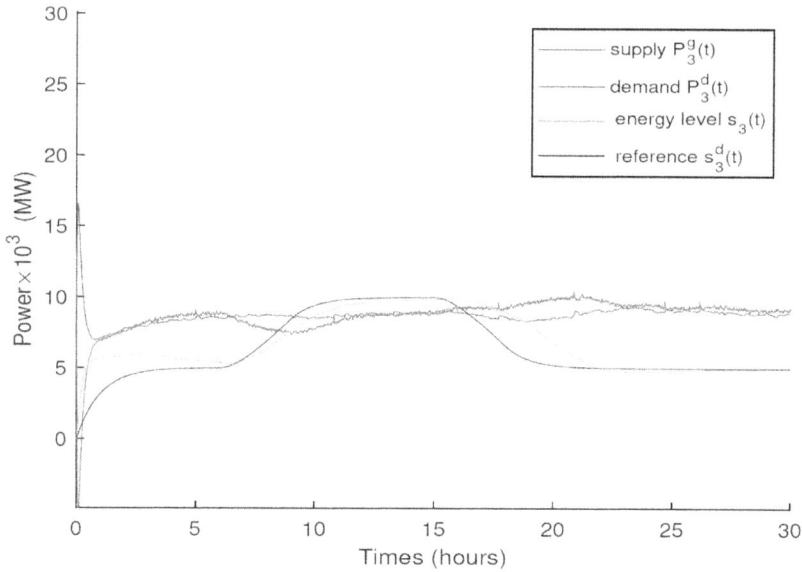

FIGURE 14.6   The distributed power, power demand, stored energy level, and desired working level for the residential area of microgrid 3.

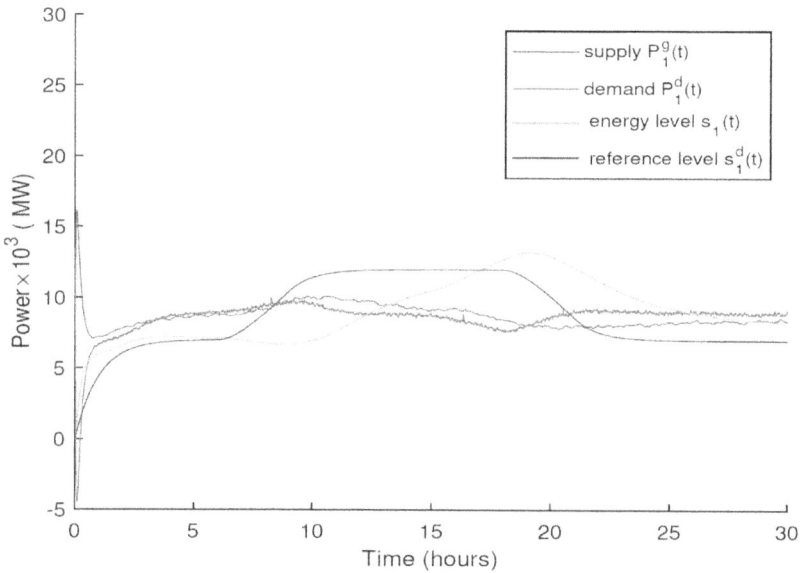

FIGURE 14.7   The distributed power, power demand, stored energy level, and desired working level for the industrial area of microgrid 1 if the weighting matrices are those designed in (14.30).

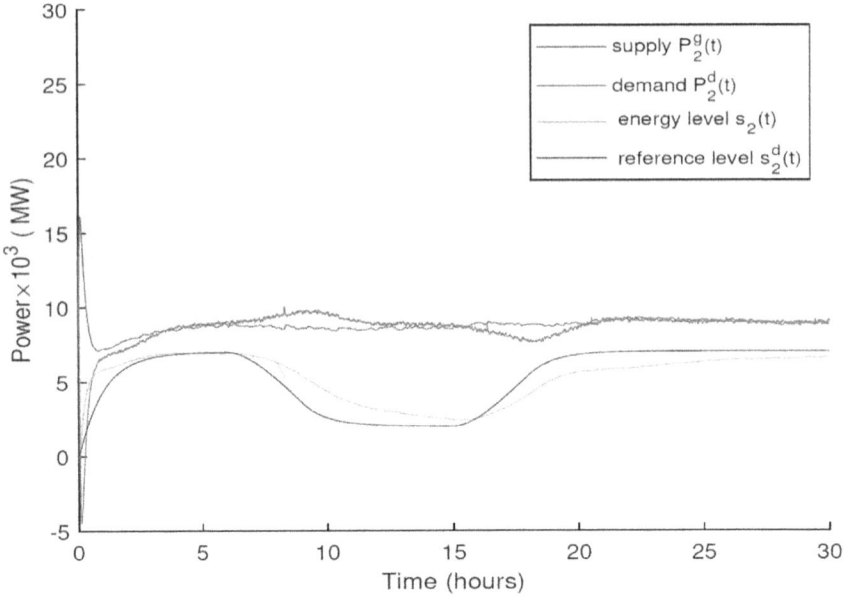

FIGURE 14.8 The distributed power, power demand, stored energy, and desired working level for residential areas of microgrid 2 if the weighting matrices are those designed in (14.30).

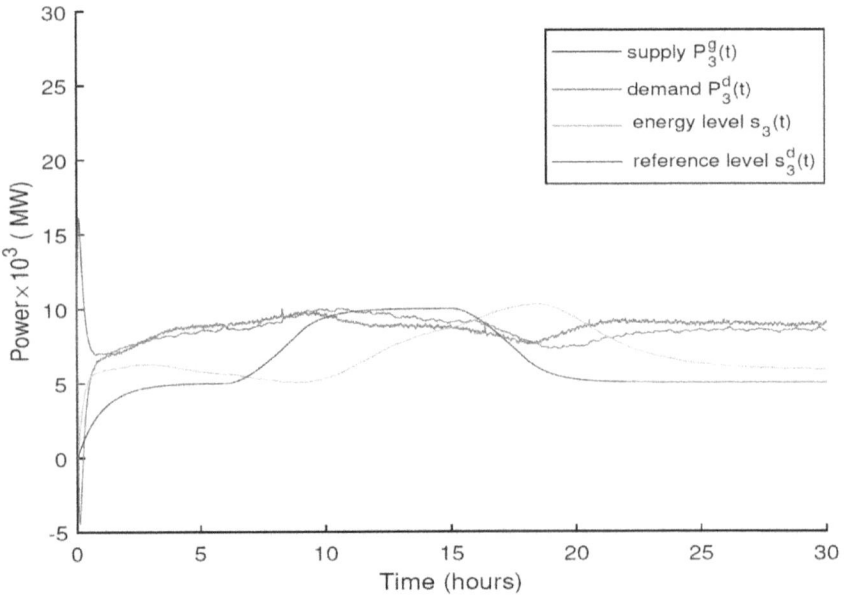

FIGURE 14.9 The distributed power, power demand, stored energy, and desired working level for commercial areas of microgrid 3 if the weighting matrices are those designed in (14.30).

the designed pricing policy focuses on the tracking performance of microgrid 2, and the desired working level in the residential area can be achieved. As a result, microgrid 2 (residential) has better tracking performance of stored energy than microgrids 1 and 3.

14.6 CONCLUSION

In this chapter, we consider a mean field stochastic smart grid network to describe the power demand behavior of microgrids, which interact through collective (mean) demand behavior with external disturbance from RESs and continuous and discontinuous random fluctuations. The managers of a smart grid network always expect that the stored energy state of each microgrid not only can track the desired working level to maintain normal operation in emergence with a parsimonious price but also attenuate the effect of the external disturbance from the RESs. Thus, a novel MO H_2/H_∞ pricing policy for energy management in a mean field smart grid network has been proposed for microgrids to achieve their own desired reference working levels. In order to effectively eliminate the effect of constants in the marginal cost and benefit function, an integrator-based state feedback control is employed for the MO H_2/H_∞ dynamic pricing policy design. An indirect method is proposed to simplify the design problem of the MO H_2/H_∞ dynamic pricing policy. Finally, the LMI-constrained MOP can be solved efficiently by the proposed LMI-constrained MOEA to find the Pareto front for the Pareto optimal solutions of the MOP. The managers of the smart grid can select their own preferred pricing policy from the set of Pareto optimal pricing policies. Finally, a simulation example of a power grid with three microgrids in different areas is given to confirm the tracking performance of the desired stored energy state with the proposed MO H_2/H_∞ pricing policy design for mean field stochastic smart grid network systems. From the simulation results, with the chosen weighting matrices, the specific microgrids in the smart grid network can access their own desired tracking performance for stored energy.

14.7 APPENDIX

14.7.1 Proof of Theorem 14.2

Let the energy function of the two subsystems $V(\bar{X}(t)) = \tilde{x}^T(t)\,P_1\tilde{x}(t) + \xi^T(t)\,P_2\xi(t) = \bar{X}^T(t)\,P\bar{X}(t)$ with $P = diag(P_1, P_2)$ be the Lyapunov function for the mean field stochastic smart grid system in (14.12), where $P = P^T > 0$ is a positive definite matrix.

We derive the sufficient condition for $J_2(u(t)) \leq \alpha$ of the MOP in (14.17):

$$J_2(u(t)) = E\{\int_0^{t_f} \bar{X}(t)^T \bar{M}^T \bar{Q}_1 \bar{M}\bar{X}(t) + \bar{u}^T(t)\bar{R}_1\bar{u}(t)\}dt$$

$$= E\{\int_0^{t_f} \bar{X}(t)^T \bar{M}^T \bar{Q}_1 \bar{M}\bar{X}(t) + \bar{u}^T(t)\bar{R}_1\bar{u}(t)dt + dV(X(t))\}$$

$$- E\{V(\bar{X}(t_f)) - V(\bar{X}(0))\}$$

$$\leq E\{V(\bar{X}(0))\} + E\{\int_0^{t_f} [\bar{X}(t)^T \bar{M}^T \bar{Q}_1 \bar{M}\bar{X}(t) + \bar{u}^T(t)\bar{R}_1\bar{u}(t)dt + dV(\bar{X}(t))\}$$

$$= E\{V(\bar{X}(0))\} + E\{\int_0^{t_f} \bar{X}^T(t)\bar{M}^T \bar{Q}_1 \bar{M}\bar{X}(t) + \bar{u}^T(t)\bar{R}_1\bar{u}(t) + 2\bar{X}^T(t)PF\bar{X}(t)$$

$$+ 2\bar{X}^T(t)PG\bar{u}(t) + \sum_{q=1}^{n} \bar{X}^T(t)H_r^T PH_r\bar{X}(t) + \sum_{q=1}^{n} \lambda_q(\bar{X}^T(t)T_q^T P\bar{X}(t)$$

$$+ \bar{X}^T(t)PT_q\bar{X}(t) + \bar{X}^T(t)T_q^T PT_q\bar{X}(t))]dt\}$$

By (14.13), we get

$$J_2(u(t)) = E\{V(\bar{X}(0))\} + E\{\int_0^{t_f} \bar{X}^T(t)\bar{M}^T \bar{Q}_1 \bar{M}\bar{X}(t) + \bar{X}^T(t)K^T \bar{R}_2 K\bar{X}(t)$$

$$+ 2\bar{X}^T(t)PF\bar{X}(t) + 2\bar{X}^T(t)PGK\bar{X}(t) + \sum_{q=1}^{n} \bar{X}^T(t)H_r^T PH_r\bar{X}(t)$$

$$+ \sum_{q=1}^{n} \lambda_q(\bar{X}^T(t)T_q^T PT_q\bar{X}(t) + \bar{X}^T(t)PT_q\bar{X}(t) + \bar{X}^T(t)T_q^T PT_q\bar{X}(t))]dt\}$$

$$= E\{\bar{X}^T(0)P\bar{X}(0)\} + E\{\int_0^{t_f} [\bar{X}^T(t)[diag(\Psi_1, \Psi_2)]\bar{X}(t)]dt\}$$

where

$$\Psi_1 = M^T Q_1 M + K_1^T R_1 K_1 + P_1 A + A^T P_1 + P_1 B K_1 + K_1^T B^T P_1 + \sum_{r=1}^{n} C_r^T P_1 C_r$$

$$+ \sum_{q=1}^{n} \lambda_q(D_q^T P_1 D_q + P_1 D_q + D_q^T P_1)$$

$$\Psi_2 = M_\xi^T Q_1^\xi M_\xi + K_\xi^T R^\mu K_\xi + P_2 A_\xi + A_\xi^T P_2 + P_2 B_\xi K_\xi + K_\xi^T B_\xi^T P_2 + \sum_{r=1}^{n} \hat{C}_{\xi,r}^T P_1 \hat{C}_{\xi,r}$$

$$+ \sum_{q=1}^{n} \lambda_q \hat{D}_{\xi,q}^T P_1 \hat{D}_{\xi,q}$$

If the following matrix inequalities hold

$$\Psi_1 < 0 \text{ and } \Psi_2 < 0 \tag{14.31}$$

$$E\{\bar{X}^T(0)P\bar{X}(0)\} \leq \alpha \tag{14.32}$$

then we have

$$E\{\int_0^{t_f} \bar{X}(t)^T \bar{M}^T \bar{Q}_1 \bar{M}\bar{X}(t) + \bar{u}^T(t)\bar{R}_1\bar{u}(t)\}dt \leq E\{\bar{X}^T(0)P\bar{X}(0)\} \leq \alpha$$

that is, if the matrix inequalities in (14.31) and (14.32) hold, then $J_2(u(t))$ has the upper bound α.

Now, we derive the sufficient condition for $J_\infty(u_{q_i}(t)) \leq \beta$ of the MOP in (14.17).

$$E\{\int_0^{t_f} \bar{X}(t)^T \bar{M}^T \bar{Q}_2 \bar{M} \bar{X}(t) + \bar{u}^T(t)\bar{R}_2\bar{u}(t)\}dt$$

$$= E\{\int_0^{t_f} \bar{X}(t)^T \bar{M}^T \bar{Q}_2 \bar{M}\bar{X}(t) + \bar{u}^T(t)\bar{R}_2\bar{u}(t)dt + dV(X(t)\} - E\{V(\bar{X}(t_f) - V(\bar{X}(0))\}$$

$$\leq E\{V(\bar{X}(0))\} + E\{\int_0^{t_f} [\bar{X}(t)^T \bar{M}^T \bar{Q}_2\bar{M}\bar{X}(t) + \bar{u}^T(t)\bar{R}_2\bar{u}(t)dt + dV(\bar{X}(t))\}$$

$$= E\{V(\bar{X}(0))\} + E\{\int_0^{t_f} \bar{X}^T(t)\bar{M}^T\bar{Q}_2\bar{M}\bar{X}(t) + \bar{u}^T(t)\bar{R}_2\bar{u}(t) + 2\bar{X}^T(t)PF\bar{X}(t)$$

$$+ 2\bar{X}^T(t)PG\bar{u}(t) + 2\bar{X}^T(t)PL\bar{v}(t) + \sum_{q=1}^n \bar{X}^T(t)H_r^T PH_r \bar{X}(t)$$

$$+ \sum_{q=1}^n \lambda_q(\bar{X}^T(t)T_q^T PT_q\bar{X}(t) + \bar{X}^T(t)PT_q\bar{X}(t) + \bar{X}^T(t)T_q^T PT_q\bar{X}(t))dt\}$$

$$= E\{V(\bar{X}(0))\} + E\{\int_0^{t_f} \bar{X}^T(t)\bar{M}^T\bar{Q}_2\bar{M}\bar{X}(t) + \bar{X}^T(t)K^T\bar{R}_2K\bar{X}(t) + 2\bar{X}^T(t)PF\bar{X}(t)$$

$$+ 2\bar{X}^T(t)PGK\bar{X}(t) + 2\bar{X}^T(t)PL\bar{v}(t) + \sum_{q=1}^n \bar{X}^T(t)H_r^T PH_r \bar{X}(t)$$

$$+ \sum_{q=1}^n \lambda_q(\bar{X}^T(t)T_q^T PT_q\bar{X}(t) + \bar{X}^T(t)PT_q\bar{X}(t) + \bar{X}^T(t)T_q^T PT_q\bar{X}(t))dt\}$$

$$\leq E\{V(\bar{X}(0))\} + E\{\int_0^{t_f} \bar{X}^T(t)\bar{M}^T\bar{Q}_2\bar{M}\bar{X}(t) + \bar{X}^T(t)K^T\bar{R}_2K\bar{X}(t) + 2\bar{X}^T(t)PF\bar{X}(t)$$

$$+ 2\bar{X}^T(t)PGK\bar{X}(t) + \tfrac{1}{\beta}\bar{X}^T(t)PLL^TP\bar{X}(t) + \beta\bar{v}^T(t)\bar{v}(t) + \sum_{q=1}^n \bar{X}^T(t)H_r^T PH_r \bar{X}(t)$$

$$+ \sum_{q=1}^n \lambda_q(\bar{X}^T(t)T_q^T P\bar{X}(t) + \bar{X}^T(t)PT_q\bar{X}(t) + \bar{X}^T(t)T_q^T PT_q\bar{X}(t))dt\}$$

$$= E\{\bar{X}^T(0)P\bar{X}(0)\} + E\{\int_0^{t_f} \bar{X}^T(t)[diag(\Phi_1, \Phi_2)]\bar{X}(t) + \beta\bar{v}^T(t)\bar{v}(t)dt\}$$

where

$$\Phi_1 = M^T Q_2 M + K_1^T R_2 K_1 + P_1 A + A^T P_1 + P_1 B K_1 + K_1^T B^T P_1 + \tfrac{1}{\beta}P_1P_1$$

$$+ \sum_{r=1}^n C_r^T P_1 C_r + \sum_{q=1}^n \lambda_q(D_q^T P_1 D_q + P_1 D_q + D_q^T P_1)$$

$$\Phi_2 = M_\xi^T Q_1^\xi M_\xi + K_\xi^T R_2^\mu K_\xi + P_2 A_\xi + A_\xi^T P_2 + P_2 B_\xi K_\xi + K_\xi^T B_\xi^T P_2 + \tfrac{1}{\beta}P_2 SS^T P_2$$

$$+ \sum_{r=1}^n \hat{C}_{\xi,r}^T P_1 \hat{C}_{\xi,r} + \sum_{q=1}^n \lambda_q \hat{D}_{\xi,q}^T P_1 \hat{D}_{\xi,q}$$

If the following matrix inequalities hold:

$$\Phi_1 < 0 \text{ and } \Phi_2 < 0 \tag{14.33}$$

then we have

$$J_\infty(u(t)) = \frac{E\{\int_0^{t_f} \bar{X}^T(t)\bar{M}^T \bar{Q}_2 \bar{M}\bar{X}(t) + \bar{u}^T(t)\bar{R}_1\bar{u}(t)dt\}}{E\{\int_0^{t_f} \bar{v}^T(t)\bar{v}(t)dt\} + E\{\bar{X}^T(0)P\bar{X}(0)\}} \le \beta$$

that is, if the inequalities in (14.33) hold, $J_\infty(u(t))$ has an upper bound β.

14.7.2   PROOF OF THEOREM 14.3

Since the matrix inequalities in (14.18)–(14.22) are still bilinear matrix inequalities, it is still not easy to solve them. Thus, let $W = P^{-1} = diag(W_1, W_2)$ and $Y_1 = K_1 W_1$, $Y_2 = K_\xi W_2$. By pre-multiplying and post-multiplying the matrix W_1 and W_2 on both sides of the matrix inequalities in (14.19) and (14.20), respectively, we have the following linear matrix inequalities

$$\begin{aligned}
&W_1 M^T Q_1 M W_1 + Y_1^T R_1 Y_1 + A W_1 + W_1 A^T \\
&+ B Y_1 + Y_1^T B^T + \sum_{r=1}^N W_1 C_r^T W_1^{-1} C_r W_1 \\
&+ \sum_{q=1}^N \lambda_q (W_1 D_q^T W_1^{-1} D_q W_1 + D_q W_1 + W_1 D_q^T) < 0
\end{aligned} \tag{14.34}$$

$$\begin{aligned}
&W_2 M_\xi^T Q_1^\xi M_\xi W_2 + Y_2^T R_1^\mu Y_2 + A_\xi W_2 + W_2 A_\xi^T + B_\xi Y_2 + Y_2^T B_\xi^T \\
&+ \sum_{r=1}^N W_2 \hat{C}_{\xi,r}^T W^{-1} \hat{C}_{\xi,r} W_2 + \sum_{q=1}^N \lambda_q W_2 \hat{D}_{\xi,q}^T W_1^{-1} \hat{D}_{\xi,q} W_2 < 0
\end{aligned} \tag{14.35}$$

By the Schur complement, the matrix inequalities in (14.34) and (14.35) are equivalent to the following LMIs, respectively.

$$
\begin{bmatrix}
AW_1 + W_1 A^T + BY_1 + Y_1^T B^T \\ + \sum_{q=1}^n \lambda_q (D_q W_1 + W_1 D_q^T) & W_1 M^T & Y_1^T & W_1 C_r^T & \cdots & W_1 C_r^T & W_1 D_q^T & \cdots & W_1 D_q^T \\
* & -Q_1^{-1} & 0 & 0 & \cdots & 0 & 0 & \cdots & 0 \\
* & * & -R_1^{-1} & 0 & \cdots & \cdots & \cdots & \cdots & 0 \\
* & * & * & -W_1 & 0 & \cdots & \cdots & \cdots & \vdots \\
* & * & * & * & \ddots & \ddots & \ddots & \ddots & \vdots \\
* & * & * & * & * & -W_1 & \ddots & \ddots & \vdots \\
* & * & * & * & * & * & -\lambda_q^{-1} W_1 & \ddots & \vdots \\
* & * & * & * & * & * & * & \ddots & 0 \\
* & * & * & * & * & * & * & * & -\lambda_q^{-1} W_1
\end{bmatrix} < 0
$$

$$
\begin{bmatrix}
A_\xi W_\xi + W_2 A_\xi^T \\ + B_\xi Y_2 + Y_2^T B_\xi^T & W_2 M_\xi^T & Y_2^T & W_2 \hat{C}_{\xi,r}^T & \cdots & W_2 \hat{C}_{\xi,r}^T & W_2 \hat{D}_{\xi,q}^T & \cdots & W_2 \hat{D}_{\xi,q}^T \\
* & -(Q_1^\xi)^{-1} & 0 & 0 & \cdots & 0 & 0 & \cdots & 0 \\
* & * & -(R_1^\mu)^{-1} & 0 & \cdots & \cdots & \cdots & \cdots & 0 \\
* & * & * & -W_1^{-1} & 0 & \cdots & \cdots & \cdots & \vdots \\
* & * & * & * & \ddots & \ddots & \ddots & \ddots & \vdots \\
* & * & * & * & * & -W_1^{-1} & \ddots & \ddots & \vdots \\
* & * & * & * & * & * & -\lambda_q^{-1} W_1^{-1} & \ddots & \vdots \\
* & * & * & * & * & * & * & \ddots & 0 \\
* & * & * & * & * & * & * & * & -\lambda_q^{-1} W_1^{-1}
\end{bmatrix} < 0
$$

that is, if the LMIs in (14.25) and (14.26) hold, then $J_2(u(t))$ has an upper bound α.

Similarly, since the matrix inequalities in (14.18) are still BMIs, it is still not easy to solve them. Thus, let $W = P^{-1} = diag(W_1, W_2)$ and $Y_1 = K_1 W_1$, $Y_2 = K_\xi W_2$. By pre-multiplying and post-multiplying the matrix W_1 and W_2 by the matrix inequalities in (14.21) and (14.22), respectively, we have the following linear matrix inequalities

$$
W_1 M^T Q_2 M W_1 + Y_1^T R_2 Y_1 + A W_1 + W_1 A^T
$$
$$
+ B Y_1 + Y_1^T B^T + \sum_{r=1}^{N} W_1 C_r^T W_1^{-1} C_r W_1 \tag{14.36}
$$
$$
+ \sum_{q=1}^{N} \lambda_q (W_1 D_q^T W_1^{-1} D_q W_1 + D_q W_1 + W_1 D_q^T) < 0
$$

$$
W_2 M_\xi^T Q_2^\xi M_\xi W_2 + Y_2^T R_2^\mu Y_2 + A_\xi W_2 + W_2 A_\xi^T + B_\xi Y_2 + Y_2^T B_\xi^T
$$
$$
+ \tfrac{1}{\beta} S S^T + \sum_{r=1}^{N} W_2 \hat{C}_{\xi,r}^T W^{-1} \hat{C}_{\xi,r} W_2 + \sum_{q=1}^{N} \lambda_q W_2 \hat{D}_{\xi,q}^T W_1^{-1} \hat{D}_{\xi,q} W_2 < 0 \tag{14.37}
$$

By the Schur complement, the inequality in (14.36) and (14.37) are equivalent to the LMIs as follows

$$
\begin{bmatrix}
AW_1 + W_1 A^T + BY_1 + Y_1^T B^T & W_1 M^T & Y_1^T & W_1 C_r^T & \cdots & W_1 C_r^T & W_1 D_q^T & \cdots & W_1 D_q^T \\
+ \frac{1}{\beta} I + \sum_{q=1}^{n} \lambda_q (D_q W_1 + W_1 D_q^T) & & & & & & & & \\
* & -Q_2^{-1} & 0 & 0 & \cdots & 0 & 0 & \cdots & 0 \\
* & * & -R_2^{-1} & 0 & \cdots & \cdots & \cdots & \cdots & 0 \\
* & * & * & -W_1 & 0 & \cdots & \cdots & \cdots & \vdots \\
* & * & * & * & \ddots & \ddots & \ddots & \ddots & \vdots \\
* & * & * & * & * & -W_1 & \ddots & \ddots & \vdots \\
* & * & * & * & * & * & -\lambda_q W_1 & \ddots & \vdots \\
* & * & * & * & * & * & * & \ddots & 0 \\
* & * & * & * & * & * & * & * & -\lambda_q W_1
\end{bmatrix} < 0
$$

$$\begin{bmatrix} A_\xi W_\xi + W_2 A_\xi^T & W_2 M_\xi^T & Y_2^T & W_2 \hat{C}_{\xi,r}^T & \cdots & W_2 \hat{C}_{\xi,r}^T & W_2 \hat{D}_{\xi,q}^T & \cdots & W_2 \hat{D}_{\xi,q}^T & S^T \\ + B_\xi Y_2 + Y_2^T B_\xi^T & & & & & & & & & \\ * & -(Q_2^\xi)^{-1} & 0 & 0 & \cdots & 0 & 0 & \cdots & 0 & 0 \\ * & * & -(R_2^\mu)^{-1} & 0 & \cdots & \cdots & \cdots & \cdots & \cdots & 0 \\ * & * & * & -W_1 & 0 & \cdots & \cdots & \cdots & \cdots & \vdots \\ * & * & * & * & \ddots & \ddots & \ddots & \ddots & \ddots & \vdots \\ * & * & * & * & * & -W_1 & \ddots & \ddots & \ddots & \vdots \\ * & * & * & * & * & * & -\lambda_q^{-1} W_1^{-1} & \ddots & \ddots & \vdots \\ * & * & * & * & * & * & * & \ddots & \ddots & 0 \\ * & * & * & * & * & * & * & * & -\lambda_q^{-1} W_1^{-1} & 0 \\ * & * & * & * & * & * & * & * & * & -\beta \end{bmatrix} < 0$$

which are LMIs in (14.27) and (14.28), respectively.
It is clear that

$$E\{\bar{X}^T(0)P\bar{X}(0)\} \le \sigma(P)E\{tr(\bar{X}^T(0)\bar{X}(0))\} \le \alpha$$

and then the following inequality can be obtained

$$tr(E\{\bar{X}^T(0)\bar{X}(0)\})P \le \alpha I$$

Thus, we have

$$diag(W_1, W_2) \ge \alpha^{-1} tr(E\{\bar{X}^T(0)\bar{X}(0)\})I$$

15 Multi-Player Noncooperative and Cooperative Game Strategies for Linear Mean Field Stochastic Systems: Multi-Objective Optimization Evolutionary Algorithm Approach

15.1 INTRODUCTION

In recent years, mean field stochastic systems and investigations on them have gained widespread attention [422–424]. The appearance of the mean term in the system or the cost function makes the optimal design of the system more complicated than that of the classical stochastic system. Many researchers have studied the mean field stochastic system [7, 105, 107, 159, 425–427] and applied it to diverse areas like finance [428], biology [429, 430], and smart grids [431, 432]. A mean field stochastic system was used to model inter-bank borrowing and lending between N banks in [428], and the authors investigated the system risk problem and Nash equilibrium of the financial system. Algorithms to sample scale-free networks in molecular biology have been analyzed in the mean field sense [429]. The growth and evolution of a protein–protein interaction network were investigated by a mean field approach [430]. A stochastic noncooperative mean field game in a mean field stochastic system can describe competitiveness in smart grids [431]. Economic analysis of electric vehicles in a smart grid used a tool from a mean field stochastic system [432]. In most of these applications, individuals considered the effect of the mean field term as the collective behaviors resulting from all individuals' mutual interaction in the social system through the internet to be marvelous nowadays. These applications can also be classified as applications of cyber-social systems. Cyber-social systems comprise various social interactions in society, such as social welfare optimization problems,

DOI: 10.1201/9781003362142-19

markets, and transportation. However, there is not a method to consider a generalized cyber-social system by a mean field stochastic system because the category of cyber-social systems is too large.

Another common tool to consider the collective behaviors of all individuals in a large-population system is game theory. It is an effective approach to analyze the behaviors of multiple players based on their goals or interests. Game theory has been applied to diverse areas, such as the design of population biological networks [433], bandwidth allocation in 4G heterogeneous wireless access networks [434], and optimal scheduling of parking-lot electric vehicle charging systems [435]. Based on the different relations between individuals (players) in the game, games can be classified into two types: noncooperative and cooperative. In noncooperative games, players compete with each other to pursue their maximum benefits [433, 436–439]. In cooperative games, players reach a compromise due to the enforcement of rules by external authorities [440–442]. In noncooperative games, it is notoriously difficult to design optimal strategies for each player because it is hard to design strategies to control the system to achieve multiple desired goals for each player.

Further, in noncooperative games, the concept of Nash equilibrium is crucial to analyze the performance of players' strategies. A strategy profile is a Nash equilibrium solution if none of the players can obtain more benefits by unilaterally changing strategy. To find the Nash equilibrium solution for noncooperative games in stochastic systems, many iteratively updating algorithms like the extreme seeking method have been presented [443–447]. In the aforementioned updating algorithm, the time of convergence of the iterative approximation to Nash equilibrium can be extremely long if the initial approximation is far from the Nash equilibrium. Further, these iterative algorithms cannot search all Nash equilibrium points. If there is a direct method to compute all Nash equilibrium solutions, the Nash equilibrium solutions can be guaranteed after a certain amount of computational effort. There are few studies on finding Nash equilibrium solutions for noncooperative game strategy in mean field stochastic systems by direct methods.

Up to now, in the field of mean field control, many studies only consider the effect of continuous random fluctuations and external disturbance; few consider discontinuous random fluctuations. In general, the effect of discontinuous random fluctuations is not negligible in the real world. On the other hand, many researchers have studied control and applications of noncooperative and cooperative games, but noncooperative and cooperative games under collective (mean) behaviors are rarely discussed. To get a thorough understanding of the interactions of the players in a stochastic system under the influence of collective behavior, noncooperative and cooperative games for the MFSJD system are discussed.

In this chapter, stochastic noncooperative and cooperative H_∞ tracking game strategy problems are discussed for a class of MFSJD system with m players suffering from unknown external disturbance and continuous and discontinuous random fluctuations. First, a novel utility function based on minimax H_∞ model reference tracking performance is proposed for each player in the noncooperative and cooperative games to achieve their own desired model reference trajectory. If the utility functions are minimized, then the Nash equilibrium solutions of the stochastic noncooperative and cooperative H_∞ tracking games can be found. To find the Nash equilibrium solutions, an indirect approach is introduced to transform the noncooperative and cooperative H_∞ model reference tracking game strategy design problems to an LMI-constrained MOP and SOP of the MFSJD system, respectively. We also prove the Pareto optimal solutions of the LMI-constrained MOP are the Nash equilibrium solutions of the stochastic noncooperative H_∞ tracking game strategy design problems for the MFSJD system, respectively. Due to the convexity of LMIs, the Pareto optimal solutions of the LMI-constrained MOP could be obtained by a parallel search scheme via the proposed reverse-order LMI-constrained multi-objective evolution algorithm with the help of the convex optimization toolbox in MATLAB. Since the number of parameters of all noncooperative game strategies is very large, especially for a large number of players, the proposed reverse-order LMI-constrained MOEA method can efficiently solve all the Nash equilibrium solutions of the stochastic noncooperative H_∞ tracking game strategy design problem. Two simulation examples of market share allocation and network security strategy in a cyber-social system are given to illustrate the design procedure and validate the performance of the noncooperative and cooperative H_∞ tracking games in an MFSJD system, respectively.

The purposes of this chapter can be summarized as follows:

- In this chapter, the MFSJD system takes a continuous Wiener process, discontinuous Poisson process, and external disturbance into account simultaneously to mimic more realistic phenomena in mean field stochastic systems. Based on the MFSJD system, unlike the conventional stabilization game in [444–447], the multi-player noncooperative and cooperative H_∞ model reference tracking game strategy design problems are studied.
- By the proposed indirect method, the multi-player noncooperative and cooperative H∞ tracking game strategy design problem in the MFSJD system can be transformed into an equivalent LMI-constrained MOP and SOP for the MFSJD system, respectively. By solving the LMI-constrained MOP and SOP, the multi-player noncooperative and cooperative H∞ tracking game strategies in the MFSJD system can be efficiently obtained without solving the minimax cost function directly.
- Different from the previous iteratively updating algorithm to search for the Nash equilibrium solution, the reverse-order LMI-constrained MOEA is proposed to parallel search for the Pareto front to obtain all Nash equilibrium solutions of the multi-player noncooperative H∞ model reference tracking game strategies, which can be efficiently applied to the market share allocation and network security strategy in cyber-social systems.

This chapter is divided into seven sections as follows. Section 15.2 is dedicated to the noncooperative H_∞ tracking game problem formulation of the MFSJD system. The strategy design of the noncooperative H_∞ tracking game is given in Section 15.3. Section 15.4 presents the cooperative H_∞ tracking game problem formulation and its solution. Section 15.5 introduces the reverse-order LMI-constrained MOEA and its design procedure for the multi-player noncooperative H_∞ tracking game strategy of the MFSJD system. In Section 15.6, two simulation examples of noncooperative and cooperative H_∞ tracking games are given to validate the accuracy and efficiency of the proposed reverse-order LMI-constrained MOEA and design strategy. Finally, the conclusion is given in Section 15.7.

Notation: A^T: transpose of matrix A; $A \geq 0(A > 0)$: positive semi-definite (positive definite) matrix; $E\{X\}$: mean of the random variable $X(t)$; I: identity matrix with appropriate dimensions; $L_F^2(\mathbb{R}^+, \mathbb{R}^n)$: space of nonanticipative stochastic processes; $y(t) \in \mathbb{R}^n$ with respect to an increasing σ-algebras $F_t(t \geq 0)$ satisfying $\|y(t)\|_{L_F^2(\mathbb{R}^+, \mathbb{R}^n)} \triangleq E\{\int_0^\infty y(t)^T y(t)dt\}^{\frac{1}{2}} < \infty$; $\mathbb{C}^2$: the function is in class $\mathbb{C}^2$ if the first and second derivative of the function both exist and are continuous.

15.2 SYSTEM DESCRIPTION AND PROBLEM FORMULATION

Consider the following MFSJD system with m players defined on the filtration probability space (Ω, F, F_t, P) [448]:

$$dx(t) = [Ax(t) + \bar{A}Ex(t) + \sum_{i=1}^{m} B_i u_i(t) + v(t)]dt$$
$$+ (Lx(t) + \bar{L}Ex(t))dw(t) + (Nx(t) + \bar{N}Ex(t))dp(t) \qquad (15.1)$$
$$x(0) = x_0$$

where the càdlàg process $x(t) \in \mathbb{R}^n$ denotes the state vector of a mean field stochastic system; $Ex(t)$ represents the expected value of the state vector $x(t)$; $u_i(t) \in \mathbb{R}^m$ is the admissible control strategy of the ith player; $v(t) \in L_F^2(\mathbb{R}^+, \mathbb{R}^n)$ indicates the unknown finite energy external disturbance; $w(t)$ is the standard 1-dimensional Wiener process; and $p(t)$ denotes the standard 1-dimensional Poisson counting process with jump intensity λ in a unit time. The term $(Lx(t) + \bar{L}Ex(t))dw(t)$ indicates continuous state and mean state-dependent random fluctuations; the term $(Nx(t) + \bar{N}Ex(t))dp(t)$ represents discontinuous and mean state-dependent random fluctuations. The Poisson counting process $p(t)$ and the Wiener process $w(t)$ are assumed to be independent of each other. $A, B_i, \bar{A}, L, \bar{L}, N$ and $\bar{N}$ are deterministic real-value system matrices with appropriate dimensions.

Remark 15.1: The important properties of the Poisson counting process $p(t)$ and the Wiener process $w(t)$ are listed as follows [20]: $E\{dp(t)\} = \lambda dt$, where the finite scalar number $\lambda > 0$ is the Poisson jump intensity; $E\{dp(t)dt\} = 0$; $E\{dp(t)dw(t)\} = 0$; $E\{dw(t)\} = 0$; and $E\{dw(t)dt\} = 0$.

In (15.1), both the state vector $x(t)$ and the mean (collective) term $Ex(t)$ affect the system dynamic of the MFSJD system. It makes the MFSJD system more complicated to design than a conventional stochastic system because the designer has to take two dynamic system states $x(t)$ and $Ex(t)$ into account. To simplify the game strategy design procedure of the MFSJD system, the decoupling method is employed to separate the MFSJD system in (15.1) into two orthogonal subsystems: a mean subsystem $Ex(t)$ and variation subsystem $\tilde{x}(t) = x(t) - Ex(t)$.

By taking the expectation of the MFSJD system in (15.1), then the mean subsystem $Ex(t)$ is given as follows:

$$dEx(t) = [(A + \bar{A} + \lambda(N + \bar{N}))Ex(t) + Ev(t) + \sum_{i=1}^{m} B_i Eu_i(t)]dt \qquad (15.2)$$

where $Eu_i(t)$ is the mean of the control strategy $u_i(t)$ of the ith player. Then the variation subsystem $\tilde{x}(t)$ can be obtained by subtracting the MFSJD system in (15.1) by the mean subsystem in (15.2) as follows:

$$d\tilde{x}(t) = [A\tilde{x}(t) - \lambda(N + \bar{N})Ex(t) + \sum_{i=1}^{m} B_i\tilde{u}_i(t) + \tilde{v}(t)]dt$$
$$+ (Lx(t) + \bar{L}Ex(t))dw(t) + (Nx(t) + NEx(t))dp(t) \qquad (15.3)$$

where $\tilde{u}_i(t) = u_i(t) - Eu_i(t)$ and $\tilde{v}(t) = v(t) - Ev(t)$.

In the noncooperative game, the players try to apply their strategy to pursue their maximum profit. However, the objectives of the players are conflicting, and the players have to compete with each other to benefit most [436]. The players are assumed to be incapable of communicating with each other, so the strategy of each player is unavailable to their competitors. In the MFSJD system, the goal of the ith player is to drive the mean term $Ex(t)$ to track their desired trajectory and eliminate the variations around the mean in (15.3). Suppose the reference trajectory $x_{r_i}(t)$ of the ith player is generated by a reference model as follows:

$$dx_{r_i}(t) = (A_{r_i} x_{r_i}(t) + r_i(t))dt \qquad (15.4)$$

where A_{r_i} is a deterministic asymptotically stable matrix, and $r_i(t)$ is a bounded reference input. A_{r_i} and $r_i(t)$ are both given by the ith player. In general, A_{r_i} and $r_i(t)$ determine the transient and steady state of the reference model. At the steady state, the reference system state $x_{r_i}(t)$ is equal to $-A_{r_i}^{-1}r_i(t)$. For example, if we choose $A_{r_i} = -I$ in (15.4), then $x_{r_i}(t) = r_i(t)$ at the steady state. In this situation, $r_i(t)$ could be set as the desired target of the ith player to simplify the design procedure of tracking game strategy of the MFSJD system. If the reference system $x_{r_i}(t)$ in (15.4) is subtracted from mean subsystem $Ex(t)$ in (15.2), then the tracking error dynamic of mean $e_i(t) = Ex(t) - x_{r_i}(t)$ is given as follows:

$$de_i(t) = [(A + \bar{A} + \lambda(N + \bar{N}))Ex(t) + Ev(t) + \sum_{i=1}^{m} B_i Eu_i(t)$$
$$- (A_{r_i} x_{r_i}(t) + r_i(t))]dt \qquad (15.5)$$

After two subsystems, the variation subsystem $\tilde{x}(t)$ in (15.3) and the subsystem of tracking error $e_i(t)$ in (15.5), have been obtained, the noncooperative game strategies of the MFSJD system will be designed as follows: For each player, if the mean of the MFSJD system $Ex(t)$ in (15.2) can optimally track one's desired trajectory $x_{r_i}(t)$ in (15.4) as possible and the variation $\tilde{x}(t)$ in (15.3) around the mean $Ex(t)$ of the MFSJD system is simultaneously minimized despite competitive strategies and external disturbance, then each player will benefit most.

In the stochastic noncooperative game strategy design problem of the MFSJD system, to be on the safe side, the ith player designs and decomposes their strategy $u_i(t)$ into $u_i(t) - Eu_i(t)$ and $Eu_i(t)$ to minimize the model reference tracking error $e_i(t)$ and the variation state $\tilde{x}(t)$ under the worst-case effect of strategies of competitors $u_{-i}(t) = [u_1^T(t), u_2^T(t), ..., u_{i-1}^T(t), u_{i+1}^T(t), ..., u_m^T(t)]^T$, external disturbance $v(t)$, reference input $r_i(t)$, and continuous and discontinuous intrinsic random fluctuations; that is, the following m-player stochastic noncooperative minimax H_∞ tracking game is proposed for the MFSJD system in (15.3), (15.4), and (15.5).

$$
g_i^* = \min_{\substack{\tilde{u}_i(t) \\ Eu_i(t)}} \max_{\substack{\tilde{u}_{-i}(t) \\ Eu_{-i}(t) \\ v(t), Ev(t) \\ r_i(t)}} \frac{E \int_0^{t_f} [\tilde{x}^T(t) Q_{i_1} \tilde{x}(t) + e_i^T(t) Q_{i_2} e_i(t)}{E(\tilde{x}^T(0)\tilde{x}(0)) + E(e_i^T(0)e_i(0))}
$$

$$
\frac{+\tilde{u}_i^T(t) R_{i_1} \tilde{u}_i(t) + Eu_i(t)^T R_{i_2} Eu_i(t)]dt}{}
$$

$$
+ E \int_0^{t_f} \left[\begin{bmatrix} \tilde{u}_{-i}^T(t) & Eu_{-i}^T(t) \end{bmatrix} \begin{bmatrix} \tilde{u}_{-i}(t) \\ Eu_{-i}(t) \end{bmatrix} \right.
$$

$$
\left. + \begin{bmatrix} v^T(t) & Ev^T(t) & r_i^T(t) \end{bmatrix} \begin{bmatrix} v(t) \\ Ev(t) \\ r_i(t) \end{bmatrix} \right]dt \tag{15.6}
$$

for $i = 1, 2, ..., m$

where t_f denotes the terminal time; $Q_{i_1} > 0$ denotes the weighting matrix on the variation state $\tilde{x}(t)$; $Q_{i_2} > 0$ represents the weighting matrix on the tracking error $e_i(t)$; $R_{i_1} > 0$ and $R_{i_2} > 0$ denote the weighting matrices to trade off between the tracking performance and the control effort; $E\tilde{x}^T(0)\tilde{x}(0) + Ee_i^T(0)e_i(0)$ in (15.6) denotes the effect of the uncertain initial condition of variation state $\tilde{x}(t)$ and $e_i(t)$ on the tracking performance of noncooperative H_∞ tracking game; $Eu_{-i}(t)$ denotes the mean of the competitive strategies $u_{-i}(t)$; and g_i^* denotes the minimax H_∞ tracking game performance of the ith player taking their optimal strategies $\tilde{u}_i^*(t)$ and $Eu_i^*(t)$ under the worst-case combined competitive strategies $\tilde{u}_{-i}^*(t)$ and $Eu_{-i}^*(t)$, external disturbance $v^*(t)$, and reference input $r_i^*(t)$. The physical meaning of the stochastic noncooperative minimax H_∞ game of the MFSJD system in (15.6) is that the worst-case effect of competitive strategies $u_{-i}(t)$, uncertain external disturbance $v(t)$, and reference signal $r_i(t)$ should be minimized with parsimonious mean control effort $Eu_i(t)$ and variation control strategy $\tilde{u}_i(t)$.

Remark 15.2: In the conventional noncooperative dynamic game, each player considers the stabilization problem only with different weighting matrices in their cost function. For the proposed mean-field noncooperative tracking game strategy, each player has a respective desired tracking target. This means each player wants to achieve their optimal tracking control performance despite the effect of the competitive strategies of other players. Also, to be on the safe side, the worst-case effect of the competitive strategies, external disturbance, and variation between the system state and mean state on the target tracking performance is to be minimized.

For simplicity of design, the variation subsystem $\tilde{x}(t)$ in (15.3), tracking error subsystem $e_i(t)$ in (15.5), and reference system $x_{r_i}(t)$ in (15.4) can be combined as an augmented MFSJD system with the augmented state $X_i(t) \triangleq [\tilde{x}^T(t), e_i^T(t), x_{r_i}{}^T(t)]^T$ and an augmented control strategy $U_i(t) \triangleq [\tilde{u}_i^T(t), Eu_i^T(t)]^T$ as follows:

$$dX_i(t) = (F_iX_i(t) + \sum_{i=1}^{m} B_iU_i(t) + G_iV_i(t))dt \tag{15.7}$$
$$+ H_iX_i(t)dw(t) + T_iX_i(t)dp(t)$$

where $e_i(t) = Ex(t) - x_{r_i}(t)$ and

$$F_i = \begin{bmatrix} A & -\lambda(N+\bar{N}) & -\lambda(N+\bar{N}) \\ 0 & A+\bar{A}+\lambda(N+\bar{N}) & -A_{r_i}+A+\bar{A}+\lambda(N+\bar{N}) \\ 0 & 0 & A_{r_i} \end{bmatrix}, G_i = \begin{bmatrix} I & -I & 0 \\ 0 & I & -I \\ 0 & 0 & I \end{bmatrix},$$

$$H_i = \begin{bmatrix} L & L+\bar{L} & L+\bar{L} \\ 0 & 0 & 0 \\ 0 & 0 & 0 \end{bmatrix}, B_i = \begin{bmatrix} B_i & 0 \\ 0 & B_i \\ 0 & 0 \end{bmatrix}, N_i = \begin{bmatrix} N & N+\bar{N} & N+\bar{N} \\ 0 & 0 & 0 \\ 0 & 0 & 0 \end{bmatrix}, V_i = \begin{bmatrix} v(t) \\ Ev(t) \\ r_i(t) \end{bmatrix}$$

The ith player has the information of the augmented state vector $X_i(t)$ to design their strategy $U_i(t)$, but the information of augmented external disturbance $V_i(t)$ and the competitive strategies $U_{-i}(t)$ are unavailable to them. The augmented MFSJD system $X_i(t)$ in (15.7) can be reconstructed as follows:

$$dX_i(t) = (F_iX_i(t) + B_iU_i(t) + B_{-i}U_{-i}(t))dt \tag{15.8}$$
$$+ H_iX_i(t)dw(t) + T_iX_i(t)dp(t)$$

where $B_{-i} = [B_1, ..., B_{i-1}, B_{i+1}, ..., B_{m-1}, B_m, G_i]$,

$$U_{-i}(t) = [U_1^T(t), U_2^T(t), ..., U_{i-1}^T(t), U_{i+1}^T(t), ..., U_{m-1}^T(t), U_m^T(t), V_i^T(t)]^T$$

For convenience of design, based on the augmented MFSJD system in (15.8), the noncooperative stochastic minmax H_∞ reference tracking game in (15.6) can be modified as follows:

$$g_i^* = \min_{U_i(t)} \max_{U_{-i}(t)} \frac{E\int_0^{t_f} [X_i^T(t)Q_iX_i(t) + U_i^T(t)R_iU_i(t)]dt}{EX_i^T(0)X_i(0) + E\int_0^{t_f}[U_{-i}^T(t)U_{-i}(t)]dt} \tag{15.9}$$

$$Q_i = diag(Q_{i1}, Q_{i2}, 0), R_i = diag(R_{i1}, R_{i2}), i = 1, 2, ..., m$$

15.3   NONCOOPERATIVE H_∞ TRACKING GAME STRATEGY DESIGN FOR MFSJD SYSTEMS

In general, due to the fractional utility (payoff) function of the minimax noncooperative H_∞ game in (15.9), it is very difficult to solve the stochastic multi-player noncooperative game strategy design problem in (15.9) for an MFSJD system in (15.7) or (15.8) directly. Therefore, the following indirect approach or the so-called stochastic suboptimal approach is applied to simplify the design procedure of the stochastic noncooperative game strategy in (15.9). Suppose there exists an upper bound of g_i^* in the stochastic minimax noncooperative H_∞ reference tracking game as follows:

$$g_i^* = \min_{U_i(t)} \max_{U_{-i}(t)} \frac{E \int_0^{t_f} [X_i^T(t)Q_i X_i(t) + U_i^T(t)R_i U_i(t)]dt}{EX_i^T(0)X_i(0) + E \int_0^{t_f} [U_{-i}^T(t)U_{-i}(t)]dt} \le g_i \qquad (15.10)$$

where $g_i^* > 0$ denotes the upper bound of the minimax H_∞ tracking performance of a noncooperative game with the augmented strategy $U_i(t)$ being selected by the ith player. The upper bound g_i will be decreased as much as possible to approach g_i^* for all players simultaneously to attain the real m minimax solutions of the m-player noncooperative H_∞ suboptimal game in (15.10). The previous complicated m-player minimax noncooperative H_∞ game problem is equivalent to the following multi-objective optimization problem by minimizing g_i, $i = 1,2,...,m$, simultaneously.

$$(g_1^*,...,g_i^*,...,g_m^*) = \min(g_1,...,g_i,...,g_m) \qquad (15.11)$$

subject to

$$\min_{U_i(t)} \max_{U_{-i}(t)} \frac{E \int_0^{t_f} [X_i^T(t)Q_i X_i(t) + U_i^T(t)R_i U_i(t)]dt}{EX_i^T(0)X_i(0) + E \int_0^{t_f} [U_{-i}^T(t)U_{-i}(t)]dt} \le g_i \qquad (15.12)$$

$$i = 1,...,m$$

The solution of the MOP in (15.11) and (15.12) is not unique, and all solutions are called Pareto optimal solutions. The following definitions are needed to define the Pareto optimal solutions for the MOP.

Definition 15.1 (Pareto Dominance) [2, 77, 205 206, 389]: For the MOP in (15.11) and (15.12), a feasible solution $(U_1^1(t),...,U_i^1(t),...,U_m^1(t))$ correspond-
ing to the objective vector $(g_1^1,...,g_i^1,...,g_m^1)$ is said to dominate another feasible solution $(U_1^2(t),...,U_i^2(t),...,U_m^2(t))$ corresponding to the objective vector $(g_1^2,...,g_i^2,...,g_m^2)$ if $g_1^1 \le g_1^2,...,g_i^1 \le g_i^2,...,g_m^1 \le g_m^2$ and at least one of the inequalities is a strict inequality.

Definition 15.2 (Pareto Optimal Solution) [2, 77, 205 206, 389]: A feasible solution $(U_1^*(t),...,U_i^*(t),...,U_m^*(t))$ of the MOP in (15.11) and (15.12) is said to be a Pareto optimal solution if there does not exist another feasible solution $(U_1(t),...,U_i(t),...,U_m(t))$ that can dominate it.

Definition 15.3 (Pareto Optimal Set) [2, 77, 205 206, 389]: For the MOP in (15.11) and (15.12), the Pareto optimal set σ^* is defined as

$\sigma^* \triangleq \{(U_1^*(t),...,U_i^*(t),...,U_m^*(t)) \big| (U_1^*(t),...,U_i^*(t),...,U_m^*(t))$ is Pareto optimal solution in (15.11) and (15.12)}.

Definition 15.4 (Pareto Front) [2, 77, 205 206, 389]: For the MOP in (15.11) and (15.12), the Pareto front is defined as $T_F \triangleq \{(g_1^*,...,g_i^*,...,g_m^*) \big| (U_1^*(t),...,U_i^*(t),...,U_m^*(t)) \in \sigma^*\}$; that is, the Pareto front collects the objective vector of Pareto optimal solution.

Remark 15.3: The concept of the Nash equilibrium solution is introduced to consider the solution of the noncooperative minimax H_∞ game in (15.9) [436, 449]. In the game theory, if each player has chosen a strategy, and no player can benefit by changing their own strategy while other players keep their strategies unchanged, then the current set of strategies and their corresponding pay-off constitutes a Nash equilibrium solution. In general, there exist a large number of Nash equilibrium solutions. In this chapter, the Nash equilibrium game strategy $(U_1^*(t),...,U_i^*(t),...,U_m^*(t))$ constitutes a Nash equilibrium objective vector $(g_1^*,...,g_i^*,...,g_m^*)$ if and only if the following inequalities are established:

$$(g_1^*,...,g_i^*,...,g_m^*) \le (g_1^*,...,g_{i-1}^*,g_i,g_{i+1}^*,...,g_m^*), i = 1,2,...,m$$

To find the Nash equilibrium solutions $(U_1^*(t),...,U_i^*(t),...,U_m^*(t))$ and their corresponding Nash equilibrium objective vectors $(g_1^*,...,g_i^*,...,g_m^*)$ of the stochastic multi-player noncooperative H_∞ game problem in (15.9), $(g_1^*,...,g_i^*,...,g_m^*)$ can be obtained indirectly by minimizing the corresponding upper bounds g_i by solving the MOP in (15.11) and (15.12).

Theorem 15.1: The MOP in (15.11) and (15.12) is equivalent to the multi-player noncooperative minimax H_∞ reference tracking game in (15.9) for the MFSJD system in (15.7) if the MOP solution in (15.11) and (15.12) is obtained.

Proof: We only need to prove the inequality constraints in (15.12) will disappear when the Pareto optimal solutions are obtained. This can be proven by contradiction. Let $(g_1^*,...,g_i^*,...,g_m^*)$ be a Pareto optimal solution of (15.11), and suppose one of the inequality constraints in (15.12) strictly holds; that is, there exists $\overline{g}_i$ such that

$$\overline{g}_i = \min_{U_i(t)} \max_{U_{-i}(t)} \frac{E\int_0^{t_f}[X_i^T(t)Q_iX_i(t)+U_i^T(t)R_iU_i(t)]dt}{EX_i^T(0)X_i(0)+E\int_0^{t_f}[U_i^T(t)U_{-i}(t)]dt} < g_i^*, \text{ for some } i$$

Then $(g_1^*,...,\overline{g}_i,...,g_m^*)$ dominates $(g_1^*,...,g_i^*,...,g_m^*)$ However, according to Definition 15.1, the Pareto optimal objective vector $(g_1^*,...,g_i^*,...,g_m^*)$ is not dominated by another objective vector, which leads to a contradiction of the assumption. Thus, when the Pareto optimal solution of the MOP in (15.11) and (15.12) is obtained, the inequalities in (15.12) all disappear, and the solution of the MOP is equal to the solution of the multiplayer noncooperative game in (15.9).

Based on Definitions 15.1–15.4 and Theorem 15.1, the simultaneous minimization of $(g_1,...,g_i,...,g_m)$ for MOP in (15.11) is to find the Pareto optimal objective vector $(g_1^*,...,g_i^*,...,g_m^*)$. Since the minimization of the numerator in (15.12) is independent of $U_{-i}(t)$, the MOP in (15.11) with m minimax constraints in (15.12) is equivalent to the following MOP with m constrained stochastic Nash quadratic game inequalities [34, 436, 450, 451, 452]:

$$(g_1^*,...,g_i^*,...,g_m^*) = \min(g_1,...,g_i,...,g_m) \tag{15.13}$$

subject to

$$\min_{U_i(t)} \max_{U_{-i}(t)} E \int_0^{t_f} [X_i^T(t)Q_i X_i(t) + U_i^T(t)R_i U_i(t)$$
$$-g_i U_{-i}^T(t)U_{-i}(t)]dt \le g_i E X_i^T(0)X_i(0) \tag{15.14}$$
$$i = 1,2,...,m$$

Let us denote

$$J_i(U_i(t),U_{-i}(t)) = E \int_0^{t_f} [X_i^T(t)Q_i X_i(t) + U_i^T(t)R_i U_i(t)$$
$$-g_i U_{-i}^T(t)U_{-i}(t)]dt \tag{15.15}$$
$$i = 1,2,...,m$$

then two steps are required to solve the m constrained stochastic Nash quadratic games in (15.14). The first step is to solve the following stochastic Nash quadratic game problems:

$$J_i^* = \min_{U_i(t)} \max_{U_{-i}(t)} J_i(U_i(t),U_{-i}(t)), i = 1,2,...,m \tag{15.16}$$

and the second step is to solve the inequality-constrained problem

$$J_i^* \le g_i E X_i^T(0)X_i(0), i = 1,2,...,m \tag{15.17}$$

By solving (15.16) and (15.17) for the constrained stochastic Nash quadratic game in (15.14), the m-player noncooperative stochastic game strategies in (15.9) can be found by solving an MOP with the constraints of m Riccati-like inequalities in the sequel. The following two lemmas are introduced to help us solve m constrained stochastic Nash quadratic games in (15.14) or (15.16) and (15.17):

Lemma 15.1: (Itô–Lévy Formula) [20] Let $L(X_i(t))$ be the Lyapunov function of the augmented MFSJD system in (15.7) such that $L(X_i(t)) \in \mathbb{R}^{3n} \to \mathbb{R}^+$,

$L(\cdot) \in \mathbb{C}^2(\mathbb{R}^{3n})$, $L(0) = 0$, and $L(X_i(t)) \geq 0$. For the m-player MFSJD system in (7), the Itô–Lévy formula of $L(X_i(t))$ is given as follows [379, 397]:

$$
\begin{aligned}
dL(X_i(t)) = & (\tfrac{\partial L(X_i(t))}{\partial X_i(t)})^T (F_i X_i(t) + B_i U_i(t) + B_{-i} U_{-i}(t)) \\
& + \tfrac{1}{2}(H_i X_i(t))^T (\tfrac{\partial^2 L(X_i(t))}{\partial^2 X_i(t)})(H_i X_i(t)) dt \\
& + (\tfrac{\partial L(X_i(t))}{\partial X_i(t)})^T (H_i X_i(t)) dw(t) \\
& + [L(X_i(t) + T_i X_i(t)) - L(X_i(t))] dp(t)
\end{aligned}
\tag{15.18}
$$

Lemma 15.2: [212] For any matrix Z and Y with appropriate dimensions, we have

$$
Z^T Y + Y^T Z \leq Z^T P^{-1} Z + Y^T P Y
$$

where P is any positive-definite symmetric matrix.

Theorem 15.2: The MOP in (15.13) with m constrained Nash games in (15.14) can be solved for the m-player noncooperative minimax H_∞ tracking game strategies in (15.9) as follows

$$
\begin{aligned}
U_i^*(t) &= -R_i^{-1} B_i^T P^* X_i(t), \text{for } i = 1, 2, ..., m \\
U_{-i}^*(t) &= \frac{1}{g_i^*} B_{-i}^T P^* X_i(t), \text{for } i = 1, 2, ..., m
\end{aligned}
\tag{15.19}
$$

where g_i^* and P^* are the solution of the following Riccati-like inequality-constrained MOP:

$$
(g_1^*, ..., g_i^*, ..., g_m^*) = \min_{P>0}(g_1, ..., g_i, ..., g_m)
\tag{15.20}
$$

subject to

$$
\begin{aligned}
& Q_i + PF_i - PB_i R_i^{-1} B_i^T P + \tfrac{1}{g_i} PB_{-i} B_{-i}^T P + F_i^T P + H_i^T PH_i \\
& + \lambda(T_i^T PT_i + T_i^T P + PT_i) \leq 0, \text{for } i = 1, 2, ..., m
\end{aligned}
\tag{15.21}
$$

with

$$
0 \leq P \leq g_i I
\tag{15.22}
$$

Proof: Please refer to Appendix 15.8.1.

In Theorem 15.2, $(U_1^*(t), ..., U_i^*(t), ..., U_m^*(t))$ is a Pareto optimal solution with the corresponding objective vector $(g_1^*, ..., g_i^*, ..., g_m^*)$. In general, the Pareto optimal solution is not unique. There is a set of Pareto optimal solutions. To prove the Pareto optimal solutions are equal to the Nash equilibrium solutions, the following theorem is needed.

Theorem 15.3: The multi-objective optimal control strategies $(U_1^*(t), ..., U_i^*(t), ..., U_m^*(t))$ with the corresponding objective vector $(g_1^*, ..., g_i^*, ..., g_m^*)$

and P^* in (15.20) are the Nash equilibrium solutions of the m-player noncooperative stochastic game in (15.9) for the MFSJD system in (15.7).

Proof: It will be proven by contradiction. Suppose the condition of the Nash equilibrium solution in Remark 15.3 is violated for a Pareto optimal solution $(U_1^*(t),...,U_i^*(t),...,U_m^*(t))$ in (12) with the objective vector $(g_1^*,...,g_i^*,...,g_m^*)$ for the ith inequality; that is, there exists a $\bar{g}_i$ such that

$$(g_1^*,...,\bar{g}_i,...,g_m^*) \leq (g_1^*,...,g_{i-1}^*,g_i^*,g_{i+1}^*,...,g_m^*)$$

Then, there exists a solution $(U_1^*(t),...,\bar{U}_i(t),...,U_m^*(t))$ in (15.12) with objective vector $(g_1^*,...,\bar{g}_i,...,g_m^*)$ that dominates the Pareto optimal solution $(U_1^*(t),...,U_i^*(t),...,U_m^*(t))$. However, it reveals $(U_1^*(t),...,U_i^*(t),...,U_m^*(t))$ is not the Pareto optimal solution in (15.12), and it leads to a contradiction to the assumption. Thus, this shows that the Pareto optimal solution in (15.12) is the Nash equilibrium solution in (15.9).

Remark 15.4: To solve the Riccati-like inequalities in (15.21), we apply a Schur complement transformation [212] to transform (15.21) into the following LMIs after multiplying $W = P^{-1}$ by both sides of (15.21):

$$
\begin{bmatrix}
WF_i^T + F_iW - B_iR_i^{-1}B_i^T \\ + \frac{1}{g_i}B_{-i}B_{-i}^T + \lambda(WT_i^T + TW_i) & WH_i^T & \lambda^{\frac{1}{2}}WT_i^T & WQ_i^{\frac{1}{2}} \\
H_iW & -W & 0 & 0 \\
\lambda^{\frac{1}{2}}T_iW & 0 & -W & 0 \\
W & 0 & 0 & -I
\end{bmatrix} \leq 0
$$

$$\text{for } i = 1,2,...,m \tag{15.23}$$

and (15.22) can be transformed as follows:

$$\begin{bmatrix} g_i & I \\ I & W \end{bmatrix} \geq 0, \text{for } i = 1,2,...,m \tag{15.24}$$

Therefore, the MOP in (15.20) for the noncooperative H_∞ reference tracking game strategy in Theorem 15.2 for MFSJD system is equivalent to the following MOP

$$(g_1^*,...,g_i^*,...,g_m^*) = \min_{W>0}(g_1,...,g_i,...,g_m) \tag{15.25}$$

$$\text{subject to LMIs in (15.22) and (15.24)}$$

After solving the MOP in (15.25) for $g_i^*, i = 1,...,m$ and W^*, we obtain $P^* = W^{*-1}$ and the corresponding Nash equilibrium solutions $U_i^*(t) = -R_i^{-1}B_i^T P^* X_i(t), i = 1,2,...,m$ for the noncooperative minimax H_∞ game in the MFSJD system.

15.4   COOPERATIVE H_∞ TRACKING GAME STRATEGY DESIGN FOR MFSJD SYSTEMS

If m players in (15.1) have reached a consensus to achieve a common desired trajectory $x_r(t)$ with each other at the beginning of the game, then all players will collaborate with each other to control the mean of state vector $Ex(t)$ to track common desired trajectory $x_r(t)$, which is generated by the following reference model.

$$dx_r(t) = (A_r x_r(t) + r(t))dt \qquad (15.26)$$

In the cooperative stochastic game, players cooperate together with their strategies $u(t) = [u_1^T(t),...,u_m^T(t)]^T$ to minimize the tracking error $(Ex(t) - x_r(t))$ and the deviation $(x(t) - Ex(t))$ under the worst-case effect of external disturbance $v(t)$ and reference input $r(t)$ despite continuous and discontinuous intrinsic random fluctuations. Let us denote $\tilde{x}(t) = x(t) - Ex(t)$, $e_c(t) = Ex(t) - x_r(t)$, and $\tilde{u}(t) = u(t) - Eu(t)$. Then the stochastic cooperative minimax H_∞ tracking game of the MFSJD system is formulated as follows:

$$g^0 = \min_{\substack{\tilde{u}_i(t) \\ Eu_i(t)}} \max_{\substack{v(t), Ev(t) \\ r_i(t)}} \frac{E\int_0^{t_f} [\tilde{x}^T(t)Q_1\tilde{x}(t) + e_c^T(t)Q_2 e_c(t)}{E\tilde{x}^T(0)\tilde{x}(0) + E(e_c^T(0)e_c(0))}$$

$$+ E\int_0^{t_f} \begin{bmatrix} v^T(t) & Ev^T(t) & r_i^T(t) \end{bmatrix} \begin{bmatrix} v(t) \\ Ev(t) \\ r_i(t) \end{bmatrix} dt \qquad (15.27)$$

where $Q_1 > 0$ is the corresponding weighting matrix on the deviation $\tilde{x}(t)$, $Q_2 > 0$ is the corresponding weighting matrix on the tracking error $e_c(t)$, $R_1 > 0$ is the corresponding weighting matrix on the deviation of the strategy $\tilde{u}(t)$, and $R_2 > 0$ is the corresponding weighting matrix on the mean of the strategy $Eu(t)$. $E(\tilde{x}^T(0)\tilde{x}(0)) + E(e_c^T(0)e_c(0))$ considers the effect of the uncertain initial condition of $\tilde{x}(t)$ and $e_c(t)$ on the reference tracking performance. g^0 denotes the performance of the cooperative minimax H_∞ tracking game of the MFSJD system. The physical meaning of the stochastic cooperative minimax H_∞ tracking game in (15.27) is that the worst-case effect of uncertain external disturbance $v(t)$ and reference trajectory $r(t)$ on the MFSJD system should be minimized by the parsimonious mean strategy $Eu(t)$ and deviation strategy $\tilde{u}(t)$.

Similar to the augmented MFSJD system $X_i(t)$ in (15.7), an augmented MFSJD system for the cooperative game $X(t) \triangleq [\tilde{x}(t), e_c^T(t), x_r^T(t)]^T$ could also be obtained as follows:

$$dX(t) = (FX(t) + BU(t) + B\_V(t))dt + HX(t)dw(t)$$
$$+ TX(t)dp(t) \qquad (15.28)$$

where

$$F = \begin{bmatrix} A & -\lambda(N+\bar{N}) & -\lambda(N+\bar{N}) \\ 0 & A+\bar{A}+\lambda(N+\bar{N}) & -A_r-A-\bar{A}-\lambda(N+\bar{N}) \\ 0 & 0 & A_r \end{bmatrix}, B = \begin{bmatrix} \bar{B} & 0 \\ 0 & \bar{B} \\ 0 & 0 \end{bmatrix},$$

$$H = \begin{bmatrix} L & L+\bar{L} & L+\bar{L} \\ 0 & 0 & 0 \\ 0 & 0 & 0 \end{bmatrix}, U(t) = \begin{bmatrix} \tilde{u}(t) \\ Eu(t) \end{bmatrix}, T = \begin{bmatrix} N & N+\bar{N} & N+\bar{N} \\ 0 & 0 & 0 \\ 0 & 0 & 0 \end{bmatrix}, V(t) = \begin{bmatrix} v(t) \\ Ev(t) \\ r(t) \end{bmatrix},$$

$$\bar{B} = [B_1,...,B_m], B_- = \begin{bmatrix} I & -I & 0 \\ 0 & I & -I \\ 0 & 0 & I \end{bmatrix}$$

For convenience of design, the cooperative minimax H_∞ tracking game in (15.27) for the MFSJD system in (15.28) can be modified as follows.

$$g^0 = \min_{U(t)} \max_{V(t)} \frac{E\int_0^{t_f} [X^T(t)QX(t)+U^T(t)RU(t)]dt}{EX^T(0)X(0)+E\int_0^{t_f}[V^T(t)V(t)]dt} \tag{15.29}$$

$$Q = diag(Q_1,Q_2,0), R = diag(R_1,R_2)$$

By using the same indirect approach, the cooperative minimax H_∞ tracking game in (15.29) can be solved by minimizing its upper bound g as follows:

$$g^0 = \min_{U(t)} \max_{V(t)} \frac{E\int_0^{t_f} [X^T(t)QX(t)+U^T(t)RU(t)]dt}{EX^T(0)X(0)+E\int_0^{t_f} V^T(t)V(t)dt} \leq g \tag{15.30}$$

Instead of solving this complicated minimax H_∞ cooperative problem, we can obtain g^0 by solving the following single-objective optimization problem via minimizing the upper bound g of g^0,

$$g^0 = \min g \tag{15.31}$$

subject to

$$\min_{U(t)} \max_{V(t)} \frac{E\int_0^{t_f} [X^T(t)QX(t)+U^T(t)RU(t)]dt}{EX^T(0)X(0)+E\int_0^{t_f}[V^T(t)V(t)]dt} \leq g \tag{15.32}$$

Theorem 15.4: The SOP in (15.31) and (15.32) is equivalent to the multi-player cooperative H_∞ tracking game in (15.29) if the SOP solution in (15.31) is obtained.

Proof: The proof is similar to Theorem 15.1. Hence, we omit the proof here.

The stochastic H_∞ game constraint in (15.32) is equivalent to the following constrained stochastic quadratic game:

$$
\min_{U(t)} \max_{V(t)} E \int_0^{t_f} [X^T(t)QX(t) + U^T(t)RU(t) - gV^T(t)V(t)]dt
$$
$$
\leq gEX^T(0)X(0)
$$

(15.33)

Let us denote

$$
J(U(t), V(t)) = E \int_0^{t_f} [X^T(t)QX(t) + U^T(t)RU(t)
$$
$$
- gV^T(t)V(t)]dt
$$

(15.34)

Two steps are required to solve the constrained stochastic Nash quadratic game in (15.33). The first step is to solve the following stochastic Nash quadratic game problem.

$$
J^0 = \min_{U(t)} \max_{V(t)} J(U(t), V(t))
$$

(15.35)

and the second step is to solve the inequality constraint.

$$
J^0 \leq EX^T(0)X(0)
$$

(15.36)

Then the main theorem of cooperative stochastic game is given as follows:

Theorem 15.5: Based on the constrained Nash games in (15.35) and (15.36) for the stochastic H_∞ game constraint in (15.32), the SOP in (15.31) and (15.32) can be solved for the m-player cooperative tracking game strategy in (15.29) as follows:

$$
U^0(t) = -R^{-1}B^T P^0 X(t), \quad V^0(t) = \frac{1}{g^0} B_-^T P^0 X(t)
$$

(15.37)

where g^0 and P^0 are the solutions of the following SOP

$$
g^0 = \min_{P>0} g
$$

subject to

$$
Q + 2PF - PBR^{-1}B^T P + \frac{1}{g} PB_- B_-^T P + H^T PH
$$
$$
+ \lambda(T^T PT + T^T P + PT) \leq 0
$$

(15.38)

$$
0 \leq P \leq gI
$$

(15.39)

Proof: The proof is similar to Theorem 15.2 and is omitted.

Based on Theorem 15.5, $U^0(t)$ and $V^0(t)$ in (15.37) are the solutions of the stochastic multi-player cooperative H_∞ game in (15.29). Therefore, by similar techniques in Remark 15.3, the SOP of solving g^0 in (15.31) for the MFSJD system can be transformed into the following LMI-constrained SOP.

$$g^0 = \min_{W>0} g \tag{15.40}$$

subject to

$$\begin{bmatrix} WF^T + FW - BR^{-1}B^T \\ + \frac{1}{g_i}B_-B_-^T + \lambda(WT^T + TW) & WH^T & \lambda^{\frac{1}{2}}WT^T & WQ^{\frac{1}{2}} \\ HW & -W & 0 & 0 \\ \lambda^{\frac{1}{2}}TW & 0 & -W & 0 \\ W & 0 & 0 & -I \end{bmatrix} \le 0 \tag{15.41}$$

$$\begin{bmatrix} g_0 & I \\ I & W \end{bmatrix} \ge 0 \tag{15.42}$$

After solving $W^0 = P_0^{-1}$ and g^0 from the LMI-constrained SOP in (15.40)–(15.42), the m-player cooperative minimax H_∞ game strategy of MFSJD in (15.37) can be obtained as $U^0(t) = -R^{-1}B^T P^0 X(t)$.

15.5 LMI-CONSTRAINED MOEA OF NONCOOPERATIVE MINMAX H_∞ GAME STRATEGY FOR MULTI-PLAYER TARGET TRACKING OF MFSJD SYSTEMS

The MOEA has long been a popular algorithm to solve MOPs [453, 454]. It is a stochastic algorithm inspired by biological evolution, such as reproduction, mutation, recombination, and selection. Based on the MOEA, we can search the global optimal solutions with multiple conflicting objectives at the same time and do not need to divide the original problem into several sub-problems to parallel search [436, 450]. There are many MOEAs adapted from NSGA-II, but most of them have algebraic functional constraints [133, 455, 456]. With conventional MOEAs, we need to search the parameters of $W^* > 0$ to solve the MOP in (15.25). For the noncooperative H_∞ reference tracking game strategy, the Pareto optimal solutions $U_i^*(t) = -R_i^{-1}B_iP^*X_i(t), i = 1,2,...,m, P^* = W^{*-1}$. Since $W \in R^{3n \times 3n}$, it is very difficult to search all components of the $W^* > 0$ from the MOP in (15.25) by conventional MOEAs when n becomes large. To simplify the solving procedure of the MOP in (15.25), a reverse-order LMI-constrained MOEA is proposed to simplify the MOP in (15.25) by searching $(g_1,...,g_i,...,g_m)$ indirectly and then solving W from LMIs in (15.23) and (15.24) via the LMI toolbox in MATLAB instead of searching W directly to solve $(g_1,...,g_i,...,g_m)$. After solving $(g_1^*,...,g_i^*,...,g_m^*)$, we can solve the corresponding $W^* > 0$ from (15.23) and (15.24) with the LMI toolbox in MATLAB indirectly. The detailed design procedure of the reverse-order LMI-constrained MOEA approach for the MOP in (15.25) of the m-player noncooperative minimax H_∞ game strategy design problem in the MFSJD system in (15.1) is given as follows:

Reverse-Order LMI-Constrained MOEA for the m-Player Noncooperative Reference Tracking Game Strategy of the MFSJD System

Step 1: Select the search range $S = [g_1^L, g_1^U] \times \cdots \times [g_i^L, g_i^U] \times \cdots \times [g_m^L, g_m^U]$, where g_i^L and $g_i^U, i = 1, 2, ..., m$ represent the lower bound and upper bound of the Pareto optimal solution $(g_1^*, ..., g_i^*, ..., g_m^*)$. Choose the population number N_p, iteration number N_I, crossover rate R_c, and mutation rate R_m, and set the initial iteration number $I = 1$.

Step 2: Choose N_p feasible individuals as the initial populations K_1 from the search regions.

Step 3: Employ the mutation and crossover operators to generate another N_p feasible individuals and add to the initial population by checking the LMIs in (15.23) and (15.24) to see if their corresponding $(g_1, ..., g_i, ..., g_m)$ are feasible.

Step 4: Choose N_p elite individuals from $2N_p$ feasible individuals generated in step 3 via the nondominated sorting scheme and the crowded comparison method. Place the iteration number $I = I + 1$ and update the population as K_I.

Step 5: Repeat steps 3 and 4 until $I > N_I$; then set the final population $K_I = T_F$ as the Pareto front and stop the iteration.

Step 6: Choose a desired feasible objective vector $(g_1^*, ..., g_i^*, ..., g_m^*) \in T_F$ according to the designer's preference with the optimal $W^* = (P^*)^{-1}$, where W^* is the preferable optimal solution of the MOP in (15.25). Moreover, the corresponding control strategy $(U_1^*, ..., U_i^*, ..., U_m^*)$, $U_i^*(t) = -R_1^{-1} B_i^T P^* X_i(t)$ is obtained for the noncooperative minimax H_∞ reference tracking game strategy in (15.6) for an MFSJD system in (15.1).

Remark 15.5: For the proposed MOEA, the search region S can be chosen as $S = [g_1^L, g_1^U] \times \cdots \times [g_m^L, g_m^U]$. In general, the lower bound $\{g_j^L\}_{i=1}^m$ can be obtained by solving the SOP, which is constructed by selecting one objective with corresponding constraints of the MOP in (15.23)–(15.25). On the other hand, the upper bound $\{g_j^U\}_{i=1}^m$ can be chosen as several times of $\{g_j^L\}_{i=1}^m$; that is, $g_j^U = \alpha g_j^L$, for some $\alpha > 1$.

Remark 15.6: In the proposed reverse-order LMI-constrained MOEA, the constraints are limited to a class of linear matrix inequality constraints. Hence, before using the proposed algorithm for nonlinear constrained MOPs, the nonlinear constraints should be linearized as a set of linear inequality constraints in different local regions. Recently, in the field of control system design, several local linearization interpolation methods have been utilized for controller synthesis of nonlinear systems, such as the fuzzy interpolation method, global linearization method, and so on. Thus, based on these local linearization interpolation methods, the proposed LMI-constrained MOEA can be further applied to solve the multi-objective controller design problem of nonlinear systems.

15.6 SIMULATION EXAMPLES IN CYBER-SOCIAL SYSTEMS

In this section, to demonstrate the effectiveness of the proposed LMI-constrained MOEA for the following MOP of the noncooperative minimax H_∞ tracking game strategy and the SOP of the cooperative minimax H_∞ tracking game strategy for the MFSJD system, we give two simulation examples of cyber-social system based on previous research.

15.6.1 SIMULATION EXAMPLE OF MARKET SHARE ALLOCATION PROBLEM

(1) Noncooperative H_∞ Reference Tracking Game Strategy: Consider a financial resource allocation problem for market share adapted from [457], in which two firms and the government compete with each other for the market share. For such a noncooperative game, the two firms can be seen as players 1 and 2, and the government can be seen as player 3. Players hope they can get the maximum profit by controlling the share price via their purchase or sale of the share. However, the market share suffers from the competitive strategy of the other players, external disturbance, and continuous and discontinuous random events like rumors in the market, changes of policy, tariff impositions, and oil price changes. The market share is also influenced by the average behavior and collective predictions about the market share. A noncooperative minimax H_∞ tracking game in (15.6) with three players on the MFSJD system is used to describe the competitiveness of the market share. The MFSJD system of the market share in [457] is given as follows:

$$dx(t) = [Ax(t) + \bar{A}Ex(t) + \sum_{i=1}^{3} B_i u_i(t) + v(t)]dt$$
$$+ (Lx(t) + \bar{L}Ex(t))dw(t) + (Nx(t) + \bar{N}Ex(t))dp(t) \qquad (15.43)$$
$$x_0 = [20,15]^T, Ex_0 = [23,14.5]^T$$

where the system state $x(t) = \left[x_1(t), x_2(t)\right]^T$ denotes the share price of firms 1 and 2. The mean term $Ex(t)$ denotes the average effect of collective prediction on the market share. The term $(Lx(t) + \bar{L}Ex(t))dw(t)$ can be regarded as the continuous intrinsic fluctuation due to random incidents like persistently fluctuating unemployment in the market share. The term $(Nx(t) + \bar{N}Ex(t))dp(t)$ can be regarded as discontinuous intrinsic fluctuation due to abrupt events like a change of tax rate in the market share. The trajectories of the Wiener process and the Poisson counting process are shown in Figure 15.1. The external disturbance vector $v(t) = 0.01\left[n_1(t), n_2(t)\right]^T$ denotes the effect of environmental events like economic recession and recovery, and $n_k(t)$ represents white Gaussian noise. $A, \bar{A}, L, \bar{L}, N, \bar{N}, B_1, B_2, B_3$ are deterministic real matrices, as follows [457]:

$$A = \begin{bmatrix} -0.6 & -0.2 \\ -0.8 & -0.5 \end{bmatrix}, \bar{A} = \begin{bmatrix} -0.025 & -0.01 \\ -0.06 & -0.03 \end{bmatrix}, L = diag([-0.01, -0.05]),$$
$$\bar{L} = 0.1L, N = diag([-0.03, -0.1]), \bar{N} = 0.1N, B_1 = B_2 = B_3 = I_2$$

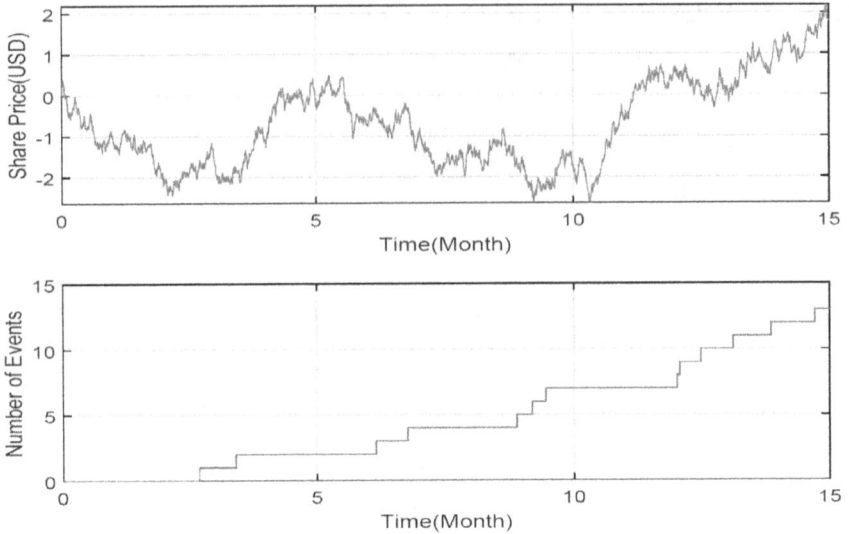

FIGURE 15.1 The Wiener process $w(t)$ and Poisson counting process $p(t)$ in the MFSJD financial system.

The objectives of the firms and the government are to control the share price to track their desired trajectories. The desired trajectory is represented by $x_{r_i}(t)$, generated by the following reference model:

$$dx_{r_i}(t) = (A_{r_i}x_{r_i}(t) + r_i(t))dt, i = 1,...,3$$
$$A_{r1}(t) = A_{r2}(t) = A_{r3}(t) = -I_2, x_{r_i}(0) = x(0), \qquad (15.44)$$
$$r_1(t) = \begin{bmatrix} 150 \\ 15 \end{bmatrix}, r_2(t) = \begin{bmatrix} 15 \\ 150 \end{bmatrix}, r_3(t) = \begin{bmatrix} 50 \\ 50 \end{bmatrix}$$

In Figure 15.2, the desired transient and steady state trajectories of player 1, player 2, and player 3 in (15.44) are presented. Players 1 and 2 both want to increase their own share price but for their share price to be higher than each other's, and player 3 wants players 1 and 2 both to have the same share price.

For the MFSJD financial system in (15.43), suppose the noncooperative minimax H_∞ tracking game strategy design $u_i(t)$ is selected by player i with a parsimonious effort to minimize the worst-case effect of other competitive strategies to control the share price. The payoff function of the noncooperative minimax H_∞ tracking game strategy is given in (15.9) with the weighting matrices as follows:

$$Q_1 = diag(1, 0.1, 1, 0.1, 1, 0.1), R_1 = diag(0.1, 1, 0.1, 1)$$
$$Q_2 = diag(0.1, 1, 0.1, 1, 0.1, 1), R_2 = diag(1, 0.1, 1, 0.1) \qquad (15.45)$$
$$Q_3 = 0.5I_6, R_3 = 0.5I_4$$

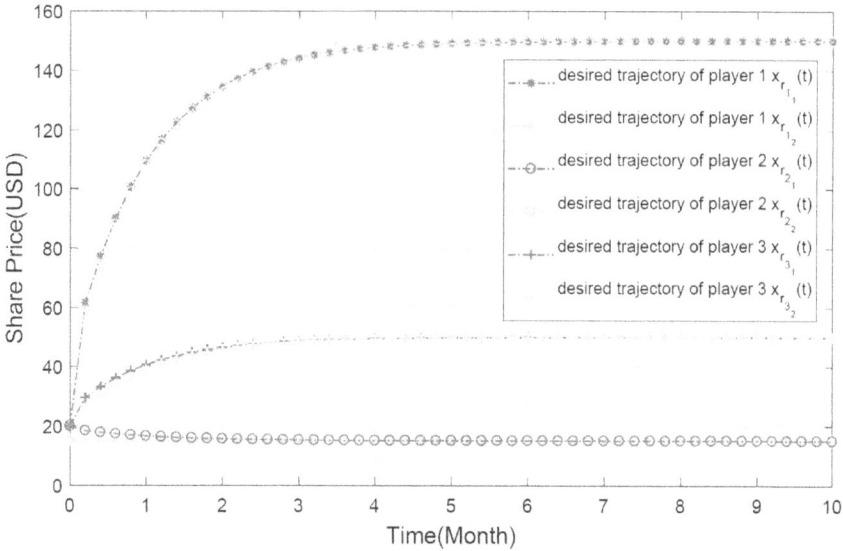

FIGURE 15.2 The desired trajectories $x_{r_i}(t)$ of each player in the noncooperative H_∞ tracking game in the MFSJD financial system.

where the diagonal weightings in Q_i and R_i mean the punishment on the corresponding tracking error and cost of the ith strategy, respectively. If a weighting in Q_i increases, the punishment on the corresponding tracking error increases, and vice versa. If a weighting in R_i decreases, the punishment on the corresponding cost of the ith strategy decreases, and vice versa. The players have to trade off between the tracking error and the cost of a control effort. In general, a high weighting in Q_i and a low weighting in R_i will lead player i to a lower tracking error but higher control effort. In this noncooperative H_∞ tracking game strategy, two firms pay most of their funds to focus on their own share price, and the government is fair with both firms.

By setting the search range $0 \le g_i \le 100, i = 1,...,3$, the population number $N_p = 1500$, the iteration number $N_I = 300$, the crossover rate $C_r = 0.9$, and the mutation rate $m_r = 0.1$, the MOP in (15.25) can be solved via the proposed reverse-order LMI-constrained MOEA. Once the iteration number $N_I = 300$ is achieved, all the Nash equilibrium solutions are on the Pareto front in Figure 15.3. Then, we select $(g_1^*, g_2^*, g_3^*) = (2.221, 2.109, 2.028)$ at the knee solution as preferable solution. Under the concept of Pareto optimality, there are many solutions in a multi-objective problem, and it is difficult to pick the best solution among these Pareto optimal solutions. Within the Pareto optimal solutions of the MOP, the knee solution is a solution that has balanced performance on each objective. Further, the corresponding Pareto optimal strategy $U_i^*(t) = -R_i^{-1} B_i^T P^* X_i(t)$ in (15.19) of the MOP in (15.25) can be obtained for the noncooperative minimax H_∞ tracking game strategy $u_i^*(t)$ of the financial resource allocation in the market share.

Figure 15.4 shows the trajectory and mean state of the original MFSJD system of the market share in (15.43) without the noncooperative minimax H_∞ tracking strategy

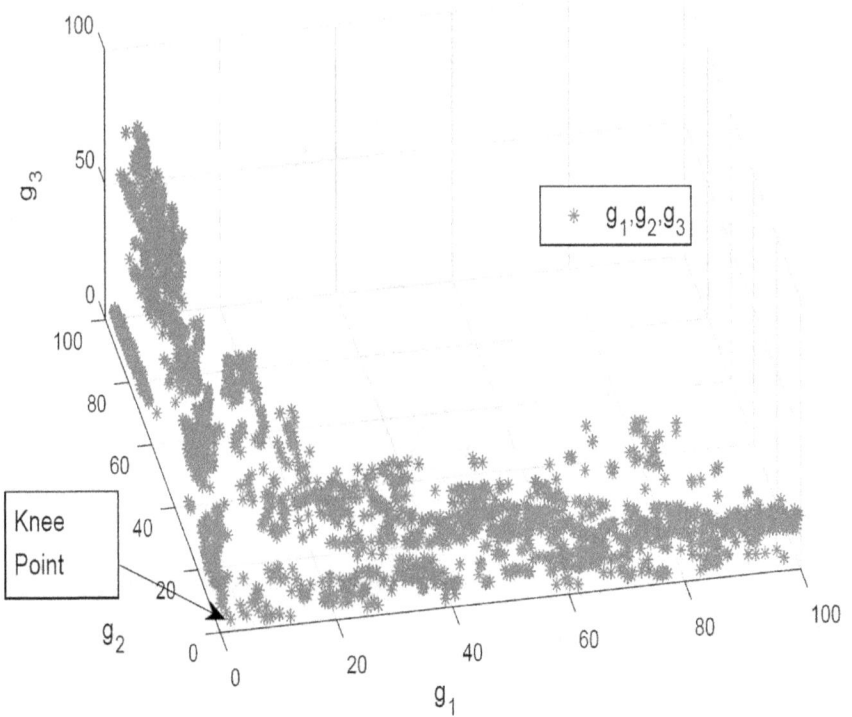

FIGURE 15.3 Pareto front by the proposed LMI-constrained MOEA for the three-player noncooperative minimax H_∞ tracking game in the MFSJD financial system.

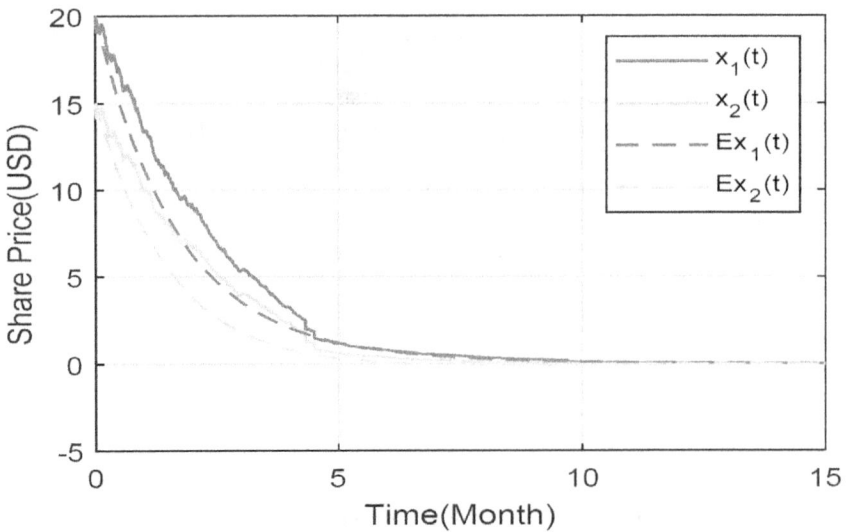

FIGURE 15.4 The trajectories of the MFSJD financial system without the control strategies of the noncooperative minimax H_∞ tracking game in the market share allocation system.

$u_i^*(t)$ but with intrinsic fluctuations due to the Wiener process and Poisson process in Figure 15.1. It can be seen that the trajectories of the original MFSJD system decay to zero. This means that the share price decreases with no investment from the firms and the government.

The trajectories of the MFSJD system of the market share under the noncooperative minimax H_∞ reference tracking game strategy $u_i^*(t)$ are shown in Figure 15.5. However, since none of players have the same goal, none of them can achieve their own target, but they achieve a compromise result because of the competitive strategy. The trajectories of the noncooperative minimax H_∞ reference tracking game strategy $u_i^*(t)$ are in Figure 15.6.

$$dx_{r_i}'(t) = (A_{r_i}'x_{r_i}'(t) + r_i'(t))dt, i = 1,...,3$$

$$A_{r1}'(t) = A_{r2}'(t) = A_{r3}'(t) = -I_2, x_{r_i}(0) = x(0) \qquad (15.46)$$

$$r_1(t) = \begin{bmatrix} 150 \\ 120 \end{bmatrix}, r_2(t) = \begin{bmatrix} 120 \\ 150 \end{bmatrix}, r_3(t) = \begin{bmatrix} 140 \\ 140 \end{bmatrix}$$

From the steady states of the reference model in (15.46), it is obvious that these players have less suppression on their competitive players than in (15.44). The trajectories of the MFSJD system of the market share by the noncooperative minimax H_∞ tracking game strategy are shown in Figure 15.7. It can be seen that the players still cannot achieve their desired goals, but the share prices of their firms are closer to their desired trajectories compared to the noncooperative minimax H_∞ tracking game

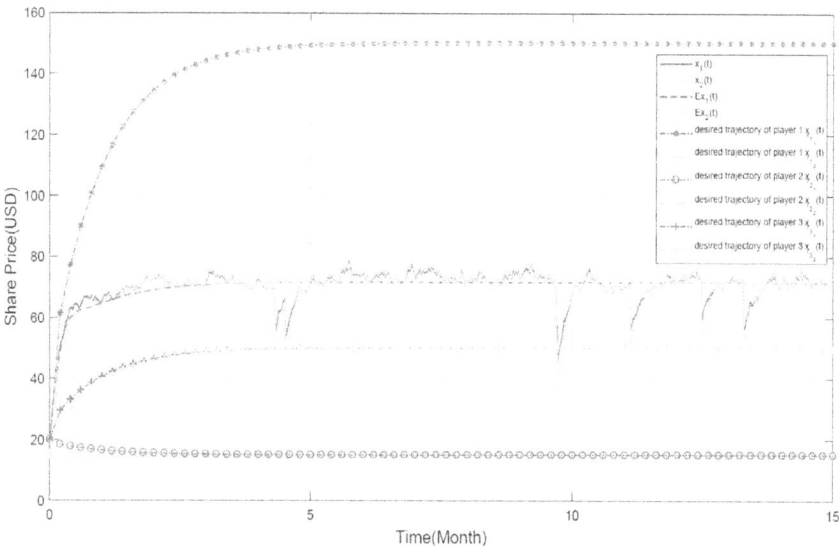

FIGURE 15.5 The trajectories of the MFSJD financial system of market share with desired trajectories $x_{r_i}(t)$ in (15.44) by the noncooperative minimax H_∞ tracking game strategy $u_i^*(t)$ for the resource allocation of market share.

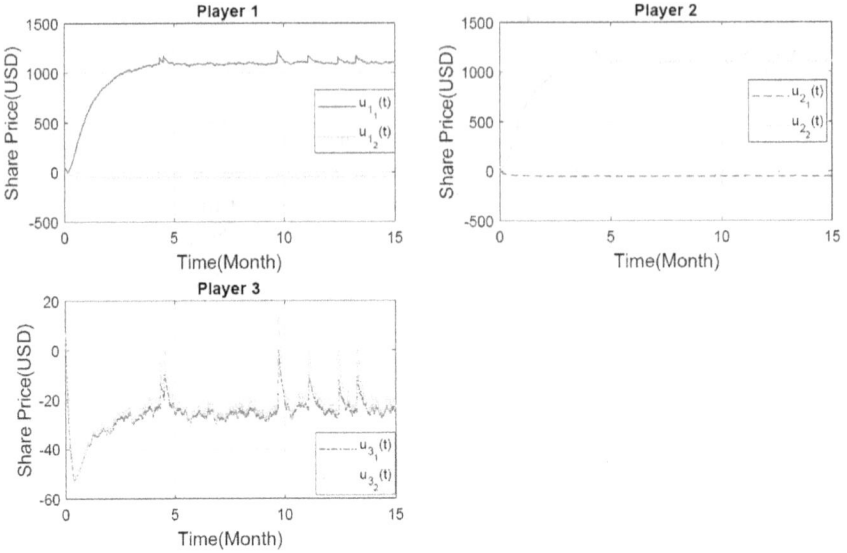

FIGURE 15.6 The noncooperative minimax H_∞ tracking game strategies of each player in the market share allocation system.

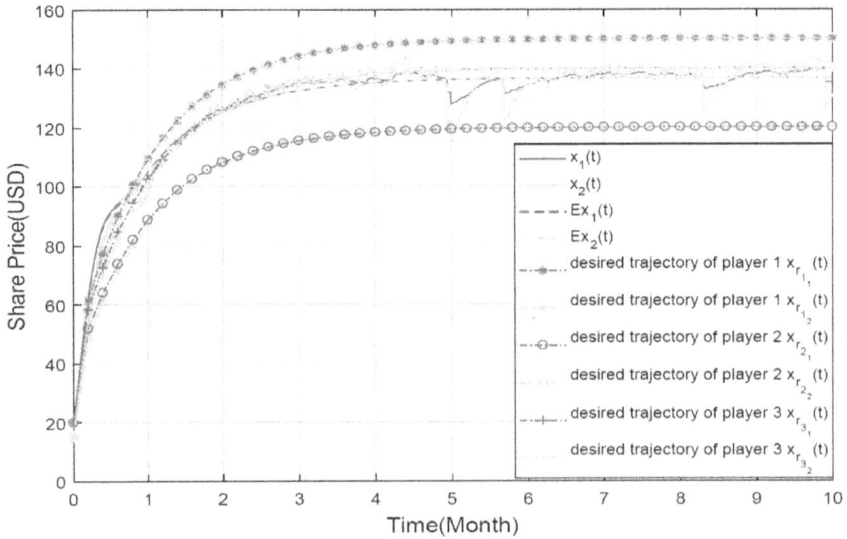

FIGURE 15.7 The trajectories of the MFSJD financial system of market share with different desired trajectories $x_{r_i}'(t)$ in (15.45) by the noncooperative minimax H_∞ tracking game strategy $u_i^*(t)$ for the resource allocation of market share.

with the desired trajectories $x_{r_i}(t)$ in (15.44): this means if these players are willing to suppress on each other less, they will gain more profit.

(2) Cooperative H_∞ Reference Tracking Game Strategy:

Consider the same MFSJD financial system in the previous three-player noncooperative game. Suppose firms 1 and 2 and the government are able to communicate with each other, and they had compromised with a common reference target before the design of the cooperative minimax H_∞ tracking game strategy. Firms 1 and 2 will have the same share price, and the three players will collaborate with each other to reach the common reference target. Since we consider the cooperative minimax H_∞ reference tracking game on the same MFSJD financial resource allocation system in market share, the MFSJD system of market share in (15.43) is modified as follows.

$$
\begin{aligned}
dx(t) &= [Ax(t) + \bar{A}Ex(t) + Bu(t) + v(t)]dt \\
&+ (Lx(t) + \bar{L}Ex(t))dw(t) \\
&+ (Nx(t) + \bar{N}Ex(t))dp(t) \\
x_0 &= [20,15]^T, Ex_0 = [23,14.5]^T
\end{aligned}
\tag{15.47}
$$

The objective of the two firms and the government is to control the share price to track the common trajectory $x_r(t)$, which is generated by the following reference model:

$$
\begin{aligned}
dx_r(t) &= (A_r x_r(t) + r(t))dt \\
x_r(0) &= x(0), A_r(t) = -I_2, r(t) = \begin{bmatrix} 150 \\ 150 \end{bmatrix}
\end{aligned}
\tag{15.48}
$$

The payoff function of the cooperative H_∞ tracking game strategy is given in (15.29), and the weighting matrices are given as $Q = I_6$, $R = I_{12}$.

Compared to the noncooperative minimax H_∞ tracking game, all players choose the same Q and R because these players will collaborate with each other to increase the share of both firms to the same price. We obtain the optimal solution $g^0 = 2.0506$ from the SOP in (15.40)–(15.42). The corresponding trajectory of the MFSJD system in market share by the proposed cooperative minimax H_∞ tracking game strategy $u^0(t)$ is given in Figure 15.8. The cooperative minimax H_∞ tracking game strategy $u^0(t)$ of the MFSJD system of market share is shown in Figure 15.9.

It can be seen that the share prices of both firms in the cooperative minimax H_∞ tracking game strategy are higher than those in the noncooperative minimax H_∞ tracking game strategy. The control effort of the cooperative minimax H_∞ tracking game strategy in Figure 15.9 is apparently less than that of the noncooperative minimax H_∞ tracking game strategy in Figure 15.6. This means if players collaborate with each other, they will gain more profit with less control effort. However, how to compromise on a common reference target is a complicated and time-consuming process because some players need to suffer a loss to achieve a compromise on a common target.

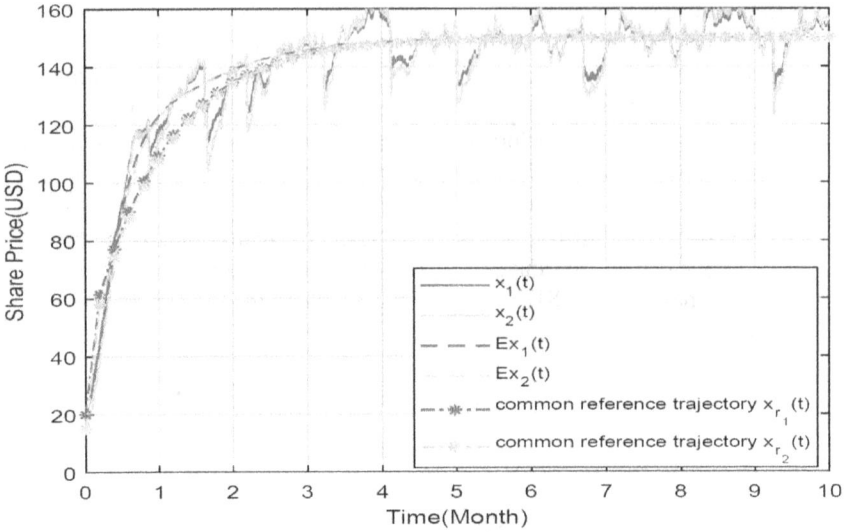

FIGURE 15.8 The trajectories of the MFSJD financial system by cooperative H_∞ tracking game strategy $u_0(t)$ in the resource allocation problem of market share.

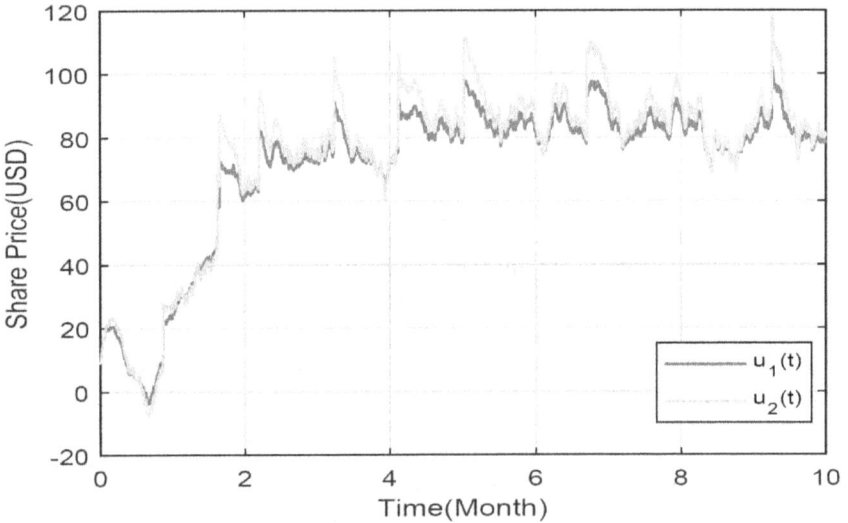

FIGURE 15.9 The cooperative H_∞ tracking game strategy $u_0(t)$ of the resource allocation problem in market share.

15.6.1.1 Non-Cooperative Minmax H_∞ Tracking Game Strategy

Since the beginning of the 21st century, the rapid growth of the internet has changed the world and has dramatically revolutionized many different fields. However, cyber attacks on networks and platforms also invade our privacy, destroy our properties, and even put our lives in danger. Therefore, network security has become an active research field [458, 459]. In the real world, systems like networks and platforms often suffer external disturbances and continuous and discontinuous parameter fluctuations due to attacks by hackers, the persistent interference of computer worms, or sudden slowdowns of the computers in the network. In a noncooperative game strategy, users attack each other to benefit. Individual users can invest money, time, or human resources in their own computer security like anti-virus software or attacks on specific users like denial-of-service attacks. In this noncooperative game strategy, we assume five users are attacking each other to steal each others' properties and protecting their properties. If the security level increases, the properties will be more safe. The MFSJD system of network security is given as follows [460]:

$$dx(t) = [Ax(t) + \bar{A}Ex(t) + \sum_{i=1}^{5} B_i u_i(t) + v(t)]dt$$

$$+ (Lx(t) + \bar{L}Ex(t))dw(t) + (Nx(t) + \bar{N}Ex(t))dp(t) \qquad (15.49)$$

$$x_0 = [5, 4, 6, 7.2, 1]^T, Ex_0 = [4.7, 6.3, 6.8, 2.4, 2]^T$$

where $x(t) = [x_1(t), x_2(t), x_3(t), x_4(t), x_5(t)]^T$ denotes the security level of users 1, 2, 3, 4, and 5, respectively. The mean term $Ex(t)$ denotes the average effect of collective prediction of $x(t)$ by all players. The term $(Lx(t) + \bar{L}Ex(t))dw(t)$ can be regarded as the continuous intrinsic fluctuation due to hidden computer worms or persistent signal interference inside the network. The term $(Nx(t) + \bar{N}Ex(t))dp(t)$ can be regarded as the discontinuous intrinsic change due to abrupt computer slowdown events. The Wiener process and the Poisson counting process in the network security system are given in Figure 15.10. The external disturbance vector $v(t) = [0.01n_1(t), 0.01n_2(t), 0.01n_3(t), 0.01n_4(t), \ 0.01n_5(t)]^T$ denotes unexpected attacks by other hackers outside the network. $A, \bar{A}, L, \bar{L}, N, \bar{N}, B_1, B_2, B_3, B_4 B_5$ are deterministic real matrices as follows [460]:

$$A = -\begin{bmatrix} 0.8 & 0.01 & 0.05 & 0.01 & 0 \\ 0.01 & 0.75 & 0.01 & 0.02 & 0.0043 \\ 0 & 0.1 & 0.1 & 0 & 0.002 \\ 0.1 & 0 & 0.02 & 0.64 & 0.1 \\ 0.01 & 0.01 & 0 & 0.0.1 & 0.72 \end{bmatrix},$$

$$\bar{A} = -\begin{bmatrix} 0.025 & 0.01 & 0.005 & 0.001 & 0 \\ 0.06 & 0.06 & 0.02 & 0.01 & 0.0003 \\ 0.05 & 0.02 & 0.05 & 0.01 & 0.001 \\ 0.03 & 0.01 & 0.005 & 0.03 & 0.04 \\ 0.003 & 0.02 & 0 & 0.2 & 0.5 \end{bmatrix},$$

$$L = -diag(0.03, 0.15, 0.09, 0.06, 0.15), \bar{L} = 0.1L$$

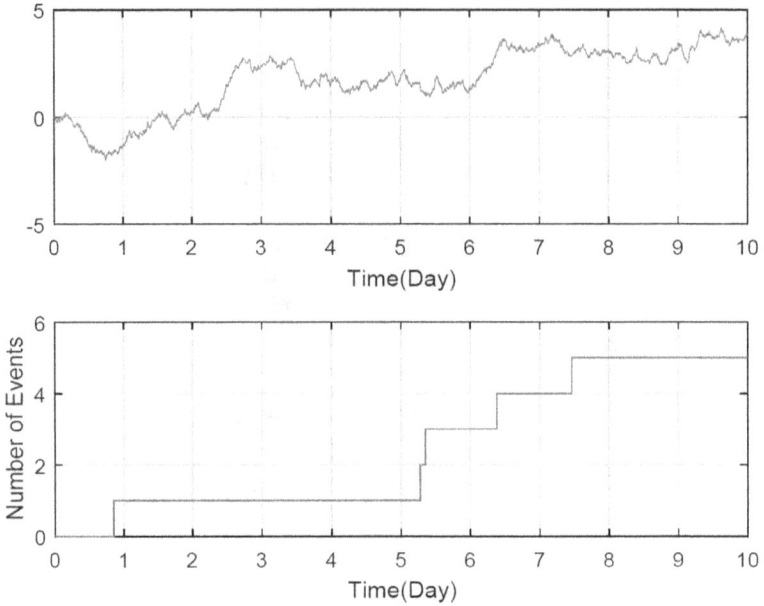

FIGURE 15.10 The Wiener process $w(t)$ and Poisson counting process $p(t)$ in the MFSJD network security system.

$$N = -diag(0.003, 0.009, 0.006, 0.012, 0.09), \; \bar{N} = 0.1N$$

$$B_1 = B_2 = B_3 = B_4 = B_5 = I_5$$

The objective of each user is to control the security level $x(t)$ to track their own desired security level $x_{r_i}(t), i = 1,...,5$. The desired security levels $x_{r_i}(t), i = 1,...,5$ of each user are generated by the following models:

$$dx_{r_i}(t) = (A_{r_i} x_{r_i}(t) + r_i(t))dt$$

$$A_{r_1}(t) = A_{r_2}(t) = A_{r_3}(t) = A_{r_4}(t) = A_{r_5}(t) = -I_5$$

$$r_1(t) = \begin{bmatrix} 150 \\ 15 \\ 15 \\ 15 \\ 15 \end{bmatrix}, r_2(t) = \begin{bmatrix} 15 \\ 150 \\ 15 \\ 15 \\ 15 \end{bmatrix}, r_3(t) = \begin{bmatrix} 15 \\ 15 \\ 150 \\ 15 \\ 15 \end{bmatrix}, r_4(t) = \begin{bmatrix} 15 \\ 15 \\ 15 \\ 150 \\ 15 \end{bmatrix},$$

$$r_5(t) = \begin{bmatrix} 15 \\ 15 \\ 15 \\ 15 \\ 150 \end{bmatrix}, x_{r_i}(0) = x(0)$$

(15.50)

The desired security levels generated by the reference model in (15.50) selected by these players are shown in Figure 15.11. In the noncooperative minmax H_∞ tracking game strategy, each player is an attacker and defender at the same time, so they will try to decrease competitive players' security levels and achieve their desired security levels.

For the MFSJD network security system, the noncooperative minmax H_∞ tracking game strategy design could be seen as the Pareto optimal strategy $u_i^*(t)$ to reach the desired security level despite the worst-case effect of competitive strategy and external disturbance. The payoff function of the cooperative game strategy is given in (15.9), and the weighting matrices are chosen as follows:

$$Q_1 = diag(1,0.1,0.1,0.1,0.1,1,0.1,0.1,0.1,0.1,1,0.1,0.1,0.1,0.1)$$
$$R_1 = diag(0.1,1,1,1,1,0.1,1,1,1,1)$$
$$Q_2 = diag(0.1,1,0.1,0.1,0.1,0.1,1,0.1,0.1,0.1,0.1,1,0.1,0.1,0.1)$$
$$R_2 = diag(1,0.1,1,1,1,1,0.1,1,1,1)$$
$$Q_3 = diag(0.1,0.1,1,0.1,0.1,0.1,0.1,1,0.1,0.1,0.1,0.1,1,0.1,0.1)$$
$$R_3 = diag(1,1,0.1,1,1,1,1,0.1,1,1)$$
$$Q_4 = diag(0.1,0.1,0.1,1,0.1,0.1,0.1,0.1,1,0.1,0.1,0.1,0.1,1,0.1)$$
$$R_4 = diag(1,1,1,0.1,1,1,1,1,0.1,1)$$
$$Q_5 = diag(0.1,0.1,0.1,0.1,1,0.1,0.1,0.1,0.1,1,0.1,0.1,0.1,0.1,1)$$
$$R_5 = diag(1,1,1,1,0.1,1,1,1,1,0.1)$$

$$(15.51)$$

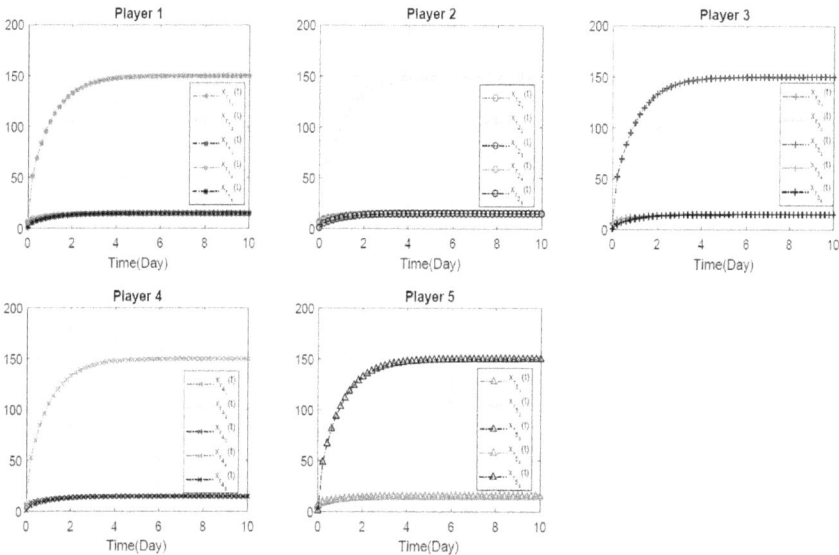

FIGURE 15.11   The desired security levels $x_{r_i}(t)$ of each user in the noncooperative minmax H_∞ tracking game of the MFSJD network security system.

By solving the MOP in (15.25) subject to (15.23) and (15.24) via the proposed LMI-constrained MOEA, we get the Nash equilibrium solutions for the noncooperative minmax H_∞ tracking strategy of the MFSJD network security system. We set the search range $0 \leq g_i \leq 100, i = 1,...,5$, the population number $N_p = 6000$, iteration number $N_i = 600$, crossover rate $C_r = 0.9$, and mutation rate $m_r = 0.1$. Once the iteration number $N_i = 600$ is achieved, we get $(g_1^*, g_2^*, g_3^*, g_4^*, g_5^*)$, P^*, the noncooperative minmax H_∞ tracking strategy $u_i^*(t)$ of player i, and the strategy $u_{-i}^*(t)$ of the other competitive players.

In Table 15.1, D denotes the 2-norm between five Pareto optimal solutions $(g_1^*, g_2^*, g_3^*, g_4^*, g_5^*)$ and the utopia point $(0,0,0,0,0)$. From Table 15.1, we choose the knee point $(g_1^*, g_2^*, g_3^*, g_4^*, g_5^*) = (4.9468, 9.8909, 4.8636, 4.1209, 4.8251)$ as the preferred Pareto optimal solution because its D is the smallest. Figure 15.12 shows the security levels of the MFSJD system without the noncooperative game strategy $U_i^*(t)$ but with the Wiener process and Poisson process. In Figure 15.12, it can be seen that the security levels of the original MFSJD network security system decay to a low security level and suffer from continuous and discontinuous random fluctuations. Figure 15.13 shows the security levels of the MFSJD network security system couldn't achieve a desired security level but a compromised security level when all players employ a noncooperative minmax H_∞ tracking game strategy $u_i^*(t)$. Figure 15.14 shows the noncooperative minmax H_∞ tracking game strategy $u_i^*(t)$ employed by players.

If the reference models of the desired security levels $x_{r_i}(t)$ of each user are changed as follows:

$$dx_{r_i}'(t) = (A_{r_i}' x_{r_i}'(t) + r_i'(t))dt, i = 1,...,5$$
$$A_{r_1}'(t) = A_{r_2}'(t) = A_{r_3}'(t) = A_{r_4}'(t) = A_{r_5}'(t) = -I_5$$

$$r_1(t) = \begin{bmatrix} 150 \\ 15 \\ 15 \\ 15 \\ 15 \end{bmatrix}, r_2(t) = \begin{bmatrix} 15 \\ 150 \\ 15 \\ 15 \\ 15 \end{bmatrix}, r_3(t) = \begin{bmatrix} 15 \\ 15 \\ 150 \\ 15 \\ 15 \end{bmatrix}, r_4(t) = \begin{bmatrix} 15 \\ 15 \\ 15 \\ 150 \\ 15 \end{bmatrix},$$

$$r_5(t) = \begin{bmatrix} 15 \\ 15 \\ 15 \\ 15 \\ 150 \end{bmatrix}, x_{r_i}(0) = x(0)$$

$$(15.52)$$

The security levels of the MFSJD network security system to track the desired security level $x_{r_i}'(t)$ by the noncooperative minmax H_∞ tracking game strategy are shown in Figure 15.15. Similar to the previous example, users still cannot achieve their desired goals, but the security level of each user is closer to their desired security level compared to the noncooperative minmax H_∞ tracking game with the desired

TABLE 15.1

Five Pareto Optimal Solutions ($g_1^*, g_2^*, g_3^*, g_4^*, g_5^*$) and Their 2-Norm to Utopia Point

g_1^*	g_2^*	g_3^*	g_4^*	g_5^*	DD
4.9468	9.8909	4.8636	4.1209	4.8251	13.6462
8.8841	8.0943	4.2297	6.2092	3.8510	14.6874
9.0738	8.1948	4.2541	6.3419	3.8241	14.9141
9.6244	5.6546	8.8048	8.4733	4.3032	17.1009
6.6749	7.9190	9.9911	8.7724	7.3318	18.3793

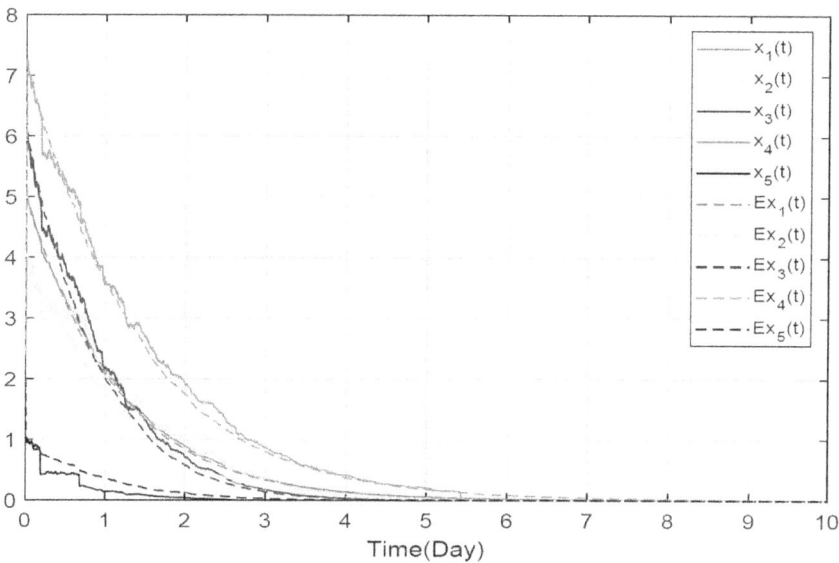

FIGURE 15.12 The security levels of MFSJD network security system without the noncooperative minmax H_∞ tracking game strategy $u_i^*(t)$.

security level $x_{r_i}(t)$ in (15.50). This means if all players are willing to attack each other less, they will be more safe.

15.6.1.2 Cooperative Minmax H_∞ Tracking Game Strategy

Consider the cooperative minmax H_∞ tracking game strategy with the same MFSJD network security (15.49). Users may cooperate with each other to protect the network. The users are interdependent, so if an individual user's security level increases, the other players will also benefit. If an individual user's security level decreases, the probability of other players being infected with a virus will increase. However, individual users may tend to shirk from their duties of maintaining the security level,

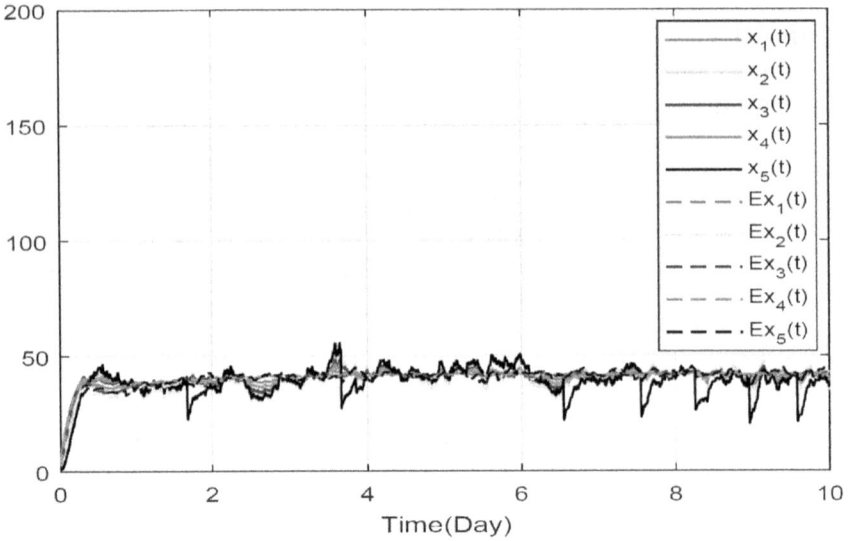

FIGURE 15.13 The security levels of MFSJD network security system with the noncooperative minmax H_∞ tracking game strategy $u_i^*(t)$.

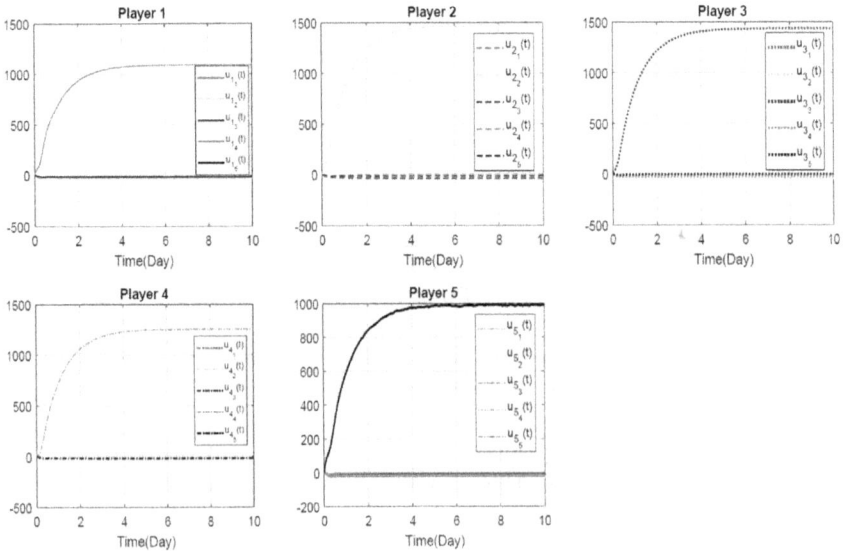

FIGURE 15.14 The noncooperative minmax H_∞ tracking game strategy $u_i^*(t)$ of each user in the MFSJD network security system.

which results in vulnerability of the network. To secure the basic level of network security, we assume the users in the network communicate with each other at first to reach a compromise on the security level; then they are obligated to protect the

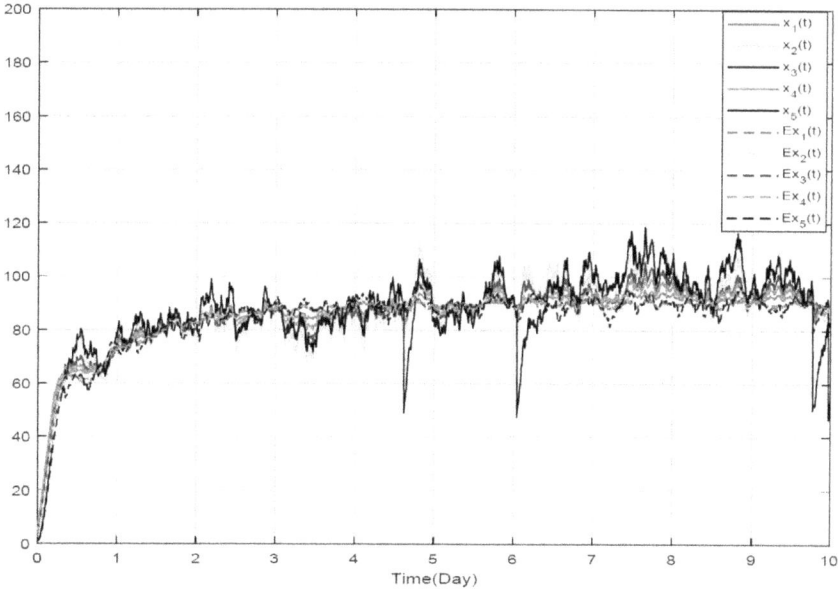

FIGURE 15.15 The security levels of the MFSJD network security system with different desired security levels $x_{r_i}'(t)$ in the noncooperative minmax H_∞ tracking game of the network security problem.

network system. In this situation, the MFSJD network security system in (15.49) is modified as follows:

$$dx(t) = [Ax(t) + \bar{A}Ex(t) + Bu(t) + v(t)]dt$$
$$+ (Lx(t) + \bar{L}Ex(t))dw(t) + (Nx(t) + \bar{N}Ex(t))dp(t) \qquad (15.53)$$
$$x_0 = [5,4,6,7.2,1]^T, Ex_0 = [4.7,6.3,6.8,2.4,2]^T$$

All users with the cooperative minmax H_∞ tracking game strategy are defending each other, so they want to have the same security level. For the MFSJD network security system, the cooperative minmax H_∞ tracking game strategy design could be seen as the optimal strategy $u(t)$ to reach the same security level despite external disturbance and continuous and discontinuous intrinsic fluctuations. The payoff function of the cooperative H_∞ tracking game strategy is given in (15.29) with the weighting matrices as $Q = I_{15}, R = I_{50}$.

We solve the optimal solution $(g^0) = 2.8674$ for the cooperative minmax H_∞ tracking game from the LMI-constrained SOP in (15.40). Figure 15.16 shows the security levels of the MFSJD network security system by the cooperative minmax H_∞ tracking game strategy $U^0(t)$. The cooperative minmax H_∞ tracking game strategy $U^0(t)$ employed by players is shown in Figure 15.17. Similar to the previous example, the cooperative H_∞ tracking game can have a higher security level with less control effort. This means the users in the network can be safer with less investment.

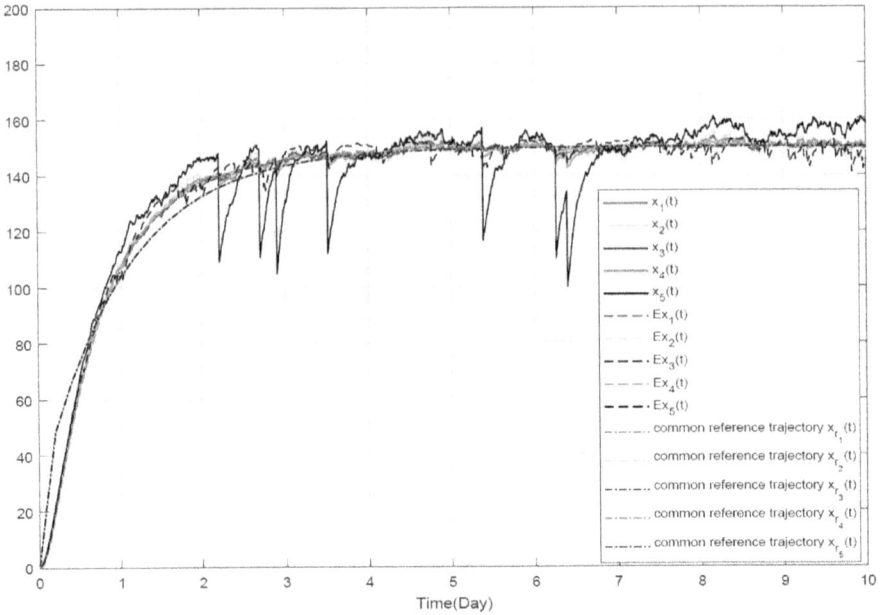

FIGURE 15.16 The security levels of the MFSJD network security system with the cooperative minmax H_∞ tracking game strategy $U_0(t)$.

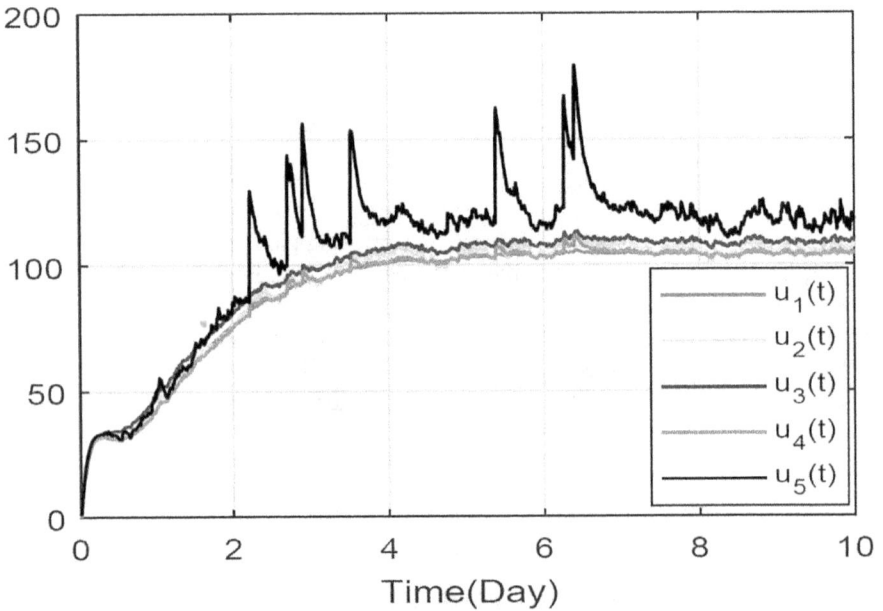

FIGURE 15.17 Cooperative minmax H_∞ tracking game strategy $U_0(t)$ of the network security problem.

15.7 CONCLUSION

This chapter has investigated multi-player stochastic noncooperative minimax H_∞ tracking games with a partially conflicting target strategy and cooperative minimax H_∞ tracking game with a common target strategy in MFSJD systems with external disturbance. With the proposed indirect method, the multi-player noncooperative and cooperative minimax H_∞ tracking game strategy design problems in the MFSJD system can be transformed into an equivalent LMI-constrained MOP and SOP for the MFSJD system, respectively. To efficiently search the Pareto optimal solutions of the LMI-constrained MOP for all the Nash equilibrium solutions of the multi-player noncooperative H_∞ tracking game, the LMI-constrained MOEA is also proposed. From the simulation results of two cyber-social systems, the proposed Pareto optimal solution is effective, and all players in the minimax H_∞ noncooperative and cooperative tracking games have a reasonable result. In the future, due to the large number agents in cyber-social systems, the behavior of large cyber-social systems can be described as a mean-field system and should be considered in the system design of the individual agent. From a practical point of view, the proposed stochastic mean-field system can be used to describe a more realistic cyber-social system with a large population (agents), such as a power grid system, autonomous vehicle network system, and so on. Meanwhile, the proposed noncooperative and cooperative tracking strategy design can be directly applied to the investment and management of these realistic large-scale cyber-social systems. Future research will focus on the mean-field game-based control strategy design for these large-scale cyber-social systems.

15.8 APPENDIX

15.8.1 PROOF OF THEOREM 15.2

Because of the variation subsystem $\tilde{x}(t)$ in (15.3) and the tracking error dynamic $e_i(t)$ of the mean subsystem in (15.5), the Lyapunov energy function of the augmented linear MFSJD system in (15.7) is united by energy functions of three subsystems as $L(X_i(t)) = \tilde{x}^T(t)P_1\tilde{x}(t) + e_i^T(t)P_2e_i(t) + x_{r_i}^T(t)P_3x_{r_i}(t) = X_i^T(t)PX_i(t)$ where $P_1 > 0$, $P_2 > 0$, and $P_3 > 0$ are positive-definite symmetric matrices and $P = diag(P_1, P_2, P_3)$. Then, from (15.15), the function $J_i(U_i(t), U_{-i}(t))$ can be rewritten as follows:

$$
\begin{aligned}
&J_i(U_i(t), U_{-i}(t)) \\
&= E\int_0^{t_f}[X_i^T(t)Q_iX_i(t) + U_i^T(t)R_iU_i(t) - g_iU_{-i}^T(t)U_{-i}(t)]dt \\
&= E\int_0^{t_f}[X_i^T(t)Q_iX_i(t) + U_i^T(t)R_iU_i(t) - g_iU_{-i}^T(t)U_{-i}(t)]dt \\
&\quad + dL(X_i(t)) + E[X_i^T(0)PX_i(0)] - E[X_i^T(t_f)PX_i(t_f)]
\end{aligned}
\tag{15.54}
$$

By applying the Itô–Lévy lemma in (15.18), we get:

$$d[L(X_i(t))]$$
$$= \{X_i^T(t)P[F_iX_i(t) + B_iU_i(t) + B_{-i}U_{-i}(t)]$$
$$+ [F_iX_i(t) + B_iU_i(t) + B_{-i}U_{-i}(t)]^T PX_i(t)$$
$$+ X_i^T(t)H_i^T PH_iX_i(t)\}dt + 2X_i^T(t)P[H_iX_i(t)]dw(t)$$
$$+ \{[X_i(t) + T_iX_i(t)]^T P[X_i(t) + T_iX_i(t)] - X_i^T(t)P_i(t)]\}dp(t)$$

(15.55)

Then, by integrating (15.55) from $t = 0$ to $t = t_f$ and substituting it into (15.54), we have the equation:

$$J_i(U_i(t), U_{-i}(t))$$
$$= E\int_0^{t_f} [[X_i^T(t)Q_iX_i(t) + U_i^T(t)R_iU_i(t) - g_iU_{-i}^T(t)U_{-i}(t)$$
$$+ X_i^T(t)P(F_iX_i(t) + B_iU_i(t) + B_{-i}U_{-i}(t))$$
$$+ (F_iX_i(t) + B_iU_i(t) + B_{-i}U_{-i}(t))^T PX_i(t)$$
$$+ X_i^T(t)H_i^T PH_iX_i(t)]dt + 2X_i^T(t)P(H_iX_i(t))dw(t)$$
$$+ [(X_i(t) + T_iX_i(t))^T P(X_i(t) + T_iX_i(t)) - X_i^T(t)P_iX_i(t))]dp(t)]$$
$$+ E[X_i^T(0)PX_i(0)] - E[X_i^T(t_f)PX_i(t_f)]$$

(15.56)

By the fact that $E[dw(t)] = 0$ and $E[dp(t)] = \lambda dt$, (15.56) can be reformulated as:

$$U_i^T(t)R_iU_i(t) + 2X_i^T(t)PB_iU_i(t)$$
$$= (U_i(t) + R_i^{-1}B_i^T PX_i(t))^T R_i(U_i(t) + R_i^{-1}B_i^T PX_i(t))$$
$$- X_i^T(t)PB_iR_i^{-1}B_i^T PX_i(t)$$

By completing the square of the terms $U_i(t)$ and $U_{-i}(t)$, we have

$$U_i^T(t)R_iU_i(t) + 2X_i^T(t)PB_iU_i(t)$$
$$= (U_i(t) + R_i^{-1}B_i^T PX_i(t))^T R_i(U_i(t) + R_i^{-1}B_i^T PX_i(t))$$
$$- X_i^T(t)PB_iR_i^{-1}B_i^T PX_i(t)$$

(15.57)

$$-g_iU_{-i}^T(t)U_{-i}(t) + 2X_i^T(t)PB_{-i}U_{-i}(t)$$
$$= -g_i(U_{-i}(t) - \tfrac{1}{g_i}B_{-i}^T PX_i(t))^T (U_{-i}(t) - \tfrac{1}{g_i}B_{-i}^T PX_i(t))$$
$$+ \tfrac{1}{g_i}X_i^T(t)PB_{-i}B_{-i}^T PX_i(t)$$

(15.58)

By substituting (15.57) and (15.58) into (15.56), (15.56) can be rewritten as:

$$
\begin{aligned}
J_i&(U_i(t), U_{-i}(t)) \\
&= E\int_0^{t_f} [X_i^T(t)[Q_i + \lambda(T_i^T PT_i + PT_i + T_i^T P) + F_i^T P + PF_i \\
&\quad + H_i^T PH_i]X_i(t) + (U_i(t) + R_i^{-1}B_i^T PX_i(t))^T R_i(U_i(t) + R_i^{-1}B_i^T PX_i(t)) \\
&\quad - g_i(U_{-i}(t) - \tfrac{1}{g_i}B_{-i}^T PX_i(t))^T (U_{-i}(t) - \tfrac{1}{g_i}B_{-i}^T PX_i(t)) \\
&\quad - X_i^T(t)PB_iR_i^{-1}B_i^T PX_i(t) + \tfrac{1}{g_i}X_i^T(t)PB_{-i}B_{-i}^T B_{-i}^T PX_i(t)]dt \\
&\quad + E[X_i^T(0)PX_i(0)] - E[X_i^T(t_f)PX_i(t_f)]
\end{aligned}
\tag{15.59}
$$

Then by selecting strategies in (15.19), we have

$$
\begin{aligned}
J_i^* &= \min_{U_i(t)} \max_{U_{-i}(t)} J_i(U_i(t), U_{-i}(t)) \\
&= E\int_0^{t_f} [X_i^T(t)[Q_i + \lambda(T_i^T PT_i + PT_i + T_i^T P) + F_i^T P + PF_i + H_i^T PH_i - PB_iR_i^{-1}B_i^T P \\
&\quad + \tfrac{1}{g_i}PB_{-i}B_{-i}^T P]X_i(t)]dt + E[X_i^T(0)PX_i(0)] - E[X_i^T(t_f)PX_i(t_f)]
\end{aligned}
$$

Next, if the Riccati-like inequalities in (15.21) and matrix inequalities in (15.22) hold, we have

$$
J_i^* \le E[X_i^T(0)PX_i(0)] - E[X_i^T(t_f)PX_i(t_f)] \le E[X_i^T(0)PX_i(0)] \le g_iX_i^T(0)PX_i(0)
$$

References

[1] K. Deb, *Multi-Objective Optimization Using Evolutionary Algorithms*, Wiley, New York, NY and Chichester, 2001.

[2] K. Deb, A. A. Pratap, S. Agarwal and T. Meyarivan, "A fast and elitist multi-objective genetic algorithm: NSGA-II," *IEEE Trans. Evol. Comput.*, vol. 6, no. 2, pp. 182–197, 2002.

[3] A. Abraham and R. Goldberg, *Evolutionary Multi-Objective Optimization: Theoretical Advances and Applications*. Springer-Verlag, London, 2005.

[4] C. A. C. Coello, G. B. Lamont, D. A. Van Veldhuizen et al., *Evolutionary Algorithms for Solving Multi-Objective Problems*, Springer, Berlin, 2007.

[5] E. Zitzler, K. Deb and L. Thiele, "Comparison of multi-objective evolutionary algorithms: Empirical results," *Evol. Comput.*, vol. 8, no. 2, pp. 173–195, Jun., 2000.

[6] B. S. Chen and S. J. Ho, "Multi-Objective tracking control design of T-S fuzzy systems: Fuzzy Pareto optimal approach," *Fuzzy Sets Syst.*, vol. 290, pp. 39–55, 2016.

[7] C. F. Wu, B. S. Chen and W. Zhang, "Multi-Objective H2/H∞ control design of the nonlinear mean-field stochastic jump-diffusion systems via fuzzy approach," *IEEE Trans. Fuzzy Syst.*, vol. 27, no. 4, pp. 686–700, 2019.

[8] B. S. Chen, M. Y. Lee, W. Y. Chen and W. Zhang, "Reverse-order multi-objective evolution algorithm for multi-objective observer-based fault-tolerant control of T-S fuzzy systems," *IEEE Access*, vol. 9, pp. 1556–1574, 2020.

[9] B. S. Chen, W. H. Chen and H. L. Wu, "Robust H2/H∞ global linearization filter design for nonlinear stochastic systems," *IEEE Trans. Circuits Syst. I, Reg. Papers*, vol. 56, no. 7, pp. 1441–1454, 2009.

[10] B. S. Chen and C. F. Wu, "Robust scheduling filter design for a class of nonlinear stochastic signal systems," *IEEE Trans. Signal Process.*, vol. 63, no. 23, pp. 6245–6257, 2015.

[11] C. F. Wu, B. S. Chen and W. H. Zhang, "Multi-Objective investment policy for a nonlinear stochastic financial system: A fuzzy approach," *IEEE Trans. Fuzzy Syst.*, vol. 25, no. 2, pp. 460–474, 2017.

[12] C. F. Wu, B. S. Chen and W. H. Zhang, "Multi-Objective control for nonlinear stochastic Poisson jump-diffusion systems via T-S fuzzy interpolation and Pareto optimal scheme," *Fuzzy Sets Syst.*, vol. 385, pp. 148–168, 2020.

[13] B. S. Chen, C. T. Yang and M. Y. Lee, "Multiplayer noncooperative and cooperative minimax H∞ tracking game strategies for linear mean-field stochastic systems with applications to cyber-social systems," *IEEE Cybern.*, no. 9216585, pp. 1–13, 2020.

[14] C. Lin and B. S. Chen, "Achieving pareto optimal power tracking control for interference limited wireless systems via multi-objective H2/H∞ optimization," *IEEE Trans. Wireless Commun.*, vol. 12, no. 12, pp. 6154–6165, 2013.

[15] B. S. Chen, H. C. Lee and C. F. Wu, "Pareto optimal filter design for nonlinear stochastic fuzzy systems via multi-objective H2/H∞ optimization," *IEEE Trans. Fuzzy Syst.*, vol. 23, no. 2, pp. 387–399, 2015.

[16] B. S. Chen, M. Y. Lee and X. H. Chen, "Security-enhanced filter design for stochastic systems under malicious attack via smoothed signal model and multi-objective estimation method," *IEEE Trans. Signal Process.*, vol. 68, pp. 4971–4986, 2020.

[17] W. Y. Chen, P. Y. Hsieh and B. S. Chen, "Multi-objective power minimization design for energy efficiency in multicell multiuser MIMO beamforming system," *IEEE Trans. Green Commun. Netw.*, vol. 4, no. 1, pp. 31–45, 2019.

[18] W. Y. Chen, B. S. Chen and W. T. Chen, "Multi-Objective beamforming power control for robust SINR target tracking and power efficiency in multicell MU-MIMO wireless system." *IEEE Trans. Vehicular Technology*, vol. 69, no. 6, pp. 6200–6214, 2020.

[19] S. P. Boyd, L. El Ghaoui, E. Feron and V. Balakrishnan, *Linear Matrix Inequalities in Systems and Control Theory*, SIAM Press, Philadelphia, PA, 1994.

[20] B. Øksendal and A. Sulem, *Applied Stochastic Control of Jump Diffusions*, vol. 498, Springer, Berlin, 2005.

[21] D. Applebaum, *Lévy Processes and Stochastic Calculus*, Cambridge University Press, Cambridge, 2009.

[22] Y. Aït-Sahalia, J. Cacho-Diaz and R. J. A. Laeven, "Modeling financial contagion using mutually exciting jump processes," *J. Financ. Econ.*, vol. 117, no. 3, pp. 585–606, 2015.

[23] A. Angius, G. Balbo, M. Beccuti, E. Bibbona, A. Horvath and R. Sirovich, "Approximate analysis of biological systems by hybrid switching jump diffusion," *Theor. Comput. Sci.*, vol. 587, pp. 49–72, 2015.

[24] P. Parpas and M. Webster, "A stochastic minimum principle and an adaptive pathwise algorithm for stochastic optimal control," *Automatica*, vol. 49, no. 6, pp. 1663–1671, 2013.

[25] E. Todorov, "Stochastic optimal control and estimation methods adapted to the noise characteristics of the sensorimotor system," *Neural Comput.*, vol. 17, no. 5, pp. 1084–1108, 2005.

[26] H. Wang and H. Zhang "LQ control for Itô-type stochastic systems with input delays," *Automatica*, vol. 49, no. 12, pp. 3538–3549, 2013.

[27] S. Chae and S. K. Nguang, "SOS based robust H_∞ fuzzy dynamic output feedback control of nonlinear networked control systems," *IEEE Trans. Cybern.*, vol. 44, no. 7, pp. 1204–1213, 2014.

[28] S. Saat and S. Nguang, "Nonlinear H_∞ output feedback control with integrator for polynomial discrete-time systems," *Int. J. Robust Nonlinear Control*, vol. 25, no. 7, pp. 1051–1065, 2015.

[29] W. Zhang and B. S. Chen, "State feedback H_∞ control for a class of nonlinear stochastic systems," *SIAM J. Control Optim.*, vol. 44, no. 6, pp. 1973–1991, 2006.

[30] R. Rishel, *A Minimum Principle for Controlled Jump Processes*, Springer, Berlin, 1975, pp. 493–508.

[31] S. Tang and X. Li, "Necessary conditions for optimal control of stochastic systems with random jumps," *SIAM J. Control Optim.*, vol. 32, no. 5, pp. 1447–1475, 1994.

[32] X. Lin and R. Zhang, "H_∞ control for stochastic systems with Poisson jumps," *J. Syst. Sci. Complexity*, vol. 24, no. 4, pp. 683–700, 2011.

[33] X. Lin, W. Zhang and X. Wang, "The output feedback H_∞ control design for the linear stochastic system driven by both Brownian motion and Poisson jumps: A nonlinear matrix inequality approach," *Asian J. Control*, vol. 15, no. 4, pp. 1139–1148, 2013.

[34] B. S. Chen and W. Zhang, "Stochastic H_2/H_∞ control with state-dependent noise," *IEEE Trans. Autom. Control*, vol. 49, no. 1, pp. 45–57, 2004.

[35] C. S. Tseng and B. S. Chen, "A mixed H_2/H_∞ adaptive tracking control for constrained non-holonomic systems," *Automatica*, vol. 39, no. 6, pp. 1011–1018, 2003.

[36] W. Zhang, Y. Huang and L. Xie, "Infinite horizon stochastic H_2/H_∞ control for discrete-time systems with state and disturbance dependent noise," *Automatica*, vol. 44, no. 9, pp. 2306–2316, 2008.

[37] P. P. Khargonekar and M. A. Rotea, "Mixed H_2/H_∞ control: A convex optimization approach," *IEEE Trans. Autom. Control*, vol. 36, no. 7, pp. 824–837, 1991.

[38] Z. Feng, Q. Zhang, Q. Tang, T. Yang and J. Ge, "Control-structure integrated multi-objective design for flexible spacecraft using MOEA/D," *Structural Multidisciplinary Optimization*, vol. 50, no. 2, pp. 347–362, 2014.

[39] M. Klug, E. B. Castelan, V. J. S. Leite and L. F. P. Silva, "Fuzzy dynamic output feedback control through nonlinear Takagi–Sugeno models," *Fuzzy Sets Syst.*, 263, pp. 92–111, 2015.

[40] W. H. Chen and B. S. Chen, "Robust stabilization design for nonlinear stochastic system with Poisson noise via fuzzy interpolation method," *Fuzzy Sets Syst.*, 217, pp. 41–61, 2013.

[41] P. Tankov and R. *Cont, Financial Modelling with Jump Processes*, Chapman and Hall/CRC, Boca Raton, FL, 2003.

[42] G. Reynoso-Meza, X. Blasco, J. Sanchis and M. Martínez, "Controller tuning using evolutionary multi-objective optimisation: Current trends and applications," *Control Eng. Pract.*, vol. 28, no. 1, pp. 58–73, 2014.

[43] C. F. Wu and B. S. Chen, "Multi-Objective H_2/H_∞ control for nonlinear stochastic jump diffusion systems," *2014 Proceedings of the SICE Annual Conference (SICE)*, Sapporo, 2014.

[44] R. Khasminskii, *Stochastic Stability of Differential Equations*, vol. 66, Springer Science and Business Media, New York, NY, 2011.

[45] L. Sheng, M. Gao, W. Zhang and B. S. Chen, "Infinite horizon H_∞ control for nonlinear stochastic Markov jump systems with (x, u, v)-dependent noise via fuzzy approach," *Fuzzy Sets Syst.*, vol. 273, pp. 105–123, 2015.

[46] S. Boyd, L. El Ghaoui, E. Feron and V. Balakrishnan, "Linear matrix inequalities in system and control theory," *SIAM*, vol. 15, 1994.

[47] J. Moon, K. Kim and Y. Kim, "Design of missile guidance law via variable structure control," *J. Guid. Control Dyn.*, vol. 24, no. 4, pp. 659–664, 2001.

[48] J. J. E. Slotine and W. Li, *Applied Nonlinear Control*, Prentice Hall, Englewood Cliffs, NJ, 1991.

[49] H. K. Khalil, *Nonlinear Systems, 2nd ed.*, Prentice Hall, Upper Saddle River and Englewood Cliffs, NJ, 1996.

[50] F. H. Hsiao, J. D. Hwang, C. W. Chen and Z. R. Tsai, "Robust stabilization of nonlinear multiple time-delay large-scale systems via decentralized fuzzy control," *IEEE Trans. Fuzzy Syst.*, vol. 13, no. 1, pp. 152–163, 2005.

[51] F. H. Hsiao, S. D. Xu, C. Y. Lin and Z. R. Tsai, "Robustness design of fuzzy control for nonlinear multiple time-delay large-scale systems via neural-network-based approach," *IEEE Trans. Syst. Man Cybern., Part B, Cybern.*, vol. 38, no. 1, pp. 244–251, 2008.

[52] B. S. Chen, C. H. Lee and Y. C. Chang, "H_∞ tracking design of uncertain nonlinear SISO systems: Adaptive fuzzy approach," *IEEE Trans. Fuzzy Syst.*, vol. 4, no. 1, pp. 32–43, 1996.

[53] C. S. Tseng, B. S. Chen and H. J. Uang, "Fuzzy tracking control design for nonlinear dynamic systems via T–S fuzzy model," *IEEE Trans. Fuzzy Syst.*, vol. 9, no. 3, pp. 381–392, 2001.

[54] B. S. Chen, H. J. Uang and C. S. Tseng, "Robust tracking enhancement of robot systems including motor dynamics: A fuzzy-based dynamic game approach," *IEEE Trans. Fuzzy Syst.*, vol. 6, no. 4, pp. 538–552, 1998.

[55] W. J. Wang and H. R. Lin, "Fuzzy control design for the trajectory tracking on uncertain nonlinear systems," *IEEE Trans. Fuzzy Syst.*, vol. 7, no. 1, pp. 53–62, 1999.

[56] W. Y. Wang, C. M. L. Chan, C. C. J. Hsu and T. T. Lee, "H_∞ tracking-based sliding mode control for uncertain nonlinear systems via an adaptive fuzzy-neural approach," *IEEE Trans. Syst. Man Cybern., Part B, Cybern.*, vol. 32, no. 4, pp. 483–492, 2002.

[57] W. Y. Wang, M. L. Chan, T. T. Lee and C. H. Liu, "Adaptive fuzzy control for strict-feedback canonical nonlinear systems with H_∞ tracking performance," *IEEE Trans. Syst. Man Cybern., Part B, Cybern.*, vol. 30, no. 6, pp. 878–885, 2000.

[58] S. F. Su, J. C. Chang and S. S. Chen, "The study on direct adaptive fuzzy controllers," *Int. J. Fuzzy Syst.*, vol. 8, no. 3, pp. 150–159, 2006.

[59] Y. T. Chang and B. S. Chen, "A fuzzy approach for robust reference-tracking-control design of nonlinear distributed parameter time-delayed systems and its application," *IEEE Trans. Fuzzy Syst.*, vol. 18, no. 6, pp. 1041–1057, 2010.

[60] T. S. Li, S. C. Tong and G. Feng, "A novel robust adaptive-fuzzy-tracking control for a class of nonlinear multi-input/multi-output systems," *IEEE Trans. Fuzzy Syst.*, vol. 18, no. 1, pp. 150–160, 2010.

[61] Y. C. Hsueh, S. F. Su, C. W. Tao and C. C. Hsiao, "Robust L_2-gain compensative control for direct-adaptive fuzzy-control-system design," *IEEE Trans. Fuzzy Syst.*, vol. 18, no. 4, pp. 661–673, 2010.

[62] R. Huang, Y. Lin and Z. Lin, "Robust H_∞ fuzzy observer-based tracking control design for a class of nonlinear stochastic Markovian jump systems," *Asian J. Control*, vol. 14, no. 2, pp. 512–526, 2012.

[63] B. S. Chen, Y. P. Lin and Y. J. Chuang, "Robust H_∞ observer-based tracking control of stochastic immune systems under environmental disturbances and measurement noises," *Asian J. Control*, vol. 13, no. 5, pp. 667–690, 2011.

[64] S. Wen, Z. Zeng and T. Huang, "Robust H_∞ output tracking control for fuzzy networked systems with stochastic sampling and multiplicative noise," *Nonlinear Dyn.*, vol. 70, no. 2, pp. 1061–1077, 2012.

[65] C. Lin, Q. G. Wang, T. H. Lee, Y. He and B. Chen, "Observer-based H_∞ fuzzy control design for T-S fuzzy systems with state delays," *Automatica*, vol. 44, no. 3, pp. 868–874, 2008.

[66] S. K. Nguang and P. Shi, "H_∞ fuzzy output feedback control design for nonlinear systems: An LMI approach," *IEEE Trans. Fuzzy Syst.*, vol. 11, no. 3, pp. 331–340, 2003.

[67] W. Assawinchaichote and S. K. Nguang, "H_∞ fuzzy control design for nonlinear singularly perturbed systems with pole placement constraints: An LMI approach," *IEEE Trans. Syst. Man Cybern., Part B, Cybern.*, vol. 34, no. 1, pp. 579–588, 2004.

[68] J. B. Qiu, G. Feng and H. J. Gao, "Static-output-feedback H_∞ control of continuous-time T-S fuzzy affine systems via piecewise Lyapunov functions," *IEEE Trans. Fuzzy Syst.*, vol. 21, no. 2, pp. 245–261, 2013.

[69] J. B. Qiu, G. Feng and H. J. Gao, "Observer-based piecewise affine output feedback controller synthesis of continuous-time T-S fuzzy affine dynamic systems using quantized measurements," *IEEE Trans. Fuzzy Syst.*, vol. 20, no. 6, pp. 1046–1062, 2012.

[70] B. S. Chen, C. S. Tseng and H. J. Uang, "Mixed H_2/H_∞ fuzzy output feedback control design for nonlinear dynamic systems: An LMI approach," *IEEE Trans. Fuzzy Syst.*, vol. 8, no. 3, pp. 249–265, 2000.

[71] W. Paszke, K. Gatkowski, E. Rogers and J. Lam, "H_2 and mixed H_2/H_∞ stabilization and disturbance attenuation for differential linear repetitive processes," *IEEE Trans. Circuits Syst. I, Reg. Papers*, vol. 55, no. 9, pp. 2813–2826, 2008.

[72] L. Q. Zhang, B. Huang and J. Lam, "LMI synthesis of H2and mixed H2/H∞ controllers for singular systems," *IEEE Trans. Circuits Syst. II, Analog Digit. Signal Process.*, vol. 50, no. 9, pp. 615–626, 2003.

[73] C. H. Wu, W. Zhang and B. S. Chen, "Multi-Objective H_2/H_∞ synthetic gene network design based on promoter libraries," *Math. Biosci.*, 233(2) pp. 111–125, 2011.

[74] C. L. Hwang, "Decentralized fuzzy control of nonlinear interconnected dynamic delay systems via mixed H_2/H_∞ optimization with Smith predictor," *IEEE Trans. Fuzzy Syst.*, vol. 19, no. 2, pp. 276–290, 2011.

[75] A. M. Stoica, "Mixed H_2/H_∞ performance analysis and state-feedback control design for networked systems with fading communication channels," *Math. Probl. Eng.*, vol. 2012, pp. 1–16, 2012.

[76] C. A. C. Coello, D. A. Van Veldhuizen and G. B. Lamont, *Evolutionary Algorithms for Solving Multi-Objective Problems*, Springer, New York, 2002.

[77] K. Deb, *Multi-Objective Optimization using Evolutionary Algorithms, 1st ed.*, John Wiley & Sons, Chichester, 2001.

[78] K. Miettinen, *Nonlinear Multi-Objective Optimization*, Springer, New York, 1998.

[79] H. Li and Q. Zhang, "Multi-Objective optimization problems with complicated Pareto sets, MOEA/D and NSGA-II," *IEEE Trans. Evol. Comput.*, vol. 13, no. 2, pp. 284–302, 2009.

[80] M. Fazzolari, R. Alcala, Y. Nojima, H. Ishibuchi and F. Herrera, "A review of the application of multi-objective evolutionary fuzzy systems: Current status and further directions," *IEEE Trans. Fuzzy Syst.*, vol. 21, no. 1, pp. 45–65, 2013.

[81] K. W. Schmidt and Y. S. Boutalis, "Fuzzy discrete event systems for multi-objective control: Framework and application to mobile robot navigation," *IEEE Trans. Fuzzy Syst.*, vol. 20, no. 5, pp. 910–922, 2012.

[82] A. B. Cara, C. Wagner, H. Hagras, H. Pomares and I. Rojas, "Multi-Objective optimization and comparison of nonsingleton type-1 and singleton interval type-2 fuzzy logic systems," *IEEE Trans. Fuzzy Syst.*, vol. 21, no. 3, pp. 459–476, 2013.

[83] T. Takagi and M. Sugeno, "Fuzzy identification of systems and its applications to modeling and control," *IEEE Trans. Syst. Man Cybern.*, vol. 15, no. 1, pp. 116–132, 1985.

[84] H. O. Wang, K. Tanaka and M. F. Griffin, "An approach to fuzzy control of nonlinear systems: Stability and design issues," *IEEE Trans. Fuzzy Syst.*, vol. 4, no. 1, pp. 14–23, 1996.

[85] K. Tanaka, T. Ikeda and H. O. Wang, "Fuzzy regulators and fuzzy observers: Relaxed stability conditions and LMI-based designs," *IEEE Trans. Fuzzy Syst.*, vol. 6, no. 2, pp. 250265, 1998.

[86] B. S. Chen, C. S. Tseng and H. J. Uang, "Robustness design of nonlinear dynamic systems via fuzzy linear control," *IEEE Trans. Fuzzy Syst.*, vol. 7, no. 5, pp. 571–585, 1999.

[87] C. S. Tseng and B. S. Chen, "H_∞ decentralized fuzzy model reference tracking control design for nonlinear interconnected systems," *IEEE Trans. Fuzzy Syst.*, vol. 9, no. 6, pp. 795–809, 2001.

[88] E. Kim, M. Park, S. Ji and M. Park, "A new approach to fuzzy modeling," *IEEE Trans. Fuzzy Syst.*, vol. 5, no. 3, pp. 328–337, 1997.

[89] S. P. Boyd and L. Vandenberghe, *Convex Optimization*, Cambridge University Press, Cambridge, 2004.

[90] S. Boyd, L. El Ghaoui, E. Feron and V. Balakrishnan, *Linear Matrix Inequalities in System and Control Theory*, SIAM, Philadelphia, PA, 1994.

[91] G. M. Siouris, *Missile Guidance and Control Systems*, Springer, New York, 2004.

[92] C. F. Lin, *Modern Navigation, Guidance, and Control Processing*, Prentice-Hall, Upper Saddle River, NJ, 1991.

[93] A. S. Locke, *Principles of Guided Missile Design—Volume 1: Guidance*, D. Van Nostrand Company, New York, 1955.

[94] Z. Xiong, J. Chen, Q. Li and Z. Ren, "Time-varying LQR on hypersonic vehicle profile-following," *53rd IEEE Conference on Decision and Control*, pp. 994–998, Los Angeles, CA, 2014.

[95] G. Hexner and H. Weiss, "Stochastic approach to optimal guidance with uncertain intercept time," *IEEE Trans. Aeros. Elec. Syst.*, vol. 46, no. 4, pp. 1804–1820, 2010.

[96] N. Harl and S. N. Balakrishnan, "Impact time and angle guidance with sliding mode control," *IEEE Trans. Control Syst. Technol.*, vol. 20, no. 6, pp. 1436–1449, 2012.

[97] F. K. Yeh, "Adaptive-sliding-mode guidance law design for missiles with thrust vector control and divert control system," *IET Control Theory Appl.*, vol. 6, no. 4, pp. 552–559, 2012.

[98] A. Zhurbal and M. Idan, "Effect of estimation on the performance of an integrated missile guidance and control system," *IEEE Trans. Aeros. Elec. Syst.*, vol. 47, no. 4, pp. 2690–2708, 2011.

[99] H. J. Uang and B. S. Chen, "Robust adaptive optimal tracking design for uncertain missile systems: A fuzzy approach," *Fuzzy Sets Syst.*, vol. 126, no. 1, pp. 63–87, 2002.

[100] B. S. Chen, Y. Y. Chen and C. L. Lin, "Nonlinear fuzzy H_∞ guidance law with saturation of actuators against maneuvering targets," *IEEE Trans. Control Syst. Technol.*, vol. 10, no. 6, pp. 76–779, 2002.

[101] C. L. Lin, H. Z. Hung, Y. Y. Chen and B. S. Chen, "Development of an integrated fuzzy-logic-based missile guidance law against high speed target," *IEEE Trans. Fuzzy Syst.*, vol. 12, no. 2, pp. 157–169, 2004.

[102] C. L. Lin and C. L. Hwang, "A dynamically fuzzy gain–scheduled design for missile autopilot," *Aeronautical J.*, vol. 107, no. 1076, pp. 599–606, 2003.

[103] M. Kac, "Foundations of kinetic theory," *Proceedings of the 3rd Berkeley Symposium on Mathematical Statistics and Probability*, vol. 3, pp. 171–197, 1956.

[104] J. Tao, R. Lu, H. Su, P. Shi and Z. G. Wu, "Asynchronous filtering of nonlinear Markov jump systems with randomly occurred quantization via TS fuzzy models," *IEEE Trans. Fuzzy Syst.*, vol. 26, no. 4, pp. 1866–1877, 2018.

[105] T. Frank, A. Daffertshofer and P. Beek, "Multivariate Ornstein-Uhlenbeck processes with mean-field dependent coefficients: Application to postural sway," *Phys. Rev. E*, vol. 63, no. 1 Pt. 1, p. 011905, 2000.

[106] C. Di Guilmi, *The Generation of Business Fluctuations: Financial Fragility and Mean-Field Interactions*, Peter Lang Pub Inc., Bern, 2008.

[107] R. Carmona, F. Delarue and A. Lachapelle, "Control of Mckean-Vlasov dynamics versus mean field games," *Math. Financ. Econ.*, vol. 7, no. 2, pp. 131–166, 2013.

[108] C. Di Guilmi, M. Gallegati and S. Landini, "Ch. 13: Financial fragility, mean-field interaction and macroeconomic dynamics: A stochastic model," in *Institutional and Social Dynamics of Growth and Distribution*, N. Salvadori, Ed., Edward Elgar Publishing, Cheltenham, 2010.

[109] J. Li, "Stochastic maximum principle in the mean-field controls," *Automatica*, vol. 48, no. 2, pp. 366–373, 2012.

[110] N. Ahmed, "Nonlinear diffusion governed by McKean-Vlasov equation on Hilbert space and optimal control," *SIAM J. Control Optim.*, vol. 46, no. 1, pp. 356–378, 2007.

[111] J. Yong, "Linear-quadratic optimal control problems for mean-field stochastic differential equations," *SIAM J. Control Optim.*, vol. 51, no. 4, pp. 2809–2838, 2013.

[112] G. P. Liu, J.-B. Yang and J. F. Whidborne, *Multi-Objective Optimisation and Control*, Research Studies Press, Baldock and London, 2003.

[113] A. Sharifi, K. Sabahi, M. A. Shoorehdeli, M. Nekoui and M. Teshnehlab, "Load frequency control in interconnected power system using multi- objective PID controller," *IEEE Conf. Soft Comput. Ind. Appl.*, pp. 217–221, 2008.

[114] C. L. Lin, H. Y. Jan and N. C. Shieh, "GA-based multi-objective PID control for a linear brushless DC motor," *IEEE/ASME Trans. Mechatronics*, vol. 8, no. 1, pp. 56–65, 2003.

[115] D. Limebeer, B. D. Anderson and B. Hendel, "A Nash game approach to mixed H_2/H_∞ control," *IEEE Trans. Autom. Control*, vol. 39, no. 1, pp. 69–82, 1994.

[116] K. Tanaka and H. O. Wang, *Fuzzy Control Systems Design and Analysis: A Linear Matrix Inequality Approach*, Wiley, Hoboken, NJ, 2004.

[117] A. Chibani, M. Chadli, P. Shi and N. B. Braiek, "Fuzzy fault detection filter design for T-S fuzzy systems in the finite-frequency domain," *IEEE Trans. Fuzzy Syst.*, vol. 25, no. 5, pp. 1051–1061, 2017.

[118] H. Ying, "Sufficient conditions on uniform approximation of multivariate functions by general Takagi-Sugeno fuzzy systems with linear rule consequent," *IEEE Trans. Syst. Man Cybern. A Syst. Hum.*, vol. 28, no. 4, pp. 515–520, 1998.

[119] Y. Wei, J. Qiu, H. K. Lam and L. Wu, "Approaches to T-S fuzzy-affine- model-based reliable output feedback control for nonlinear Itô stochastic systems," *IEEE Trans. Fuzzy Syst.*, vol. 25, no. 3, pp. 569–583, 2017.

[120] H. D. Choi, C. K. Ahn, P. Shi, L. Wu and M. T. Lim, "Dynamic output- feedback dissipative control for T-S fuzzy systems with time-varying input delay and output constraints," *IEEE Trans. Fuzzy Syst.*, vol. 25, no. 3, pp. 511–526, 2017.

[121] J. Liu, C. Wu, Z. Wang and L. Wu, "Reliable filter design for sensor networks using type-2 fuzzy framework," *IEEE Trans. Ind. Inform.*, vol. 13, no. 4, pp. 1742–1752, 2017.

[122] W. Y. Chiu, B. S. Chen and H. V. Poor, "A multi-objective approach for source estimation in fuzzy networked systems," *IEEE Trans. Circuits Syst. I, Reg. Papers*, vol. 60, no. 7, pp. 1890–1900, 2013.

[123] W.-Y. Chiu, "Multi-Objective controller design by solving a multi-objective matrix inequality problem," *IET Control Theory Appl.*, vol. 8, no. 16, pp. 1656–1665, 2014.

[124] P. Baranyi et al., *TP-model Transformation-Based-Control Design Frame-works*, Springer, New York, NY, 2016.

[125] P. Baranyi, Y. Yam and P. Varlaki, *Tensor Product Model Transformation in Polytopic Model-Based Control*, CRC Press, Boca Raton, FL, 2013.

[126] A. Szollosi and P. Baranyi, "Influence of the tensor product model representation of qLPV models on the feasibility of linear matrix inequality," *Asian J. Control*, vol. 18, no. 4, pp. 1328–1342, 2016.

[127] L. Ma, T. Zhang and W. Zhang, "H_∞ control for continuous-time mean- field stochastic systems," *Asian J. Control*, vol. 18, no. 5, pp. 1630–1640, 2016.

[128] L. Y. Sun, S. Tong and Y. Liu, "Adaptive backstepping sliding mode H_∞ control of static Var compensator," *IEEE Trans. Control Syst. Technol.*, vol. 19, no. 5, pp. 1178–1185, 2011.

[129] B. K. Øksendal and A. Sulem, *Applied Stochastic Control of Jump Diffusions, 2nd ed.*, Springer, New York, NY and Berlin, 2007.

[130] F. B. Hanson, *Applied Stochastic Processes and Control for Jump-Diffusions: Modeling, Analysis, and Computation*, SIAM, Philadelphia, PA, 2007.

[131] K. Deb, S. Agrawal, A. Pratap and T. Meyarivan, "A fast elitist non-dominated sorting genetic algorithm for multi-objective optimization: NSGA-II," *Int. Conf. Parallel Problem Solving Nature*, pp. 849–858, 2000.

[132] J. Ma and Y. Chen, "Study for the bifurcation topological structure and the global complicated character of a kind of nonlinear finance system (II)," *Appl. Math. Mech.*, vol. 22, no. 12, pp. 1375–1382, 2001.

[133] M. Abido, "A niched Pareto genetic algorithm for multi-objective environmental/economic dispatch," *Int. J. Electr. Power Energy Syst.*, vol. 25, no. 2, pp. 97–105, 2003.

[134] W.-Y. Chiu, "Method of reduction of variables for bilinear matrix inequality problems in system and control designs," *IEEE Trans. Syst. Man Cybern. Syst.*, vol. 47, no. 7, pp. 1241–1256, 2017.

[135] S. H. Strogatz, *Nonlinear Dynamics and Chaos: With Applications to Physics, Biology, Chemistry, and Engineering*, Westview, Boulder, CO, 2014.

[136] N. C. Framstad, B. Øksendal and A. Sulem, "Optimal consumption and portfolio in a jump diffusion market with proportional transaction costs," *J. Math. Econ.*, vol. 35, no. 2, pp. 233–257, 2001.

[137] I. Pan, S. Das and S. Das, "Multi-objective active control policy design for commensurate and incommensurate fractional order chaotic financial systems," *Appl. Math. Model.*, vol. 39, no. 2, pp. 500–514, 2015.

[138] R. J. Patton, "Fault-tolerant control systems: The 1997 situation," *Proceedings of IFAC Symposium Safeprocess*, pp. 1033–1054, 1997.

[139] M. Blanke, R. Izadi-Zamanabadi, S. A. Bøgh and C. P. Lunau, "Fault- tolerant control systems—A holistic view," *Control Eng. Pract.*, vol. 5, no. 5, pp. 693–702, 1997.

[140] A. Shui, W. Chen, P. Zhang, S. Hu and X. Huang, "Review of fault diagnosis in control systems," *Proceedings of IEEE 21st Chinese Control and Decision Conference*, pp. 5324–5329, New York, NY, 2009.

[141] H. Lee and Y. Kim, "Fault-tolerant control scheme for satellite attitude control system," *IET Control Theory Appl.*, vol. 4, no. 8, pp. 1436–1450, 2010.

[142] S. Wang and D. Dong, "Fault-tolerant control of linear quantum stochastic systems," *IEEE Trans. Autom. Control*, vol. 62, no. 6, pp. 2929–2935, 2017.

[143] S. Kawahata, M. Deng and S. Wakitani, "Operator theory based nonlinear fault tolerance control for MIMO micro reactor," *Proceedings of UKACC 11th International Conference on Control (CONTROL)*, pp. 1–6, Belfast, Aug. 2016.

[144] F. Chen, J. Niu and G. Jiang, "Nonlinear fault-tolerant control for hypersonic flight vehicle with multi-sensor faults," *IEEE Access*, vol. 6, pp. 25427–25436, 2018.

[145] N. Li, X. Y. Luo, K. Xu and X. P. Guan, "Robust H_∞ fault-tolerant control for nonlinear time-delay systems against actuator fault," *Proceedings 31st Chinese Control and Decision Conference*, pp. 5265–5270, Hefei, Jul. 2012.

[146] K. Tanaka and H. O. Wang, *Fuzzy Control Systems Design and Analysis*, Wiley, Hoboken, NJ, 2001.

[147] D. Zhang, Q.-L. Han and X. Jia, "Network-based output tracking control for a class of T-S fuzzy systems that can not be stabilized by nondelayed output feedback controllers," *IEEE Trans. Cybern.*, vol. 45, no. 8, pp. 1511–1524, 2015.

[148] D. Zhang, Q.-L. Han and X. Jia, "Network-based output tracking control for T–S fuzzy systems using an event-triggered communication scheme," *Fuzzy Sets Syst.*, vol. 273, pp. 26–48, 2015.

[149] E.-H. Guechi, J. Lauber, M. Dambrine, G. Klančar and S. Blažič, "PDC control design for non-holonomic wheeled mobile robots with delayed outputs," *J. Intell. Robotic Syst.*, vol. 60, no. 3–4, pp. 395–414, 2010.

[150] R.-E. Precup and M. L. Tomescu, "Stable fuzzy logic control of a general class of chaotic systems," *Neural Comput. Appl.*, vol. 26, no. 3, pp. 541–550, 2014.

[151] Y.-X. Li and G.-H. Yang, "Fuzzy adaptive output feedback fault-tolerant tracking control of a class of uncertain nonlinear systems with nonaffine nonlinear faults," *IEEE Trans. Fuzzy Syst.*, vol. 24, no. 1, pp. 223–234, 2016.

[152] P. Li and G. Yang, "Backstepping adaptive fuzzy control of uncertain nonlinear systems against actuator faults," *J. Control Theory Appl.*, vol. 7, no. 3, pp. 248–256, 2009.

[153] X. Su, F. Xia, L. Wu and C. L. P. Chen, "Event-triggered fault detector and controller coordinated design of fuzzy systems," *IEEE Trans. Fuzzy Syst.*, vol. 26, no. 4, pp. 2004–2016, 2018.

[154] Y. Li, K. Sun and S. Tong, "Observer-based adaptive fuzzy fault-tolerant optimal control for SISO nonlinear systems," *IEEE Trans. Cybern.*, vol. 49, no. 2, pp. 649–661, 2019.

[155] Z. Wang, L. Liu, Y. Wu and H. Zhang, "Optimal fault-tolerant control for discrete-time nonlinear strict-feedback systems based on adaptive critic design," *IEEE Trans. Neural Netw. Learn. Syst.*, vol. 29, no. 6, pp. 2179–2191, 2018.

[156] S. K. Kommuri, S. B. Lee and K. C. Veluvolu, "Robust sensors-fault- tolerance with sliding mode estimation and control for PMSM drives," *IEEE/ASME Trans. Mechatronics*, vol. 23, no. 1, pp. 17–28, 2018.

[157] E. Kamal, A. Aitouche, R. Ghorbani and M. Bayart, "Robust fuzzy fault-tolerant control of wind energy conversion systems subject to sensor faults," *IEEE Trans. Sustain. Energy*, vol. 3, no. 2, pp. 231–241, 2012.

[158] K. Zhou, K. Glover, B. Bodenheimer and J. Doyle, "Mixed H_2/H_∞ performance objectives I: Robust performance analysis," *IEEE Trans. Autom. Control*, vol. 39, pp. 1564–1575, 1994.

[159] Y. Lin, T. Zhang and W. Zhang, "Pareto-based guaranteed cost control of the uncertain mean-field stochastic systems in infinite horizon," *Automatica*, vol. 92, pp. 197–209, 2018.

[160] K. Li, R. Chen, G. Min and X. Yao, "Integration of preferences in decom- position multi-objective optimization," *IEEE Trans. Cybern.*, vol. 48, no. 12, pp. 3359–3370, 2018.

[161] C. S. Tseng and B. S. Chen, "Multi-Objective PID control design in uncertain robotic systems using neural network elimination scheme," *IEEE Trans. Syst. Man Cybern. A Syst. Hum.*, vol. 31, no. 6, pp. 632–644, 2001.

[162] X. S. Yang, *Nature-Inspired Optimization Algorithms, 1st ed.*, Elsevier, Amsterdam, Netherlands, Mar. 2014.

[163] H. Zapata, N. Perozo, W. Angulo and J. Contreras, "A hybrid swarm algorithm for collective construction of 3D structures," *Int. J. Artif. Intell.*, vol. 18, no. 1, pp. 1–18, 2020.

[164] R. E. Precup and R. C. David, *Nature-Inspired Optimization Algorithms for Fuzzy Controlled Servo Systems*, Butterworth, London, 2019.

[165] B. S. Chen and M. Y. Lee, "Noncooperative and cooperative strategy designs for nonlinear stochastic jump diffusion systems with external disturbance: T-S fuzzy approach," *IEEE Trans. Fuzzy Syst.*, vol. 28, no. 10, pp. 2437–2451, 2020.

[166] R. S. Varga, "Extrapolation methods: Theory and practice," *Numer. Algorithms*, vol. 4, no. 2, pp. 305, 1993.

[167] J. D. Stefanovski, "Fault tolerant control of descriptor systems with disturbances," *IEEE Trans. Autom. Control*, vol. 64, no. 3, pp. 976–988, 2019.

[168] S. J. Huang and G. H. Yang, "Fault tolerant controller design for T-S fuzzy systems with time-varying delay and actuator faults: A K-step fault-estimation approach," *IEEE Trans. Fuzzy Syst.*, vol. 22, no. 6, pp. 1526–1540, 2014.

[169] B. S. Chen, M. Y. Lee, W. Y. Chen and W. H. Chang. *Supplementary Materials*. Hsin-Chu, Taiwan. Accessed: Dec. 24, 2020. [Online]. Available: www.dropbox.com/s/egwxkyozsiomvof/Silumation%20Parameters.pdf?dl=0

[170] B. S. Chen, M. Y. Lee, W. Y. Chen and W. H. Chang, *Supplementary Materials*. Hsin-Chu, Taiwan. Accessed: Dec. 24, 2020. [Online]. Available: www.dropbox.com/s/dq8sc9gp8vl7swe/code.rar?dl=0

[171] S. H. Zak, *Systems and Control*, Oxford University Press, Oxford, London and New York, NY, 2003.

[172] G. Minkler and J. Minkler, *Theory and Application of Kalman Filtering*, Magellan, Palm Bay, FL, 1993.

[173] F. Zheng, "A robust H_2 filtering approach and its application to equalizer design for communication systems," *IEEE Trans. Signal Process.*, vol. 53, no. 8, pp. 2735–2747, 2005.

[174] L. H. Xie, Y. C. Soh and C. E. Desouza, "Robust Kalman filtering for uncertain discrete-time-systems," *IEEE Trans. Autom. Control*, vol. 39, no. 6, pp. 1310–1314, 1994.

[175] H. P. Liu, F. C. Sun, K. Z. He and Z. Q. Sun, "Design of reduced-order H_∞ filter for Markovian jumping systems with time delay," *IEEE Trans. Circuits Syst. II-Exp. Briefs*, vol. 51, no. 11, pp. 607–612, 2004.

[176] H. J. Gao and C. H. Wang, "Robust L_2-L_∞ filtering for uncertain systems with multiple time-varying state delays," *IEEE Trans. Circuits Syst. I, Fundam. Theory Appl.*, vol. 50, no. 4, pp. 594–599, 2003.

[177] C. S. Tseng, "Robust fuzzy filter design for a class of nonlinear stochastic systems," *IEEE Trans. Fuzzy Syst.*, vol. 15, no. 2, pp. 261–274, 2007.

[178] W. H. Zhang, B. S. Chen and C. S. Tseng, "Robust H_∞ filtering for nonlinear stochastic systems," *IEEE Trans. Signal Process.*, vol. 53, no. 2, pp. 589–598, 2005.

[179] M. J. Grimble and A. Elsayed, "Solution of the H_∞ optimal linear-filtering problem for discrete-time-systems," *IEEE Trans. Acoust. Speech Signal Process.*, vol. 38, no. 7, pp. 1092–1104, 1990.

[180] C. S. Tseng and B. S. Chen, "H_∞ fuzzy estimation for a class of nonlinear discrete-time dynamic systems," *IEEE Trans. Signal Process.*, vol. 49, no. 11, pp. 2605–2619, 2001.

[181] S. K. Nguang and W. Assawinchaichote, "H_∞ filtering for fuzzy dynamical systems with D stability constraints," *IEEE Trans. Circuits Syst. I, Reg. Papers*, vol. 50, no. 11, pp. 1503–1508, 2003.

[182] W. Assawinchaichote and S. K. Nguang, "H_∞ filtering for fuzzy singularly perturbed systems with pole placement constraints: An LMI approach," *IEEE Trans. Signal Process.*, vol. 52, no. 6, pp. 1659–1667, 2004.

[183] S. Y. Xu, T. W. Chen and J. Lam, "Robust H_∞ filtering for uncertain Markovian jump systems with mode-dependent time delays," *IEEE Trans. Autom. Control*, vol. 48, no. 5, pp. 900–907, 2003.

[184] S. Zhou, J. Lam and A. Xue, "H_∞ filtering of discrete-time fuzzy systems via basis-dependent Lyapunov function approach," *Fuzzy Sets Syst.*, vol. 158, pp. 180–193, Jan. 16, 2007.

[185] H. J. Gao, Y. Zhao, J. Lam and K. Chen, "H_∞ fuzzy filtering of nonlinear systems with intermittent measurements," *IEEE Trans. Fuzzy Syst.*, vol. 17, no. 2, pp. 291–300, 2009.

[186] X. Su, P. Shi, L. Wu and Y. D. Song, "A novel approach to filter design for T-S fuzzy discrete-time systems with time-varying delay," *IEEE Trans. Fuzzy Syst.*, vol. 20, no. 6, pp. 1114–1129, 2012.

[187] L. Wu and D. Ho, "Fuzzy filter design for Itô stochastic systems with application to sensor fault detection," *IEEE Trans. Fuzzy Syst.*, vol. 17, no. 1, pp. 233–242, 2009.

[188] X. Su, P. Shi, L. Wu and S. K. Nguang, "Induced filtering of fuzzy stochastic systems with time-varying delays," *IEEE Trans. Cybern.*, vol. 43, no. 4, pp. 1251–1264, 2013.

[189] L. Wu and W. X. Zheng, "L_2-L_∞ control of nonlinear fuzzy Itô stochastic delay systems via dynamic output feedback," *IEEE Trans. Syst. Man Cybern. B Cybern.*, vol. 39, no. 5, pp. 1308–1315, 2009.

[190] K. Tanaka and H. O. Wang, *Fuzzy Control Systems Design and Analysis: A Linear Matrix Inequality Approach*, Wiley, New York, NY, 2001.

[191] S. H. Strogatz, *Nonlinear Dynamics and Chaos: With Applications to Physics, Biology, Chemistry, and Engineering*, Addison-Wesley, Reading, MA, 1994.

[192] B. S. Chen, C. H. Chiang and S. K. Nguang, "Robust H_∞ synchronization design of nonlinear coupled network via fuzzy interpolation method," *IEEE Trans. Circuits Syst. I, Reg. Papers*, vol. 58, no. 2, pp. 349–362, 2011.

[193] W. Y. Chiu and B. S. Chen, "Multisource prediction under nonlinear dynamics in WSNs using a robust fuzzy approach," *IEEE Trans. Circuits Syst. I, Reg. Papers*, vol. 58, no. 1, pp. 137–149, 2011.

[194] H. Lin, Y. Xu and Y. Zhao, "Frequency analysis of T-S fuzzy control systems," *Int. J. Innov. Comput. Inf. Control*, vol. 9, no. 12, pp. 4781–4791, 2013.

[195] X. Su, P. Shi, L. Wu and Y. Song, "A novel control design on discrete-time T-S fuzzy systems with time-varying delays," *IEEE Trans. Fuzzy Syst.*, vol. 21, no. 4, pp. 655–671, 2013.

[196] C. K. Ahn and M. K. Song, "New sets of criteria for exponential L_2-L_∞ stability of T-S fuzzy systems combined with Hopfield neural networks," *Int. J. Innov. Comput. Inf. Control*, vol. 9, no. 7, pp. 2979–2986, 2013.

[197] L. Wu, W. Xing, W. X. Zheng and H. Gao, "Dissipativity-based sliding mode control of switched stochastic systems," *IEEE Trans. Autom. Control*, vol. 58, no. 3, pp. 785–791, 2013.

[198] J. C. Hung and B. S. Chen, "Genetic algorithm approach to fixed-order mixed H_2/H_∞ optimal deconvolution filter designs," *IEEE Trans. Signal Process.*, vol. 48, no. 12, pp. 3451–3461, 2000.

[199] M. J. Grimble, "H_∞ inferential filtering, prediction and smoothing problems," *Signal Process.*, vol. 60, no. 3, pp. 289–304, 1997.

[200] B. S. Chen, C. L. Tsai and Y. F. Chen, "Mixed H_2/H_∞ filtering design in multirate trans-multiplexer systems: LMI approach," *IEEE Trans. Signal Process.*, vol. 49, no. 11, pp. 2693–2701, 2001.

[201] H. J. Gao, J. Lam, L. H. Xie and C. H. Wang, "New approach to mixed H_2/H_∞ filtering for polytopic discrete-time systems," *IEEE Trans. Signal Process.*, vol. 53, no. 8, pp. 3183–3192, 2005.

[202] Z. D. Wang and B. Huang, "Robust H_2/H_∞ filtering for linear systems with error variance constraints," *IEEE Trans. Signal Process.*, vol. 48, no. 8, pp. 2463–2467, 2000.

[203] L. Q. Zhang, B. Huang and J. Lam, "LMI synthesis of H_2 and mixed H_2/H_∞ controllers for singular systems," *IEEE Trans. Circuits Syst. I, Exp. Briefs*, vol. 50, no. 9, pp. 615–626, 2003.

[204] W. H. Chen and B. S. Chen, "Robust stabilization design for stochastic partial differential systems under spatio-temporal disturbances and sensor measurement noises," *IEEE Trans. Circuits Syst. I, Reg. Papers*, vol. 60, no. 4, pp. 1013–1026, 2013.

[205] A. Abraham, L. C. Jain and R. Goldberg, *Evolutionary Multi-Objective Optimization: Theoretical Advances and Applications*, Springer, New York, NY and Berlin, 2005.

[206] L. Rachmawati and D. Srinivasan, "Multi-Objective evolutionary algorithm with controllable focus on the knees of the Pareto front," *IEEE Trans. Evol. Comput.*, vol. 13, pp. 810–824, 2009.

[207] M. Elmusrati, H. El-Sallabi and H. Koivo, "Applications of multi-objective optimization techniques in radio resource scheduling of cellular communication systems," *IEEE Trans. Wireless Commun.*, vol. 7, no. 1, pp. 343–353, 2008.

[208] Q. L. Ma and C. H. Hu, "An effective evolutionary approach to mixed H_2/H_∞ filtering with regional pole assignment," *Proc. 6th WCICA*, vols. 1–12, pp. 1590–1593, 2006.

[209] J. H. Ryu, S. Kim and H. Wan, "Pareto front approximation with adaptive weighted sum method in multi-objective simulation optimization," *Proc. Winter Simul. Conf.*, vols. 1–4, pp. 615–625, 2009. doi: 10.1109/WSC.2009.5429562

[210] M. Oltean, C. Grosan, A. Abraham and M. Koppen, "Multi-Objective optimization using adaptive pareto archived evolution strategy," *Proceedings of the 5th International Conference on Intelligent Systems Design and Applications*, pp. 558–563, Warsaw, Poland, 2005. doi: 10.1109/ISDA.2005.69

[211] R. T. Marler and J. S. Arora, "Survey of multi-objective optimization methods for engineering," *Struct. Multidisciplinary Optimization*, vol. 26, pp. 369–395, 2004.

[212] S. P. Boyd, Linear Matrix Inequalities in System and Control Theory, Society for Industrial and Applied Mathematics, Philadelphia, PA, 1994.

[213] S. H. Strogatz, *Nonlinear Dynamics and Chaos: With Applications to Physics, Biology, Chemistry, and Engineering*. Reading, MA, USA: Addison-Wesley, 1994.

[214] E. Branke, K. Deb, H. Dierolf and M. Osswald, "Finding knees in multi-objective optimization," *Parallel Problem Solving from Nature—PPSN VIII*, pp. 722–731, Springer, New York, NY, vol. 3242, 2004.

[215] S. C. Lee and C. Y. Liu, "Trajectory estimation of reentry vehicles by use of on-line input estimator," *Guid. Control Dyn.*, vol. 22, pp. 808–815, 1999.

[216] P. Zarchan, *Tactical and Strategic Missile Guidance*, AIAA, Washington, DC, 1990.

[217] W. H. Zhang, B. S. Chen and C. S. Tseng, "Robust filtering for nonlinear stochastic systems," *IEEE Trans. Signal Process.*, vol. 53, no. 2, pp. 589–598, 2005.

[218] H. L. Dong, Z. D. Wang, X. Ding and H. J. Gao, "Event-based H_∞ filter design for a class of nonlinear time-varying systems with fading channels and multiplicative noises," *IEEE Trans. Signal Process.*, vol. 63, no. 13, pp. 3387–3395, 2015.

[219] R. van der Merwe, A. Doucet, J. F. G. de Freitas and E. Wan, "The unscented particle filter," *Proceedings of the 13th International Conference on Neural Information Processing Systems*, pp. 563–569, MIT Press, Cambridge, MA, 2000.

[220] S. Yin and X. Zhu, "Intelligent particle filter and its application on fault detection of nonlinear system," *IEEE Trans. Ind. Electron.*, vol. 62, no. 6, pp. 3852–3861, 2015.

[221] X. H. Chang, Z. M. Li and J. H. Park, "Fuzzy generalized H_2 filtering for nonlinear discrete-time systems with measurement quantization," *IEEE Trans. Syst. Man Cybern. Syst.*, vol. 48, no. 12, pp. 2419–2430, 2018.

[222] V. Dragan, S. Aberkane and I. L. Popa, "Optimal H_2 filtering for periodic linear stochastic systems with multiplicative white noise perturbations and sampled measurements," *J. Franklin Inst.*, vol. 352, no. 12, pp. 5985–6010, 2015.

[223] H. L. Dong, Z. D. Wang and H. J. Gao, "Robust H_∞ filtering for a class of nonlinear networked systems with multiple stochastic communication delays and packet dropouts," *IEEE Trans. Signal Process.*, vol. 48, no. 4, pp. 1957–1966, 2010.

[224] T. J. Ho and B. S. Chen, "Novel extended Viterbi-based multiple-model algorithms for state estimation of discrete-time systems with Markov jump parameters," *IEEE Trans. Signal Process.*, vol. 54, no. 2, pp. 393–404, 2006.

[225] P. X. Sun, Y. H. Wang and W. H. Zhang, "Robust H_∞ filtering for nonlinear uncertain stochastic time-varying delayed system," *Proceedings of Chinese Control and Decision Conference*, pp. 3697–3702, Shenyang, 2018.

[226] H. B. Zhang, H. Zhong and C. Y. Dang, "Delay-dependent decentralized H_∞ filtering for discrete-time nonlinear interconnected systems with time-varying delay based on the T–S fuzzy model," *IEEE Trans. Fuzzy Syst.*, vol. 20, no. 3, pp. 431–443, 2012.

[227] Q. Liu, Z. Wang, X. He, G. Ghinea and F. E. Alsaadi, "A resilient approach to distributed filter design for time-varying systems under stochastic nonlinearities and sensor degradation," *IEEE Trans. Signal Process.*, vol. 65, no. 5, pp. 1300–1309, 2017.

[228] S. Sun, F. Peng and H. Lin, "Distributed asynchronous fusion estimator for stochastic uncertain systems with multiple sensors of different fading measurement rates," *IEEE Trans. Signal Process.*, vol. 66, no. 3, pp. 641–653, 2018.

[229] D. R. Cox and H. D. Miller, *Theory Stochastic Processes*, Methuen and Company, London, 1965.

[230] Z. D. Wang, Y. R. Liu and X. H. Liu, "H_∞ filtering for uncertain stochastic time-delay systems with sector-bounded nonlinearities," *Automatica*, vol. 44, no. 5, pp. 1268–1277, 2008.

[231] R. M. AbuSaris and F. B. Hanson, "Computational suboptimal filter for a class of Wiener-Poisson driven stochastic processes," *Dyn. Control*, vol. 7, no. 3, pp. 279–292, 1997.

[232] J. Westman and F. Hanson, "State dependent jump models in optimal control," *Proceedings of the 38th IEEE Conference on Decision and Control*, vol. 3, pp. 2378–2383, Phoenix, AZ, Dec. 1999.

[233] R. Situ, *Theory Stochastic Differential Equations with Jumps Applications: Mathematical Analytical Techniques with Applications to Engineering*, Springer, New York, NY, 2005.

[234] X. L. Lin, C. F. Wu and B. S. Chen, "Robust H_∞ adaptive fuzzy tracking control for MIMO nonlinear stochastic Poisson jump diffusion systems," *IEEE Trans. Cybern.*, vol. 49, no. 8, pp. 3116–3130, 2019.

[235] X. M. Yao, L. G. Wu and W. X. Zheng, "Fault detection filter design for Markovian jump singular systems with intermittent measurements," *IEEE Trans. Signal Process.*, vol. 59, no. 7, pp. 3099–3109, 2011.

[236] H. Dong, Z. Wang, D. Ho and H. Gao, "Robust H_∞ filtering for Markovian jump systems with randomly occurring nonlinearities and sensor saturation: The finite-horizon case," *IEEE Trans. Signal Process.*, vol. 59, no. 7, pp. 3048–3057, 2011.

[237] E. Özkan, F. Lindsten, C. Fritsche and F. Gustafsson, "Recursive maximum likelihood identification of jump Markov nonlinear systems," *IEEE Trans. Signal Process.*, vol. 63, no. 3, pp. 754–765, 2015.

[238] D. Minoli, K. Sohraby and B. Occhiogrosso, "IoT considerations, requirements, and architectures for smart buildings—Energy optimization and next-generation building management systems," *IEEE Internet Things J.*, vol. 4, no. 1, pp. 269–283, 2017.

[239] K. Georgiou, S. Xavier-de-Souza, and K. Eder, "The IoT energy challenge: A software perspective," *IEEE Embed. Syst. Lett.*, vol. 10, no. 3, pp. 53–56, 2018.

[240] A. Al-Ali, I. A. Zualkernan, M. Rashid, R. Gupta and M. AliKarar, "A smart home energy management system using IoT and big data analytics approach," *IEEE Trans. Consum. Electron.*, vol. 63, no. 4, pp. 426–434, 2017.

[241] S. Li, L. Da Xu and S. Zhao, "5G internet of things: A survey," *J. Ind. Inf. Integr.*, vol. 10, pp. 1–9, 2018.

[242] T. Abels, R. Khanna and K. Midkiff, "Future proof IoT: Composable semantics, security, QoS and reliability," *Proceedings of IEEE Topical Conference on Wireless Sensors and Sensor Networks (WiSNet)*, Phoenix, AZ, 2017.

[243] M. A. Khan and K. Salah, "IoT security: Review, blockchain solutions, and open challenges," *Future Gener. Comput. Syst.*, vol. 82, pp. 395–411, 2018.

[244] A. Dorri, S. S. Kanhere, R. Jurdak and P. Gauravaram, "Blockchain for IoT security and privacy: The case study of a smart home," *Proceedings of IEEE International Conference on Pervasive Computing and Communications Workshops (PerCom Workshops)*, pp. 618–623, 2017. doi: 10.1109/PERCOMW.2017.7917634

[245] Y. H. Hwang, "IoT security & privacy: Threats and challenges," *Proceedings of 1st ACM Workshop IoT Privacy, Trust and Security*, ser. IoTPTS'15, ACM, New York, NY, 2015.

[246] M. Dhawan, R. Poddar, K. Mahajan and V. Mann, "SPHINX: Detecting security attacks in software-defined networks," *Proceedings of Network and Distributed System Security (NDSS) Symposium*, pp. 1–15, San Diego, CA, 2015.

[247] E. G. AbdAllah, H. S. Hassanein and M. Zulkernine, "A survey of security attacks in information-centric networking," *IEEE Commun. Surv. Tuts.*, vol. 17, no. 3, pp. 1441–1454, 3rd Quart., 2015.

[248] M. Nawir, A. Amir, N. Yaakob and O. B. Lynn, "Internet of Things (IoT): Taxonomy of security attacks," *Proceedings of 3rd International Conference on Electron Device*, pp. 321–326, Phuket, Thailand, Aug. 2016. doi: 10.1109/ICED.2016.7804660

[249] C. Mavromoustakis, G. Mastorakis and J. M. Batalla, *Internet Things (IoT) 5G Mobile Technol.*, vol. 8, Springer, Cham, 2016.

[250] Z. Wang, P. Shi and C. Lim, "$H-/H_\infty$ fault detection observer in finite frequency domain for linear parameter-varying descriptor systems," *Automatica*, vol. 86, pp. 38–45, 2017.

[251] Y. K. Wu, B. Jiang and N. Y. Lu, "A descriptor system approach for estimation of incipient faults with application to high-speed railway traction devices," *IEEE Trans. Syst. Man Cybern. Syst.*, vol. 49, no. 10, 2019.

[252] Z. W. Gao and W. C. Ho, "State/noise estimator for descriptor systems with application to sensor fault diagnosis," *IEEE Trans. Signal Process.*, vol. 54, no. 4, pp. 1316–1326, 2006.

[253] Z. W. Gao and Steven X. Ding, "State and disturbance estimator for time-delay systems with application to fault estimation and signal compensation," *IEEE Trans. Signal Process.*, vol. 55, no. 12, 2007.

[254] B. S. Chen, W. H. Chen and H. L. Wu, "Robust global linearization filter design for nonlinear stochastic systems," *IEEE Trans. Circuits Syst. I, Reg. Papers*, vol. 56, no. 7, pp. 1441–1454, 2009.

[255] W. Mao, F. Deng and A. Wan, "Robust H_2/H_∞ global linearization filter design for nonlinear stochastic time varying delay systems," *Sci. China (Inform. Sci.)*, vol. 59, no. 3, pp. 1–17, 2016.

[256] S. X. Du, A. Dekka and B. Wu, *Modular Multilevel Converters: Analysis, Control, and Applications*, Wiley-IEEE Press, Hoboken, NJ, Jan. 2018.

[257] S. R. Kou, D. L. Elliott and T. J. Tarn, "Observability of nonlinear systems," *Inf. Control.*, vol. 22, pp. 89–99, 1973.

[258] K. Deb, S. Agrawal, A. Pratap and T. Meyarivan, "A fast elitist nondominated sorting genetic algorithm for multi-objective optimization: NSGA-II," *Proceedings of International Conference on Parallel Problem Solving from Nature*, pp. 849–858, 2000.

[259] S. Mishra, S. Mondal and S. Saha, "Fast implementation of steady-state NSGA-II," *Proceedings of 2016 IEEE CEC*, pp. 3777–3784, 2016.

[260] R. Y. Zhang and J. Lavaei, "Efficient algorithm for large-and-sparse LMI feasibility problems," *IEEE Conf. Decision Control*, pp. 6868–6875, 2018.

[261] T. H. Lee, C. P. Lim, S. Nahavandi and R. G. Roberts, "Observer-based H_∞ fault-tolerant control for linear systems with sensor and actuator faults," *IEEE Syst. J.*, vol. 12, no. 2, 2019.

[262] G. M. Siouris, G. Chen and J. Wang, "Tracking an incoming ballistic missile using an extended interval Kalman filter," *IEEE Trans. Aeros. Elec. Syst.*, vol. 33, no. 1, pp. 232–240, 1997.

[263] M. I. Garcia-Planas, J. L. Dominguez-Garcia and L. E. Um, "Sufficient conditions for controllability and observability of serial and parallel concatenated linear systems," *Int. J. Circuits Syst. Signal Process.*, vol. 8, pp. 622–630, 2014.

[264] IEEE, "IEEE draft amendment standard for local and metropolitan area networks—Part 16: Air interface for broadband wireless access systems -advanced air interface," *IEEE*, pp. 1–1120, 2011.

[265] 3GPP, "GPP TS 36.211 v9.1.0, LTE standard document." Available: www.3gpp.org/ftp/Specs/html-info/36211.htm

[266] Y.-C. Liang, K.-C. Chen, G. Li and P. Mahonen, "Cognitive radio networking and communications: An overview," *IEEE Trans. Veh. Technol.*, vol. 60, no. 7, pp. 3386–3407, 2011.

[267] D. Gesbert, S. Hanly, H. Huang, S. Shamai Shitz, O. Simeone and W. Yu, "Multi-cell MIMO cooperative networks: A new look at interference," *IEEE J. Sel. Areas Commun.*, vol. 28, no. 9, pp. 1380–1408, 2010.

[268] E. Björnson, N. J. Jaldén, M. Bengtsson and B. Ottersten, "Optimality properties, distributed strategies, and measurement-based evaluation of coordinated multicell OFDMA transmission," *IEEE Trans. Signal Process.*, vol. 59, no. 12, pp. 6086–6101, 2011.

[269] L. Ventruino, N. Prasad and X.-D. Wang, "Coordinated scheduling and power allocation in downlink multicell OFDMA networks," *IEEE Trans. Veh. Technol.*, vol. 58, no. 6, pp. 2835–2848, 2009.

[270] V. Cadambe and S. Jafar, "Interference alignment and degrees of freedom of the k-user interference channel," *IEEE Trans. Inf. Theory*, vol. 54, no. 8, pp. 3425–3441, 2008.

[271] P. Karunakaran, P. Suryasarman, V. Ramaswamy, K. Kuchi, J. Milleth and B. Ramamurthi, "On pilot design for interference limited OFDM systems," *Proceedings 2011 International Symposium on Wireless Communication Systems*, pp. 522–526, 2011.

[272] B. Blaszczyszyn and M. Karray, "Quality of service in wireless cellular networks subject to log-normal shadowing," *IEEE Trans. Commun.*, vol. 61, no. 2, pp. 781–791, 2013.

[273] E. Calvo, O. Munoz, J. Vidal and A. Agustin, "Downlink coordinated radio resource management in cellular networks with partial CSI," *IEEE Trans. Signal Process.*, vol. 60, no. 3, pp. 1420–1431, 2012.

[274] C. Martinez-Sanchez, J. M. Luna-Rivera and D. Campos-Delgado, "A cross-layer power allocation scheme for CDMA wireless networks," *Proceedings of 2012 American Control Conference*, pp. 2018–2023.

[275] F. Lau and W. Tam, "Achievable-SIR-based predictive closed-loop power control in a CDMA mobile system," *IEEE Trans. Veh. Technol.*, vol. 51, no. 4, pp. 720–728, 2002.

[276] L. Mendo and J. Hernando, "System-level analysis of closed-loop power control in the uplink of DS-CDMA cellular networks," *IEEE Trans. Wireless Commun.*, vol. 6, no. 5, pp. 1681–1691, 2007.

[277] C. Li, L. Duan and X. Dong, "Robust power control for CDMA cellular communication systems via h2 optimal theory," *Proceedings of 2008 International Conference on Wireless Communications, Networking and Mobile Computing*, pp. 1–5, 2008.

[278] J. Oh, S.-J. Kim and J. Cioffi, "Optimum power allocation and control for OFDM in multiple access channels," *Proceedings of 2004 IEEE Vehicular Technology Conference*, vol. 2, pp. 774–778, 2004.

[279] A. Abrardo and D. Sennati, "On the analytical evaluation of closed loop power-control error statistics in DS-CDMA cellular systems," *IEEE Trans. Veh. Technol.*, vol. 49, no. 6, pp. 2071–2080, 2000.

[280] B.-K. Lee, Y.-H. Chen and B.-S. Chen, "Robust power control for CDMA cellular communication systems," *IEEE Trans. Signal Process.*, vol. 54, no. 10, pp. 3947–3956, 2006.

[281] B.-S. Chen, B.-K. Lee and Y.-H. Chen, "Power control for CDMA cellular radio systems via l/sub 1/ optimal predictor," *IEEE Trans. Wireless Commun.*, vol. 5, no. 10, pp. 2914–2922, 2006.

[282] K. Deb, *Multi-Objective Optimization Using Evolutionary Algorithms*, vol. 16, Wiley, New York, 2004.

[283] M. Ziolkowski and S. Gratkowski, "Weighted sum method and genetic algorithm based multi-objective optimization of an exciter for magnetic induction tomography," *Proceedings of 2009 International Symposium on Theoretical Engineering*, pp. 1–5, 2009.

[284] J.-H. Ryu, S. Kim and H. Wan, "Pareto front approximation with adaptive weighted sum method in multi-objective simulation optimization," *Proceedings of 2009 Winter Simulation Conference*, pp. 623–633, 2009.

[285] J. Horn, N. Nafpliotis and D. Goldberg, "A niched Pareto genetic algorithm for multi-objective optimization," *Proceedings of 1994 IEEE Conference on Evolutionary Computation*, vol. 1, pp. 82–87, 1994.

[286] R. Schoenen and F. Qin, "Adaptive power control for 4G OFDMA systems on frequency selective fading channels," *Proceedings of 2009 International Conference Wireless Communications, Networking and Mobile Computing*, pp. 1–6, 2009.

[287] X. Kang, H. Garg, Y.-C. Liang and R. Zhang, "Optimal power allocation for OFDM-based cognitive radio with new primary transmission protection criteria," *IEEE Trans. Wireless Commun.*, vol. 9, no. 6, pp. 2066–2075, 2010.

[288] A. Abrardo, M. Belleschi, P. Detti and M. Moretti, "A message passing approach for multi-cellular OFDMA systems," *Proceedings of 2010 International Symposium on Wireless Communication Systems*, pp. 651–655, 2010.

[289] D. Popescu, D. Joshi and O. Dobre, "Spectrum allocation and power control in OFDM-based cognitive radios with target SINR constraints," *Proceedings of 2010 Asilomar Conference Signals, Systems and Computers*, pp. 1891–1895, 2010.

[290] D. Wang, Z. Li and X. Wang, "Joint optimal subcarrier and power allocation for wireless cooperative networks over OFDM fading channels," *IEEE Trans. Veh. Technol.*, vol. 61, no. 1, pp. 249–257, 2012.

[291] E. F. S. Boyd, L. El Ghaoui and V. Balakrishnan, *Linear Matrix Inequalities in System and Control Theory*. Society for Industrial Mathematics, Philadelphia, PA, 1994.

[292] R. B. Agrawal, K. Deb, K. Deb and R. B. Agrawal, "Simulated binary crossover for continuous search space," *tech. rep.*, 1994.

[293] T. S. Rappaport and S. B. Online, *Wireless Communications: Principles and Practice*, vol. 2, Prentice Hall PTR, Upper Saddle River, NJ, 1996.

[294] S. G. Glisic, *Adaptive WCDMA*, Wiley Online Library, New York, 2003.

[295] D. Giancristofaro, "Correlation model for shadow fading in mobile radio channels," *Electron. Lett.*, vol. 32, no. 11, pp. 958–959, 1996.

[296] F. Graziosi and F. Santucci, "A general correlation model for shadow fading in mobile radio systems," *IEEE Commun. Lett.*, vol. 6, no. 3, pp. 102–104, 2002.

[297] H.-J. Su and E. Geraniotis, "Adaptive closed-loop power control with quantized feedback and loop filtering," *Proceedings of 1998 IEEE International Symposium on Personal, Indoor and Mobile Radio Communications*, vol. 2, pp. 926–931, 1998.

[298] M. Sim, E. Gunawan, C. Soh and B. H. Soong, "Characteristics of closed loop power control algorithms for a cellular DS-CDMA system," *IEE Proc.-Commun.*, vol. 145, no. 5, pp. 355–362, 1998.

[299] N. Abe and K. Yamanaka, "Smith predictor control and internal model control—A tutorial," *Proceedings of 2003 Annual Conference SICE*, vol. 2, pp. 1383–1387, 2003.

[300] B. Chen, T. Lee and J. Feng, "A nonlinear H∞ control design in robotic systems under parameter perturbation and external disturbance," *Inter. J. Control*, vol. 59, no. 2, pp. 439–461, 1994.

[301] E. Castañeda, A. Silva, A. Gameiro and M. Kountouris, "An overview on resource allocation techniques for multi-user MIMO Systems," *IEEE Commun. Surv. Tuts.*, vol. 19, no. 1, pp. 239–284, 1st Quart., 2017.

[302] D. Tse and P. Viswanath, *Fundamentals of Wireless Communication*, Cambridge University Press, New York, NY, 2005.

[303] A. J. Paulraj, D. A. Gore, R. U. Nabar and H. Bolcskel, "An overview of MIMO communications—A key to gigabit wireless," *Proc. IEEE*, vol. 92, no. 9, pp. 198–218, 2004.

[304] H. Weingarten, Y. Steinberg and S. S. Shamai, "The capacity region of the Gaussian multiple-input multiple-output broadcast channel," *IEEE Trans. Inf. Theory*, vol. 52, no. 9, pp. 3936–3964, 2006.

[305] M. Agiwal, A. Roy and N. Saxena, "Next generation 5G wireless networks: A comprehensive survey," *IEEE Commun. Surv. Tuts.*, vol. 18, no. 3, pp. 1617–1655, 3rd Quart., 2016.

[306] Q. Shi, M. Razaviyayn, Z.-Q. Luo and C. He, "An iteratively weighted MMSE approach to distributed sum-utility maximization for a MIMO interfering broadcast channel," *IEEE Trans. Signal Process.*, vol. 59, no. 9, pp. 4331–4340, 2011.

[307] D. H. N. Nguyen and T. Le-Ngoc, "Sum-rate maximization in the multicell MIMO broadcast channel with interference coordination," *IEEE Trans. Signal Process.*, vol. 62, no. 6, pp. 1501–1513, 2014.

[308] J. Park, Y. Sung, D. Kim and H. V. Poor, "Outage probability and outage-based robust beamforming for MIMO interference channels with imperfect channel state information," *IEEE Trans. Wireless Commun.*, vol. 11, no. 10, pp. 3561–3573, 2012.

[309] X. Wang and X.-D. Zhang, "Linear transmission for rate optimization in MIMO broadcast channels," *IEEE Trans. Wireless Commun.*, vol. 9, no. 10, pp. 3247–3257, 2010.

[310] H. Du, T. Ratnarajah, M. Pesavento and C. B. Papadias, "Joint transceiver beamforming in MIMO cognitive radio network via second order cone programming," *IEEE Trans. Signal Process.*, vol. 60, no. 2, pp. 781–792, 2012.

[311] Q. Zhang, C. He and L. Jiang, "Per-stream MSE based linear transceiver design for MIMO interference channels with CSI error," *IEEE Trans. Commun.*, vol. 63, no. 5, pp. 1676–1689, 2015.

[312] H. Shen, B. Li, M. Tao and X. Wang, "MSE-Based transceiver designs for the MIMO interference channel," *IEEE Trans. Wireless Commun.*, vol. 9, no. 11, pp. 3480–3489, 2010.

[313] S. M. J. Asgari-Tabatabaee and H. Zamiri-Jafarian, "Per-subchannel joint equalizer and receiver filter design in OFDM/OQAM systems," *IEEE Trans. Signal Process.*, vol. 64, no. 19, pp. 5094–5105, 2016.

[314] J. Li, D.-Z. Feng and W. X. Zheng, "Space-time semi-blind equalizer for dispersive QAM MIMO system based on modified Newton method," *IEEE Trans. Wireless Commun.*, vol. 13, no. 6, pp. 3244–3256, 2014.

[315] S. W. Peters and R. W. Heath, "Cooperative algorithms for MIMO interference channels," *IEEE Trans. Veh. Technol.*, vol. 60, no. 1, pp. 206–218, 2011.

[316] Y. Li, Y. Tian and C. Yang, "Energy-efficient coordinated beamforming under minimal data rate constraint of each user," *IEEE Trans. Veh. Technol.*, vol. 64, no. 6, pp. 2387–2397, 2015.

[317] Y. Li, P. Fan and N. C. Beaulieu, "Cooperative downlink max-min energy-efficient precoding for multicell MIMO networks," *IEEE Trans. Veh. Technol.*, vol. 65, no. 11, pp. 9425–9430, 2016.

[318] Z. Xiang, M. Tao and X. Wang, "Coordinated multicast beamforming in multicell networks," *IEEE Trans. Wireless Commun.*, vol. 12, no. 1, pp. 12–21, 2013.

[319] D. N. Nguyen and M. Krunz, "Power minimization in MIMO cognitive networks using beamforming games," *IEEE J. Sel. Areas Commun.*, vol. 31, no. 5, pp. 916–925, 2013.

[320] M. Codreanu, A. Tolli, M. Juntti and M. Latva-Aho, "Joint design of Tx-Rx beamformers in MIMO downlink channel," *IEEE Trans. Signal Process.*, vol. 55, no. 9, pp. 4639–4655, 2007.

[321] M. Moretti, L. Sanguinetti and X. Wang, "Resource allocation for power minimization in the downlink of THP-based spatial multiplexing MIMO-OFDMA systems," *IEEE Trans. Veh. Technol.*, vol. 64, no. 1, pp. 405–411, 2015.

[322] L. Lei, D. Yuan and P. Värbrand, "On power minimization for non-orthogonal multiple access (NOMA)," *IEEE Commun. Lett.*, vol. 20, no. 12, pp. 2458–2461, 2016.

[323] A. Ramezani-Kebrya, M. Dong, B. Liang, G. Boudreau and R. Casselman, "Per-relay power minimization for multi-user multichannel cooperative relay beamforming," *IEEE Trans. Wireless Commun.*, vol. 15, no. 5, pp. 3187–3198, 2016.

[324] D. W. K. Ng, Y. Wu and R. Schober, "Power efficient resource allocation for full-duplex radio distributed antenna networks," *IEEE Trans. Wireless Commun.*, vol. 15, no. 4, pp. 2896–2911, 2016.

[325] W.-Y. Chiu, H. Sun and H. V. Poor, "A multi-objective approach to multimicrogrid system design," *IEEE Trans. Smart Grid*, vol. 6, no. 5, pp. 2263–2272, 2015.

[326] D. W. K. Ng, E. S. Lo and R. Schober, "Multi-Objective resource allocation for secure communication in cognitive radio networks with wireless information and power transfer," *IEEE Trans. Veh. Technol.*, vol. 65, no. 5, pp. 3166–3184, 2016.

[327] Y. Sun, D. W. K. Ng, J. Zhu and R. Schober, "Multi-objective optimization for robust power efficient and secure full-duplex wireless communication systems," *IEEE Trans. Wireless Commun.*, vol. 15, no. 8, pp. 5511–5526, 2016.

[328] S. Leng, D. W. K. Ng, N. Zlatanov and R. Schober, "Multi-objective beamforming for energy-efficient SWIPT systems," *IEEE, Proc. Int. Conf. Comput. Netw. Commun. (ICNC)*, 2016, pp. 1–7.

[329] S. Leng, D. W. K. Ng, N. Zlatanov and R. Schober, "Multi-objective resource allocation in full-duplex SWIPT systems," *Proceedings of IEEE International Conference on Communications (ICC)*, pp. 1–7, 2016.

[330] X. Xie, H. Yang and A. V. Vasilakos, "Robust transceiver design based on interference alignment for multi-user multi-cell MIMO networks with channel uncertainty," *IEEE Access*, vol. 5, pp. 5121–5134, 2017.

[331] Y. Huang and D. P. Palomar, "Rank-constrained separable semidefinite programming with applications to optimal beamforming," *IEEE Trans. Signal Process.*, vol. 58, no. 2, pp. 664–678, 2010.

[332] N. Vucic, H. Boche and S. Shi, "Robust transceiver optimization in downlink multiuser MIMO systems," *IEEE Trans. Signal Process.*, vol. 57, no. 9, pp. 3576–3587, 2009.

[333] Y. Huang, D. P. Palomar and S. Zhang, "Lorentz-positive maps and quadratic matrix inequalities with applications to robust MISO transmit beamforming," *IEEE Trans. Signal Process.*, vol. 61, no. 5, pp. 1121–1130, 2013.

[334] K. Deb, *Multi-Objective Optimization Using Evolutionary Algorithms, 1st ed.*, Wiley, Chichester, 2009.

[335] S. Boyd and L. Vandenberghe, *Convex Optimization*, Cambridge University Press, Cambridge, 2004.

[336] M. Grant and S. Boyd, *CVX: MATLAB Software for Disciplined Convex Programming, Version 2.0 Beta*. Sep. 2013. [Online]. Available: http://cvxr.com/cvx

[337] T.-J. Ho and B.-S. Chen, "Robust minimax MSE equalizer designs for MIMO wireless communications with time-varying channel uncertainties," *IEEE Trans. Signal Process.*, vol. 58, no. 11, pp. 5835–5844, 2010.

[338] A. G. Helmy, A. R. Hedayat and N. Al-Dhahir, "Robust weighted sum-rate maximization for the multi-stream MIMO interference channel with sparse equalization," *IEEE Trans. Commun.*, vol. 63, no. 10, pp. 3645–3659, 2015.

[339] Y.-S. Jeon, Y.-J. Kim, M. Min and G.-H. Im, "Distributed block diagonalization with selective zero forcing for multicell MU-MIMO systems," *IEEE Signal Process. Lett.*, vol. 21, no. 5, pp. 605–609, 2014.

[340] M. Sadek, A. Tarighat and A. H. Sayed, "A leakage-based precoding scheme for downlink multi-user MIMO channels," *IEEE Trans. Wireless Commun.*, vol. 6, no. 5, pp. 1711–1721, 2007.

[341] D. Shen, Z. Pan, K.-K. Wong and V. O. K. Li, "Effective throughput: A unified benchmark for pilot-aided OFDM/SDMA wireless communication systems," *Proceedings of IEEE INFOCOM*, pp. 1603–1613, San Francisco, CA, 2003.

[342] M. Y. Hong, R. Y. Sun, H. Baligh and Z. Q. Luo, "Joint base station clustering and beamformer design for partial coordinated transmission in heterogeneous networks," *IEEE J. Sel. Areas Commun.*, vol. 31, no. 2, pp. 226–240, 2013.

[343] S. W. He, Y. M. Huang, L. X. Yang, B. Ottersten and W. Hong, "Energy efficient coordinated beamforming for multicell system: Duality-based algorithm design and massive MIMO transition," *IEEE Trans. Commun.*, vol. 63, no. 12, pp. 4920–4935, 2015.

[344] S. W. He, Y. M. Huang, S. Jin and L. X. Yang, "Coordinated beamforming for energy efficient transmission in multicell multiuser systems," *IEEE Trans. Commun.*, vol. 61, no. 12, pp. 4961–4971, 2013.

[345] J. Schreck, G. Wunder and P. Jung, "Robust iterative interference alignment for cellular networks with limited feedback," *IEEE Trans. Wireless Commun.*, vol. 14, no. 2, pp. 882–894, 2015.

[346] Y. M. Huang, S. W. He, S. Jin and W. Y. Chen, "Decentralized energy efficient coordinated beamforming for multicell systems," *IEEE Trans. Veh. Technol.*, vol. 63, no. 9, pp. 4302–4314, 2014.

[347] L. N. Tran, M. Juntti, M. Bengtsson and B. Ottersten, "Weighted sum rate maximization for MIMO broadcast channels using dirty paper coding and zero-forcing methods," *IEEE Trans. Commun.*, vol. 61, no. 6, pp. 2362–2373, 2013.

[348] D. H. N. Nguyen, H. Nguyen-Le and T. Le-Ngoc, "Block-diagonalization precoding in a multiuser multicell MIMO system: Competition and coordination," *IEEE Trans. Wireless Commun.*, vol. 13, no. 2, pp. 968–981, 2014.

[349] H. Shen, W. Xu, S. Jin and C. M. Zhao, "Joint transmit and receive beamforming for multiuser MIMO downlinks with channel uncertainty," *IEEE Trans. Veh. Technol.*, vol. 63, no. 5, pp. 2319–2335, 2014.

[350] F. H. Panahi, T. Ohtsuki, W. Jiang, Y. Takatori and T. Nakagawa, "Joint interference alignment and power allocation for multi-user MIMO interference channels under perfect and imperfect CSI," *IEEE Trans. Green Commun. Netw.*, vol. 1, no. 2, pp. 131–144, 2017.

[351] J. Jose, A. Ashikhmin, T. L. Marzetta and S. Vishwanath, "Pilot contamination and precoding in multi-cell TDD systems," *IEEE Trans. Wireless Commun.*, vol. 10, no. 8, pp. 2640–2651, 2011.

[352] M. F. Hanif, L. N. Tran, A. Tolli, and M. Juntti, "Computationally efficient robust beamforming for SINR balancing in multicell downlink with applications to large antenna array systems," *IEEE Trans. Commun.*, vol. 62, no. 6, pp. 1908–1920, 2014.

[353] H. Q. Du and P. J. Chung, "A probabilistic approach for robust leakage-based MU-MIMO downlink beamforming with imperfect channel state information," *IEEE Trans. Wireless Commun.*, vol. 11, no. 3, pp. 1239–1247, 2012.

[354] Z. J. Peng, W. Xu, L. C. Wang and C. M. Zhao, "Achievable rate analysis and feedback design for multiuser MIMO relay with imperfect CSI," *IEEE Trans. Wireless Commun.*, vol. 13, no. 2, pp. 780–793, 2014.

[355] E. G. Larsson and H. V. Poor, "Joint beamforming and broadcasting in massive MIMO," *IEEE Trans. Wireless Commun.*, vol. 15, no. 4, pp. 3058–3070, 2016.

[356] B. K. Chalise and L. Vandendorpe, "MIMO relay design for multipoint to-multipoint communications with imperfect channel state information," *IEEE Trans. Signal Process.*, vol. 57, no. 7, pp. 2785–2796, 2009.

[357] Z. G. Ding, F. Adachi and H. V. Poor, "The application of MIMO to non-orthogonal multiple access," *IEEE Trans. Wireless Commun.*, vol. 15, no. 1, pp. 537–552, 2016.

[358] W. Ikram, S. Petersen, P. Orten and N. F. Thornhill, "Adaptive multichannel transmission power control for industrial wireless instrumentation," *IEEE Trans. Ind. Inform.*, vol. 10, no. 2, pp. 978–990, 2014.

[359] M. Rasti, M. Hasan, L. B. Le and E. Hossain, "Distributed uplink power control for multi-cell cognitive radio networks," *IEEE Trans. Commun.*, vol. 63, no. 3, pp. 628–642, 2015.

[360] H. N. Wang, J. H. Wang and Z. Ding, "Distributed power control in a two-tier heterogeneous network," *IEEE Trans. Wireless Commun.*, vol. 14, no. 12, pp. 6509–6523, 2015.

[361] D. Catrein and R. Mathar, "Feasibility and power control for linear multiuser receivers in CDMA networks," *IEEE Trans. Wireless Commun.*, vol. 7, no. 11, pp. 4700–4709, 2008.

[362] C. Y. Yang, B. S. Chen and C. Y. Jian, "Robust two-loop power control for CDMA systems via multi-objective optimization," *IEEE Trans. Veh. Technol.*, vol. 61, no. 5, pp. 2145–2157, 2012.

[363] T. E. Bogale and L. B. Le, "Massive MIMO and MM Wave for 5G wireless HetNet: Potential benefits and challenges," *IEEE Veh. Technol. Mag.*, vol. 11, no. 1, pp. 64–75, 2016.

[364] F. G. Wang, X. J. Yuan, S. C. Liew and D. N. Guo, "Wireless MIMO switching: Weighted sum mean square error and sum rate optimization," *IEEE Trans. Inf. Theory*, vol. 59, no. 9, pp. 5297–5312, 2013.

[365] J. Rubio, A. Pascual-Iserte, D. P. Palomar and A. Goldsmith, "Joint optimization of power and data transfer in multiuser MIMO systems," *IEEE Trans. Signal Process.*, vol. 65, no. 1, pp. 212–227, 2017.

[366] Q. Z. Lin, J. Chen, Z. H. Zhan, W. N. Chen, C. A. Coello Coello, Y. Yin; C. M. Lin, J. Zhang, "A hybrid evolutionary immune algorithm for multi-objective optimization problems," *IEEE Trans. Evol. Comput.*, vol. 20, no. 5, pp. 711–729, 2016.

[367] S. Bandyopadhyay, S. Saha, U. Maulik and K. Deb, "A simulated annealing-based multi-objective optimization algorithm: AMOSA," *IEEE Trans. Evol. Comput.*, vol. 12, no. 3, pp. 269–283, 2008.

[368] K. Deb and D. Kalyanmoy, *Multi-Objective Optimization Using Evolutionary Algorithms*, Wiley, New York, NY, 2001.

[369] W. Y. Chiu, G. G. Yen and T. K. Juan, "Minimum Manhattan distance approach to multiple criteria decision making in multi-objective optimization problems," *IEEE Trans. Evol. Comput.*, vol. 20, no. 6, pp. 972–985, 2016.

[370] T. V. Chien and E. Björnson, "Massive MIMO communications," in *5G Mobile Communications*, W. Xiang et al., Eds., Springer, Berlin, 2017, pp. 77–116.

[371] A. Nemirovski, *Lectures on Modern Convex Optimization*, Class Notes Georgia Inst. Technol., 2013. [Online]. Available: http://www2.isye.gatech.edu/~nemirovs/Lect_ModConvOpt.pdf

[372] T. V. Chien, T. N. Canh, E. Björnson and E. G. Larsson, "Power control in cellular massive MIMO with varying user activity: A deep learning solution," 2019, arXiv:1901.03620.

[373] *Evolved Universal Terrestrial Radio Access (E-UTRA);* Further Advancements for E-UTRA Physical Layer Aspects (Release 9), 3GPP TR 36.814 V9.0.0 (2010-03), 2010.

[374] G. L. Stüber, *Principles of Mobile Communication*, Springer, New York, NY, 2011.

[375] J. Lofberg, "YALMIP: A toolbox for modeling and optimization in MATLAB," *Proceedings of IEEE International Symposium on Computer Aided Control System Design*, pp. 284–289, Taipei, 2004.

[376] MOSEK Aps, "The MOSEK optimization toolbox for MATLAB manual. Version 7.1 (Revision 28)," 2015. [Online]. Available: http://docs.mosek.com/7.1/toolbox/index.html

[377] H. W. Lorenz, *Nonlinear Economic Dynamics and Chaotic Motion*, Springer, New York, NY, 1993.

[378] W. C. Chen, "Nonlinear dynamics and chaos in a fractional-order financial system," *Chaos Solit. Fractals.*, vol. 36, no. 5, pp. 1305–1314, 2008.

[379] F. Hanson, *Applied Stochastic Processes and Control for Jump-Diffusions: Modeling, Analysis and Computation, 2nd ed.*, SIAM, Philadelphia, PA, 2007.

[380] N. C. Framstad, B. Øksendal and A. Sulem, "Optimal consumption and portfolio in a jump diffusion market with proportional transaction costs," *J. Math. Econ.*, vol. 35, pp. 233–257, 2001.

[381] R. Cont and P. Tankov, *Financial Modelling with Jump Processes*, CRC Press, Boca Raton, FL, 2004.

[382] H. C. Sung, D. W. Kim, J. B. Park and Y. H. Joo, "Robust digital control of fuzzy systems with parametric uncertainties: LMI-based digital redesign approach," *Fuzzy Sets Syst.*, vol. 161, pp. 919–933, 2010.

[383] D. Hinrichsen and A. Pritchard, "Stochastic H_∞," *SIAM J. ControlOptim.*, vol. 36, pp. 1504–1538, 1998.

[384] W. H. Zhang and G. Feng, "Nonlinear stochastic H_2/H_∞ control with (x,u,v)-dependent noise: Infinite horizon case," *IEEE Trans. Autom. Control*, vol. 53, no. 5, pp. 1323–1328, 2008.

[385] F. L. Lewis, L. Xie and D. Popa, *Optimal and Robust Estimation: With an Introduction to Stochastic Control Theory*, CRC Press, Boca Raton, FL, 2008.

[386] D. P. Bertsekas and S. E. Shreve, *Stochastic Optimal Control: The Discrete Time Case*, vol. 139, Academic, New York, NY, 1978.

[387] R. F. Stengel, *Stochastic Optimal Control: Theory and Application*, Wiley, Hoboken, NJ, 1986.

[388] C. S. Tseng and C. K. Hwang, "Fuzzy observer-based fuzzy control design for nonlinear systems with persistent bounded disturbances," *Fuzzy Sets Syst.*, vol. 158, no. 2, pp. 164–179, 2007.

[389] G. P. Liu, J. B. Yang and J. F. Whidborne, *Multi-Objective Optimisation & Control*, RSP Ltd, Hertfordshire, 2002.

[390] E. Branke, K. Deb, H. Dierolf and M. Osswald, "Finding knees in multi-objective optimization," *Parallel Problem Solving Nature*, vol. 3242, pp. 722–731, 2004.

[391] D. Hasan, N. Guzel and M. Sivri, "A fuzzy set-based approach to multi-objective multi-item solid transportation problem under uncertainty," *Int. J. Fuzzy Syst.*, vol. 18, pp. 716–729, 2016. [Online]. Available: http://link.springer.com/article/10.1007/s40815-015-0081-9

[392] A. Gulisashvili, *Analytically Tractable Stochastic Stock Price Models*, Springer, New York, NY, 2012.

[393] J. Stoyanov, "Stochastic Financial Models," *J. Roy. Statist. Soc.: Ser. A (Statist. Soc.)*, vol. 174, no. 2, pp. 510–511, 2011.

[394] B. S. Chen, C. S. Tseng and H. J. Uang, "Mixed H_2/H_∞ fuzzy output feedback control for nonlinear dynamic systems: An LMI approach," *IEEE Trans. Fuzzy Syst.*, vol. 8, no. 3, pp. 249–265, 2000.

[395] H. Zhang, H. Yan, F. Yang and Q. Chen, "Quantized control design for impulsive fuzzy networked systems," *IEEE Trans. Fuzzy Syst.*, vol. 19, no. 6, pp. 1153–1162, 2011.

[396] H. K. Lam, "LMI-based stability analysis for fuzzy-model-based control systems using artificial T–S fuzzy model," *IEEE Trans. Fuzzy Syst.*, vol. 19, no. 3, pp. 505–513, 2011.

[397] R. Khasminskii and G. N. Milstein, *Stochastic Stability of Differential Equations, 2nd ed.*, Springer, Berlin, 2011.

[398] T. Logenthiran, D. Srinivasan, and T. Z. Shun, "Demand side management in smart grid using heuristic optimization," *IEEE Trans. Smart Grid*, pp. 1244–1252, 2012.

[399] W. Y. Chiu, Hongjian Sun, and H. V. Poor, "Robust power flow control in smart grids with fluctuating effects," *The 1st IEEE INFOCOM Workshop on Communications and Control for Sustainable Energy Systems: Green Networking and Smart Grids*.

[400] W. Y. Chiu, "A Multi-Objective Approach to Resource Management in Smart Grid," *The 2014 International Conference on Control, Automation and Information Sciences*.

[401] H. Yang, J. Zhang, J. Qiu, S. Zhang, M. Lai, Z. Y. Dong, "A practical pricing approach to smart grid demand response based on load classification," *IEEE Trans. Smart Grid*, vol. 9, no. 1, pp. 179–190, 2018.

[402] W.-Y. Chiu, H. Sun and H. V. Poor, "An H1 design for dynamic pricing in the smart grid," *Proceedings of Asian Control Conference*, pp. 1–6, May 2015.

[403] W.-Y. Chiu, H. Sun and H. V. Poor, "Energy imbalance management using a robust pricing scheme," *IEEE Trans. Smart Grid*, vol. 4, no. 2, pp. 896–904, 2013.

[404] P. Samadi, A.-H. Mohsenian-Rad, R. Schober, V. W. Wong and J. Jatskevich, "Optimal real-time pricing algorithm based on utility maximization for smart grid," *Proceedings of IEEE International Conference on Smart Grid Communications*, pp. 415–420, Oct. 2010.

[405] A. Vojdani, "Smart integration," *IEEE Power Energy Mag.*, vol. 6, no. 6, pp. 71–79, 2008.

[406] H. Wang and J. Huang, "Incentivizing energy trading for interconnected microgrids," *IEEE Trans. Smart Grid*, vol. 9, no. 4, pp. 2647–2657, 2018.

[407] S. Belhaiza and U. Baroudi, "A game theoretic model for smart grids demand management," *IEEE Trans. Smart Grid*, vol. 6, no. 3, pp. 1386–1393, 2015.

[408] M. Nourian, P. E. Caines, R. P. Malhame and M. Huang, "Mean field LQG control in leader-follower stochastic multi-agent systems: Likelihood ratio based adaptation," *IEEE Trans. Autom. Control*, vol. 55, no. 11, pp. 2801–2816, 2012.

[409] M. Nourian, P. E. Caines, R. P. Malhame and M. Huang, "Nash, social and centralized solutions to consensus problems via mean field control theory," *IEEE Trans. Autom. Control*, vol. 58 no. 3, pp. 639–653, March, 2013.

[410] O. Gueant, J. M. Lasry and P. L. Lions, "Mean Field Games and Applications," *Paris-Princeton Lectures on Mathematical Finance* by Springer, Berlin, Heidelberg, 2011.

[411] S. Grammatico, F. Parise, M. Colombino and J. Lygeros, "Decentralized convergence to Nash Equilibria in constrained deterministic mean field control," *IEEE Trans. Autom. Control*, vol. 61, no. 11, pp. 3315–3329, Nov., 2016.

[412] A. Clemence, B. T. Imen and M. Anis, "An extended mean field game for storage in smart grid," *J. Optim. Theory Appl.*, vol. 184, pp. 644–670, 2020.

[413] A. D. Paola, D. Angeli and G. Strbac, "Distributed control of microstorage devices with mean field games," *IEEE Trans. Smart Grid*, vol. 7, no. 2, pp. 1119–1127, 2016.

[414] A. C. Kizilkale, S. Mannor and P. E. Caines, "Large scale real time bidding in the smart grid a mean field," *2012 IEEE 51st IEEE Conf. Decision and Control*, Dec., 2012.

[415] R. Couillet, S. M. Perlaza, H. Tembine and M. Debbah, "Electrical vehicles in the smart grid: A mean field game analysis," *IEEE J. Sel. Areas Commun.*, vol. 30, no. 6, pp. 1086–1096, May, 2012.

[416] P. Wang, H. Zareipour and W. D. Rosehart, "Descriptive models for reserve and regulation prices in competitive electricity markets," *IEEE Trans. Smart Grid*, vol. 5, no. 1, 2014.

[417] Z. Lu, W. Wang, G. Li and D. Xie, "Electricity market stochastic dynamic model and its mean stability analysis," *Math. Probl. Eng.*, vol. 2014, Article 207474, pp. 1–8, 2014.

[418] *PJM Interconnection LLC* [Online]. www.pjm.com

[419] Shengrong Bu, F. Richard Yu, Yegui Cai and Xiaoping P. Liu, "When the smart grid meets energy-efficient communications: Green wireless cellular networks powered by the smart grid," *IEEE Trans. Wireless Commun.*, vol. 11, no. 8, pp. 3014–3024, 2012.

[420] F. Genoese, M. Genoese and M. Wietschel, "Occurrence of negative prices on the German spot market for electricity and their influence on balancing power markets," *Proceedings of 2010 International Conference on the European Energy Market*.

[421] C. Brandstatt, G. Brunekreeft and K. Jahnke, "How to deal with negative power price spikes? Flexible voluntary curtailment agreements for large-scale integration of wind," *Energ. Policy*, vol. 39, pp. 3732–3740, 2011.

[422] D. Crisan and J. Xiong, "Approximate McKean-Vlasov representations for a class of SPDEs," *Stochastics*, vol. 82, no. 1, pp. 53–68, 2009.

[423] R. Buckdahn, J. Li, S. Peng and C. Rainer, "Mean-field stochastic differential equations and associated PDEs," *Ann. Probab.*, vol. 45, no. 2, pp. 824–878, 2017.

[424] R. Buckdahn, B. Djehiche, J. Li and S. Peng, "Mean-field backward stochastic differential equations: A limit approach," *Ann. Probab.*, vol. 37, no. 4, pp. 1524–1565, 2009.

[425] C. D. Guilmi, *The Generation of Business Fluctuations: Financial Fragility and Mean-field Interactions*. Peter Lang GmbH, Internationaler Verlag der Wissenschaften, Berlin, 2008.

[426] Y. N. Lin, X. S. Jiang and W. H. Zhang, "Necessary and sufficient conditions for Pareto optimality of the stochastic systems in finite horizon," *Automatica*, vol. 94, pp. 341–348, 2018.

[427] W. H. Zhang, Y. N. Lin and L. G. Xue, "Linear quadratic Pareto optimal control problem of stochastic singular systems," *J. Franklin Institute*, vol. 354, no. 2, pp. 1220–1238, 2017.

[428] R. Carmona, J. P. Rene and L. H. Sun, "Mean field games and systemic risk," arXiv:1308.2172 [q-fin.PR], *SSRN Elect. J.*, vol. 13, 2013. [online]. Available: https://arxiv.org/abs/1308.2172

[429] S. Konini and E. J. J. V. Rensburg, "Mean field analysis of algorithms for scale-free networks in molecular biology," *PLoS One*, vol. 12, no. 12, 2017.

[430] S. Cai, Z. R. Liu and H. C. Lee, "Mean field theory for biology inspired duplication-divergence network model," *Chaos*, vol. 25, pp. 83106-1–83106-10, 2015.

[431] A. Clemence, I. B. Tahar and M. Anis, *An extended mean field game for storage in smart grids*. [Online] arXiv.org. Available: https://arxiv.org/abs/1710.08991

[432] R. Couillet, S. M. Perlaza, H. Tembine and M. Debbah, "A mean field game analysis of electric vehicles in the smart grid," *2012 Proceedings IEEE INFOCOM Workshops*, pp. 79–84, Orlando, FL, 2012.

[433] B. S. Chen and C. H. Yeh, "Stochastic noncooperative and cooperative evolutionary game strategies of a population of biological networks under natural selection," *Syst. Evol. Bio.*, pp. 173–214, 2018.

[434] D. Niyato and E. Hossain, "WLC04-5: Bandwidth allocation in 4G heterogeneous wireless access networks: A noncooperative game theoretical approach," *IEEE Globecom* 2006, pp. 1–5, San Francisco, CA, 2006.

[435] L. Zhang and Y. Li, "A game-theoretic approach to optimal scheduling of parking-lot electric vehicle charging," *IEEE Trans. Veh. Technol.*, vol. 65, no. 6, pp. 4068–4078, 2016.

[436] T. Basar and G. J. Olsder, *Dynamic Noncooperative Game Theory*, Academic Press, London, 1982.

[437] X. N. Zhong, H. B. He, D. Wang and Z. Ni, "Model free adaptive control for unknown nonlinear zero-sum differential game," *IEEE Trans. Cybern.*, vol. 48, no. 5, 2018.

[438] H. Mukaidani, H. Xu, "Incentive Stackelberg games for stochastic linear systems with H infinity constraint," *IEEE Trans. Cybern.*, vol. 49, no. 4, 2019.

[439] M. Sobel, "Noncooperative stochastic games," *Ann. Math. Stat.*, vol. 42, no. 6, pp. 1930–1935, 1971.

[440] R. Olfati-Saber, J. A. Fax and R. M. Murray, "Consensus and cooperation in networked multi-agent systems," *Proc. IEEE*, vol. 95, no. 1, pp. 215–233, 2007.

[441] J. R. Marden, G. Arslan and J. S. Shamma, "Connections between cooperative control and potential games illustrated on the consensus problem," *2007 European Control Conference (ECC)*, Kos, pp. 4604–4611, 2007.

[442] M. U. B. Niazi, A. B. Ozguler and A. Yildiz, "Consensus as a Nash equilibrium of a dynamic game," arXiv.org, 5-Jan-2017. [Online]. Available: https://arxiv.org/abs/1701.01223

[443] M. J. Ye and G. Q. Hu, "Distributed Nash equilibrium seeking in multiagent games under switching communication topologies," *IEEE Trans. Cybern.*, vol. 48, no. 11, 2018.

[444] M. Krstic, P. Frihauf, J. Krieger and T. Bas,ar, "Nash equilibrium seeking with finitely-and infinitely-many players," *IFAC Proc. Vols.*, vol. 43, no. 14, pp. 1086–1091, 2010.

[445] P. Frihauf, M. Krstic and T. Basar, "Nash equilibrium seeking in noncooperative games," *IEEE Trans. Autom. Control*, vol. 57, no. 5, pp. 1192–1207, 2012.

[446] J. Shamma and G. Arslan, "Dynamic fictitious play, dynamic gradient play, and distributed convergence to Nash equilibria," *IEEE Trans. Autom. Control*, vol. 50, no. 3, pp. 312–327, 2005.

[447] M. Ye, G. Hu and F. Lewis, "Nash equilibrium seeking for N-coalition noncooperative games," *Automatica*, vol. 95, pp. 266–272, 2018.

[448] B. S. Chen, W. Y. Chen, C. T. Young and Z. Yan, "Noncooperative game strategy in cyber-financial systems with Wiener and Poisson random fluctuations: LMIs constrained MOEA approach," *IEEE Trans. Cybern.*, vol. 48, no. 12, pp. 3323–3336, 2018.

[449] J. Nash. "Non-cooperative games," *Ann. Math.*, vol. 54, no. 2, pp. 286–295, 1951.

[450] B. S. Chen, C. S. Tseng and H. J. Uang, "Fuzzy differential games for nonlinear stochastic systems: Suboptimal approach," *IEEE Trans. Fuzzy Syst.*, vol. 10, no. 2, pp. 222–233, 2002.

[451] W. H. Zhang and Bor-Sen Chen," State feedback H_2/H_∞ control for a class of nonlinear stochastic systems," *SIAM J. Control Optim.*, vol. 44, no. 6, pp. 1973–1991, 2006.

[452] W. H. Zhang, L. H. Xie and B. S. Chen, Stochastic H_2/H_∞ *Control: A Nash Game Approach*, CRC Press, Boca Raton, FL, 2017.

[453] C. A. C. Coello, G. B. Lamont and D. A. V. Veldhuizen, *Evolutionary Algorithms for Solving Multi-Objective Problems, 2nd ed.*, Springer, Boston, MA, 2007.

[454] K. Deb, S. Agrawal, A. Pratap and T. Meyarivan, "A fast elitist non-dominated sorting genetic algorithm for multi-objective optimization: NSGA-II," in *Parallel Problem Solving from Nature PPSN VI Lecture Notes in Computer Science* by Springer, Berlin, Heidelberg, pp. 849–858, 2000.

[455] G. C. Wang, C. H. Zhang and W. H. Zhang, "Stochastic maximum principle for mean-field type optimal control under partial information," *IEEE Trans. Autom. Control*, vol. 59, no. 2, pp. 522–528, 2014.

[456] Y. Y. Tan, Y. C. Jiao, H. Li and X. K. Wang, "A modification to MOEA/D-DE for multi-objective optimization problems with complicated Pareto sets," *Inf. Sci.*, vol. 213, pp. 14–38, 2012.

[457] Y. Lin, X. Jiang and W. Zhang, "An open-loop Stackelberg strategy for the linear quadratic mean-field stochastic differential game," *IEEE Trans. Autom. Control*, vol. 64, no. 1, pp. 97–110, 2019.

[458] Y. Z. Wang, C. Lin, Y. Wang and K. Meng, "Security analysis of enterprise network based on stochastic game nets model," *2009 IEEE International Conference on Communications*, pp. 1–5, Dresden, 2009.

[459] X. Liang and Y. Xiao, "Game theory for network security," *IEEE Commun. Surv. Tutor.*, vol. 15, no. 1, pp. 472–486, 2013.

[460] A. T. Siwe and H. Tembine, "Network security as public good: A mean-field-type game theory approach," *2016 13th International Multi-Conference on Systems, Signals & Devices (SSD)*, pp. 601–606, Leipzig, 2016.

Index

For Product Safety Concerns and Information please contact our EU
representative GPSR@taylorandfrancis.com
Taylor & Francis Verlag GmbH, Kaufingerstraße 24, 80331 München, Germany